PSYCHOLOGY
OF ENTERTAINMENT

PSYCHOLOGY OF ENTERTAINMENT

Edited by

Jennings Bryant

College of Communication & Information Sciences
The University of Alabama

Peter Vorderer

Annenberg School for Communication
University of Southern California

Routledge
Taylor & Francis Group
New York   London

Routledge is an imprint of the
Taylor & Francis Group, an informa business

First Published by Lawrence Erlbaum Associates, Inc., Publishers
10 Industrial Avenue
Mahwah, New Jersey 07430

Reprinted 2008 by Routledge

Routledge
Taylor and Francis Group
270 Madison Avenue
New York, NY 10016

Routledge
Taylor and Francis Group
2 Park Square
Milton Park, Abingdon
Oxon OX14 4RN

Library of Congress Cataloging-in-Publication Data

Psychology of entertainment / edited by Jennings Bryant, Peter Vorderer.
     p. cm.
  Includes bibliographical references and index.
  ISBN 0-8058-5237-9 (casebound : alk. paper)—ISBN 0-8058-5238-7 (pbk. : alk. paper)
1. Performing arts–Psychological aspects.   I. Bryant, Jennings. II. Vorderer, Peter.
  PN1590.P76P79   2006
  791.01′9—dc22

                         2005032343

10 9 8 7 6 5 4 3

To Plato, Aristotle, Locke, Hobbes, Hazlitt, Hume, Schiller, Freud, Darwin, and the other intellectual giants on whose shoulders all of us have been blessed to stand while studying entertainment.

Contents

Preface

The Seattle grunge band Nirvana (1991) contributed what may be the defining mantra of these postmodern times in "Smells Like Teen Spirit," when they proclaimed, "Here we are now, entertain us!" As entertainment becomes a trillion dollar a year industry worldwide (EmanuEl, 1995), as our modern era increasingly lives up to its label of "the entertainment age" (Zillmann & Vorderer, 2000, p. vii), and as economists begin to recognize that entertainment has become the driving force of the new world economy (Wolf, 1999), we can safely say with only a touch of irony that scholars are beginning to catch up with Nirvana and take entertainment seriously. The scholarly spin on entertainment has been manifested in traditional ways (e.g., courses, symposia, sessions at scholarly conferences, consulting) as well as in innovative ones (e.g., videogame design, launching new entertainment companies). Without a doubt, the accumulating empirical evidence, theoretical formulations, and practical wisdom are contributing mightily to the emerging area of entertainment theory.

But the battle is far from won. It is often noted that only by teaching a subject does one fully understand it. A corollary is that teaching makes you cognizant of a topic's limitations and weaknesses. Having taught numerous seminars in entertainment theory over the years, and having presented a plethora of lectures on various facets of entertainment on several continents, the volume co-editors were not at all surprised to learn that we concurred on most areas about which we had experienced delimiting knowledge gaps in entertainment theory and research. Moreover, independently we had come to the conclusion that the weakest links included fundamental aspects of the topic's intellectual infrastructure. Specifically, both of us had found that conceptualization and explication of key *psychological mechanisms* underlying entertainment often were inadequate, and the specific ways *entertainment processes* purportedly differed from those commonly associated with information, education, or persuasion were not always well-articulated.

Once we realized we concurred on areas in which conceptualization and explication of entertainment *mechanisms* and *processes* were underdeveloped, we found ourselves taking a positive turn and identifying scholars who were doing excellent research and theory construction in these underserved aspects of entertainment theory. Again, more often than not, we found that we agreed on the identity and scholarly ability of those who were successfully addressing these troublesome abysses.

At some point in our deliberations (i.e., that "Eureka" moment), we decided that the most productive way of advancing understanding of these psychological mechanisms and processes of entertainment would be to call upon those peers who seemed to be making the most significant progress in understanding these fundamental intellectual underpinnings of entertainment theory and ask them if they would be willing to share their insights—as well as the fruits of their scholarship—with kindred spirits. To our great pleasure, our associates were more than willing to synthesize their research, as well as the cognate scholarship of others, in the several research domains we had identified. The concrete product of our collaboration is *Psychology of Entertainment*, which is dedicated to advancing understanding of the fundamental psychological processes and mechanisms of entertainment.

Preparation and Reception Processes

The content of this volume is divided into three basic units. The first section is entitled "Preparation and Reception Processes." The six chapters in this portion of the book deal with those fundamental mechanisms and processes involved in orienting to and selecting entertainment fare, as well as receiving and processing it.

In the chapter "Motivation," Peter Vorderer, Francis F. Steen, and Elaine Chan ask why human beings from different cultures and historical periods seek out and enjoy the experience of entertainment. They also productively address the question, "what is entertainment?," which they consider from both intentionalist and objectivist stances.

Jennings Bryant and John Davies examine a second precursor to entertainment by considering theories and models of the ways modern consumers select entertainment fare from the plethora of choices available to them in today's digital media environment. In "Selective Exposure Processes," after considering the intellectual history of selective exposure theory, the authors focus on the processes and outcomes of the theory and critically examine its effectiveness in terms of predicting and explaining selectivity in the information age.

One of the most productive research traditions in media psychology involves the examination of "Attention and Television." Daniel R. Anderson and Heather L. Kirkorian describe, integrate, and synthesize the several conceptual, methodological, and empirical dimensions of attention and indicate just how and why attention affords a necessary, but not sufficient, condition for entertainment.

L. J. Shrum tackles the elusive concept of "Perception" and its manifold roles as a mechanism of entertainment. The literatures from which this explication and extension are derived are extremely diverse, and their successful integration provides novel insights into the ways we perceive entertainment media.

Nothing is more fundamental to entertainment experiences than the ways in which we encode and store media messages. Richard Jackson Harris, Elizabeth Tait Cady, and Tuan Quoc Tran serve as guides through these processes and mechanisms in their chapter on "Comprehension and Memory." The chapter concludes with a model of comprehension and memory of entertainment.

The final chapter in this first section is "Media Information Processing" by Robert H. Wicks. By nature, and by intention, this chapter integrates a great deal of the material from the previous five chapters, plus it offers a number of novel discussions, such as its treatment of schemas and framing.

Reaction Processes

The mechanisms and processes by which we are entertained by the media messages we select and receive are explored in the unit on "Reaction Processes." This is the largest section of *Psychology of Entertainment* and contains 14 chapters.

Patti M. Valkenburg and Jochen Peter consider the role of "Fantasy and Imagination" before, during, and after exposure to entertainment. They advance and consider a number of thoughtful hypotheses for clarifying the role of fantasy and imagination in various entertainment phenomena.

In "Attribution and Entertainment: It's Not Who Dunnit, It's Why," Nancy Rhodes and James C. Hamilton examine the role causal thinking plays in the process of entertainment. In advancing a new thesis of the role of attributional thinking in the entertainment experience, the authors explore how various aspects of quality fiction are comparable to effective attributional thinking.

One of the theories most often utilized by entertainment researchers is disposition theory. In "The Psychology of Disposition-Based Theories of Media Enjoyment," Arthur A. Raney critically examines the various assumptions and mechanisms of disposition theory and articulates six principles and features that are common to all disposition-based theories.

Dolf Zillmann tackles one of the most complex yet fundamental processes associated with media entertainment in "Empathy: Affective Reactivity to Others' Emotional Experiences." This chapter takes us through various theories of empathy, offers a three-factor theory of empathy, and examines the role of empathy in the changing media landscape.

One of the thorniest topics in entertainment theory is the concept of identification, which Jonathan Cohen addresses in the chapter "Audience Identification with Media Characters." Perspectives as diverse as literary analysis, critical studies, and traditional social science are integrated in order to comprehensively explore the mechanisms and processes of identification.

Werner Wirth examines "Involvement" and its relationship to the entertainment experience. Beginning with the diverse characterizations and conceptualizations of involvement, this chapter plumbs issues such as the effects of intensity and valence of involvement on entertainment, and it offers an integrated conceptualization of involvement as it relates to entertainment theory.

A second chapter by Dolf Zillmann focuses on "Dramaturgy for Emotions from Fictional Narration." After critically analyzing aspects of the "willing suspension of disbelief," this chapter focuses on the roles of excitatory processes in emotional experience and cognitive processes in the mediation of excitatory reactions.

Silvia Knobloch-Westerwick provides a systematic portrayal and assessment of one of the more prominent media theories in "Mood Management: Theory, Evidence, and Advancements." In addition to carefully profiling and integrating the extant empirical research on the topic, several ideas on further development of mood-management research are provided.

In "Social Identity Theory," Sabine Trepte examines the ways in which groups to which one belongs affect reception processes and the enjoyment of consuming media with diverse attributes. Also considered are the effects of media entertainment on social identity—the flip side of enjoyment—and how entertainment in the virtual environment may well alter social-identity processes.

One underresearched topic in entertainment theory is the consideration of "Equity and Justice" in media enjoyment. Manfred Schmitt and Jürgen Maes examine the justice motive, demonstrate its role in various entertainment genres, and examine several hypotheses regarding the effects on entertainment value of consideration of equity and justice.

A perennially hot topic in entertainment-theory circles is parasocial interaction. Christoph Klimmt, Tilo Hartmann, and Holger Schramm explicate the concepts of "Parasocial Interactions and Relationships" and examine their role and function in the entertainment process.

They conclude with a process-oriented model of parasocial interaction with media characters and posit its impact on entertainment responses.

In "Why Horror Doesn't Die: The Enduring and Paradoxical Effects of Frightening Entertainment," Joanne Cantor examines the interplay between fright reactions and the entertainment experience. Developmental differences are considered in correspondence with media characteristics in examining the lingering mixture of pain and pleasure.

Mary Beth Oliver, Jinhee Kim, and Meghan S. Sanders explore the complex role of "Personality" in media entertainment. One distinctive aspect of their chapter comes from their division of entertainment-related variables into "selectivity" versus "enjoyment," and the articulation of quite distinct personality patterns associated with these different dimensions of the entertainment experience. Personality is also considered as a moderator of media effects and as a dependent variable.

Relying heavily on scholarship from the area of affective neuroscience, Dorina Miron examines the role of "Emotion and Cognition in Entertainment." In considering the role of pleasure seekers at several different levels, this chapter not only integrates the cognitive and affective sides of the emotional equation, it also raises many new issues and questions for future generations of entertainment theorists to explore.

Application of Psychological Theories and Models to Entertainment Theory

The final section of *Psychology of Entertainment* provides an opportunity for the application of some well-established—as well as some emerging—psychological and psychobiological theories to be applied to the study of entertainment in ways that seldom have been utilized previously. Four chapters form this section and complete the volume.

In "Sensation Seeking in Entertainment," Marvin Zuckerman first examines the formulation of the popular personality dimension of sensation seeking, and then explains how high- versus low-sensation seekers select and enjoy various forms of entertainment. Of special interest is the examination of environmental and biological influences of sensation seeking and considerations of how various dimensions of personality underlie different entertainment preferences.

Hedonic psychology has come of age, and Margrit Schreier examines entertainment aspects of this tradition in "(Subjective) Well-Being." The relationships between psychological well-being, media consumption, and the entertainment experience are shown to be both complex and fascinating, and a number of additional variables are required to begin to understand their interrelationships.

Just when we thought it was safe to put the study of catharsis on the shelf, Brigitte Scheele and Fletcher DuBois provide a radical reformulation of the concept and its intellectual history in "Catharsis as a Moral Form of Entertainment." Offering developments as different as a critical re-examination of Aristotelian catharsis and a complex remodeling of emotions, this chapter offers rich new possibilities for an expanded understanding of entertainment.

Finally, principles of evolutionary biology are brought to bear on the study of entertainment by Peter Ohler and Gerhild Nieding in "An Evolutionary Perspective on Entertainment." From entertainment's origins around Pleistocene campfires to its place in sexual preference and creative intelligence, fresh perspectives on the study of entertainment are provided.

In Sum

As is typically the case in mediated communication, this volume is the work of teams and institutions. We are most grateful to our talented contributors and to the universities that support all of us. Although we often forget to say it or even think it, we really do appreciate the freedom of expression they uphold and foster.

To our good friends at Lawrence Erlbaum Associates, Publishers, it is both a pleasure and an honor to collaborate with you again. To Linda Bathgate, the public face of LEA's values in most of the disciplines we represent, your guidance and friendship are invaluable, and you make those ubiquitous conferences so much more fun. To the other members of the Erlbaum family who toil behind the scenes, thank you.

To our mentors, particularly Dolf Zillmann and Norbert Groeben, we owe you so much for stimulating our interest in entertainment theory and research. We are equally grateful to you for embodying and modeling the purest possible scholarly zeal.

To our students, who are becoming both legion and legend, it really is true that you taught us more than we taught you. As our intellectual family tree grows, because of you we have great confidence in the future of entertainment theory: The best is yet to be.

It has been this pair of editors' great pleasure to learn so much from the fertile minds of our generous collaborators. It is with considerable pride—and with the sort of enjoyment that only satisfying intellectual rumination and discovery can provide—that we remand this product of their creative genius to your edification, enlightenment, and enjoyment.

—Jennings Bryant
—Peter Vorderer

REFERENCES

Emanuel, E. F. (1995). *Action & idea: The roots of entertainment.* Dubuque, IA: Kendall/Hunt Publishing Company.

Nirvana. (1991). Smells like teen spirit. On *Nervermind* (CD). London: DGC Records.

Wolf, M. J. (1999). *The entertainment economy: How mega-media forces are transforming our lives.* New York: Times Books.

Zillmann, D., & Vorderer, P. (2000). *Media entertainment: The psychology of its appeal.* Mahwah, NJ: Lawrence Erlbaum Associates.

Contributors

Daniel R. Anderson
University of Massachusetts at Amherst

Jennings Bryant
University of Alabama

Joanne Cantor
University of Wisconsin-Madison

Elaine Chan
University of Southern California

Jonathan Cohen
University of Haifa

John Davies
University of North Florida

Fletcher Dubois
University of Cologne

James C. Hamilton
University of Alabama

Richard J. Harris
Kansas State University

Tilo Hartmann
Hanover University of Music and Drama

Christoph Klimmt
Hanover University of Music and Drama

Jinhee Kim
Pennsylvania State University

Heather L. Kirkorlaw
University of Massachusetts, at Amherst

Silvia Knobloch-Westerwick
Ohio State University

Dorina Miron
University of Alabama

Gerhild Nieding
University of Wuerzburg

Jurgen Mass
University of Trier

Peter Ohler
University of Technology Chemnitz

Mary Beth Oliver
Pennsylvania State University

Jochen Peter
University of Amsterdam

Arthur A. Raney
Florida State University

Nancy Rhodes
University of Alabama

Meghan S. Sanders
Pennsylvania State University

Brigitte Scheele
University of Cologne

Manfred Schmitt
University of Koblenz-Landau

Holger Schramm
University of Zurich

Margrit Schreier
International University Bremen

L. J. Shrum
University of Texas, at San Antonio

Francis F. Steen
University of California, Los Angeles

Sabine Trepte
University of Hamburg

Patti M. Valkenburg
University of Amsterdam

Peter Vorderer
University of Southern California

Rob Wicks
University of Arkansas

Werner Wirth
University of Zurich

Dolf Zillmann
University of Alabama

Marvin Zuckerman
University of Delaware

PSYCHOLOGY OF ENTERTAINMENT

PREPARATION
AND RECEPTION PROCESSES

1

Motivation

Peter Vorderer
University of Southern California

Francis F. Steen
University of California, Los Angeles

Elaine Chan
University of Southern California

When we approach entertainment as an experience that is sought after and enjoyed, we encounter the enduring questions of its psychological cause. Why do human beings, across a range of different cultures and historical periods, seek out and enjoy the experience of entertainment? Why do they select and create certain types of situations—and not others—to entertain themselves? Why do they seek entertainment so often, for such long periods of time, and in so many different situations and settings? To ask these questions is to adopt the perspective that entertainment is a response to a certain set of opportunities rather than a feature of a particular media product itself (Bosshart & Macconi, 1998; Bryant & Miron, 2003; Vorderer, 2001; Zillmann & Bryant, 1994).

The *motivational* question in entertainment research tries to reach beyond a simple description of "who is doing what in which kind of situation" and attempts to explain *why* things are done as they are. In this chapter, we try to systematize various answers to the question of why people perceive entertainment to constitute a good—why different people, in a wide range of circumstances, want to be entertained and find the experience of being entertained rewarding. However before we do so, we would like to bring a typical, everyday entertainment situation into focus, so that we may capture the complexity of what entertainment can mean to an individual.

It is Thursday night. Rowan, an undergraduate student, is at home in her apartment. She thinks about spending the evening in, rather than heading out to a party with friends. She is not in the mood for that sort of excitement tonight. She brings her laptop computer to the couch and sits. Rowan grabs the remote control and turns on the television, then uses her laptop to instant message her friend that she would rather stay at home. Flipping through the channels, Rowan narrows the viewing alternatives to a cooking special, the second half of Ferris Bueller's Day Off, *and a reality show on MTV. She has seen* Ferris Bueller's Day Off *before, but settles on watching a bit until her favorite show,* ER, *comes on. She gets up and goes to the kitchen to find something to eat, and returns with a bowl of potato chips. Next, she checks the website for ER to make sure this week's episode will be a new one. She laughs at Bueller's antics and forgets about ER until the commercial break. Remembering, however, Rowan flips to the appropriate channel, then returns to the kitchen to replenish her chip supply. The TV plays the familiar ER theme song as Rowan returns and again sits on the couch.*

We take the sort of situation portrayed in this imaginary scenario to be commonplace, particularly among young people in Western societies who rather frequently find themselves in a broad entertainment context like the one described. Applying a motivational perspective to entertainment and to this specific situation raises the following questions: Why is Rowan (or anybody similar to her) doing this? What prompted her to initiate and arrange this experience? Why does she sustain it for a certain length of time, and what causes her to terminate it? What, if anything, does she gain from it?

What Is Entertainment?

Recently, the academic study of entertainment has been identified as one of the most important challenges facing communication theory and research (Bryant, 2004). Since the systematic investigations in the 1980s and 1990s, in particular those by Zillmann and his collaborators in the United States (e.g., Raney & Bryant, 2002; Zillmann & Bryant, 1994; Zillmann & Vorderer, 2000) and by Bosshart and others in Europe (Bosshart, 1979; Bosshart & Hoffman-Riem, 1994; Bosshart & Macconi, 1998), entertainment has often been conceptualized to be an affective response to entertainment products such a movies, TV, music, or books. Similarly, researchers have regarded entertainment as a human activity that might be influenced, triggered, and maybe even shaped by the media product that is selected. Still, it is the individual who deliberately and voluntarily controls entertainment; entertainment is not determined by the product. As a human activity, it includes various physiological, cognitive, affective, and behavioral components. Therefore, entertainment can and should be described and explained by a discipline concerned with human thinking, feeling, and behavior. Psychology, particularly motivational psychology, seeks to answer the questions explicated above. In fact, most researchers in the tradition of Zillmann refer to psychological theories, and have even developed and explicated their own theories that dwell on psychological processes, assumptions, and models (cf., Bryant, Roskos-Ewoldson, & Cantor, 2003). Zillmann himself has explicated a number of different theories, such as mood-management theory, affective-disposition theory, or excitation transfer theory, all of which originate from a psychological understanding of human functioning (cf., Vorderer, 2003). The challenge facing us in the attempt to deepen our understanding of the motivational causes of entertainment is to continue to update our models in light of ongoing advances in psychological theory and research into the causes of human action and behavior. The explanations currently available for why people seek out and enjoy entertainment do not make full use of major theoretical achievements' made more recently within particular areas of psychology.

In this chapter, we focus our attention on two key areas of new work within the field of psychology to provide a richer and better-grounded causal theory of entertainment. The first of these areas studies human activities that are characterized as being intrinsically motivated and experienced as being ends within themselves. Although the study of intrinsic motivation reaches back to the origins of psychology (e.g., Spencer, 1872–73), the area has in recent years been revitalized by a concerted effort to identify the actual needs that people seek to satisfy with their behavior, to relate the dynamics of intrinsic to extrinsic forms of motivation, and to understand what makes people change their levels of motivation and preferences over time. The second area is evolutionary psychology, which examine the role of natural selection in the design of human cognitive and emotional abilities and proclivities (Chapter 24 by Ohler & Nieding). Following previous work (Steen & Owens, 2001; Tooby & Cosmides, 2001), we develop the proposal that a suite of adaptations tied to the natural phenomenon of play provides an additional gateway to a better understanding of important aspects of entertainment.

By drawing on these two new areas of work, we wish to emphasize that a coherent theory of entertainment must be able to handle theoretical innovations and new data from two very different perspectives, or stances, that we term intentionalist and objectivist. Adopting an intentionalist stance (Dennett, 1987) is to look for the causes of entertainment in people's subjective mental states. Mood-management theory, for instance, implies an intentionalist stance, in which the subjective mental state, "mood", and the intention to manage it are attributed to the participants of the study, and by induction to people in general. Intentionalist theories utilize a general understanding of agents to formulate causal theories.

In contrast, an objectivist stance assumes that causal relations between material, physiological processes can explain the phenomena (Leslie, 1994). In these theories, people's motivations are understood to be the result of these processes. What links the adaptive design and the physiological effect is the notion that the brain is processing information, and that natural selection over long periods of time has the effect of creating organic structures that are optimized for certain kinds of information processing. From this point of view it might be argued, for instance, that the reason people devote time and resources to multiplayer online games is that reward circuits in the nucleus accumbens are activated during this activity. Because major structures in the brain are common to all, this argument can be extended to suggest that natural selection for some other activity, such as rough-and-tumble play (Bjorklund & Pellegrini, 2002) has created a connection between certain forms of play and the brain's reward structures that society can to tap into and activate with the design of dedicated technologies.

The intentionalist and the objectivist stances rely on different assumptions, argue on different levels of abstraction, and have therefore long been treated as rivaling explanations, their boundaries delineating the split between the "two cultures" (Snow, 1959/1993), humanities and the sciences. Nevertheless, we will here present a general framework here suggesting that both types of theories are necessary and complement each other in important ways.

What Is Motivation? Seeking An Intentional Approach to Entertainment

Most theories of motivational psychology have distinguished between potential causes of human activities as originating either from within an individual or from external sources (Heider, 1958). The differentiation between intrinsically and extrinsically motivated behavior (or action) that developed later stems from this dichotomy. According to this view, external causes initiate and shape extrinsically motivated behavior, particularly with rewards and punishments. Behaviorism, as a school of thought and a research paradigm, dominated psychology for decades of the 20th century, and successfully demonstrated how specific rewards and punishments given at various points in time could influence almost all dimensions of human functioning. Intrinsically motivated behavior, on the other hand, is seen as something that has its end in itself. An intrinsically motivated individual behaves or acts for the satisfaction inherent in the behavior he or she performs. Satisfaction may come from positive feelings of being effective (cf., White, 1959), or simply from being the origin of behavior (cf., deCharms, 1968). Following this dichotomy, A person may devote time to learning a particular subject because he or she wants to do well on an exam (extrinsically motivated) *or* may do so because the subject is fascinating in itself and learning about it is inherently enjoyable (intrinsically motivated).

Although some psychologists have more recently suggested that researchers further differentiate between a "structural sense" (the relation between the activity and its goal) and a "substantive sense" (the type of goal the activity is meant to attain) of intrinsic motivation (cf., e. g., Shah & Kruglanski, 2000), we will use the simple dichotomy here to describe *entertainment as an intrinsically motivated response* to certain media products. Cases in which one must entertain him- or herself in order to attain a goal that lies outside the activity itself also exist (e.g., a student who must watch a movie in order to criticize it for class, a professor who must play a video game so that he or she can teach about it at school, etc.), but these certainly are not prototypical cases.

Someone who is seeking entertainment usually does so for its own sake, that is , in order to experience something positive, such as enjoyment, suspense, amusement, serenity, and so on (cf., Vorderer, Klimmt, & Ritterfeld, 2004). As a person prepares to be entertained, and when he or she chooses a specific activity or selects a particular product for this purpose, however, the desired experience still lies in the future. Take, for example, the previously outlined scenario in which Rowan gets ready for some entertainment. She sits in front of her TV with a laptop and a supply of chips, not because of an entertainment experience she is already having, but because of one she is expecting to have. In other words, she directs her activity toward a future psychological state that, so far, is only represented in her mind. Representations like this future state are usually considered goals, and individuals appear to be able to regulate their activities toward these goals.

What sorts of activities hold intrinsic interest for people? According to Ryan and Deci (2000), these activities must have the appeal of novelty (cf., also Berlyne, 1971; 1974), challenge (Csikszentmihalyi, 1975), or aesthetic value. Following the observation that, from the time of birth, children are active, curious, and playful, even in the absence of rewards, Ryan and Deci (2000; see also Ryan, 1995) regard intrinsic motivation as a construct that "describes this natural inclination toward assimilation, mastery, spontaneous interest, and exploration" (p. 70; also see Kelly, 1955, for an early example of such a model). This cognitive evaluation theory (Deci & Ryan, 1985) aims to specify factors that explain variability in intrinsic motivation. The theory is based on the assumption that humans have three fundamental needs, and that the satisfaction of these needs is essential and crucial not only for an individual's intrinsic motivation, but for a person's well-being and mental health generally, as well. They claim a need for *competence* and a need for *autonomy*. In fact, empirical research in line with this theory shows that social-contextual factors (e.g., feedback) that support feelings of competence during a given action enhance intrinsic motivation for that action, as much as optimal challenges and freedom from demeaning evaluations facilitate it (Fisher, 1978; Ryan, 1982; deCharms, 1968). They conclude that, in order to maintain intrinsic motivation and perform well, people must perceive their behavior to be self-determined (Reeve, 1996) and autonomous.

Ryan and Deci's so-called self-determination theory (into which later on the previously mentioned cognitive evaluation theory seems to have merged) has identified *relatedness* as a third human need that, when satisfied, helps intrinsic motivation flourish. That is, a secure relational base between an individual and another person (parent, teacher, peer, friend, etc.) is not only important so that infants can explore their early environments, but it also impacts (intrinsic) activities over their life spans. Ryan and Deci (2000) stipulate that the three needs are universal and developmentally persistent, although their relative salience and the ways these needs can be satisfied may change across the life span. Also, there is little doubt that the modes of expression of these needs may vary in different cultures: "The very fact that need satisfaction is facilitated by the internalization and integration of culturally endorsed values and behaviors suggests that individuals are likely to express their competence, autonomy, and relatedness differently within cultures that hold different values" (Ryan & Deci, 2000, p. 75).

Entertainment As Intrinsically Motivated Experience

We use self-determination theory to explain people's overall interest in entertainment by suggesting that media consumption in general, and the use of entertainment media products in particular, provide specific ways to satisfy the fundamental psychological needs proposed. First, exposure to entertainment products is usually an activity that has an end in itself, and thereby qualifies as being intrinsically motivated. It serves all three fundamental needs of competence, autonomy, and relatedness, although these needs materialize differently over a life span, in different cultures, situations, and even personalities. However, entertainment products are suitable to use nearly anywhere and anytime.

As far as the need for *competence* is concerned, it is interesting to note that television use in particular often has been considered to be "easy," while reading frequently is regarded by many as "tough," or requiring effort. Salomon (1984) has shown that the cognitive activity invested in following TV programs is rather low, and that most audiences select programs that are not overly challenging (see also Weidenmann, 1989). Henning and Vorderer (2001) demonstrated that massive consumption of TV is correlated with a rather low need for cognition. Groeben and Vorderer (1988) have followed Berlyne's (1971; 1974) motivational theory of aesthetic appreciation and argued that readers of literary texts usually select and particularly enjoy texts that challenge them—but only up to the point where they can still master the challenge. In sum, entertainment has almost always been described as an activity where there is rather little challenge, or only as much challenge as the media user can still handle successfully. This, of course, would be the optimal challenge, the level that allows people to feel the greatest sense of competence. The feeling of competence is therefore almost guaranteed, and it can be created without much effort. Where else can somebody feel competent so often, so easily, and so profoundly? Video games, which are among the most appealing facilitators of entertainment these days, particularly among younger males (as an overview: Vorderer & Bryant, in press), provide an excellent example of this. The level of complexity, difficulty, and challenge of a given game varies and is dependent on the settings that are either chosen by the player or automatically set by the game. The game itself may choose an optimal difficulty level based on the amount of skill or expertise a player has demonstrated previously. In other words, a video game guarantees the player to be challenged at a level he or she can master without becoming bored or overwhelmed. The importance of competition and challenge for a video game player's entertainment experience is often mentioned (e.g., Vorderer, Bryant, Weber, Pieper, in press), which seems to support the notion that individuals often seek to experience competence in entertainment. As in the above mentioned example of Rowan's entertainment experience, we can consider how carefully she prepares for a situation in which she will be able to avoid boredom and find some challenge (which is why she chooses to use different media at the same time) while making sure that this situation will not overwhelm her or give her a sense of being incompetent.

With respect to the need for *autonomy*, it is also obvious that exposure to a particular entertainment program typically fits the description of an activity that is not forced, influenced, or triggered by others. While there is little doubt that entertainment users may be influenced by their peers' attitudes, preferences, and values in selecting specific entertainment products, most users do not consider themselves to be subject to such influences. Indeed, media users typically overestimate their independence from outside influences and see their choices as directed by their individual interests and preferences. This biased self-view has consequences for an entire research paradigm on uses and gratifications (Rosengren, Wenner, & Palmgreen, 1985; Rubin, 2002), in which the motivational question is answered by summarizing the various responses that are given by research participants. The uses-and-gratifications-paradigm assumes that respondents are fully aware of their own motives and able and willing to express them accurately. Verbal reports provided by the subjects are thus considered valid and sufficient evidence. Although this approach has been questioned and criticized for a range of reasons (e.g., Zillmann, 1985), what matters in the present context is that the approach will perpetuate an exaggerated assessment of a person's autonomy. Certainly, it confirms that people tend to see themselves as highly self-directed. The availability of media entertainment products, almost anywhere and at any time, provides users with the objective circumstances that make the experience of autonomy easy to sustain. The perception of unconstrained choice also gives users a sense that the activity is absent outside control. In our original example, it appears that Rowan believes she is selecting *ER* on TV as her favorite show, whereas positive appraisals of the show by her peers may in fact be highly influential.

Finally, as far as *relatedness* is concerned—that is, the need to feel in touch with somebody else—an extensive body of literature shows that TV especially provides its users with a sense of relatedness, even in situations when they are watching alone. The connection that some viewers feel with hosts, anchormen and -women, or other media personae who appear to be addressing a single member of the audience, gave way to the study of so-called parasocial interactions and parasocial relations (originally: Horton & Wohl, 1956; see Chapter 17 Klimmt, Hartmann & Schramm). This research shows that media users do not sense that they are alone. They usually feel connected and related to characters in movies, shows, novels, and even video games. The sense of parasocial interaction, or having a relationship with a media figure, accounts for much of the interest in specific programs. Some users may go so far as to admire and adore some persona and select a movie, a book, or a game primarily—or even exclusively—because of the person or character's role in it. Parasocial interaction explains the prominence of celebrities in entertainment products, which leads many to consume media for the purpose of connecting or relating to a celebrity.

Affective disposition-theory also explains how the audience's moral judgments of the characters' behavior lead to either hopes or fears about the characters' fortune and subsequently to empathy or counterempathy with them. Again, applied to Rowan's situation described above, we understand that she intends to relate to a friend through messaging, presumably about the TV show she is going to watch, and maybe also about one or more personae of this show. This could be a character for instance, Dr. Carter, who has been part of this show from its very beginning in 1994, or it could be Noah Wyle, the actor playing this character. Such examples suggest the impact on media users of real or fictional personae in the media, and the effectiveness of entertainment products in creating the impression of being related, thereby satisfying the third fundamental need proposed by Ryan and Deci (2000).

So far, we have focused on recent work in the psychology of intrinsically motivated behavior. We now propose to broaden the perspective by situating intrinsically motivated behaviors within an evolutionary perspective, thus adopting an objectivist stance in which causal material processes are invoked to explain the design and functioning of the mind.

What Is Function? Seeking An Objectivist Approach to Entertainment

At first blush, a theory of human behavior that holds that the behavior is engaged in for its own sake may appear to pose a challenge to an evolutionary account. Evolution is the story of functional outcomes, and an activity that is indifferent to its own consequences would appear to elude the dragnet of biology altogether. We will show that this reasoning is fallacious, and the task of clearing up the fallacy will allow us to characterize in a more precise manner what is distinctive about entertainment.

As the French materialist philosopher Maupertuis (1745) realized more than a century before Darwin, organic design can be explained by a dynamic process of incremental change from generation to generation (cf., Glass, 1959). Over a multiplicity of individual events, organisms with a physiology that provide a better fit with their environment have a greater chance of surviving to leave offspring. An organism without a mouth, Maupertuis reasoned, would be unable to eat, and thus die without issue. Central to Maupertuis' argument, later formalized by Darwin, is the notion of a biological function. A theory of the biological function of an organ, such as a mouth, explains the precise organic design of the mouth—the presence of an opening, of teeth, of a tongue, of saliva glands, and so on—as a result of a historical process over a large number of generations. Those individuals who were endowed with a well-engineered mouth, given their local environment, tended on average to leave more offspring, thus making that particular design more common in the population.

Within evolutionary theory, the generalization of physiological design to behavior is not unproblematic. It does not follow, for instance, from the notion that organic design is the outcome of a long period of natural selection, that an organism's behavior is similarly determined by this history. In the case of human beings, for instance, our full range of possible activities cannot be determined by an examination of our history, however exhaustive. There is a very simple reason for this: our physiology, including that of our brain and nervous system, enable us to engage in a far larger number of activities than natural selection has acted on. We must therefore distinguish between the *biological* function of a particular organ and its *actual* function. While the biological function of the mouth, in Millikan's term, is the function the mouth was designed to solve, by virtue of its past successes (Millikan, 1984, 1993), its actual function may include such evolutionarily novel activities as sucking on a cigarette, inflating a balloon, and playing a harmonica.

In a word, the causal link between natural selection and behavior is attenuated by the intermediation of psychology. It is here we encounter the first clue to the misunderstanding that evolutionary theories cannot comprehend—intrinsically rewarding activities. Yet it is not altogether simple to navigate the strait between the Scylla of a total evolutionary reductionism, on the one hand, and the Charybdis of a vision of human psychology untouched by biological history on the other.

Consider the case of a series of activities that human beings find intrinsically enjoyable, yet would not characterize as entertainment—activities such as eating, sleeping, and sexual intercourse. To engage in these activities, we do not typically need an extrinsic reward; they are experienced as ends in themselves. We sit down and eat when we are in a physiological and mental state that makes food taste good, and we stop eating when that state has passed. Now, does this mean that our behavior was caused by our biological history?

In part, we would have to reply "yes"—in consonance with Maupertuis, the individual who lacked the psychological states of hunger, who did not experience a desire for the food placed in front of him, and who did not persist in the activity of eating until he had acquired an adequate amount of nourishment, would surely die. Hunger has a biological function: to ensure that we eat and nourish ourselves. Thus, the explanation that was used to explain the design of the mouth can be extended to include the design of the appetites and the preferences that regulate what goes into the mouth. Yet, at the same time, the food we put on the table today consists largely of items that are evolutionarily novel—that, as it were, natural selection has never seen. Potato chips and sugar-frosted doughnuts did not form part of our ancestors' diet, and our preferences cannot therefore have evolved to prefer them. In fact the food we just ate may not be nourishing us at all; on the contrary, it might, just as likely, be slowly killing us off, if we participate in the unprecedented national epidemic of obesity.

Although natural selection built the structures that underpin our motivational systems, those structures operate correctly only in an environment that resembles, or approximately reproduces, the environment in which these structures themselves evolved. It is as if nature designed a tricycle, and culture built a freeway for it. Or to invert the metaphor: as if nature constructed a supercomputer, and culture posed it as a problem of which color soap to buy. There is a radical disjunction between the environment in which most of our evolution took place— what Bowlby (1969; cf., Foley, 1995) called "the environment of evolutionary adaptedness"— and our present socially and technologically transformed reality. This disjunction, referred to as adaptive or phenotypic lag, is a general result of changing environmental conditions; in the human case, the effect is vastly speeded up and magnified by our capacity to imagine and create realities without precedence in nature. To claim that the mental state of hunger and the act of eating has a biological function, then, is not to claim that, in any given meal, they actually provide the benefits that the adaptation was designed to provide. Human psychology is optimized for an environment that no longer exists, and culture generates products that target psychological adaptations in a manner that need not be advantageous to the individual, either in terms of survival or reproduction.

Although natural selection operates on outcomes, behavior itself cannot be inherited. Rather, what can be passed on through genetic material is the ability and proclivity to engage in particular types of behavior under perceived types of circumstances. Such abilities, proclivities, and perceptions belong to the domain of psychology. Evolution, then, generates a psychology that has its own priorities, only indirectly connected to the logic of natural selection. In a celebrated and maligned formulation, Herbert Spencer compressed Darwin's theory into a single phrase, "the survival of the fittest" (1861). He used this phrase to justify a certain vision of society, Social Darwinism, where nature was seen to bestow a normative blessing on successful competitive social climbers. Among other things, Spencer's motto inadvertently and inappropriately collapsed the logic of natural selection onto human motivational psychology. Just because a biologist finds the concept of fitness useful in understanding the emergence of organic design doesn't mean that human beings are psychologically motivated to maximize this quality. If human beings truly were fitness maximizers, following the supposed call of nature to leave as many descendants as possible, they would be paying their last savings to be allowed to donate their eggs and sperm to childless couples and sperm banks. Clearly this is not happening: There is nothing in human psychology committed to maximizing what a biologist terms fitness.

In the case of eating, sleeping, and sexual intercourse, it is not difficult to demonstrate that these activities fit the evolutionary paradigm, with the qualification that psychology must be accorded a degree of autonomy. Although these behaviors are subjectively experienced as intrinsically rewarding, an evolutionary account would point out that this psychological intermediation is itself designed by natural selection to encourage and sustain the biological functions of acquiring nourishment, rest, and producing offspring. In this view, these behaviors are intrinsically rewarding precisely because, over evolutionary history, they on average promoted an underlying biological function.

This line of reasoning nevertheless leaves the question this chapter addresses unanswered: What, if any, is the biological function of entertainment? Returning to Rowan's situation described earlier, does her very modern behavior have an ancient antecedent, and does her appetite for television and her preference for certain programs reflect the operation, in a novel environment, of an ancestral adaptation? One can certainly be forgiven for thinking that natural selection simply cannot be invoked to explain her enjoyment of *Ferris Bueller's Day Off*, in particular, or her enjoyment of television and movies in general. To show how our evolutionary history is relevant to contemporary forms of entertainment, we need a better model of what is distinctive about the activity of being entertained.

We propose that what distinguishes entertainment from other forms of intrinsically motivating activities is that in the case of activities such as eating, sleeping, and sexual intercours
the purpose of the activity is to change the state of the world. When somebody is hungry, he
or she enjoys eating, and the act of eating moves nourishment from outside the body to the
inside. If that somehow didn't happen, the behavior would not be fully satisfied. Imagine, for
instance, that somebody was given delicious food, and that this person enjoyed the activity of
having it in the mouth and chewing it, but was not allowed to swallow. Such an experience
might well be extremely frustrating, and we would ascribe this frustration to a design feature,
whose purpose it is to ensure that eating actually delivers its biological result of nourishment.

What is puzzling about entertainment is that it appears to fail to deliver the real thing,
and to fail by design. A key feature of entertainment is pretense: In the case of Rowan, for
instance, she treats the flickering lights on the screen in some way *as if* they were real events
she was observing, and the actors and actresses in the movie merely pretend that they have
the concerns and the emotions they appear to have. In fact, even the buildings, streets, and
natural surroundings may be pretend-houses of cardboard, fake studio set streets, and painted
or digitally added backdrops. When a patient is rushed to the operating table in *ER,* he or
she doesn't actually undergo surgery at all; there is just the pretense of cutting and sutures.
Yet, in contrast to the case of other intrinsically motivating activities, when somebody is
being entertained, this person appears to be satisfied by make-believe, and indeed to prefer a
simulacrum over reality itself.

These features of entertainment indicate that its biological function is of a character distinct
from that of other forms of intrinsically motivated behaviors. To understand what this function
might be, we need to develop a more detailed understanding of what the mind is doing when
it is engaged in an act of entertainment—we need a cognitive model.

A Cognitive Model for the Study of Entertainment

The first coherent theory of fiction-based forms of entertainment is Aristotle's (350 BC) notion
that the verbal arts and music are forms of *mimesis*, traditionally translated as 'imitation':

> Epic poetry and Tragedy, Comedy also and Dithyrambic poetry, and the music of the
> flute and of the lyre in most of their forms, are all in their general conception modes of
> imitation (*Poetics*, I.i).

Following Stephenson (1967) and Oatley (1994), we argue that modern technology has provided us with a more potent concept for the arts, namely that of simulation. In a simulation,
substitute objects are used to enact the core causal relations of a target phenomenon. Climate
modeling, for instance, uses computers to generate a virtual earth with a digital atmosphere
warmed by a make-believe sun, in such a manner that the causal relations of the real earth
and sun can be systematically investigated. The model is necessarily partial, providing only a
selective congruence of entailment structures, or it would be useless, like the map in Borges'
story, which entirely covers the territory.

Because simulations selectively preserve causal relations, they are useful for training
purposes, allowing you to acquire an understanding of a significant subset of causal relations without incurring the cost of experimenting with the phenomenon itself. An F16 flight
simulator, for instance, allows a novice to acquire experience and practical skills in manipulating a single-seat airplane without risking the loss of life and a multimillion-dollar
fighter jet.

Similarly, fiction-based forms of entertainment may be coherently understood as a species
of simulation. The *ER* that Rowan watches, for instance, is not a literal recording of an actual

emergency room, but a choreographed simulation using substitute agents and objects—actors instead of doctors, nurses, and injury victims; stage sets instead of real hospitals. The entertaining simulation is superior to an actual recording for two reasons. First of all, a literal recording would be hard to understand. Patients and people would come go, and a live recording would lack the background and contextual information that allows an outsider to make sense of what is happening. A fictional recreation of the events, in contrast, will typically focus on a manageable number of patients and doctors, and provide the viewer with the contextual information. In fact, fictive entertainment not only portrays the core causal relations of the target phenomenon; it makes these relations hyperintelligible (cf. Steen, 2005). In *The Distinction of Fiction*, Cohn (1999) argues that fictional narrative characteristically employs "narrative situations that open to inside views of the characters' minds" (vii), rendering the relations between perception, thought, emotion, and action transparent and visible to the reader or spectator. For a striking technologically updated instantiation of this general feature of fiction, consider the dramatic visualizations of the heroine's mental states in the television series *Ally McBeal*. Such a feature of fiction-based forms of entertainment is consistent with the notion that the events portrayed are simulations rather than imitations of events: through the use of sets, actors, and scripts, they enact the core relations that characterize the target phenomenon.

The second reason a fictional enactment is superior to a literal, historical recording is that the latter would in most cases be mind-numbingly boring. Fiction is a simulation crafted not only to be hyperintelligible, but to reveal a range of different aspects and possibilities of the phase space modeled. There is a constant premium on novelty at the level of the core relations modeled, typically achieved by introducing some new element in every episode and exploring the entailments of this new element from different angles. A continuous generation of this kind of novelty is closely tied to the experience of entertainment as intrinsically rewarding, as argued by Ryan and Deci (2000).

Our sketch of a cognitive model of entertainment, then, identifies it as a form of simulation, a dynamic model relying on substitute agents and objects that maintains a selective congruence of entailments. The simulation is prototypically focused on the domain of human action and characterized by a level of intelligibility of human relations exceeding those of real-life interactions, a constant exploration of variations in the core relations modeled, and an intrinsic motivation. From an evolutionary perspective, the question can now be asked: Why do human beings engage in an activity structured in this manner at all, and why is it subjectively experienced as intrinsically motivating? Why would a skillfully created fiction-based presentation of an emergency room be *entertaining*?

Entertainment As Play

The key to answering this question lies in the phenomenon of play (for an overview, see Vorderer, 2001; 2003). Aristotle (350 BC) traces mimesis back to a natural proclivity in animals. "The instinct of imitation," he argues, "is implanted in man from childhood, one difference between him and other animals being that he is the most imitative of living creatures, and through imitation learns his earliest lessons; and no less universal is the pleasure felt in things imitated" (*Poetics* I.iv). Aristotle here identifies four natural dimensions of play: its presence in several species of animals, its reliable expression in development, its pedagogical effect, and its inherent motivation. In contemporary terms, play has been found to be primarily a mammalian adaptation (Burghardt, 1984), and to be particularly important in infancy (Fagen, 1981). Its main pedagogical effect may be behavioral flexibility (Fagen, 1975). A tradition from Groos (1898) onward has attempted to make sense of its apparent frivolity or purposelessness; a prominent definition of play is that it encompasses "all motor activity performed postnatally that *appears* to be purposeless, in which motor patterns from other contexts may often be used in modified forms and altered temporal sequencing" (Bekoff and Byers (1981, pp. 300-301). As Rubin, Fein, & Vandenberg (1983) put it, "Play is intrinsically motivated."

Play is a credible developmental and evolutionary antecedent to the more sophisticated forms of entertainment we engage in today. Clear continuities exist between animal and human play. In a study of elementary forms of children's play, Steen & Owens (2001) seek to show that the activity of chase play is structured as a form of predator-evasion training, a form of play ubiquitous in mammals. They argue that play is an evolved pedagogy making use of simulations to lower the cost of learning. Children's play possesses the core features of simulations, such as the congruence of entailments with the situation modeled. Kavenaugh and Harris (1994) demonstrated in a series of experiments that children engaging in play are reliably able to track the causal consequences of pretend manipulations, so that, for instance, if you immerse a teddy bear into a pretend bath, it becomes wet (cf., Harris, 2000). The complex cognitive machinery required to understand an episode of *ER* has a long biological and developmental history.

The evolutionary argument is that biology would favor, through natural selection, cognitive adaptations supporting the capacity for and proclivity towards simulation in what was at first a relatively narrow range of situations, namely in cases where high levels of skill are required and the risk of failure is high (Steen & Owens, 2001). The prototypical situation is predation, or, more generally, high-stakes adversarial encounters. In brief, natural selection may have favored individuals who engaged in forms of play that trained them in skills relevant for predation. Such play would have functioned to build skills without having to incur the cost of actually encountering the predator.

An adaptation for learning from play would need to be intrinsically motivating. The virtues of practicing behavioral simulations for an event that has not yet taken place would be lost on young animals and children; they lack the cognitive apparatus for imagining the potentially lethal consequences of not being prepared when the predator arrives. This is why play is inherently gratifying without care about the outcome: Natural selection has built into the mind a natural delight in a range of activities, for reasons the individual need not concern herself with. The natural pedagogy of play and entertainment is no less effective though we do not realize that the biological function of the activity is learning.

An evolutionary theory of play, drawing on a cognitive model of entertainment as a form of simulation, yields a very simple proposition. It suggests that the biological function of entertainment is learning, that such learning is accomplished by means of cognitive adaptations for running, understanding, and parsing simulations, and that the pleasure and enjoyment we feel in fiction-based forms of entertainment is a necessary design feature of this natural pedagogical system. Since the *actual* function of an adaptation can differ greatly from its *biological* or evolved function, an evolutionary theory of entertainment does not amount to a claim that modern forms of entertainment, in fact, deliver genuine benefits. The theoretical significance of an evolutionary model is, above all, that it alerts us to the possibility that an underlying biological function is regulating entertainment preferences, even in a world that has long since outrun the environment in which human being evolved.

Toward an Integrated Model of Entertainment

In this chapter, we have introduced two broad types of models to explain what motivates people to seek out entertainment. Entertainment theory, we suggest, will benefit from adopting a mix of intentional and objectivist stances. Recent work in the psychology of intrinsically motivating behaviors allows us to situate entertainment in terms of an intrinsically motivated response to a set of entertainment opportunities. Specifically, we apply and extend Ryan and Deci's (2000) self-determination theory to the phenomenon of entertainment, emphasizing the centrality of autonomy, competence, and relatedness. We would like to end by demonstrating how these dimensions of intrinsically motivating responses to entertainment can both illuminate and be illuminated by our understanding of the evolutionary ancestry of entertainment in play.

In order to meaningfully relate the intentional and objectivist stances, we have sought to develop a core cognitive model of entertainment. We build on the work of Aristotle (350 BC), Stephenson (1967), and Oatley (1994) to suggest that entertainment can coherently be understood as a form of simulation. A simulation, we now argue, is uniquely suited to facilitating the experience of autonomy and competence that Ryan and Deci (2000) propose are core characteristics of intrinsically motivating experiences. Simulations model a phenomenon by the use of substitute agents and objects, selectively conserving causal relations. This loose and facultative relation between reality and entertainment provides both creators and audiences with a vast opportunity space. Because causal relations are selectively imported, agents retain an autonomy over the model that is unattainable in real life. Moreover, this autonomy can be utilized to fine-tune the selective importation of causal relations, so as to produce optimal challenges that generate the gratifying experience of competence.

These features of entertainment, we suggest, are in turn recognizable as design features of an evolved system whose biological function is learning. By creating a make-believe scenario in which the child can pretend to be, in certain controlled respects, a fearsome lion, or a prey animal escaping a monster, the child is able to assert his autonomy over the situation and indeed to direct the behavior of adults who wish to join in the play (Steen & Owens, 2001). In so doing, the child is also able and motivated to create challenges that are precisely optimized for his or her personal and temporally specific appetite for skill acquisition. This provides the child with an unparalleled opportunity to create a sustained and intensely gratifying experience of competence and mastery. As Csikszentmihalyi (1975) points out, the enjoyment derived from mastery is not merely an affirmation of a routine and tested ability; it is a successful movement into a new area of competence. In terms of biological function, the subjective phenomenology of competence is a sign to the organism that it has mastered an adaptively significant challenge.

The third cardinal element of Ryan and Deci's (2000) self-determination theory is that of relatedness, and we have argued in some detail prior that entertainment provides a rich opportunity for and experience of becoming closely related to others. In terms of the evolutionary model, the dimension of relatedness prompts us to supplement the pan-mammalian model described herein with a distinctively human extension. For the case of animal play, we argued that a core adaptive target was high-stakes adversarial encounters. While the theme of chase and mortal conflict is certainly a key theme in entertainment—ER, for instance, may heighten interest by dealing with life and death—this is clearly not in itself sufficient to make a television series attractive to watch as entertainment, nor is it central to what makes the series entertaining. In fact human play, starting with young children, adds a dramatically different dimension to the underlying primate and mammalian base. From an early age, children's play is primarily focused on human relations, and specifically on the complex relations between perception, thought, and action, represented in social roleplay and collaborative narrative play. In adult forms of entertainment, elementary forms of make-believe are elaborated into myths, tragedies, and Hollywood blockbusters (cf., Goldman, 1998). Entertainment, then, involves the exploration of relationships through simulations that permit individuals to *identify* with substitute agents (cf., Steen, Greenfield, Davies, & Tynes, in press) and thus create the subjective experience of relationships.

We suggest that the model we have sketched in this chapter, uniting work in the psychology of intrinsic motivation, the cognitive analysis of fiction-based forms of entertainment, and the evolutionary and developmental psychology of play, provides an integrated causal model for the study of entertainment. This model opens for a systematic experimental study of entertainment motivation. The utility of such an overarching theory is, above all, to help us formulate coherent research hypotheses about what types of information viewers and audiences may be searching for when then set themselves down before a television, movie screen, theatrical performance, or perform a skit or play before their friends one night when they are free to do exactly as they please.

REFERENCES

Aristotle (350 BC). *Poetics*. Trans. S. H. Butcher, 1955. Available at http://classics.mit.edu/Aristotle/poetics.1.1.html

Baron-Cohen, S. (1995). *Mindblindness: An essay on autism and theory of mind*. Cambridge, MA: MIT.

Bekoff, M., and Byers, J. (1981). A critical reanalysis of the ontogeny and phylogeny of mammalian social and locomotor play: An ethological hornet's nest. In K. Immelmann, G. W. Barlow, L. Petrinovich, & M. Main (Eds.), Behavioral *Development: The Bielefeld Interdisciplinary Project*. Cambridge, UK: Cambridge. 269–337.

Berlyne, D. E. (1971). *Aesthetics and psychobiology*. New York: Appleton-Century-Crofts.

Berlyne, D. E. (1974). *The new experimental aesthetics: Steps toward an objective psychology of aesthetic appreciation*. Washington, D.C.: Hemisphere.

Bjorklund, D. J., & Pellegrini, A. D. (2002). *Phylogeny and ontogeny: The emergence of evolutionary developmental psychology*. Washington, D.C.: American Psychological Association.

Bosshart, L., & Hoffmann-Riem, W. (Eds.). (1994). *Medienlust und Mediennutz. Unterhaltung als oeffentliche Kommunikation. Muenchen*: Verlag Oelschlaeger GmbH.

Bosshart, L. (1979). *Dynamik der Fernseh-Unterhaltung. Eine kommunikationswissenschaftliche Analyse und Synthese*. Freiburg: Universitätsverlag Freiburg Schweiz.

Bosshart, L., & Macconi, I. (1998). Defining "Entertainment." *Communication Research Trends, 18*(3), 3–6.

Bowlby, J. (1969). *Attachment and loss*. London, UK: Hogarth.

Bryant, J., & Miron, D. (2003). Excitation-transfer theory and three-factor theory of emotion. In J. Bryant, D. Roskos-Ewoldson, & J. Cantor (Eds.), *Communication and emotion. Essays in honor of Dolf Zillmann* (pp. 31–59). Mahwah, NJ: Lawrence Erlbaum Associates.

Bryant, J. (2004). Presidential address: Critical communication challenges for the international era. *Journal of Communication, 54*, 389–401.

Bryant, J., Roskos-Ewoldson, D., & Cantor, J. (Eds.). (2003). *Communication and emotion. Essays in honor of Dolf Zillmann*. Mahwah, NJ: Lawrence Erlbaum Associates.

Burghardt, G. M. (1984). On the origins of play. In P. K. Smith (Ed), *Play In Animals and Humans* (pp. 5–41). Oxford: Blackwell.

Cohn, D. (1999). *The distinction of fiction*. Baltimore, MD: Johns Hopkins.

Csikszentmihalyi, M. (1975). *Beyond boredom and anxiety*. San Francisco: Jossey-Bass.

deCharms, R. (1968). *Personal causation*. New York: Academic Press.

Deci, E. L., & Ryan, R. M. (1985). *Intrinsic motivation and self-determination in human behavior*. New York: Plenum.

Dennett. D. (1987). *The intentional stance*. Cambridge, MA: Bradford.

Fagen, R. (1975). Modelling how and why play works. In J. S. Bruner, A. Jolly, & K. Sylva (Eds.), Play: *Its role in development and evolution*. New York, NY: Penguin.

Fagen, R. (1981). *Animal play behavior*. New York, NY: Oxford.

Fisher, C. D. (1978). The effects of personal control, competence, and extrinsic reward systems on intrinsic motivation. *Organizational Behavior and Human Performance, 21*, 273–288.

Foley, R. (1995). The adaptive legacy of human evolution: A search for the environment of evolutionary adaptedness. *Evolutionary Anthropology, 4*, 194–203.

Glass, B. (1959). Maupertuis, pioneer of genetics and evolution. In B. Glass, O. Temkin, & W. Straus (Eds.), *Forerunners of Darwin: 1745–1859* (pp. 51-83). Baltimore, MD: Johns Hopkins.

Goldman, L. R. (1998). *Child's play. Myth, mimesis, and make-believe*. New York: Berg.

Groeben, N., & Vorderer, P. (1988). *Leserpsychologie: Lesemotivation—Lektuerewirkung*. Muenster: Aschendorffsche Verlagsbuchhandlung.

Groos, K. (1898). *The play of animals*. (E. L. Baldwin, Trans.). New York: Appleton.

Heider, F. (1958). *The psychology of interpersonal relations*. New York: Wiley.

Harris, P. L. (2000). The work of the imagination. Malden, MA: Blackwell.

Henning, B., & Vorderer, P. (2001). Psychological escapism: Predicting the amount of television viewing by need for cognition. *Journal of Communication, 51*, 100–120.

Horton, D., & Wohl, R. (1956). Mass communication and para-social interaction: Observation on intimacy at a distance. *Psychiatry, 19*, 215–229.

Kavenaugh, R. D. & Harris, P. L. (1994). Imagining the outcome of pretend transformations: Assessing the competence of normal children and children with autism. *Developmental Psychology, 30*, 847–54.

Kelly, G. A. (1955). *The psychology of personal constructs*. New York: Norton.

Leslie, A. M. (1987). Pretense and representation: The origins of "theory of mind." *Psychological Review, 94*, 412–426.

Leslie, A. M. (1994). ToMM, ToBy, and Agency: Core architecture and domain specificity. In L. A. Hirschfeld and S. A. Gelman (Eds.), *Mapping the mind. Domain specificity in cognition and culture*. New York: Cambridge.

Maupertuis, P. L. M. de (1745). *Venus physique*. La Haye.

Millikan, R. G. (1984). *Language, thought, and other biological categories: New foundations for realism*. Cambridge, MA: MIT Press.

Millikan, R. G. (1994). A common structure for concepts of individuals, stuffs, and real kinds: More mama, more milk and more mouse. *Behavioral and Brain Sciences, 9*, 55–100.

Oatley, K. (1994). A taxonomy of the emotions of literary response and a theory of identification in fictional narrative. *Poetics, 23*, 53–74.

Rosengren, K. E., Wenner, L. A., & Palmgreen, P. (Eds.). (1985). *Media gratifications research. Current perspectives*. Beverly Hills, CA: Sage.

Raney, A. A., & Bryant, J. (2002). An integrated theory of enjoyment. *Journal of Communication, 52*, 402–415.

Reeve, J. (1996). *Motivating others*. Needham Heights, MA: Allyn & Bacon.

Rubin, A. M. (2002). The uses-and-gratifications perspective of media effects. In J. Bryant & D. Zillmann (Eds.), *Media effects. Advances in theory and research* (pp. 525–548). Mahwah, NJ: Lawrence Erlbaum Associates.

Rubin, K. H., Fein, G., & Vanderberg, B. (1983). Play. In P. Mussen & E. M. Hetherington (Eds.), *Handbook of child psychology: Vol. 4. Socialization, personality, and social development* (pp. 693–774). New York: Wiley.

Ryan, R. M., & Deci, E. L. (2000). Self-determination theory and the facilitation of intrinsic motivations, social development, and well-being. *American Psychologist, 1*, 68–78.

Ryan, R. M. (1982). Control and information in the intrapersonal sphere: An extension of cognitive evaluation theory. *Journal of Personality and Social Psychology, 43*, 450–461.

Ryan, R. M. (1995). Psychological needs and the facilitation of integrative processes. *Journal of Personality, 63*, 397–427.

Shah, J. Y., & Kruglanski, A. W. (2000). The structure and substance of intrinsic motivation. In C. Sansone & J. M. Harackiewicz (Eds.), *Intrinsic and extrinsic motivation. The search for optimal motivation and performance* (pp. 105–127). San Diego, CA: Academic Press.

Snow, C. P. (1959/1993). *The two cultures*. New York: Cambridge.

Salomon, G. (1984). Television is 'easy' and print is 'tough': The differential investment of mental effort in learning as a function of perceptions and attributions. *Journal of Educational Psychology, 4*, 647–658.

Spencer, H. (1872-3). *The Principles of Pychology*. New York, NY: Appleton.

Steen, F. F. (2005). The paradox of narrative thinking. *Journal of Cultural and Evolutionary Psychology, 3*(1), 87–105.

Steen, F. F. & Owens, S. A. (2001). Evolution's pedagogy: An adaptationist model of pretense and entertainment. *Journal of Cognition and Culture, 1*, 289–321.

Steen, F. F., Greenfield, P. M., Davies, M. S., & Tynes, B. (in press). What went wrong in The Sims Online? Cultural learning and barriers to identification in a MMORPG. In P. Vorderer and J. Bryant (Eds.), *Playing computer games: Motives, responses, and consequences*. Mahwah, NJ: Lawrence Erlbaum Associates.

Stephenson, W. (1967). *The play theory of mass communication*. Chicago, IL: University of Chicago.

Tooby, J. & Cosmides, L. (2001). Does beauty build adapted minds? Toward an evolutionary theory of aesthetics, fiction and the arts. *SubStance, 30*, 6–27.

Vorderer, P. (2001). It's all entertainment—sure. But what exactly is entertainment? Communication research, media psychology, and the explanation of entertainment experiences. *Poetics. Journal of Empirical Research on Literature, Media and the Arts, 29*, 247–261.

Vorderer, P. (2003). Entertainment theory. In J. Bryant, D. Roskos-Ewoldsen, & J. Cantor (Eds.), *Communication and emotion: Essays in honor of Dolf Zillmann* (pp. 131–153). Mahwah, NJ: Lawrence Erlbaum Associates.

Vorderer, P., & Bryant, J. (Eds.). (in press). *Playing computer games—Motives, responses, and consequences*. Mahwah, NJ: Lawrence Erlbaum Associates.

Vorderer, P., Bryant, J., Pieper, K., & Weber, R. (in press). Playing computer games as entertainment. In P. Vorderer & J. Bryant (Eds.), *Playing computer games—motives, responses, and consequences*. Mahwah, NJ: Lawrence Erlbaum Associates.

Vorderer, P., Klimmt, C., & Ritterfeld, U. (2004). Enjoyment: At the heart of media entertainment. *Communication Theory, 4*, 388–408.

Weidenmann, B. (1989). Der mentale Aufwand beim Fernsehen. In J. Groebel & P. Winterhoff-Spurk (Eds.), *Empirische Medienpsychologie* (pp. 134–149). Muenchen: PVU.

White, R. W. (1959). Motivation reconsidered: The concept of competence. *Psychological Review, 66*, 297–333.

Zillmann, D., & Bryant, J. (1994). Entertainment as media effect. In J. Bryant & D. Zillmann (Eds.), *Media effects: Advances in theory and research* (pp. 437-461). Hillsdale, NJ: Lawrence Erlbaum Associates.

Zillmann, D., & Vorderer, P. (Eds.). (2000). *Media entertainment: The psychology of its appeal.* Mahwah, NJ: Lawrence Erlbaum Associates.

Zillmann, D. & Bryant, J. (1985). Affect, mood, and emotion as determinants of selective exposure. In D. Zillmann & J. Bryant (Eds.), *Selective exposure to communication* (pp. 157-190). Hillsdale, NJ: Lawrence Erlbaum Associates.

2

Selective Exposure Processes

Jennings Bryant
University of Alabama

John Davies
University of North Florida

The so-called Sovereign Consumer was the archetype of the innocent 20th century portion of the early information age. Ready access to vast amounts of information and entertainment was perceived to be empowering and wholly satisfying. However, as that optimistic era evolves into the more realistic, if not pessimistic, 21st century information age, the Sovereign Consumer is in danger of being replaced by the Overwhelmed Consumer, as the explosion in modern media offers an excess of abundance of every conceivable type of entertainment message and medium. For example, with television having achieved its 500-channel enhanced-TV promise (albeit still with "nothing on"), and with on-screen search-and-access systems falling woefully behind channel availability, viewers are increasingly spending more time searching for something to watch and less time engaged in pleasurable viewing. Satellite radio and Internet radio have replaced the 20th century's limited local dial with a virtually unlimited national and international digital keypad that puts consumers in touch with any type of audio message they can desire anyplace, anytime. Today's incredible array of Netpapers, which allows consumers to access with a click of the mouse daily and weekly newspapers of all types from almost any locale around the globe, has usurped the one-newspaper-per-town model of the recent past. Virtually endless racks of specialized magazines and a staggering array of e-zines provide modern periodical readers with an immense storehouse of information and entertainment riches. As for movies, box theaters and video stores are giving way to 24-screen multiplexes and internet DVD rental stores and the like, which operate in tandem with PPV cable or satellite systems to provide consumers with more movies than even the most ardent film buff can possibly watch.

This overwhelming abundance of media channels and messages issues a clarion call for veridical models and theories of consumer selectivity. Sorely needed are theoretical models that can help us understand and predict consumer choice behaviors for modern fishers in the overstocked Myriad Media Sea. It is crucial that communication scholars arrive at a more comprehensive and deeper understanding of consumer selectivity if we are to have any hope of mastering entertainment theory in the next iteration of the information age. Essentially, understanding selective-exposure theory is a prerequisite for constructing a useful psychology of entertainment.

The purpose of this chapter is to examine selective-exposure phenomena by explicating the processes through which selective exposure occurs. First, we sketch a brief history of selective-exposure research, including important precursor theories. Next, we outline some of the basic processes associated with how people learn the emotional and cognitive outcomes of media use in the service of entertainment. Then, we provide a survey of contemporary selective-exposure literature, focusing primarily on research involving Zillmann & Bryant's (1985) affect-dependent theory of stimulus arrangement. Finally, we offer some concluding assessments on the state of selective-exposure theory.

History

The roots of selective-exposure research are tightly interwoven with the history of the study of propaganda. By the time television had become a regular fixture in American homes, in the early 1950s, the notion that people avoided messages that conflicted with their opinions was widely held by psychologists (Cooper & Jahoda, 1947).

It was also believed that when people did encounter discordant messages, they reacted with various strategies to reconcile the discomfort those messages produced. A series of studies conducted following World War II by the Bureau of Applied Social Research at Columbia University demonstrated this phenomenon. The studies revealed that prejudiced individuals interpreted a cartoon designed to combat prejudicial attitudes differently than did non-prejudiced persons (Cooper & Jahoda, 1947). The researchers concluded that readers who harbored prejudiced views chose to perceive the meaning of the cartoon in accordance with their beliefs. These propaganda studies supported the idea that the same mass-mediated message can mean different things to different people.

The question that is a precursor to this finding is: Do different people even always *receive* the same message? Obviously, they do not, but those who espouse the idea of selective exposure assert that reception of a message can be predicted by various characteristics of the individuals who make up the audience.

In mass communications, one of the earliest studies to provide solid empirical support for the selective-exposure hypothesis was the classic Erie County study of the 1940 presidential election (Lazarsfeld, Berelson, & Gaudet, 1968). That investigation employed a panel technique that, according to Lazarsfeld et al., was a superior method because "we did not describe opinion; we studied it *in the making.*" (p. xxii, emphasis in original). However, the study revealed very little opinion-making as a result of the media campaigns. Rather, the media appeared to reinforce intentions that constituents already held. Evidence for this conclusion is found by noting that approximately two-thirds of decided voters saw and heard more of their own party's publicity than the opposition's, and that the campaigns convinced very few residents to change their voting intentions. This latter phenomenon was even more pronounced among voters who reported the strongest party affiliation.

Of course, during the early 1940s, at the time of the Lazarsfeld et al. (1968) study, radio and newspapers were the most universal media; television penetration was not substantial until after World War II (Comstock & Scharrer, 1999). Perhaps different results would have been found had the presidential hopefuls had use of modern television technology. A study of a televised political campaign by Schramm and Carter (1959), however, casts some doubt on this speculation. Their study focused on the effects of a marathon media event staged by then-Senator William Knowland in an effort to persuade voters. Knowland, a Republican, appeared on the air continuously for a period of over twenty hours from the evening of October 31, 1958 until 7:00 p.m. the following evening. Telephone interviews revealed that Republicans were nearly twice as likely as Democrats to view the telecast. Of the 65 viewers interviewed by Schramm and Carter, only two said Knowland's program helped them make up their minds how to vote, and only one said it convinced him to change his mind.

Studies such as these undoubtedly influenced Klapper's (1960) survey of mass media research entitled *The Effects of Mass Communication.* A major theme that runs through the book is the media's inability to persuade, on the one hand, and its capacity to reinforce, on the other. After an exhaustive review of the literature, Klapper wrote, "mass communication ordinarily does not serve as a necessary and sufficient cause of audience effects, but rather functions among and through a nexus of mediating factors and influences" (p. 8). It was Klapper's contention that some of the primary mediators of effects were the processes of selective exposure, selective perception, and selective retention. For selective exposure, he offered the following definition: "the tendency of people to expose themselves to mass communications in accord with their existing opinions and interests and to avoid unsympathetic material" (p. 19).

Klapper's (1960) definition of selective exposure is firmly grounded in Festinger's (1957) influential theory of cognitive dissonance. Briefly, that theory asserts that mental discomfort is produced in individuals when they simultaneously hold two conflicting thoughts, attitudes, or beliefs. When the individual becomes aware of the contradiction, dissonance is created, and the person will seek to reduce it, either through behavior or by modifying one of the discordant elements. One method of reducing dissonance is either avoiding information that produces it, or seeking information sympathetic to one's beliefs.

Because mass media are major sources of information in modern society, Festinger's (1957) theory appeared to hold considerable promise for the study of mass communication processes and effects. The theory, however, failed to live up to expectations—at least according to some researchers. Freedman and Sears (1965), for example, claimed that field research and experimental studies produced a puzzling contradiction. They acknowledged that the composition of audiences for mass communications generally included more individuals sympathetic to the message expressed through the medium than individuals opposed to the message. However, laboratory research did not provide sufficient evidence to confirm the theory. Freedman and Sears asked how it could be possible that audiences in general clearly demonstrated selective-exposure tendencies, but individuals in the laboratory failed to seek out supportive information? They reasoned that other alternative motives, such as informational utility, might be underlying selective-exposure phenomena.

Other researchers were not quite so pessimistic as Freedman and Sears (1965). Katz (1968) speculated that the supportive-selectivity hypothesis, although not definitively sustained by the research of the time, nevertheless would eventually be proven valid. In the meantime, he advocated exploring the idea that utility and/or interest could serve as a basis for selectivity.

In fact, earlier studies of television use among children had identified several factors motivating television use in general. Schramm, Lyle, and Parker (1961) observed, for example, that the desire to be entertained was a primary motive for choosing to watch television. This finding, although seemingly obvious, typifies the focus of research in the 1960s. Katz (1968) accurately noted that the study of mass communications had shifted from "what the media *do to* people" to "what people *do with* the media" (p. 88, emphasis in original).

The shift in focus from cognitive factors as a basis for selectivity to that of interest and entertainment marked the beginning of a stream of research in communication that centered on the emotional antecedents of media use.

Learning the Emotional Outcomes of Media Use

Most media scholars—and laypersons as well—are concerned about whether exposure to media messages causes social or psychological effects. One aspect of that concern is expressed by the questions: "Do we learn from the media, and, if so, *what* do we learn?" The first portion of the question has received considerable empirical attention, and the answer consistently has proven to be "Yes, all media use entails learning" (e.g., Bandura, 2002). Moreover, research in cognitive psychology has taught us that repeated exposure to a particular stimulus strengthens the mental connections between that stimulus and its related constructs (e.g., Higgins, 1996). Apparently, the mere activation of a construct increases the probability that the same or related construct will come to mind. The obvious implication of this phenomenon is that memory is a dynamic rather than a static system, because every encounter with any stimulus, no matter how trivial, serves to reinforce or fortify some mental connections. (Similarly, other connections will be weakened, in a relative sense, by mere virtue of their failure to be reinforced or fortified.) Therefore, all media use (indeed, all sensory experience) is a learning experience in the sense that we are chronically activating mental representations in memory.

What, then, do we learn from media use? To a large extent, the answer to that question depends on the content we choose to consume. That is not to say, however, that all media use is so idiosyncratic that we cannot identify some very basic processes involved with most, if not all, mediated experiences. In the next section, we identify some of these processes.

Classical Conditioning. About the time mass communication scholars began to focus on "what people do with media" (Katz, 1968, p. 88), many psychologists were taking an opposing approach to the study of human behavior. Behavioral psychology assumed that environmental factors determined behavior, that the human mind was largely an impenetrable black box, and that by observing an organism's reactions to the environment we could make inferences about the nature of the human condition. The key to predicting human behavior, therefore, was observing how human beings responded to the presentation of various stimuli. Although many of the assumptions made by early behaviorists have fallen out of favor, much of the learning paradigm they created is still applicable and in use today. Some clinical therapies, for example, rely on classical conditioning techniques in treating phobias. In any case, given the circumstances under which most media consumption takes place, it is hard to imagine that people do not acquire some sort of learning (in the classical conditioning sense) from media.

According to the classical conditioning paradigm, exposure to some stimulus (called an unconditioned stimulus) naturally elicits a response (called an unconditioned response). When the unconditioned stimulus is repeatedly paired with some other stimulus (conditioned stimulus) the conditioned stimulus may come to elicit a response (conditioned response) that is very similar to the unconditioned response. In simpler terms, we learn to react in the same way to two stimuli that have been repeatedly associated with each other. For example, a child may react with fear upon seeing a snake. If the child also hears a bell every time he sees the snake, the child will likely learn to react with fear every time the bell is rung, even if no snake is present.

The same process may characterize media use. In our earliest experiences with media we may see a mediated image, or hear a mediated voice, and we respond to that stimulation. What is the nature of that response? It almost certainly has an emotional component. Much media research has confirmed that emotional reactions to media are part and parcel of the experience of media consumption (e.g., Bradley & Lang, 2000; Reeves & Nass, 1996). Research has shown that exposure to different structural features of media (i.e., pacing, editing, message complexity) results in different emotional outcomes (e.g., Grabe, Zhou, Lang, & Bolls, 2000). Eventually, after repeated association, certain structural features come to be associated with emotional outcomes. The same process may occur with more complex stimuli (i.e., media message content), and likely both structural features and content come to be associated with emotional outcomes. Thus, initially, media use in general is associated with emotional consequences. Eventually, specific content comes to be associated with specific emotional results.

Excitation Transfer. Beyond classical conditioning, other processes ensure that the associations between media and emotional outcomes are generally positive; otherwise, there would be no compelling motive to repeatedly seek entertainment experiences. Most mediated entertainment, and television in particular, is particularly adept at inducing arousal in media consumers (Zillmann, 1991). Perhaps not by conscious design of its creator, but just by its very nature, a good yarn gets the audience excited, and television is the central source of stories in contemporary culture (Gerbner & Gross, 1976). In addition to the arousal-inducing properties that arise from the storyteller function of media, structural features (so-called "formal features," e.g., editing, sound effects) contribute to the stimulating experience of media use. Other entertainment media rely on stimuli other than stories to entertain audiences, but the consequences are similar—excitation. By excitation we mean, in biological terms, an increase in sympathetic activity of the autonomic nervous system.

How does a good story produce excitation? Typically, attention is captured by some sort of conflict or obstruction between a protagonist's current state and his or her desired goals. By its very nature, conflict is an unpleasant experience, and it is precisely this unpleasantness that induces arousal in the listener. Conflict is a staple of mediated entertainment, and we should therefore expect the consumer of such fare to be in some state of unpleasant distress. This raises an interesting paradox. How does a person in a state of unpleasant distress come to associate that experience with positive emotional outcomes? This seeming contradiction is eloquently resolved by excitation-transfer theory (Zillmann, 1971; see also Bryant & Miron, 2003; Zillmann, 2003). Essentially, the arousal induced by witnessing an unpleasant conflict intensifies the feelings of relief that are experienced when the conflict is overcome. The resulting positive feelings are misattributed to the enjoyment of the experience. This process occurs naturally in many entertainment experiences, and, in most cases, without our conscious awareness. Indeed, lack of awareness is essential to the transfer process (Cantor, Zillmann, & Bryant, 1975). Typically, we are too cognitively preoccupied with the storyline to pay much attention to subtle physiological cues furnished by emotional fluctuations. The result is that we come away from a potentially distressing experience feeling satisfied and entertained.

Of course, conflict is only one of several domains of media content in which excitation transfer operates. The arousal-inducing stimuli need not be negative. Research has also shown transfer effects in the context of comedy (Cantor, Bryant, & Zillmann, 1974) as well as erotic stimuli (Cantor et al., 1975). Thus, excitation-transfer theory is one explanation for how we come to associate media experiences with positive emotional outcomes.

Mere Exposure. Another mechanism that possibly underlies the connection between exposure to media and positive emotional outcomes is the mere-exposure effect (Zajonc, 1968). In a seminal study, Zajonc demonstrated that merely exposing participants to a stimulus (Turkish adjectives, Chinese characters, or photographs) caused an increase in liking for that stimulus. Since then, more than 200 published studies have confirmed that the effect is both robust and reliable (Bornstein, 1989). From these studies, psychologists have slowly pieced together the conditions that are most likely to yield a mere-exposure effect.

One of the most important of these conditions, is the duration for which a stimulus is presented. A typical mere-exposure experiment consists of an exposure phase in which participants view some stimuli for a brief time, and a test phase during which the participants rate the same stimuli for liking, typically using a seven-point scale anchored by "like" and "dislike." Generally, shorter familiarization phases tend to yield stronger effects. In fact, several researchers have found that exposures at the subliminal level produced stronger effects than stimuli that could be consciously perceived (Kunst-Wilson & Zajonc, 1980; Seamon, Marsh, & Brody, 1984). Bornstein's (1989) meta–analysis concluded that studies of subliminal mere exposure had a combined effect size of $r = .53$, whereas experiments in non-subliminal (or conscious) mere exposure had a combined effect size of $r = .14$—a substantial difference.

One interesting exception to this trend is the case of studies conducted outside of a laboratory with naturally occurring stimuli. These natural experiments do not employ a familiarization phase and so duration of exposure is not controlled. Even so, these studies tend to yield strong effect sizes comparable to subliminal mere-exposure investigations (Bornstein, 1992).

Advertisers and marketers, in an effort to cultivate positive attitudes toward brands, regularly employ a similar approach. That is, an advertisement is repeatedly broadcast or printed in the hopes that consumers will develop positive feelings for the brand. This strategy does not always work, and at least in the case of attitudes toward the advertisement, consumers initially may have positive feelings about the ad, but with repeated exposures those feelings may sour via a so-called "wear-out" effect.

Several reasons could account for the relative failure of advertising to consistently produce mere-exposure effects. First and foremost, research in the laboratory generally utilizes much less complex stimuli (e.g., names, Chinese characters, photographs of faces) than do advertisers. Second, the ultimate purpose of advertising is to persuade, and people may naturally be resistant to attempts to manipulate their attitudes or behaviors. Indeed, several mere-exposure studies have demonstrated that the mere-exposure effect diminishes or disappears altogether if participants are consciously aware that the familiarization phase of an experiment is linked to the subsequent test phase (Bornstein & D'Agostino, 1992, 1994; Murphy & Zajonc, 1993). Likewise, if viewers are aware that advertisers' attempts to familiarize consumers with brands are linked to ulterior goals (i.e., liking of the brand, awareness of the brand) the persuasive objectives of the advertisement fail, or backfire. Advertisers know that a successful ad is one that is entertaining enough to capture attention, but not so entertaining that it draws attention to itself and causes consumers to question the motives of the advertiser. Such phenomena have been well researched under the theoretical framework of the elaboration-likelihood model (Petty & Cacioppo, 1986).

One purpose of entertainment media is the stimulation of emotions, and audiences generally are not resistant to such attempts, unless the plot is so over-the-top, or the content so blatantly panders, that it calls attention to itself. Thus, entertainment media are potentially excellent stimuli for naturalistic experiments in mere-exposure effects. Potentially, exposure to naturally occurring stimuli in media produces liking for those stimuli. That is, with repeated exposures to similar media content, we develop affinity for that content. The mere-exposure effect is generalizable as long as some overlap of features among stimuli exists (Gordon & Holyoak, 1983; Monahan, Murphy, & Zajonc, 2000), so media consumers do not have to see exactly the same stimuli in order for liking to develop.

The question arises: What would constitute an appropriate stimulus to yield a mere-exposure effect? Something as abstract as a theme, good triumphs over evil, for instance, might not be concrete enough for viewers to make a connection. Recall that most stimuli in mere-exposure experiments are much more tangible. However, in any typical program viewers are repeatedly shown shots of characters that appear in the episode. At this level, mere-exposure effects could occur. Several experiments have demonstrated mere-exposure effects with photographs of human faces (e.g., Bornstein & D'Agostino, 1994; Harmon-Jones & Allen, 2001; Zajonc, 1968). One study even found that subliminally exposed photographs of a confederate had an influence on how participants interacted with the confederate in a later phase of the experiment, although no differences were found in actual liking of the confederates (Bornstein, Leone, & Galley, 1987). Indeed, the original intent of Zajonc's (1968) seminal study was to understand the contribution exposure makes in dynamic social processes of communication and cooperation.

If mere exposure to mediated characters increases liking of those characters, then we would expect that viewers would hold stronger affective dispositions for characters that they see repeatedly, compared to other less frequently appearing characters. This phenomenon has a chain of implications that starts with the suspense we might feel as a liked character is endangered, and ends with the processes of excitation transfer as we feel a heightened sense of relief when the liked character escapes unharmed.

Unfortunately, to our knowledge no study has ever tested whether mere exposure contributes to increased liking of characters and whether this has any influence on empathy for a character. Still, theoretically it seems plausible that mere exposure is one process at work as people learn to associate media with emotional outcomes, and it is this learning process that plays a crucial role in selective exposure.

Implicit Memory. Research on mere exposure yields one more potentially important factor in learning. The finding that subliminal exposure effects are more robust than supraliminal effects has led some researchers to argue that the subliminal mere-exposure effect is best understood in terms of implicit memory (Seamon, Williams, Crowley, Kim, Langer, Orne, & Wishengrad, 1995). Implicit memory, according to Schacter (1987), is information encoded in a particular episode that is subsequently expressed without deliberate or conscious recollection. Thus, participants in subliminal mere-exposure studies express greater liking for stimuli that they have seen in the familiarization phase of an experiment, even though they cannot consciously recognize having seen them. Therefore, implicit memory and liking apparently have an important relationship.

The nature of this relationship is clarified by recent research using psychophysiological and neurological approaches. Lieberman (2000) argued that the basal ganglia are the neurological basis for implicit learning. He further claimed that the basal ganglia function in the evaluation of positive-affective stimuli and positive emotional experience. He stated, "the basal ganglia learn temporal patterns that are predictive of events of significance, regardless of conscious intent to learn the predictor–reward relationship, as long as exposure is *repeatedly instantiated*" (p. 116, emphasis added). Repeated exposure is essential because the recognition of reward-producing stimuli apparently develops slowly over time. Support for these arguments is found, in part, by mere-exposure studies. Harmon-Jones and Allen (2001) found that familiar stimuli elicited more zygomatic activity, which indicates positively valenced arousal. On the other hand, unfamiliar stimuli failed to elicit increased corrugator activity, which suggests that the mere-exposure effect enhances positive affect rather than decreases negative affect. Elliot and Dolan (1998) utilized positron emission tomography and found that different brain regions were more activated when participants in a mere-exposure paradigm preformed liking judgments than when they were engaged in recognition judgments. Specifically, they found caudate activation in the brain. The caudate is a component of the basal ganglia.

These studies suggest that brain structures associated with implicit memory also function in the registration of positive reward. Lieberman (2000) used this evidence to argue that intuition is associated with implicit learning of positive rewards. These studies provide an explanation for an area of selective-exposure research largely taken for granted. Until now, research that utilizes selective-exposure paradigms has assumed that media consumers learn from past experience which stimuli provide the most effective means to achieve a desired mood state. When a person sits down to watch television, for example, he or she may flip through the channels, sampling the available options, until the individual see something of interest and settles down to watch. Often, this sampling behavior entails exposure to a channel for less than 500 milliseconds (Davies, 2004). In that short period of time, how does a viewer judge that a program will or will not be an effective means to manage mood? Intuitively, we seem to know that certain content will lead to reward. In fact, expression of behavior in this manner is indicative of learning in implicit memory.

If implicit memory is indeed the mechanism by which we learn to associate positive emotional reward with media, then several media-related behaviors ought to be apparent. First, when presented with novel stimuli (i.e., never-before-seen programs, websites, music) heavy viewers (in the case of television programs) should be able to make a media-use decision more quickly than could light viewers. Second, children should be less adept than adults at selecting media content that appropriately alters mood in a desired direction. Third, media consumers ought to be able to make rapid media-use decisions even when exposure to media stimuli occurs for very brief durations.

To summarize this section, selective-exposure theory assumes that we learn through prior experience to associate media with positive emotional outcomes. The processes and mechanisms by which this learning takes place include classical conditioning, excitation transfer, mere exposure, and implicit memory. We now turn to a survey of current research that supports selective-exposure theory, and detail the specific processes involved in media-use decision-making.

Selective-Exposure Processes

A large body of literature suggests that individuals select media content because of the effect it has on their emotional state. The most well-developed theory in this area is Zillmann and Bryant's (1985) affect-dependent theory of stimulus arrangement (see also chapter 14 by Knobloch-Westerwick on mood management, this volume). Two basic premises anchor this theory. First, the theory assumes, on the one hand, that people are motivated to minimize exposure to negative, noxious, or aversive stimuli, and second, on the other hand, are motivated to maximize exposure to positive, pleasurable stimuli. The theory further maintains that individuals will try to arrange external stimuli to maximize the chances of achieving these goals. Often, this goal is achieved through selecting a variety of affect-inducing programs, music, stories, or other entertaining fare commonly found in the mass media. However, the affect-dependent theory is broad enough to include means of mood management other than mediated entertainment.

This theory considers four key aspects of using media as a mood manager: excitatory homeostasis, intervention potential of a message, message-behavioral affinity, and hedonic valence. Several studies have found support for these assumptions.

Excitatory Homeostasis. Excitatory homeostasis refers to the tendency of individuals to choose entertainment to achieve an optimal level of arousal (i.e., if affective state is considered aversive, understimulated persons tend to choose arousing material, overstimulated persons choose calming fare). This phenomenon was clearly observed in a study by Bryant and Zillmann (1984). In that study, when individuals were placed in a state of boredom, they selected arousing media messages and avoided relaxing fare when given the opportunity to do so. On the other hand, individuals who experienced stress chose to view relaxing programs more than did bored individuals. Mastro, Eastin, and Tamborini (2002) observed a similar pattern of results among Internet users. In a pretest, they observed that rapidly surfing a Website was more likely to be associated with stress than slowly surfing a Website. In contrast, participants found the experience of slowly surfing a Website to be more boring than rapidly surfing the Website. In the experiment proper, Mastro et al. found that the Web-surfing behaviors of stressed individuals were associated with a slow-surfing pattern (which produced boredom in the pretest), and bored individuals tended to surf the Web more rapidly (which was associated more closely with stress than slow-surfing). In another study, researchers used a series of surveys and noted associations between selective exposure to television and stressful life events (Anderson, Collins, Schmitt, & Jacobvitz, 1996). Overall, the surveys found support for mood management by television but revealed some gender differences. As predicted, stressed men paid greater attention to television than did men who were under less stress, but both stressed women and women under less stress paid equal attention to television. Stressful life events were associated with TV addiction for women, but not for men. Both stressed men and women viewed more comedy, as expected, but stressed men also watched more violent/action/horror programs than did men under less stress.

Intervention Potential.   The intervention potential of a message refers to the ability of a message to engage or absorb an aroused individual's attention or cognitive-processing resources. Highly engaging messages supposedly prevent an individual from dwelling on a particular affective state and consequently diminish its perceived intensity. Bryant and Zillmann (1977) reported that the intervention potential of a message moderated subjects' retaliation against a person who annoyed them. That is, highly absorbing messages reduced retaliation, and minimally absorbing messages failed to reduce retaliation. Another study found that highly involving TV programs impeded memory of experiences that occurred prior to viewing (Davies, 2004).

Message-Behavioral Affinity.   Communication that has a high degree of similarity with affective state is said to have message-behavioral affinity. Generally, the greater the degree of similarity between a message and a individual's mood, the less potential the message has of altering or diminishing affective state. This explanation accounts for the finding by Zillmann, Hezel, and Medoff (1980) that participants who were insulted in an experiment tended to avoid hostile comedy when given the opportunity to view television. Medoff's (1982) results provided strong support for the prediction that annoyed people would be motivated to avoid comedy programs that remind them of their emotional state. Bryant and Zillmann's (1977) investigation had initially demonstrated a similar phenomenon. Wakshlag, Vial, and Tamborini (1983) manipulated apprehension about crime in film viewers by showing them either a graphic crime documentary or a control film. When given the choice to view another film, apprehensive viewers chose films that strongly portrayed the fulfillment of justice.

Hedonic Valence.   Presumably, all communication messages can be classified on a continuum of positive and negative. Positive messages might be described as uplifting, reassuring, amusing, happy, and so on. The list of adjectives that describe negative messages includes threatening, noxious, distressing, sad, and the like. Hedonic valence refers to the positive or negative nature of a message. Affect-dependent theory of stimulus arrangement posits that people in a bad mood will prefer media (or experiences) with a positive valence. Consistent with this prediction, Knobloch and Zillmann (2002) demonstrated that when students in a negative mood were given the opportunity to listen to music, they opted to listen to joyful music for longer periods than did students who were in a good mood. Generally, in cases where the hedonic valence of a message is opposite to the valence of an individual's affective state, the message will effectively diminish the intensity of the affective state. However, this relationship is true only under certain conditions. The hedonic valence in combination with the excitatory potential of a message yields different effects. For example, a meta-analysis of the effects of pornography on aggression revealed differently valenced correlations between aggression and specific types of erotic stimuli (Allen, D'Allessio, & Brezgel, 1995). In that study, exposure to nonpornographic nudity was associated with lower levels of aggression, whereas exposure to nonviolent or violent pornography yielded positive associations with aggression. Another study produced analogous findings—messages with a positive valence diminished the effects of a negative mood if the message had low excitatory potential, but exacerbated aggressive reactions if the excitatory potential was high (Zillmann, Bryant, Comisky, & Medoff, 1981).

Studies that have investigated the relationship between a women's menstrual cycle and mood management have yielded particularly strong evidence in favor of the notion that people manage moods through media. The literature indicates that women are more likely to view comedies when their menstrual cycle makes them most likely to be depressed (Helgrel & Weaver, 1989; Meadowcroft & Zillmann, 1987), and they are more likely to choose programs with higher sexual and romantic content when their cycle makes them more likely to have an increased libido (Weaver & Baird, 1995).

The vast majority of mood management studies has focused on the effects of aversive affective states on selectivity. In survey research, the focus is on correlations between life events or situations and television viewing. In experimental studies, the researcher induces a "bad" mood in participants and then gives them the opportunity to view television programs. In general, the literature has found support for mood management through media. People have been observed to use media to alleviate boredom (Bryant & Zillmann, 1984), stress (Anderson, Collins, Schmitt, & Jacobvitz, 1996), apprehension (Wakshlag, Vial, & Tamborini, 1983; Wakshlag, Bart, Dudley, Groth, McCuthcheon, & Rolla, 1983), annoyance (Medoff, 1982), and depression (Dittmar, 1994; Helgrel & Weaver, 1989; Meadowcroft & Zillmann, 1987; Potts & Sanchez, 1994, Weaver & Baird, 1995). Other research has shown selective avoidance of television in response to annoyance (Christ & Medoff, 1984).

In contrast, very few studies have investigated the effects of positive affect on selectivity. When given the choice, first and second graders were more likely to select programs with fast-paced humorous inserts (Wakshlag, Day, & Zillmann, 1981). Similar results were found with the same age group in a study of the effects of tempo of background music (Wakshalg, Reitz, & Zillmann, 1982). Perse's (1998) investigation of channel changing revealed that positive affect did not deter people from switching programs, but negative affect increased the likelihood a viewer would change the channel. However, these studies only considered emotional reactions to television programs; none of them considered initial mood state.

Critique of the Theory

A large body of literature supports the basic tenets of the affect-dependent theory of stimulus arrangement. This does not mean that the theory has already achieved its full potential, however. For example, by design, the theory has focused on micro-psychological processes to explain behavior. Thus, social-psychological and sociological influences on the same behavior typically have been neglected. In context of the theory, we know relatively little about how small groups or dyads make decisions to consume media, or how the presence of others influences individual media-use decisions. Likewise, we know relatively little about how broader, macrosocial forces, institutions, or social structures interact with individuals in making media-use decisions.

At the heart of the affect-dependent theory is the assumption that individuals will maximize psychological gratifications through media use. Experimental studies have demonstrated that selectivity is characterized by media choices that have the potential to minimize negative moods. To a lesser extent, studies have also shown that individuals select media to maximize gratifications associated with positive moods. Nevertheless, selectivity might be influenced by factors other than a person's private emotional state. For example, individuals likely shift internal definitions of what constitutes psychological gratification in the presence of others. For example, when viewing TV alone, individuals may strive to maximize pleasure by seeking out programs according to the hedonic valence of the program—according to the theory. However, the presence of others may cause the individual to consider social norms of behavior. If another person is present and is of equal status, social norms dictate that the individual consider the other person's affective state as well as his or her own. Thus, the definition of what constitutes psychological gratifications shifts from a personal level to a group level, and behavior is in terms of maximizing the group's psychological gratifications rather than an individual's. In this scenario, viewing choice might not align with theory at all, because maximizing pleasure is now in terms of hedonic valence relative to the group's overall mood. The situation may be further complicated if the group's mood is uncertain or difficult to gauge, if group cohesion is more important than media gratification, if the group is extremely heterogeneous, and so on. Furthermore, individuals may obtain gratification from solving the problem of uncertainty; that is, an individual may achieve the goal of mood optimization solely from finding a program that is most appropriate for the other person—an effect independent of the applicability of the media content to the individual.

Another assumption made by the affect-dependent theory is that individuals will maximize pleasure and minimize aversive states. Behavior may be totally different if we look at short-term versus long-term behavior. Generally, the theory, and research on the theory, has looked at short-term effects and consequences, but individuals may adopt (or have greater ability to adopt) long-term strategies. Consequently, the theory may predict short-term behavior, but individuals may act according to the dictates of a long-term strategy. To be fair, later versions of the theory have introduced spontaneous versus telic hedonism (Zillmann, 2000), but little research has been conducted in this area.

The maximize/minimize assumption also dictates that individuals are optimizers in terms of mood management, when in fact they may be satisfiers. That is, it may require too much effort to balance a mood perfectly, and individuals may opt for a reduction in aversive state rather than a complete reversal of it. Complicating factors, such as the amount of choice, may also exist. That is, a person may have to flip through 500 channels to find an optimally satisfying solution, and rather than expend the effort to sample each individual channel, she or he may select the first channel that approximates her or his ultimate mood management goal.

Perhaps, in the interest of parsimony, the theory focuses on emotional states, and, other than the notion of intervention potential, has little to offer in terms of cognitive states and selective exposure. The theory also operates from the perspective that media choice is a function of affective state. Consequently, most research in this area has induced emotion and then measured media consumption choices. Cognitive processes play a small role in this paradigm, but in real life may play an important role. For example, Festinger's (1957) theory of cognitive dissonance proposes that mental discomfort is produced in individuals when they simultaneously hold two conflicting thoughts, attitudes, or beliefs. When the individual becomes aware of the contradiction, dissonance is created, and the person will seek to reduce it, either through behavior or by modifying one of the discordant elements. Selective exposure to information is a branch of dissonance research that argues that people will seek out information that supports the decisions they have made. Cotton (1985) has provided a review of earlier research in selective exposure to information, and he concluded that although initial research was disappointing and contradictory, subsequent research has shown that dissonance reduction is a phenomenon that motivates exposure to communication. Recent research has borne out these observations in some instances (e.g., D'Alessio & Allen, 2002; Jonas, Greenberg, & Frey, 2003).

Survey research suggests that preferences for certain genres, such as crime dramas, may be motivated by a desire to manage moods, but a desire for cognitive stimulation is a primary motivator for viewing news magazine shows (Hawkins et al., 2001).

The cognitive role may also have been manifest in the control groups of mood-management experiments. For these participants, mood state was not induced, and therefore they have been considered to be in a relatively neutral emotional state. In such cases, consumption may have been the result, for example, of priming a mental model of media effects (i.e., if I view this program I will experience arousal and positive feelings). Thus, media choice may have been made on the basis of cognition and the potential for affect, rather than on the basis of affect alone. (This is not necessarily problematic for the theory; it simply has not incorporated such processes).

The affect-dependent theory is particularly suited to mood management of entertainment media. It is less clear how this theory might relate to other types of media, which have a primary purpose to inform or educate. The concept of informational utility contributes to the understanding of media use behavior, but research has not yet separated the effects of hedonism and informational utility on selective exposure. Given the rise of so-called "infotainment" and so-called "edutainment," this is an important issue to address.

Moreover, as noted earlier, very little research has considered the effect of positive mood states on selective exposure to media. Because of the ambiguity found in these few investigations, clarifying research is needed.

Finally, the notion of intervention potential assumes that media users will want to disrupt cognitive rehearsal of negative experiences and maintain rehearsal of positive experiences. However, there are times when it is beneficial to dwell on negative experiences in the interest of making sense of an experience, or in finding solutions to a problem. Some personality types may simply prefer to dwell on negative experiences; others may chronically avoid such thoughts. For instance, neuroticism is associated with a tendency to ruminate on negative moods and experiences (Roberts, Gilboa, & Gotlib, 1998). Use of the media for its intervention potential may be more of a function of problem-solving strategies and personality type than mood management per se. Indeed, Davies (2004) reported that neuroticism negatively influenced perceptions of a stressful experience and predicted exposure to media with low intervention potential, but only when negative mood was relatively less intense. Under extreme duress, mood—rather than personality—predicted media exposure.

Conclusions

At first glance, the basic premise of selective-exposure theory seems so simple; we watch what makes us feel good. Our personal experiences as users of media seem to confirm such a conclusion. As researchers, however, we realize that such a simple conclusion is merely the proverbial tip of the iceberg. Clearly, the processes involved in making a decision to consume media are complex and involved. Throughout this chapter we have drawn upon existing research to identify processes of selective exposure that are reasonably well-established. Where such research has not yet been conducted, we have attempted to offer our own speculation as to the nature of these processes, and to identify areas where future research is needed.

In terms of addressing areas of needed research, it may suffice to offer a general approach to advancing selective exposure. That is: What factors predict the use of media to manage moods? In other words, what are the antecedents of mood management behavior? Which individuals are more likely to prefer media as a solution to emotional distress, and which individuals regulate their moods through other activities? Answering these questions would go far toward advancing our understanding of the processes that serve as the foundations of selective-exposure behavior, and for our ability to predict dysfunctional behavior or harmful effects.

We suggested that implicit memory might play an important role as an antecedent of selective exposure. At this point, our speculation remains untested, but if verified, suggests a means by which greater control can be achieved in selective-exposure experiments. Typically, experimental stimuli are pretested to ensure that the stimuli differ along one crucial dimension—for instance, excitatory potential. Although this is an appropriate step to ensure that the media choices offered to viewers differ along a key dimension, it does not directly address the actual phenomenon of selectivity. That is, there is no guarantee that the content offered to viewers provides the cues that are required to connect media content with a positive outcome. However, if researchers could demonstrate that certain media content is preferred, using tests of implicit memory in addition to the typical pretest of key dimensions, then we could be much more confident that behavior is due to selective-exposure processes and not some other factor.

In terms of advancing other relatively unestablished theoretical notions, it would seem important that we gain a better understanding of how groups make the decision to use media. Increasingly, media use is becoming a narrowly individual pursuit, yet so many of our media-use decisions take place in the presence of others—families, friends, and so on. Do the processes of group decision-making mirror those of individuals? Or do we rely on entirely different processes? This chapter has mainly focused on psychological processes, but sociological processes surely also play a role in media-use decisions.

Likewise, it would seem important to understand the processes underlying so-called telic hedonism, mood satisfaction (as opposed to mood optimization), and when, how, and even if, cognitive factors become more important to media-use decisions than do affective components. In addressing questions such as these, it would seem most profitable to consider how personality factors and individual differences may influence how individuals approach the decision to use media.

It is also imperative that we modify our research paradigms to more viridically match the real-world situations of manifold media choice. Our experimental treatments and assessments typically have included a limited constellation of viewing, listening, or reading choices. Obviously this model has little external validity today, because the world in which we live is typified by media abundance, if not excess. This may well indicate that the next generation of selective-exposure research should utilize field experiments or other forms of controlled naturalistic inquiry, if it proves untenable to accommodate ecological validity in our laboratories.

In sum, we already know something of the mental decision making that underlies media-use decisions, and we can predict, with some degree of confidence, media uses and effects. We are equally confident, however, that what looms below the iceberg's tip is far greater than what is immediately apparent. Indeed, a far richer understanding of the processes of selective exposure, one that more closely approximates our actual information-age choice behaviors, may well be one of the most fundamental challenges we face in developing a comprehensive psychology of entertainment.

REFERENCES

Allen, M., D'Alessio, D., & Brezgel, K. (1995). A meta-analysis summarizing the effects of pornography II: Aggression after exposure. *Human Communication Research, 22*, 258–283.

Anderson, D. R., Collins, P. A., Schmitt K. L., & Jacobvitz, R. S. (1996). Stressful life events and television viewing. *Communication Research, 23*, 243–260.

Bandura, A. (2002). Social cognitive theory of mass communication. In J. Bryant & D. Zillmann (Eds.), *Media effects: Advances in theory and research* (2nd ed., pp. 121–153). Mahwah, NJ: Lawrence Erlbaum Associates.

Bradley, M. M., & Lang, P. J. (2000). Measuring emotion: Behavior, feeling, and physiology. In R. D. Lane & L. Nadel (Eds.), *Cognitive neuroscience of emotion* (pp. 242–276). New York: Oxford University Press.

Bornstein, R. F. (1989). Exposure and affect: Overview and meta-analysis of research 1968-1987. *Psychological Bulletin, 106*, 265–289.

Bornstein, R. F. (1992). Subliminal mere exposure effects. In R. F. Bornstein & T. S. Pittman (Eds.), *Perception without awareness: cognitive, clinical, and social perspectives* (pp. 191–210). New York: Guilford Press.

Bornstein, R. F., & D'Agostino, P. R. (1992). Stimulus recognition and the mere exposure effect. *Journal of Personality and Social Psychology, 63*, 545–552.

Bornstein, R. F., & D'Agostino, P. R. (1994). The attribution and discounting of perceptual fluency: Preliminary tests of a perceptual fluency/attributional model of the mere exposure effect. *Social Cognition, 12*, 103–128.

Bornstein, R. F., Leone, D. R., & Galley, D. J. (1987). The generalizability of subliminal mere exposure effects: Influence of stimuli perceived without awareness on social behavior. *Journal of Personality and Social Psychology, 56*, 1070–1079.

Bryant J., & Miron, D. (2003). Excitation-transfer theory and three-factor theory of emotion. In J. Bryant, D. Roskos-Ewoldsen, & J. Cantor (Eds.), *Communication and emotion: Essays in honor of Dolf Zillmann* (pp. 31–59). Mahwah, NJ: Lawrence Erlbaum Associates.

Bryant, J., & Zillmann, D. (1977). The mediating effect of the intervention potential of communications on displaced aggressiveness and retaliatory behavior. In B. D. Ruben (Ed.), *Communication Yearbook 1* (pp. 291–306). New Brunswick, NJ: Transaction.

Bryant, J., & Zillmann D., (1984). Using television to alleviate boredom and stress: Selective exposure as a function of induced excitational states. *Journal of Broadcasting, 28*, 1–20.

Cantor, J. R., Bryant, J., & Zillmann, D. (1974). Enhancement of humor appreciation by transferred excitation. *Journal of Personality and Social Psychology, 92*, 231–244.

Cantor, J. R., Zillmann, D., & Bryant, J. (1975). Enhancement of experienced sexual arousal in response to erotic stimuli through misattribution of unrelated residual excitation. *Journal of Personality and Social Psychology, 32*, 69–75.

Christ, W. G., & Medoff, N. J. (1984). Affective state and the selective use of television. *Journal of Broadcasting, 28*, 51–63.

Comstock, G., & Scharrer, E. (1999). *Television: What's on, who's watching, and what it means.* New York: Academic Press.

Cooper, E., & Jahoda, M. (1947). The evasion of propaganda: How prejudiced people respond to anti-prejudice propaganda. *Journal of Psychology, 23*, 15–25.

Cotton, J. L. (1985). Cognitive dissonance in selective exposure. In D. Zillmann & J. Bryant (Eds.). *Selective exposure to communication* (pp. 11–34). Hillsdale, NJ: Lawrence Erlbaum Associates.

D'Alessio, D., & Allen, M. (2002). Selective exposure and dissonance after decisions. *Psychological Reports, 91*, 527–532.

Davies, J. J. (2004). The effects of neuroticism, mood, and the intervention potential of media messages on selective exposure to television. Unpublished doctoral dissertation, University of Alabama, Tuscaloosa.

Dittmar, M. L. (1994). Relations among depression, gender, and television viewing of college students. *Journal of Behavior and Personality, 9*, 317–328.

Elliot, R., & Dolan, R. J. (1998). Activation of different anterior cingulated foci in association with hypothesis testing and response selection. *Neuroimage, 8*, 17–29.

Festinger, L. (1957). *A theory of cognitive dissonance.* Evanston, IL: Row & Peterson.

Freedman, J. L., & Sears, D. O. (1965). Selective exposure. In L. Berkowitz (Ed.), *Advances in experimental social psychology* (Vol 2, pp. 58–98). New York: Academic Press.

Gerbner, G., & Gross. L. (1976). Living with television: The violence profile. *Journal of Communication, 26(2)*, 173–199.

Grabe, M. E., Zhou, S., Lang, A., & Bolls, P. D. (2000). Packaging television news: The effects of tabloid on information processing and evaluative responses. *Journal of Broadcasting and Electronic Media, 44*, 581–598.

Gordon, P. C., & Holyoak, K. J. (1983). Implicit learning and generalization of the mere exposure effect. *Journal of Personality and Social Psychology, 45*, 492–500.

Harmon-Jones, E., & Allen, J. J. B. (2001). The role of affect in the mere exposure effect: Evidence from psychophysiological and individual differences approaches. *Personality and Social Psychology Bulletin, 27*, 889–898.

Hawkins, R. P., Pingree, S., Hitchon, J., Gorham, B. W., Kannaovakun, P. Gilligan, E., Radler, B., Kolbeins, G. H. & Schmidt, T., (2001). Predicting selection and activity in television genre viewing. *Media Psychology, 3*, 237–263.

Helgrel, B. K., & Weaver, J. B. (1989). Mood management during pregnancy through selective exposure to television. *Journal of Broadcasting and Electronic Media, 33*, 15–33.

Higgins, E. T. (1996). Knowledge activation: Accessibility, applicability, and salience. In E. T. Higgins & A. W. Kruglanski (Eds.), *Social psychology: Handbook of basic principles* (pp. 133–168). New York: Guilford Press.

Jonas, E., Greenberg, J., & Frey, D. (2003) Connecting terror management and dissonance theory: evidence that mortality salience increases the preference for supporting information after decisions. *Personality and Social Psychology Bulletin, 29*, 1181–1189.

Katz, E. (1968). On reopening the question of selectivity in exposure to mass communication. In R. P. Abelson, E. Aronson, W. J. McGuire, T. M. Newcomb, M. J. Rosenberg, & P. H. Tannenbaum (Eds.), *Theories of cognitive consistency: A sourcebook* (pp. 788–796). Chicago: Rand McNally and Company.

Klapper, J. T. (1960). *The effects of mass communication.* New York, Free Press.

Knobloch, S., & Zillmann, D. (2002). Mood management via the digital jukebox. *Journal of Communication, 52*, 351–366.

Kunst-Wilson, W. R., & Zajonc, R.B. (1980). Affective discrimination of stimuli that cannot be recognized. *Science, 207*, 557–558.

Lazarsfeld, P. F., Berelson, B., & Gaudet, H. (1968) *The people's choice: How the voter makes up his mind in a presidential campaign* (3rd ed.). New York: Columbia University Press.

Lieberman, M.D. (2000). Intuition: A social, cognitive, neuroscience approach. *Psychological Bulletin, 126*, 109–137.

Mastro, D. E., Eastin, M. S., & Tamborini, R. (2002). Internet search behaviors and mood alterations: A selective exposure approach. *Media Psychology, 4*, 157–172.

Meadowcroft, J. M., & Zillmann, D. (1987). Women's comedy preferences during the menstrual cycle. *Communication Research, 14*, 204–218.

Medoff, N. J. (1982). Selective exposure to televised comedy programs. *Journal of Applied Communication Research, 10*, 117–132.

Monahan, J. L., Murphy, S. T., & Zajonc, R. B. (2000). Subliminal mere exposure: Specific, general and diffuse effects.*Psychological Science, 11*, 462–466.

Murphy, S. T., & Zajonc, R. B. (1993). Affect, cognition and awareness: Affective priming with optimal and suboptimal exposures. *Journal of Personality and Social Psychology, 64*, 723–739.

Perse, E. M. (1998). Implications of cognitive and affective involvement for channel changing. *Journal of Communication, 48*(3), 49–68.

Petty, T. E., & Cacioppo, J. T. (1986). *Communication and persuasion: Central and peripheral routes to attitude change*. New York: Springer-Verlag.

Potts, R., & Sanchez, D. (1994). Television viewing and depression: No news is good news. *Journal of Broadcasting & Electronic Media, 38*, 79–90.

Reeves, B., & Nass, C. (1996). *The media equation*. Stanford, CA: CLSI Publications.

Roberts, J. E., Gilboa, E., & Gotlib, I. H. (1998). Ruminative response style and vulnerability to episodes of dysphoria: Gender, neuroticism, and episode duration. *Cognitive Therapy and Research, 22*, 401–423.

Schacter, D. L. (1987). Implicit memory: History and current status. *Journal of Experimental Psychology: Learning, Memory, and Cognition, 13*, 501–518.

Schramm, W., & Carter, R. F. (1959). Effectiveness of a political telethon. *Public Opinion Quarterly, 23*, 121–126.

Schramm, W., Lyle, J., & Parker, E. (1961). *Television in the lives of our children*. Palo Alto, CA: Stanford University Press.

Seamon, J. G., Marsh, R. L., & Brody, N. (1984). Critical importance of exposure duration for affective discrimination of stimuli that are not recognized. *Journal of Experimental Psychology: Learning, Memory, and Cognition, 10*, 465–469.

Seamon, J. G., Williams, P. C., Crowley, M. J., Kim, I. J., Langer, S. A., Orne, P. J., & Wishengrad, D. L. (1995). The mere exposure effect is based on implicit memory: Effects of stimulus type encoding conditions, and number of exposures on recognition and affect judgments. *Journal of Experimental Psychology: Learning Memory, and Cognition, 21*, 711–721.

Wakshlag, J., Day, K., & Zillmann, D. (1981). Selective exposure to educational television programs as a function of differently paced humorous inserts. *Journal of Educational Psychology, 73*, 27–32.

Wakshlag, J. J., Bart, L., Dudley, J., Groth, G., McCuthcheon, J., & Rolla, C. (1983). Viewer apprehension about victimization and crime drama programs. *Communication Research, 10*, 195–217.

Wakshlag, J., Reitz, R., & Zillmann, D. (1982). Selective exposure to and acquisition of information from educational television programs as a function of appeal and tempo of background music. *Journal of Educational Psychology, 74*, 666–677.

Wakshlag, J., Vial, V., & Tamborini, R. (1983). Selecting crime drama and apprehension about crime. *Human Communication Research, 10*, 227–242.

Weaver, J. B. III, & Baird, E. A. (1995). Mood management during the menstrual cycle through selective exposure to television. *Journalism and Mass Communication Quarterly, 72*, 139–146.

Zajonc, R. B. (1968). Attitudinal effects of mere exposure. *Journal of Personality and Social Psychology Monographs, 9*(2 Pt. 2), 1–27.

Zillmann, D. (1971). Excitation transfer in communication-mediated aggressive behavior. *Journal of Experimental Social Psychology, 7*, 419–434.

Zillmann, D. (1991). Television viewing and physiological arousal. In J. Bryant & D. Zillmann (Eds.), *Responding to the screen: Reception and reaction processes* (pp. 103-133). Hillsdale, NJ: Lawrence Erlbaum Associates.

Zillmann, D. (2000). Mood management in the context of selective exposure theory. In M. E. Roloff (Ed.), *Communication Yearbook 23* (pp. 103–123). Thousand Oaks, CA: Sage.

Zillmann, D. (2003). Theory of affective dynamics: Emotions and moods. In J. Bryant, D. Roskos-Ewoldsen, & J. Cantor, (Eds.), *Communication and emotion: Essays in honor of Dolf Zillmann* (pp. 533–567). Mahwah, NJ: Lawrence Erlbaum Associates.

Zillmann, D., & Bryant, J. (1985). Affect, mood, and emotion as determinants of selective exposure. In D. Zillmann and J. Bryant (Eds.), *Selective exposure to communication* (pp. 157–190). Hillsdale, NJ: Lawrence Erlbaum Associates.

Zillmann, D., Bryant, J., Comisky, P. W., & Medoff, N. J., (1981). Excitation and hedonic valence in the effect of erotica on motivated intermale aggression. *European Journal of Social Psychology, 11*, 233–252.

Zillmann, D., Hezel, R. T., & Medoff, N. J. (1980). The effect of affective states on selective exposure to televised entertainment fare. *Journal of Applied Social Psychology, 10*, 323–339.

3

Attention and Television

Daniel R. Anderson and Heather L. Kirkorian
University of Massachusetts at Amherst

INTRODUCTION

To be entertaining, a production, regardless of the medium, must get and keep the attention of its audience. This is not to say that attention is synonymous with entertainment. For example, we might pay rapt attention to a tornado headed our way without feeling the least bit entertained. On the other hand, we might find that a special effects movie based on tornadoes would both get our attention and be entertaining. Attention is probably best seen as a necessary but not sufficient condition for entertainment.

There is no simple definition of attention that is inclusive of all the ways in which the term is used. In general, though, the term attention refers to a psychological process by which information, usually from the external environment, is made available for cognitive and emotional analysis. Attention is overtly manifest by orientation of the body, and especially the head and eyes, toward a source of stimulation, and covertly by a variety of physiological activities that serve the purpose of intensifying the processing of the stimulation. Although cognitive theorists make a variety of distinctions between types of attention such as vigilance, orientation, attention to objects, attention to location, and others, media researchers have generally been concerned more simply with attention onset and offset, as well as the intensity of attentional engagement. Cognitive researchers, moreover, have generally studied the role of attention in motivated adult subjects who are instructed to be attentive and who are assigned task situations that are not intrinsically entertaining. Furthermore, cognitive scientists have usually been interested in detailed aspects of attention as it is deployed over time courses of less than a second in duration. Cognitive science, therefore, has had a limited impact on research on attention to media, and that impact has primarily been on studies of intensity of attention.

There has been very little research that directly connects attention to the psychological experience of being entertained, *per se*. Rather, the extant research examines attention to entertainment media, usually television, with the goal of developing rules that predict attention onset, continuation, intensity, and offset. In this chapter we summarize the research and theory concerning attention to television. Although there is a small scatter of studies concerning attention to interactive entertainment media, the literature is not yet sufficiently coherent to be worth reviewing here.

The television research generally falls into two classes, each with its own methodologies. Research with children has overwhelmingly focused on visual orientation toward television with looks (visual orientations toward the TV screen) being the primary behaviors measured. Research with adults has more typically focused on issues concerning intensity of attention and has utilized methodologies that are thought to measure intensity, such as the secondary reaction time task and some physiological measures. Not surprisingly, there is an overlap of methodologies such that intensity measures are used in some studies with children, and looks are measured in some studies with adults.

Although there have been a few field investigations, most studies have been laboratory experiments. Child studies have focused on infants and preschoolers, and adult studies have focused on college students. In this chapter, updated from the Anderson and Burns review of 1991, we begin with a brief description of common methodologies, followed by summaries of the research on looks and intensity. We try to characterize our current state of knowledge and theory and suggest significant gaps that should be addressed in future research.

METHODOLOGY

Looks

Even though they may not be aware of it, when people watch TV they look away from the TV screen with some frequency. Looks are usually measured from video recordings of a person as that person watches television. An experimenter watching the viewer records the video frame number at the beginning of the look and the video frame number at the end of the look. In our laboratory, this procedure is currently accomplished by using the tape-logging utility of the video-editing program, Adobe Premiere. The observer presses an appropriate key at the judged beginning of a look and presses another at the end of the look. The video frames for each event are automatically recorded in a computer file that can be synchronized with respect to events that occur on the TV. A continuous time record of the vicissitudes of looking at and away from the TV is thus obtained. Inter-observer reliability is typically quite high with very little disagreement between observers (e.g., Anderson & Levin, 1976).

Visual Fixations

Until recently, given typical screen sizes, the distances viewers sit from them, and the relatively poor detail resolution of standard American video, viewers could pretty well take in the screen image with little necessity for visual scanning (cf, Nathan et al., 1985). Perhaps because of this, very few published studies have attempted to identify *where* on the TV screen a look is directed. Such studies require the use of an apparatus that can determine the precise direction of gaze. Although such apparatus is now readily available, use of it by media researchers has been limited to special situations such as the use of text in the context of a TV program (e.g., Flagg, 1978). There are, for example, no normative studies of how often viewers move their eyes in the course of watching a TV program, much less any detailed information as to what they look at. This is an area well worth future research, especially as large-screen TV sets become increasingly common, making extensive visual scanning likely.

Secondary Task Reaction Times

Another method used to assess attention during television viewing is to measure a viewer's reaction time to a secondary task. The use of this method stems from the theory that there is a limited pool of cognitive resources available to devote to processing. The more that resources are tied up by a primary task, the longer should be the latency of response to a secondary task (e.g., Navon & Gopher, 1979). Secondary tasks in studies of attention to television often entail orienting toward another, intermittent visual stimulus (e.g., distractor slides) or pressing a button in response to a tone that occurs randomly throughout the television-viewing session. This method has been used to measure attention to television in infants and toddlers (e.g., Richards & Turner, 2001), preschoolers (e.g., Lorch & Castle, 1997), and, more often, adults (e.g., Geiger & Reeves, 1993).

Physiological Measures

The most common physiological measures of attention in media research focus on heart rate and brain electrical activity. In both cases, surface electrodes are placed on the chest or on the scalp. Volume-conducting electrical potentials are recorded, amplified, digitized, and subjected to a variety of analyses.

When a person encounters a novel or changed stimulus, that person may react to the stimulus with an orienting response that affects numerous physiological systems, including the cardiac system. Probably reflecting changes in blood flow in the brain, heart rate measurably decelerates. As attention is sustained, the slowed heart rate continues with reduced variability between beats. The heart rate response has been used as a measure of attention both in child studies (e.g., Richards & Casey, 1992) and in research with adults (e.g., Lang, 1990).

The electrical activity of the brain is measured by scalp electrodes in an array ranging from a few to more than a hundred electrodes, depending on the study. Studies of brain electrical activity during media use have generally employed alpha-suppression as the primary measure of attention. The alpha wave is a rhythmic frequency of about 8 Hz and reflects a relaxed state of inattention. Absence of alpha is inferred as indicating focused attention to the medium, especially if alpha blocking is time-locked to some event in the program such as a scene change.

LOOKING AT TELEVISION

General Characteristics

Much of the research on looking at television has been done by our research group. In our typical laboratory study with children or adults, a person is videotaped in a comfortably furnished room containing a TV set, snacks, and objects that afford alternative activities, such as toys for children or magazines and hand-held non-electronic games for adults. In our field research, automated video equipment was set up in homes in the rooms containing TV sets. The equipment automatically began recording the viewing area when the TV set was turned on, and stopped recording when the TV set was turned off. In both cases, the videotapes were coded for looks at the TV.

A few observations are common across all studies of looking at television. Unless viewers are in a dark, barren room, they look at and away from the TV set numerous times over the course of an hour. At home, or in the typical laboratory study, they look at and away from the TV 120 to 150 times an hour (e.g., Anderson, 1985; Anderson & Levin, 1976; Burns & Anderson, 1993). Other research groups report the same phenomenon (e.g., Hawkins et al.,

2002; Richards & Cronise, 2000). The look lengths are not normally distributed; they are skewed such that there are many short looks of a few seconds' duration, interspersed with fewer long looks which may last as much as 10 minutes or more. The distribution of look lengths can be quantitatively characterized: Both group and individual data from infants to adults closely conform to the lognormal distribution (for a detailed review and analysis see Richards & Anderson, 2004).

When a TV program receives high levels of looking (about 90% of program time), compared to a program that receives a modest amount of looking (about 50% of program time), there are still many looks at and away from the TV, but intervals between the looks become much shorter, and the long looks become much longer. Programs that receive low levels of looking (about 20% of program time) still receive many short looks but there are few, if any, long looks, and the intervals between looks become much longer as the viewer becomes engaged in an alternative activity, such as toy play or reading a magazine (unpublished analyses from our laboratory; also see Hawkins et al., 2005). Fundamentally, high attention to a TV program is reflected in more long looks with only brief pauses between looks.

The Development of Looking at Television

Prior to the 1990s, research from both laboratory and field studies found that looking increased from very low levels during infancy to relatively high levels by age 5 years, finally peaking at about age 12 years, with some decline in looking among adults (see Anderson, Lorch, Collins, Field & Nathan, 1986, for field research; Anderson & Smith, 1984, for a summary of laboratory studies). Absolute levels of looking depend on the types of programs on the TV in relation to the age of the viewer. Not surprisingly, children pay greatest attention to children's programs, and adults pay greatest attention to adult fare (Bechtel, Achelpohl & Akers, 1972; Schmitt, Anderson & Collins, 1999). For the most part, at home the increase in looking with age occurs because children pay attention to a greater range of programs as they get older.

Anderson and Lorch (1983) argued that this increase with age reflects greater comprehension ability. They hypothesized that the comprehensibility of programming is the single greatest factor that drives attention by young children to TV. Infants and toddlers under 3 years of age paid little attention to TV because there was relatively little they could understand. This hypothesis was supported by a series of studies relating program comprehensibility to children's attention to TV. In several experiments, for example, preschool children were shown *Sesame Street* segments that were either the normal segments, or that were reduced in comprehensibility by using either foreign language or backward language, or with the shots re-edited in random order. In the reduced comprehensibility conditions, the children looked less than they did at the normal segments (Anderson, Lorch, Field & Sanders, 1981; Lorch & Castle, 1997; Pingree, 1986).

Although historically infants and toddlers looked relatively little at television under normal viewing conditions at home, that began to change in the 1990s with the success of *Teletubbies,* a TV program directed at toddlers, as well as the *Baby Einstein* series of videos directed at infants. Since then, the entertainment industry has discovered that infants and toddlers constitute a lucrative market for video production. Infants pay substantial amounts of attention to baby videos. Rachel Barr and her colleagues (Barr et al., 2003) found that looking at these videos averaged about 60% in a study conducted in the infants' homes.

The fact that infants may pay relatively high levels of attention to television raises the question of why they pay attention. There are basically two possibilities: the first is that they are watching an essentially kaleidoscopic display with little comprehension. The second is that, like older children, their attention is being driven at least in part by comprehension. Huston and Wright (1983; 1989) argued that early television viewing is driven passively by orienting to salient formal features such as cuts, movement, and auditory changes, but that with age and experience attention is more systematically and cognitively driven in the service of comprehension and entertainment. If so, it may be that during infancy viewing is relatively free of comprehension, but that with age it gradually becomes more cognitive in nature.

In a complex experiment, Richards and Cronise (2000) showed infants aged 6-, 12-, 18-, and 24-months a clip from the movie *Follow that Bird*, and examined their looking at the clip as compared to computer-generated, randomly moving forms with coordinated sounds. In the most straightforward comparison, 6- and 12-month-olds looked equally at the two displays, but the 18- and 24-month-olds looked substantially more at *Follow that Bird*. This experiment is suggestive that by 18 months infants begin to prefer a structured video that features animate characters. Because there were many differences between the two displays, rendering a specific interpretation impossible, we are currently replicating the experiment in collaboration with John Richards. In this experiment we are using *Teletubbies* in three versions: normal video, randomized shots, and backward speech. Because the shots and sound envelopes are identical in all three versions, the primary differences are in sequential or linguistic comprehensibility. We have presented the findings for the 18- and 24-month-olds at an international infancy conference (Frankenfield et al., 2004). Briefly, in keeping with Richards' earlier findings, there are clear differences such that both age groups have longer looks at the normal version of *Teletubbies* than at either of the distorted versions. This suggests that by 18 months of age, if not earlier, infants' attention is at least partially guided by their cognitive processing of the content.

There is some research indicating that looking is responsive to content much earlier. From about 6 months of age, infants who are placed in front of side-by-side TV screens drawing images of their mother and father will selectively look at the image that matches a narrator saying "mama" or "papa" or other term used in the family to identify the mother and father (Tincoff & Jusczyk, 1999). By 6 months, they also prefer to look at television programs with coordinated audio as compared to images with mismatched audio tracks (Hollenbeck & Slaby, 1979). By 14 months of age, infants prefer to look at a video that has a voiceover narrator correctly describing an action compared to the same video with the same voiceover narrator incorrectly describing the action (Hirsh-Pasek & Gollinkoff, 1999). These simple examples indicate that, at least in principle, infants can attend to videos based on content, not just on the basis of salient formal features.

Baby videos and shows such as *Teletubbies*, however, are far more complex than most of the stimuli used in experimental studies with infants. *Teletubbies*, for example, is an edited program with numerous cuts and other transitions. Processing such transitions with understanding takes considerable perceptual and inferential capacity in order to comprehend the continuity of content across the transitions (cf, Anderson & Smith, 1984). Although our research (Frankenfield et al., 2004) suggests that infants are sensitive to the canonical order of transitions by 18 months of age, it is not yet clear that younger infants are capable of processing visual transitions at all. Moreover, the evidence for comprehension of videos by children under 2 years of age is very meager, and suggestive of much poorer comprehension of equivalent real-life displays (for a review see Anderson & Pempek, 2005). Overall, it is not known how much of infant or toddler attention to baby videos is based on content and how much is based on a kaleidoscopic effect. Nevertheless, infants and toddlers do look at baby videos and TV programs that are made for them. It is an important task for future research to clarify why they do so.

Formal Features

Formal features are characteristics of television programs that can be deployed across many types of content and are used to convey content. An obvious example is the cut. Television programs are typically shot as multiple streams of continuous video which edited and joined by cuts (or other devices such as dissolves, wipes, fades, and the like). Cuts can be used in multiple ways to convey content. Sometimes, they simply provide change of camera angle or distance. Other times, they show who is speaking in a conversation, or they may show a listener's reactions. At yet other times, they may convey the illusion of a single continuous action sequence, as in a car chase scene, even though the chase may have been filmed in multiple locations over many days of shooting.

Other characteristics of television are not as clearly part of the form, rather than the content, of television; nevertheless, they have generally been grouped under the rubric of formal features. These include animation, visual movement, character type (man, woman, boy, girl, puppet, animal), and audio features such as sound effects, applause, and voice type (man, woman, child, peculiar voice).

A television producer may employ formal features as necessary devices to convey content (for instance, the use of cuts in a chase scene), or she may employ cuts to provide visual or auditory change and variation (for instance, the use of cuts during a monologue). The question is whether formal features influence attention independently of their role in conveying content.

At this point, we do not know of any research that has systematically studied the effects of formal features on looking independently of their use to convey content. Rather, the approach in the three studies that have examined an array of formal features has been to examine the relationship of formal feature occurrence to attention in normal videos. The results of these studies have been remarkably consistent.

Across age groups and across studies, the most consistent formal feature is movement, which is related to enhanced looking (Alwitt, Anderson, Lorch & Levin, 1980; Anderson & Levin, 1976; Schmitt, et al., 1999). Viewers look more in the presence of movement from infancy through adulthood (Schmitt et al., 1999). Furthermore, if a viewer is not looking at the screen at the time movement occurs, the viewer is likely to begin looking, probably because of the acute sensitivity of peripheral vision to movement (Alwitt et al., 1980). The response to cuts is also consistent, insofar as looking is greater in an interval following a cut compared to intervals that do not follow cuts. Despite their attention-maintaining capacity, however, cuts do not elicit looking from inattentive viewers as consistently as movement (Alwitt et al., 1980).

The results for character types depend on whether the viewers are adults or children, and the effects are stronger based on voices than on characters' visual presence. This stronger effect of voices is primarily due to viewers' ability to hear voices even when they are not currently looking at the TV; consequently, a particular type of voice can suppress or elicit looking, as well as maintain or extinguish looks in progress at the time of voice onset. For children, all three studies find less looking in the presence of men on the screen or sound track as compared to the absence of men (Alwitt et al., 1980; Anderson & Levin, 1976; Schmitt et al., 1999). This effect does not hold for adult viewers (Schmitt et al., 1999). Woman characters are neutral to positive for child viewers, and are positive for adult female viewers. Child characters are very positive for child viewers in all three studies, but there are not enough data on child characters viewed by adults to come to any certain conclusion. Puppets and animated characters are consistently positive for child viewers.

With respect to audio features such as sound effects and applause, the results are consistently positive. Music is inconsistently related to looking; the effects on attention appear to depend on the type of music and context.

Huston and Wright (1983; 1989) provide a lucid theory on the impact of formal features on attention, and one with which we agree. They note the power of salient formal features to elicit orienting reactions, and suggest that this may be the fundamental effect of formal features in very young viewers. As a child becomes more cognitively mature and experienced in watching television, however, the child becomes increasingly aware of the role of formal features in conveying content. Formal features thus come to gain significance in relation to content. Children recognize that adult men are typically associated with adult-level content on television; such content is largely incomprehensible and uninteresting to children. Adult men on television thus come to signal content that is not interesting. Children learn that features such as peculiar voices, brightly colored sets, puppets, and other features combine to signal child content. Boys recognize that certain abrupt editing and sound effect styles signal content that is intended for them, whereas girls recognize that softer editing styles and music signal content intended for them.

The Huston and Wright research group experimentally tested aspects of the theory. Campbell, Wright, and Huston (1987) presented public service announcement-type videos in either a child format (using formal features that are characteristic of children's programs) or an adult format. Despite making the content as similar between the two formats as possible, children paid substantially more attention to the video presented in a child format. Huston, Greer, Wright, Welch, and Ross (1984) presented abstract video images to children produced and edited with formal features that are characteristic of commercials directed at boys or with formal features that are characteristic of commercials directed at girls. The children distributed their looking at the videos in ways completely consistent with their gender and the production technique.

Pacing and Formal Features

Television shows vary considerably in the density of formal features. Generally speaking, the more formal features per unit time, the more rapidly paced is the show. While it is generally assumed that rapid pacing produces greater attention, especially in children, this has not been subjected to systematic research. Potts, Huston, and Wright (1986), in a study focused on violence, showed a nonviolent children's TV show that was rapidly paced, a violent children's show that was rapidly paced, and violent and nonviolent children's TV shows that were slowly paced. They found greater attention to the rapidly paced shows than to the slowly paced shows with no effect of violence on attention. It should be noted that the differences in attention could have been due to the varying content of the shows rather than the pacing. Nevertheless, their results are consistent with the notion that more rapidly paced programs gain greater attention.

While it may be the case that increased pacing, other things being equal, leads to increased attention, it should also be pointed out that increased pacing tends to increase the information processing burden on the viewer. If the burden becomes too great, the program becomes incomprehensible and for that reason may lose attention. Taking the information processing burden into account, recent successful TV programs for preschoolers have been deliberately slowly paced in terms of the density of formal features (Anderson, 2004). These programs, such as *Blue's Clues,* get consistently high levels of attention from preschool viewers (Crawley, Anderson, Wilder, Williams, and Santomero, 1999; Crawley et al., 2002).

Content and Individual Differences

We have already noted that although formal features play an important role in attention to television, attention is primarily in service of the viewer's processing of the content. For example, children pay much more attention to children's programs than they do to adult programs. They also pay more attention to children's programs than they do to commercials for children's products (Schmitt, Woolf & Anderson, 2003). This latter observation is particularly important insofar as commercials tend to be rapidly paced and dense in formal features.

Techniques for conveying content, such as the use of humor, can also increase attention to programs (Zillmann, Williams, Bryant, Boynton, & Wolf, 1980). On the whole, attention to television appears to be more driven by content than by formal features and pacing.

The meaning of content, however, varies with the individual so that individual needs, interests, capabilities, and other psychological factors determine the attention value of any particular program. Although individual psychological characteristics other than age presumably drive specific patterns of attention to television, there have been only a few studies that have examined this.

One such psychological characteristic is gender. Striking gender differences in program and character preferences begin to emerge in the early elementary school years (age 5 years and older). In particular, there is a shift to preference for same-sex characters, particularly in boys. One development that occurs at the time of this shift is the acquisition of gender constancy. Gender constancy is marked by the onset of understanding that one's own biological sex and gender are not determined by superficial characteristics such as clothing or hairstyle but are relatively immutable (Slaby & Frey, 1975). With this understanding, one might suppose that children would begin to take a great interest in same-sex characters insofar as such characters may provide information about how to behave relative to cultural expectations of gender roles.

In the home television viewing study of Anderson et al. (1985), participating children near the fifth birthday were administered tests of gender constancy. Luecke-Aleksa, Anderson, Collins, and Schmitt (1995) examined home-recorded videotapes of the children's TV viewing and coded the gender of the character (if any) on the screen when the child was looking and when the child was present but not looking at the TV. Both boys and girls who had not achieved gender constancy looked more at female characters than at male characters regardless of whether the characters were adult humans, children, or nonhumans such as puppets. This was consistent with findings for younger children (e.g., Anderson & Levin, 1976). Children who had achieved gender constancy, on the other hand, looked more at same-sex characters, also regardless of type. That is, gender-constant girls maintained their preference for female characters, but gender-constant boys consistently paid greater attention to male characters. Such a shift in attention with the acquisition of gender constancy was predicted by Slaby and Frey (1975) and was also found by them when children watched a movie of a male and female each assembling a bicycle.

Recently, in one of the few studies to employ visual fixation methodology, Linebarger and Chernin (2004, unpublished) found results consistent with those of Luecke-Aleksa et al. (1995) and Slaby and Frey (1975). When gender-constant children look at the screen, they preferentially fixate on same-sex characters when there are multiple characters present. Pre-constant children do not have this preference; rather, they tend to fixate on female characters.

Relatively temporary psychological factors may also induce some attentional bias to content. Anderson, Collins, Schmitt, and Jacobvitz (1996) examined adult TV viewers from the Anderson et al. (1985) home-viewing study. In particular the investigators studied looking at television in relation to stress as measured by the standard Life Events Scale. Not only did they find that program preference was related to stress (for example, stressed viewers gravitated toward comedy), but they also found that attention was related to stress as well. When they watched TV in their homes, stressed men actually looked more at the screen than did men who had experienced less stress. The results were interpreted as being consistent with mood-management theory (e.g., Zillmann & Bryant, 1994), which hypothesizes that media are used to displace anxious thought patterns.

Hawkins et al. (2005) showed several TV programs from different genres to college students. Although they found that the pattern of looking varied according to genre, they found little relationship to individual differences in a measure of the students' uses and gratifications for media. Thus, not all individual difference characteristics influence looking at TV.

Look Termination

There has been rather little work devoted to the issue of why attention terminates. We do know that looks at television tend to end at content boundaries, other things being equal (e.g., Alwitt et al., 1980), consistent with the notion that comprehension of content is a primary driver of looking. That said, Richards found that sustained heart rate decelerations in infants and toddlers tend to terminate several seconds before the sustained looks end, suggesting some kind of internal state changes before the overt looking behavior is terminated (e.g., Richards & Gibson, 1997). It is not known whether this phenomenon is present in older viewers whose attention may be more schema-driven and therefore more sensitive to content boundaries.

In one experiment (Anderson, Lorch, Smith, Bradford & Levin, 1981), analyses compared the temporal similarities in look onsets and offsets across 3- and 5-year-old children. The age differences were revealing: 3-year-olds were less similar to each other than were 5-year-olds as to when looks began, but 5-year-olds were less similar to each other than were 3-year-olds as to when looks ended. Look onsets were in general more similar across children than were look offsets. The authors interpreted these findings as indicating that 1) look onsets and look offsets were determined by different factors; 2) with experience, children become more stereotyped in their look onsets which are probably cued by formal features as discussed above; 3) look offsets probably reflect idiosyncratic interests in the content of TV. Younger children are more like each other in experience and interests and thus are more likely to lose interest in the content at the same time. Older children are less like each other, both developmentally and in terms of experience, and so are more likely to have different levels of interest in the content thus terminating looks at different times.

Children reliably look away from the TV during extended zoom shots (Alwitt et al., 1980; Anderson & Levin, 1976; Susman, 1978), perhaps because they become disoriented by the constantly changing spatial frame of reference. In addition, adult male TV characters tend to terminate young children's looks at TV as discussed above (Alwitt et al., 1980; Anderson & Levin, 1976). Of course, distractions external to the TV set can terminate looks (Anderson, Choi & Lorch, 1987), but we know little beyond these findings, even though attention offsets are ultimately as important as attention onsets.

Repetition and Familiarity

With the advent of home videocassette recorders it became possible to watch a TV program or movie repeatedly. There is evidence that children have a great tolerance for repetition of favored videos and watch them many times (Mares, 1998). Rachel Barr and her colleagues (Barr et al., 2003) showed familiar or unfamiliar videos in the homes of 12- to 15-month-old infants. They found greater looking at the familiar videos (67%) as compared to the unfamiliar videos (50%). They estimated that the familiar baby videos had been watched an average of 30 times prior to the study, suggesting 1) that familiarity is an important determinant of attention by infants, and 2) that infants tolerate an enormous amount of repetition.

Crawley et al. (1999) showed an episode of *Blue's Clues* to 3-, 4-, and 5-year-old children who had not previously seen the series. One group of children saw the same episode on five consecutive days. Only 5-year-old boys showed a slight drop in looking at the program over the five days. On closer examination, the investigators distinguished between educational versus entertainment content within the program. Looking was initially greater to educational content but after three repetitions looking declined to the same level as entertainment content. Crawley et al. (1999) interpreted the decline as being due to mastery of the educational content after three showings so that the educational content eventually acquired the same status as non-demanding entertainment content. It should be noted that although attention to the episode dropped little, audience participation with this participatory program steadily increased over the five days.

Crawley et al. (2002) extended this line of research by examining familiarity with the whole series *Blue's Clues*. They compared children who had been watching the program for two years to children who had rarely if ever seen it (usually because Nickelodeon was not offered as part of the basic cable package). When shown a new episode of *Blue's Clues* that neither group had previously seen, the experienced viewers paid less overall attention to the episode than did the inexperienced viewers. This effect was primarily due to recurrent portions of the program, that is, parts of the program that are highly similar from episode to episode. Experienced and inexperienced viewers paid equivalent levels of attention to portions of the episode that were unique to that episode.

As noted above, *Blue's Clues* invites audience participation and preschool children readily shout out answers, point to the screen, dance with music, and so on. Crawley et al. (2002), in analyzing patterns of looking, audience participation, and comprehension across the Crawley et al. (1999; 2002) studies, argued that high levels of looking indicate that the children are heavily engaged in information processing and learning. To some extent during these periods of high attention, audience participation is relatively low. As the young audience masters the material and becomes familiar with the episode (with repetition) or with the series as a whole, attention drops while audience participation increases. They argued that audience participation reflects mastery and knowledge, whereas high levels of attention reflect information processing and learning.

LISTENING TO TELEVISION

It is clear that the auditory component of a program is an important predictor of attention to television, even in infants. As noted earlier, infants are responsive to language and the congruency of the audio with the video channel. Also, as noted earlier, infants and preschoolers look less at the TV if the audio track is distorted through reversed speech or foreign language. Furthermore, auditory formal features are highly effective at gaining looking by inattentive viewers. Clearly, listening is an important aspect of attention to television.

That noted, there is little research on auditory attention to television. The primary reason is the difficulty in satisfactorily measuring auditory attention. Unlike visual attention, which has a clear behavioral correlate (i.e., gaze), auditory attention cannot be monitored directly. There have been a few indirect studies that have allowed some inferences.

In one study, five-year-olds either had a variety of toys available while watching an episode of *Sesame Street* or watched the program without toys available. Not having toys doubled the amount of time that children spent looking at the program but, surprisingly, subsequent tests of comprehension yielded results that were identical for the two groups. This was true for program information presented visually as well as that presented purely through the audio channel. Furthermore, considering the exact time when information critical to answering a question was presented, looking predicted whether the question was later answered correctly, but only for the group with toys (Lorch, Anderson & Levin, 1979). These findings suggest that the children who played with toys, unlike those in the no-toys group, selectively looked at the screen only when critical information necessary for comprehension was presented. The authors concluded that by five years of age, children develop a strategy for dual-tasking during television viewing in that they monitor the audio track at a superficial level for cues of comprehensible central information, at which point they visually orient toward the screen. This strategy was highly effective insofar as doubling looking (the no-toys condition) did not further increase comprehension. The findings were also consistent with the hypothesis that young children tend to listen at a level of verbal comprehension primarily when they are also looking at the screen.

A later study by Field and Anderson (1985) replicated the finding that children's looking predicted their comprehension of purely auditory information. These researchers also found that this relationship was stronger for 5-year-olds than for 9-year-olds, suggesting that the visual and auditory modalities are more interdependent for younger children. That is, 9-year-olds appeared to be better able to listen to the TV when they were not actually looking at it. The authors hypothesized that looking and listening are strongly linked in young children but that, with maturity, auditory attention can be increasingly deployed independently of the focus of visual attention.

This is not to say that there is no linkage in adults. Burns and Anderson (1993) examined the ability of college students to recognize brief (3- to 4-second) audiovisual clips taken from TV shows that the students had just viewed. Not surprisingly, if they were looking at the time the clip occurred, recognition accuracy was high. If they were not looking, however, recognition had to be based on the audio alone. Not only was recognition lower, but it progressively declined as time elapsed between the students' last look and when the clip occurred. Burns and Anderson (1993) interpreted this result as indicating that adult viewers progressively withdraw auditory attention from the TV the longer they maintain a period of not looking at the TV. This presumably reflects increased engagement in some other non-TV viewing activity (such as looking at a magazine).

Friedlander and Cohen de Lara (1973) investigated individual differences in 5- to 8-year-old children's ability to select clear, intelligible dialogue by using a two-position switch allowing the children to select between video with normal and noise-degraded soundtracks. Subjects were instructed to hold the switch to one side for up to 15 seconds to select a particular soundtrack at which point they would have to release the switch and press it again to maintain the same soundtrack. Seventy-five percent of children in this study clearly preferred normal over degraded dialogue using a selectivity criterion of 65% attention to the normal soundtrack.

Rolandelli, Wright, Huston, and Eakins (1991) adopted a similar technique to study both visual and auditory attention to television in children. Observing 5- and 7-year-olds watching a TV program, they randomly degraded the video, the audio, or both, and informed the children that they could press a lever to "fix" the television when it was not working correctly. The audio was degraded by gradually introducing white noise while video was degraded by gradually fading in "salt-and-pepper." There was also an audio-visual degrade condition in which both modalities were degraded simultaneously. Latency to respond to the degraded stimuli was used as a measure of attention. The results indicated that overall looking at the screen was related to reaction time for all three types of degrades. The only exception was for girls' latency to audio degrades, suggesting that girls at this age attend to the auditory modality more independently than boys. Not surprisingly, the shortest latency was to audio-visual degrades because children could attend to either or both modalities to detect the change. The slowest response time was to audio-only degrades, suggesting there was less overall attention to the audio. More importantly, latency to respond was shorter when looking at the screen at degrade onset for audio-visual and video-only degrades but not for audio-only degrades, demonstrating that these children were able to detect a change in the audio track regardless of visual attention to the program. Furthermore, 7-year-olds' latency to respond to audio degrades was shorter than that for 5-year-olds when the video was narrated, suggesting that the older children are more likely to attend to the audio track than younger children, at least when the stimulus contains language.

Consistent with earlier research, looking at the screen was predictive of comprehension for auditory material for the younger age group but not for the older children, further supporting the position that auditory and visual attention become less interdependent with age.

INTENSITY OF ATTENTION

There is substantially less known about the intensity of attentional engagement with television than about looking at the screen in general. The most common procedure to estimate intensity of engagement, the secondary task reaction time, is relatively simple but is an indirect measure of attention. Conversely, physiological measures may be more direct measures of attentional engagement, but have yet to be extensively employed. This is partly due to controversies as to what the best physiological measures are, and also to the fact that experienced investigators who use these measures have not, by and large, been interested in attention to media. What research does exist has focused heavily on the distinction between automatic orienting responses elicited by the medium and more sustained attention. Experiments involve varying content, such as emotionally evocative or arousing material, and also examine the degree to which attention is related to developing story structures. A substantial amount of the more recent research in this area has emphasized the interactive effects of form and content on the allocation of attention.

Secondary Task Reaction Time Studies

Secondary task reaction time (STRT) studies rely on the assumption that as attention to a primary task (e.g., watching television) increases, so will response time to a secondary task (e.g., pressing a button in response to a tone) due to limited cognitive resources available to process incoming stimuli at any given moment. As such, research presented in this section defined increased intensity of attention as slower reaction times to orient toward distracters or to respond to secondary tasks.

Formal features, such as cuts, elicit attention, at least at the automatic level of an orienting response (e.g., Geiger & Reeves, 1993). Although the onset of a particular structural feature is known to elicit attention at that moment, Geiger and Reeves (1993) found that increasing the number of cuts did not increase overall attention to the television messages as measured by STRT. One study utilizing STRT to measure attention found that subjects were more deeply engaged with structurally simple video than complex (Thorson, Reeves, & Schleuder, 1985). Complexity in this study was defined by the number of movements, pans, zooms, edits, and cuts, rather than the content of the program. Although seemingly counterintuitive, this finding is consistent with earlier studies of attention suggesting that simple or meaningful materials can be more engaging (e.g., Britton, Glynn, Meyer, & Penland, 1982; Britton, Holdredge, Curry, & Westbrook, 1979). Furthermore, this effect seems to be modality-dependent. The density of propositions or meaningful units in the script did not influence reaction time when the secondary task involved responding to a tone. A replication of this study with a visual secondary task (responding to a flash) showed the opposite effect: Subjects were more deeply engaged with commercials with few auditory propositions, but structural complexity of visual features had no effect. These authors reasoned that the subjects were able to "borrow" resources from one modality when the other was consumed by the secondary task.

Engagement also appears to change over the course of a television-viewing session. Using STRT to measure attention, Geiger and Reeves (1993) found that the act of processing cuts became easier over time into the viewing session, consuming fewer resources, so that STRTs immediately following cuts became shorter. In contrast, Lorch and Castle (1997), in a study of

5-year-olds, found that the longest reaction times to random probes were during the second half of the viewing session. Perhaps increased facility in processing cuts allows deeper engagement with the content over time.

With respect to content, several studies have investigated what types of programming and information are most likely to engage viewers. Material presented on television can be considered related or unrelated to immediately previous information and can be categorized by its relevance to the ongoing story structure or plot. Adult subjects are more intensely attentive to content that is central to the plot rather than to incidental detail (Meadowcroft & Reeves, 1989). Other studies investigating attention as a function of content compare overall attention to arousing or emotionally evocative material to calm or neutral content. Lang, Bolls, Potter, and Kawahara (1999) found that arousing scenes received greater attention than calm ones.

Lorch and Castle (1997) found that 5-year-olds had shorter reaction times when not looking at the screen than when looking and that, when looking, reaction times were longer for relatively comprehensible versions of *Sesame Street* as compared to language-distorted (backward speech) versions of the program. This finding demonstrates that cognitive capacity is more engaged during comprehensible material, lending further support to the position that relatively understandable material is more engaging.

Several STRT studies have investigated the interaction between formal features and content. One line of work considered the effects of related versus unrelated cuts. Lang, Geiger, Strickwerda, and Sumner (1993) demonstrated that unrelated cuts required more capacity to process than related cuts. Geiger and Reeves (1993) examined the changes in attention immediately after related and unrelated cuts. They found that the overall capacity required to process unrelated cuts remained constant and high compared to the gradually decreasing attention devoted to related cuts. More specifically, these researchers found that attention to related cuts increased very briefly but then decreased within one second following the cut. Conversely, unrelated cuts led to a very brief and initial decrease in attention, signifying a "release" from the previous content, followed by a sharp increase in attention within one second of the cut associated with attending to the new content. Essentially, these researchers found that the impact of formal features on viewers' attention depends, at least in part, on the overarching content. For instance, there is an automatic orienting response to all cuts, but unrelated cuts require more controlled attention to process and comprehend (Geiger & Reeves, 1993).

Finally, Lang and colleagues (1999) investigated the interaction between arousal and pacing on attention to television messages. Their results indicated that for calm messages, overall attention increased with more rapid pacing. Conversely, attention decreased for arousing content with more rapid pacing. In other words, the least attention was found for slow, calm messages and fast, arousing ones. On the other hand, the greatest amount of cognitive resources was allocated toward fast, calm messages and slow, arousing ones. This finding suggests that arousal and pacing both influence resource allocation and that attention can be maximized by an optimal combination of arousal and pacing that does not overtax cognitive resources. Taken together, STRT studies have demonstrated that automatic and controlled attention is related to both form and content, but to fully predict the intensity of engagement consideration must be given as to how the two interact.

Heart Rate Studies

One line of research has investigated the influence of structural features of television, namely cuts, on attention as measured by heart rate. For instance, the onset of a commercial is known to elicit the orienting response (OR), a component of which is a decrease in heart rate (Lang, 1990). Heart rate decelerates for approximately four seconds after a cut on screen (Lang et al.,

1993). Richards and Gibson (1997) also found decelerations in heart rate in 3- to 6-month-olds in response to scene changes.

Although some research suggests that overall pacing does not influence broad changes in heart rate over the viewing session (Lang et al., 1999), Lang, Zhou, Schwartz, Bolls, & Potter, (2000) found greater decreases in heart rate in the second half of the viewing session, particularly for fast-paced messages. This is consistent with the STRT results by Lorch and Castle (1997) described above.

An interesting feature of entertainment media that is relevant to attention is screen size. Reeves, Lang, Kim, & Tatar (1999) demonstrated that adults' heart rate decreased significantly more when viewing on large-format (56") screens as compared to moderate (13") or small (2") screens. This effect held regardless of arousal or emotional valence of content. These researchers posit that several known effects of media could be accentuated on large screens, and recommend that screen size be included in future research.

Studies examining the effects of program content on heart rate have focused largely on emotional valence. Heart rate increases over a viewing session for emotional content but decreases for messages rated as rational or a combination of rational and emotional (Lang, 1990). Furthermore, Lang (1990) found that emotional messages increase the intensity of evoked cardiac responses. That is, orienting occurred in response to changes on screen but the intensity of the OR increased with emotionality. Reeves et al. (1999) found that negative content produced greater decreases in heart rate than positively valenced messages, although this effect was only marginal. Lang et al. (2000) posit that related cuts do not tax cognitive capacity, unlike unrelated cuts. Consistent with this position, these researchers found that both related and unrelated cuts led to decreased heart rate but only related cuts showed a complementary increase in memory for the content presented. Furthermore, the production implications of these findings are that producers can maximize learning from information presented in the program by adding attention-getting features that do not tax cognitive capacity. In many ways findings from heart rate studies parallel those exhibited by secondary task reaction time studies.

Electrophysiological Studies

There are a few studies that have used EEG to examine attention to television. Rothschild, Thorson, Reeves, Hirsch, and Goldstein (1986) demonstrated that EEG can be used to discriminate scenes that gain and hold attention, those that gain but then lose attention, and those that never succeed in gaining attention. Epochs of alpha suppression (indicative of attention) often begin with some visual change, such as a cut, movement, or commercial onset. Superimposed voiceover also elicits alpha suppression. Rothschild et al. (1986) identified "points of interest" in their television stimuli where the most alpha blocking was observed in adult subjects. Lang (1990), using the same stimuli, found that these points of interest also elicited ORs according to heart rate measures, but she was unable to identify specific formal features common to all or most of these points. Rothschild et al. (1986) suggest that alpha may drop quickly during an OR to a visual or auditory cue, but that the maintenance of attention may depend on the motivation of the viewer or interest in the content. To our knowledge, no studies have since been done to investigate the effects of program content on EEG, but this prediction is consistent with research cited earlier in the present chapter utilizing other measures of attention.

Simons, Detenber, and Cuthbert (2003) showed adults either stills or brief video clips of stimuli that had been known to evoke emotional responses. They found alpha attenuation, indicative of attention, in parietal regions. The attention was produced by both positive and negative emotional clips and, as with other measures of attention, by movement as compared to stills.

Smith and Gevins (2004), studying EEG responses to television commercials, found that posterior alpha attenuation (indicative of orienting) was associated with scene changes. Frontal alpha attenuation in the lower frequencies of the alpha range was associated with the subjects' finding the commercials interesting, and frontal alpha attenuation in the higher frequencies of the alpha range was associated with commercials being recalled. Like the studies based on looking, on STRT, and on heart rate, the EEG studies, taken together, suggest that there are distinct attentional processes associated with initial orienting as compared to more sustained attention. Orienting is elicited by formal features and sustained attention by cognitive engagement with the content.

Developmental Issues in the Intensity of Attention

By and large the research cited in this section was conducted with adult participants. With few exceptions, most of what we know about the intensity of attention is from research with adults. Only very recently have there been studies recording STRT (e.g., Richards & Turner, 2001), heart rate (e.g., Richards & Cronise, 2000), and brain activity (e.g., Richards, 2003) in infants with regard to television viewing. Only a couple of studies have used any measure of intensity of attention, namely STRT, in older children (e.g., Lorch & Castle, 1997). These studies contribute mostly to the literature on attentional inertia and, as such, are described in the following section.

ATTENTIONAL INERTIA

On the whole, research on looking at television has addressed somewhat different questions than has research on intensity of attention. There is one phenomenon of attention to television, however, where both lines of research converge. This phenomenon is referred to as *attentional inertia*. Attentional inertia is defined with respect to the onset of an episode of attention (usually a look): At episode onset the intensity of attention is low and the episode is fragile and easily disrupted. As an episode is sustained, however, it becomes increasingly robust, less vulnerable to disruption, and information processing becomes more intense. Metaphorically, as an episode of attention to television is sustained, it develops its own inertia.

The term originated with an analysis reported by Anderson, Alwitt, Lorch, and Levin (1979). They plotted look length data from children as a type of hazard function. Hazard functions describe the probability of a failure of some kind of an entity conditional upon the time it has already survived. Hazard functions are used, for example, as part of life insurance calculations concerning the probability of death within the next year as a function of how old the person is currently. Anderson et al. (1979) applied a version of the analysis to looks at television. For a group of 5-year-olds who watched *Sesame Street* in a room equipped with attractive toys, the hazard of a look ending within 3 seconds from its onset was about .57. Given that the look survived to being 3 seconds old, the hazard of it ending before it became 6 seconds old dropped to .34. Given that the look survived to be 6 seconds, the hazard of termination further dropped to .24 and so on. The entire curve, based on 3 second intervals, showed a smoothly decelerating negative function over time. In other words, the longer a look survived, the lower was its hazard of ending over each succeeding time interval.

Anderson et al. (1979) showed that this hazard function characterized the data of individual children as well as group data and that the function was found with children as young as 12 months of age as well as adults. Subsequently, Richards and his colleagues showed that this function characterizes the television look data of infants ranging from 6 weeks to 24 months (Richards & Cronise, 2000; Richards & Gibson, 1997; Richards & Turner, 2001). Attentional inertia is not limited to television viewing. The characteristic hazard function is also found for episodes of preschool toy play (Choi & Anderson, 1991) and bouts of music listening by preschoolers (Sims, 2001).

A series of investigations has explored the nature of attentional inertia. One line of research has found that the longer a look at television has been sustained, the less distractible the viewer becomes (Anderson et al., 1987; Richards & Turner, 2001), with a parallel finding for toy play (Choi & Anderson, 1991; Ruff, Cappazolli & Salterelli, 1996). This reduction in distractibility is accompanied by an increase in attentional intensity or engagement as indexed by reaction times to distractors or secondary tasks (Anderson et al., 1987; Choi & Anderson, 1991; Lorch & Castle, 1997). In toy play, as episodes are sustained, the child is more likely to enter a state of deep concentration known as focused attention (Oakes, Ross-Sheehy & Kanass, 2004; Ruff, et al., 1996). Progressively deepened engagement as a look at TV is sustained has also been found in analyses of heart rate by Richards and his colleagues (Richards & Cronise, 2000; Richards & Gibson, 1997; Richards & Turner, 2001). Information processing of television is enhanced as a look is sustained. Burns and Anderson (1993) found that recognition memory for program content is increased the longer a look has been sustained at the time that the tested content appeared. Lorch et al. (2004) found that children better understood causal connections in television narratives the longer a look had been in progress when the information about those connections was presented.

Attentional inertia appears to be a general property of attention to entertainment media including TV, toys, and music. A question arises as to what use is attentional inertia. A number of theorists, prior to the discovery of attentional inertia, have argued that the attentional system must have a way to maintain attention to a task or to a source of discourse regardless of momentary fluctuations in interest value, comprehensibility, or other aspects of the source (Hebb, 1949; Hochberg & Brooks, 1978; James, 1890). Hochberg and Brooks (1978), for example, argued that in order to watch a movie, something has to drive attention across the visual content boundaries defined by cuts. In other words, in temporally structured entertainment media, as well as in play activities and perhaps other kinds of activities, there must be a "glue" that maintains attention across content or action boundaries.

Anderson and his colleagues showed that the longer a look has been in progress prior to a complete change in content (as in a shift from program to commercial), the longer the look will remain in progress after the change (Anderson & Lorch, 1983; Burns & Anderson, 1993). Attentional inertia thus serves to sustain attention to completely new content. Anderson and his colleagues have argued that because attentional inertia is found in young infants and across unrelated pieces of content, it is probably a fundamental, biologically based process underlying sustained attention.

Attentional inertia is found in infants' TV viewing before they could possibly comprehend much on television, and it is also found when infants view stimuli such as computer-generated random forms and sounds. In quantitative analyses of look distributions, however, beginning at about 18 months, infants show greater inertia if the stimuli are structured, normal television sequences (Richards & Anderson, 2004). Inertia is also greater in older than younger children. Together, these results indicate that attentional inertia is enhanced by cognitive comprehension processes. This conclusion is also consistent with studies on the intensity of attention, reviewed above, that find greater attentional engagement with related, as compared to unrelated, edited sequences.

Hawkins and his colleagues have explored this phenomenon more fully, finding that the degree to which looks are driven across content boundaries varies with the type of boundary (Hawkins, et al., 2002; Hawkins, Tapper, Bruce & Pingree, 1995). They argue that there are learned, strategic processes that modify the strength of attentional inertia and that emotional processes may modify it as well. As examples, expecting a transition to be from an entertaining program to a commercial will produce less inertia than the reverse but in addition a highly entertaining program produces a greater level of attentional inertia than a less entertaining program.

THEORY AND FUTURE DIRECTIONS

Across all the areas of research that we have reviewed, there has been a consistent distinction between the television characteristics that produce attentional onset or, if already paying attention, produce transient increases in attentional intensity, and those that produce sustained attention. Certain formal features elicit looks, produce transient increases in secondary task reaction times, produce orienting response heart rate decelerations, and attenuate alpha in parietal regions. These are the fairly automatic responses. The factors that produce sustained attention are less well characterized, but are primarily factors of content. These include comprehensibility, interest value, emotional valence, and the personal significance of the content to the viewer. In addition, as attention is sustained, engagement progressively deepens as indicated by attentional inertia. The latter factor, while apparently produced by the processes of comprehension of content, and which may be strategically modified in adults, also appears to be an automatic process insofar as attentional inertia sustains attention into new and unrelated content, occurs in pre-cognitive infants, and can occur with non-meaningful stimuli.

In reading for this chapter, we have been struck by the general lack of theory of how attention is related to entertainment. In particular, there is little theory as to how sustained attention in task situations differs from sustained attention in entertainment situations. Why is sustained attention in certain task situations seemingly effortful whereas sustained attention in entertainment situations seems effortless? Is sustained attention in these contrasting situations fundamentally different in kind? Do they call on different underlying neural mechanisms of attention? We are not aware of published studies that have attempted to systematically explore this issue, although many of the tools described above could be used to do so. This is not the only issue that has received little research.

We also have noted that there is, as yet, little research on auditory attention to entertainment media even though the audio clearly plays an important role. Similarly, although we know that children are tolerant of—and even demand repetition—we know little about the factors that underlie this phenomenon. There is some evidence that repeated viewing of videos declines as children age beyond the preschool years (Mares, 1998), but there have been no systematic investigations beyond the studies cited earlier.

Although the research on attention to television has reached a critical mass, even allowing the development of systematic principles for program design (Anderson, 2004), there has been no parallel research literature on other entertainment media. Even though there is a vast literature on eye movements in reading, for example, there is no literature on the factors that cause a person to start reading, read for some period of time, and then stop reading (analogous to looking at TV). We do not know, for example, whether attentional inertia applies to entertainment reading. Nor is there such a parallel literature on video game play.

Finally, American television is on the verge of becoming a digital medium, with large-screen high-definition television and surround sound becoming common. Digital TV, moreover, affords the possibility of being interactive. While much of what we have learned about attention to television may be applicable to this evolved medium, much may change. On the other end of the spectrum, cell phone users will be able to receive television on two-inch screens. What will be the nature of attention to this evolved medium? Whatever these media may bring to entertainment, attention will still be necessary. Systematic study of attention to the new media will allow us to better understand their impact and help with the rational design of content.

REFERENCES

Alwitt, L. F., Anderson, D. R., Lorch, E. P., & Levin, S. R. (1980). Preschool children's visual attention to attributes of television. *Human Communication Research, 7*, 52–67.

Anderson, D. R. (1985). On-line cognitive processing of television. In A. Mitchell & L. Alwitt (Eds.), *Psychological processes and advertising effects: Theory, research and application*. Hillsdale, N.J.: Lawrence Erlbaum Associates.

Anderson, D. R. (2004). Watching children watch television and the creation of *Blue's Clues*. In H. Hendershot (Ed.), *Nickelodeon nation: The history, politics, and economics of America's only TV channel for kids* (pp. 241–268). New York: New York University Press.

Anderson, D. R., Alwitt, L. F., Lorch, E. P., & Levin, S. R. (1979). Watching children watch television. In G. Hale & M. Lewis (Eds.), *Attention and cognitive development* (pp. 331–361). New York: Plenum.

Anderson, D. R., & Burns, J. (1991). Paying attention to television. In D. Zillman & J. Bryant (Eds.), *Responding to the screen: Perception and reaction processes*. pp. 3–26. Hillsdale, NJ: Lawrence Erlbaum Associates.

Anderson, D. R., Choi, H. P., & Lorch, E. P. (1987). Attentional inertia reduces distractibility during young children's television viewing. *Child Development, 58*, 798–806.

Anderson, D. R., Collins, P. A., Schmitt, K. L., & Jacobvitz, R. S. (1996). Stressful life events and television viewing. *Communication Research, 23*, 243–260.

Anderson, D. R., Field, D. E., Collins, P. A., Lorch, E. P., & Nathan, J. G. (1985). Estimates of young children's time with television: A methodological comparison of parent reports with time-lapse video home observation. *Child Development, 56*, 1345–1357.

Anderson, D. R., & Levin, S. R. (1976). Young children's attention to Sesame Street. *Child Development, 47*, 806–811.

Anderson, D. R., & Lorch, E. P. (1983). Looking at television: Action or reaction? In J. Bryant & D. R. Anderson (Eds.), *Children's understanding of TV: Research on attention and comprehension* (pp. 1–34). New York: Academic Press.

Anderson, D. R., Lorch, E. P., Collins, P. A., Field, D. E., & Nathan, J. G. (1986). Television viewing at home: Age trends in visual attention and time with TV. *Child Development, 57*, 1024–1033.

Anderson, D. R., Lorch, E. P., Field, D. E., & Sanders, J. (1981). The effect of television program comprehensibility on preschool children's visual attention to television. *Child Development, 52*, 151–157.

Anderson, D. R., Lorch, E. P., Smith, R. Bradford, R., & Levin, S. R. (1981). The effects of peer presence on preschool children's television viewing behavior. *Developmental Psychology, 17*, 446–453.

Anderson, D. R., & Pempek, T. A. (2005). Television and very young children. *American Behavioral Scientist, 46*, 505–522.

Anderson, D. R., & Smith, R. N. (1984). Young children's television viewing: The problem of cognitive continuity. In F. Morrison, C. Lord, & D. Keating (Eds.), *Advances in applied developmental psychology* (pp 116–165). New York: Academic Press.

Barr, R., Chavez, M., Fujimoto, M., Garcia, A., Muentener, P., & Strait, C. (2003, April). *Television exposure during infancy: Patterns of viewing, attention, and interaction*. Poster presented at the Biennial Meeting of the *Society for Research in Child Development*, Tampa, FL.

Bechtel, R., Achelpohl, C., & Akers, R. (1972). Correlates between observed behavior and questionnaire responses on television viewing. In E.A. Rubinstein, G.A. Comstock, & J.P. Murray (Eds.), *Television and social behavior* (Vol. 4), *Television in day to day life: Patterns of use* (pp. 274–344). Washington, DC: U.S. Government Printing Office.

Britton, B. K., Glynn, S. M., Meyer, B. J. F., & Penland, M. F. (1982). Effects of text structure on use of cognitive capacity during reading. *Journal of Educational Psychology, 74*, 51–61.

Britton, B. K., Holdredge, T. S., Curry, C., & Westbrook, R. D. (1979). Use of cognitive capacity in reading identical texts with different amounts of discourse level meaning. *Journal of Experimental Psychology: Human Learning and Memory, 5*, 262–270.

Burns, J. J., & Anderson, D. R. (1993). Attentional inertia and recognition memory in adult television viewing. *Communication Research, 20*, 777–799.

Campbell, T. A., Wright, J. C., & Huston, A. C. (1987). Form cues and content difficulty as determinants of children's cognitive processing of televised educational messages. *Journal of Experimental Child Psychology, 43*, 311–327.

Choi, H. P., & Anderson, D. R. (1991). A temporal analysis of toy play and distractibility in young children. *Journal of Experimental Child Psychology, 52*, 41–69.

Crawley, A. M., Anderson, D. R., Wilder, A., Williams, M., & Santomero, A. (1999). Effects of repeated exposures to a single episode of the television program *Blue's Clues* on the viewing behaviors and comprehension of preschool children. *Journal of Educational Psychology, 91*, 630–637.

Crawley, A. M., Anderson, D. R., Santomero, A., Wilder, A., Williams, M., Evans, M. K., & Bryant, J. (2002). Do children learn how to watch television? The impact of extensive experience with *Blue's Clues* on preschool children's television viewing behavior. *Journal of Communication, 52,* 264–280.

Field, D. E., & Anderson, D. R. (1985). Instruction and modality effects on children's television attention and comprehension. *Journal of Educational Psychology, 77,* 91–100.

Flagg, B. N. (1978). Children and television: Effects of stimulus repetition and eye activity. In J. W. Senders, D. F. Fisher & R. A. Monty (Eds), *Eye movements and the higher psychological functions* (pp. 279–291). Hillsdale, NJ: Lawrence Erlbaum Associates.

Frankenfield, A. E., Richards, J. R., Lauricella, A. R., Pempek, T. A., Kirkorian, H. L., & Anderson, D. R. (2004, May). *Looking at and interacting with comprehensible and incomprehensible* Teletubbies. Poster session to be presented at the biennial International Conference for Infant Studies, Chicago, IL.

Friedlander, B. Z., & Cohen de Lara, H. C. (1973). Receptive language anomaly and language/reading dysfunction in "normal" primary-grade school children. *Psychology in the Schools, 10,* 12–18.

Geiger, S., & Reeves, B. (1993). The effects of scene changes and semantic relatedness on attention to television. *Communication Research, 20,* 155–175.

Hawkins, R. P., Pingree, S., Hitchon, J. B., Gilligan, E., Kahlor, L., Gorham, B. W., Radler, B., Kannaovakun, P., Schmidt, T., Kolbeins, G. H., Wang, C.-I., & Serlin, R. C. (2002). What holds attention to television? *Communication Research, 29,* 3–30.

Hawkins, R. P., Pingree, S., Hitchon, J., Radlor, B., Gorham, B. W., Kahlor, L., Gilligan, E., Serlin, R. C., Schmidt, T., Kannaovakun, P., & Kolbeins, G. H. (2005). What produces television attention and attention style? *Human Communication Research, 31,* 162–187.

Hawkins, R. P., Tapper, J., Bruce, L., & Pingree, S. (1995). Strategic and non-strategic explanations for attentional inertia. *Communication Research, 22,* 188–206.

Hebb, D. O. (1949). *The organization of behavior.* New York: Wiley.

Hirsh-Pasek, K., & Golinkoff, R. N. (1999). *Origins of grammar: Evidence from early language comprehension.* Cambridge, MA: MIT Press.

Hochberg, J., & Brooks, V. (1978). Film cutting and visual momentum. In J. W. Senders, D. F. Fisher & R. A. Monty (Eds), *Eye movements and the higher psychological functions.* (pp. 293–316). Hillsdale, NJ: Lawrence Erlbaum Associates.

Hollenbeck, A., & Slaby, R. (1979). Infant visual responses to television. *Child Development, 50,* 41–45.

Huston, A. C., Greer, D., Wright, J. C., Welch, R., & Ross, R. (1984). Children's comprehension of televised formal features with masculine and feminine connotations. *Developmental Psychology, 20,* 707–716.

Huston, A. C., & Wright, J. C. (1983). Children's processing of television: The informative functions of formal features. In J. Bryant & D. R. Anderson (Eds.), *Children's understanding of television: Research on attention and comprehension* (pp. 35–68). New York: Academic Press, Inc.

Huston, A. C., & Wright, J. C. (1989). The forms of television and the child viewer. In G. Comstock (Ed.), *Public communication and behavior: Volume 2* (pp. 103–158). New York: Academic Press, Inc.

James, W. (1890). *Principles of psychology.* New York: Holt.

Lang, A. (1990). Involuntary attention and physiological arousal evoked by structural features and emotional content in TV commercials. *Communication Research, 17,* 275–299.

Lang, A., Bolls, P., Potter, R. F., & Kawahara, K. (1999). The effects of production pacing and arousing content on the information processing of television messages. *Journal of Broadcasting & Electronic Media, 43,* 451–475.

Lang, A., Geiger, S., Strickwerda, M., & Sumner, J. (1993). The effects of related and unrelated cuts on television viewers' attention, processing capacity, and memory. *Communication Research, 20,* 4–29.

Lang, A., Zhou, S., Schwartz, N., Bolls, P. D., & Potter, R. F. (2000). The effects of edits on arousal, attention and memory for television messages: When an edit is an edit can an edit be too much? *Journal of Broadcasting & Electronic Media, 44,* 94–109.

Linebarger, D. L., & Chernin, A. R. (2004). *Gender Differences in Young Children's Eye Movements While Watching Television: Implications for Attention and Comprehension.* Unpublished manuscript, University of Pennsylvania.

Lorch, E. P., Anderson, D. R., & Levin, S. R. (1979). The relationship of visual attention and comprehension of television by preschool children. *Child Development, 50,* 722–727.

Lorch, E. P., & Castle, V. J. (1997). Preschool children's attention to television: Visual attention and probe response times. *Journal of Experimental Child Psychology, 66,* 111–127.

Lorch, E. P., Eastham, D., Milich, R., Lemberger, C. C., Sanchez, R. P., & Welsh, R. (2004). Difficulties in comprehending causal relations among children with ADHD: The role of attentional engagement. *Journal of Abnormal Psychology, 113,* 56–63.

Luecke-Aleksa, D., Anderson, D. R., Collins, P. A., & Schmitt, K. L. (1995). Gender constancy and television viewing. *Developmental Psychology, 31,* 773–780.

Mares, M. L. (1998). Children's use of VCRs. *Annals of the American Academy of Political and Social Science, 557,* 120–131.

Meadowcroft, J. M., & Reeves, B. (1989). Influence of story schema development on children's attention to television. *Communication Research, 16,* Special issue: Social cognition and communication. 352–374.

Nathan, J. G., Anderson, D. R., Field, D. E., & Collins, P. A. (1985). Television viewing at home: Distances and viewing angles of children and adults. *Human Factors, 27,* 467–476.

Navon, D., & Gopher, D. (1979). On the economy of the human-processing system. *Psychological Review, 86,* 214–255.

Oakes, L. M., Ross-Sheehy, S., & Kanass, K. M. (2004). Attentional engagement in infancy: The interactive influence of attentional inertia and attentional state. *Infancy, 5,* 239–252.

Pingree, S. (1986). Children's activity and television comprehensibility. *Communication Research, 13,* 239–256.

Potts, R., Huston, A. C., & Wright, J.C. (1986). The effects of television form and violent content on boys' attention and social behavior. *Journal of Experimental Child Psychology, 41,* 1–17.

Reeves, B., Lang, A., Kim, E. Y., & Tatar, D. (1999). The effects of screen size and message content on attention and arousal. *Media Psychology, 1,* 49–67.

Richards, J. E. (2003). Attention affects the recognition of briefly presented visual stimuli in infants: an ERP study. *Developmental Science, 6,* 312–328.

Richards, J. E., & Anderson, D. R. (2004). Attentional inertia in children's extended looking at television. In R. V. Kail (Ed.), *Advances in child development and behavior* (Vol. 32, pp. 163–212). Amsterdam: Academic Press.

Richards, J. E., & Casey, B. J. (1992). Development of sustained visual attention in the human infant. In B. A. Campbell, H. Hayne, & R. Richardson (Eds.), *Attention and information processing in infants and adults* (pp. 30–60). Hillsdale, NJ: Lawrence Erlbaum Associates.

Richards, J. E., & Cronise, K. (2000). Extended visual fixation in the early preschool years: Look duration, heart rate changes, and attentional inertia. *Child Development, 71,* 602–620.

Richards, J. E., & Gibson, T. L. (1997). Extended visual fixation in young infants: Look distributions, heart rate changes, and attention. *Child Development. 68,* 1041–1056.

Richards, J. E., & Turner, E. D. (2001). Distractibility during the extended viewing of television during the early preschool years. *Child Development, 68,* 963–972.

Rolandelli, D. R., Wright, J. C., Huston, A. C., & Eakins, D. (1991). Children's auditory and visual processing of narrated and nonnarrated television programming. *Journal of Experimental Child Psychology, 51,* 90–122.

Rothschild, M. L., Thorson, E., Reeves, B., Hirsch, J. E., & Goldstein, R. (1986). EEG activity and the processing of television commercials. *Communication Research, 13,* 192–220.

Ruff, H. A., Cappozzolli, M., & Salterelli, L. M. (1996). Focused visual attention and distractibility in 10-month-old infants. *Infant Behavior and Development, 19,* 281–293.

Schmitt, K. L., Anderson, D. R., & Collins, P. A. (1999). Form and content: Looking at visual features of television. *Developmental Psychology, 35,* 1156–1167.

Schmitt, K. L., Woolf, K. D. & Anderson, D. R. (2003). Viewing the viewers: Viewing behaviors by children and adults during television programs and commercials. *Journal of Communication, 53,* 265–281.

Simons, R. F., Detenber, B. H., & Cuthbert, B. N. (2003). Attention to television: Alpha power and its relationship to image motion and emotional content. *Media Psychology, 5,* 283–301.

Sims, W. L. (2001). Characteristics of preschool children's individual music listening during free choice time. *Bulletin of the Council for Research in Music Education, 149,* 53–63.

Slaby, R. G., & Frey, K. S. (1975). Development of gender constancy and selective attention to same-sex models. *Developmental Psychology, 46,* 849–856.

Smith, M. E., & Gevins, A. (2004). Attention and brain activity while watching television: Components of viewer engagement. *Media Psychology, 6,* 285–305.

Susman, E. J. (1978). Visual and verbal attributes of television and selective attention in preschool children. *Developmental Psychology, 14,* 565–566.

Thorson, E., Reeves, B., & Schleuder, J. (1985). Message complexity and attention to television. *Communication Research, 12,* 427–454.

Tincoff, R., & Jusczyk, P. W. (1999). Some beginnings of word comprehension in 6-month-olds. *Psychological Science, 10,* 172–175.

Zillmann, D., & Bryant, J. (1994). Entertainment as media effect. In J. Bryant, & D. Zillmann (Eds.), *Media effects: Advances in theory and research.* Hillsdale, NJ: Lawrence Erlbaum Associates.

Zillmann, D., Williams, B., Bryant, J., Boynton, K., & Wolf, M. (1980). Acquisition of information from educational television programs as a function of differently paced humorous inserts. *Journal of Educational Psychology, 72,* 170–180.

4

Perception

L. J. Shrum
University of Texas at San Antonio

Perception is an elusive concept. In my readings and conversations in preparation of this chapter, I've seldom heard a consistent definition. Perhaps that is understandable, given that the usage of the term and its cognates differs across fields. Precept, perceptual field, and perceptual fluency (cognitive psychology), person perception, social perception, and selective perception (social psychology), perceived reality and perceptions of social reality (communications), are just a few examples of different usages. The progression of terms, from precept through perceptions of social reality, actually suggests a progression through stages of information processing, from the categorization and encoding of basic stimuli to the formation of trait (person) inferences to the construction of more elaborate inferences and judgments about complex social stimuli such as groups, society, and events.

Regardless of which of the fields, terms, or stages best captures the consensus of what perception entails, each is clearly fundamental to communication in general, and the processing of entertainment in particular. Moreover, as will (hopefully) be clear throughout the chapter, these processes are dynamic and reciprocal: Person and situation factors influence the perception of people and events, and the frequent processing of people and events influences which person and situation factors are employed in subsequent perceptions and judgments.

In the next section, I provide a general overview of cognitive psychology's view of perception. The purpose is to begin laying the groundwork for understanding how the processes that comprise perception influence communication processes and how communication processes also influence perceptual processes. These dynamic and reciprocal relations between perception and communication form the basis for how media in general, and entertainment media in particular, are perceived and for how the consumption of entertainment media influences perceptions.

A COGNITIVE PSYCHOLOGY PERSPECTIVE ON PERCEPTION

One might think that focusing on a particular field's conceptualization of perception would alleviate ambiguity, particularly when that field (cognitive psychology) is the one with which perception is most closely associated. One would be wrong. Although there is considerable agreement on certain aspects of the definition, that consensus breaks down at the boundaries between perception and other processes, such as when perception becomes memory (Erdelyi, 1992). However, as with the ambiguity across fields mentioned earlier, this ambiguity is also understandable. The current view of perception is not one of a "locus," or unitary event, but of a "vast processing region" comprised of multiple stages (Erdelyi, 1974, p. 14). This focus on stages of processing and the dynamic nature of information processing in general was initially fostered by what has become known as the "New Look" in perception (Bruner, 1957), and the reformulations of this early research into information processing models have become known as the "New Look 2" (cf. Erdelyi, 1974; Greenwald, 1992). It is difficult to overestimate the impact of this seminal research on both the cognitive and social psychology fields of today, because it laid the foundation for the information processing revolution in psychology.

The "New Look" in Perception

Perception at its most basic level is a process of categorization (Bruner, 1957). When stimulus information is received as a sensory input, the perceptual process attempts to make sense out of this information by placing it into a category of things (e.g., fruit, animal, woman, etc.). Note that these inputs could come from various senses (visual, aural, etc.), and for the former, could take the form of pictures or words. The process of perceiving involves taking the surface features of the stimulus (e.g., colors, tones, shapes, letters) and placing them into a semantic category.

Prior to the New Look movement, perception research embodied the positivist perspective that there was an objective reality (a "pure precept," Bruner, 1992, p. 780) that was processed by the senses in a relatively passive manner and was generally affected only by external factors (e.g., intensity, novelty). The New Look took a more constructivist perspective. It suggested that perception was an adaptive process, and, as such, was influenced by internal constructs such as expectancies and motivations. In a series of studies, Bruner, Postman, and colleagues demonstrated that these internal constructs affected "perceptual readiness," or the ease with which stimuli could be categorized (for a review, see Bruner, 1957; Erdelyi, 1974). For example, Postman, Bruner, and McGinnies (1948) showed that the speed with which participants in their study recognized words corresponding to the Allport-Vernon values list (Allport & Vernon, 1931) was a function of the place of those values within each individual's value hierarchy. Values ranked as more important by an individual tended to be recognized faster than those ranked as less important. Bruner (1951) later showed that categorization (interpretation) of an ambiguous picture (e.g., a man bending over) varied as a function of the importance of the Allport-Vernon values. Thus, participants who held strong religious values tended to describe the man as praying and those with strong economic values tended to describe the man as working.

Other internal constructs unrelated to values, motivations, or needs have produced results similar to Postman et al. (1948). Bruner and Postman (1949) showed that past experience can result in expectancies, which in turn influence the perceptual process of categorization. Playing cards were presented to participants tachistoscopically; some of the cards were normal (e.g., a red eight of hearts) and others were not (e.g., a black three of diamonds). Recognition thresholds were substantially longer for the anomalous cards than for the normal cards. However, once the anomalous cards had been initially presented, recognition thresholds for all other anomalous cards were lowered, but still not to the levels of the ordinary cards. Bruner (1958) framed these results in terms of an expectancy, or hypothesis, theory of perception. Over time, people learn "what goes with what." The strength of the hypothesis also increases with consistent results over time, which in turn influences the perceptual process embodied by the recognition thresholds in Bruner and Postman (1949). Moreover, once confidence was reduced by showing participants anomalous cards, participants showed less of an effect of expectancies.

The enhanced ability of people to perceive (categorize) stimuli as a function of their internal states was referred to as "perceptual vigilance" (Bruner & Postman, 1947b). However, there was another set of findings that was equally novel: People also showed a *decreased* ability to recognize certain emotional stimuli (termed "perceptual defense"), and these were invariably related to "taboo" words (e.g., bitch, penis, death; Bruner & Postman, 1947a).

Both of these concepts (vigilance and defense) had a profound influence on psychology, from cognitive, to social, to clinical. First, the notion of defense mechanisms provided a clear link between the clinical nature of Freudian psychology and the more scientifically inclined experimental work in cognitive psychology. That is, the clinical findings on such mechanisms as repression that had heretofore been subjected to criticisms of lack of scientific rigor and falsifiability now had just the evidence for which critics had called. Moreover, the selectivity of perception under conditions in which participants were not aware of any selective processes clearly suggested the operation of an active subconscious. Indeed, as Bruner (1992) noted, it was not long before the psychoanalytical aspects of the debate began to dominate, and links between Freudian and cognitive psychology became more formalized (cf. Lazarus, Erikson, & Fonda, 1951; Erdelyi, 1974, 1983). Second, the notion of perceptual vigilance, or readiness, provided the impetus for the explosion of research on construct accessibility (for reviews, see Higgins, 1996; Wyer & Srull, 1989). Specifically, people are always ready to perceive (categorize), and they may do so by choosing among a number of categorization possibilities. Which category is chosen is a function of the accessibility of that category, and category accessibility can be a function of internal (e.g., motivation, needs, attitudes) or external (e.g., environmental, situational) factors. Thus, a Rorschach inkblot might be more likely to be interpreted in terms of food-related objects for those who are hungrier, or an ambiguous behavior (sitting on a park bench) might be interpreted negatively because of a recent exposure to a racial or gender stereotype.

The "New Look 2"

The findings of Bruner, Postman, and colleagues posed significant problems for the then-current view of cognitive processes. According to the extant theory of that time, the perceptual system was posited to operate between the stimulus and response systems, with a clear sequential progression from stimulus to perception to response, at which point information could be stored in long-term memory. Yet Bruner and colleagues clearly showed that at some point, long-term memory must be exerting an influence on the perceptual process. That is, perceptual readiness was enhanced by past experience with covariation. Perceptual defense was enhanced by past experience with unpleasant stimuli.

This conundrum was addressed by Erdelyi (1974), who conceptualized perception as a multistage process that is under cognitive control (via long-term memory) but occurs outside of awareness. Thus, individuals are generally (but not always) unaware of why particular categories or constructs are activated in the process of perception/categorization of stimuli and they are also generally unaware of any perceptual defense mechanisms that might not let particular stimuli be perceived and thus brought into awareness. In other words, people are not typically aware of why particular categories get "selected in" (perceptual vigilance) or why they get "selected out" (perceptual defense), even though long-term memory is monitoring the sensory inputs in order to perform these selection tasks in the most adaptive manner.

The general notion of perceptual selectivity provides the foundation for the transition to a social psychological perspective on perception. As noted earlier, the work of Bruner and colleagues was instrumental to the development of theories of social perception, particularly as they relate to construct accessibility. In addition, although to my knowledge it has not been directly applied in this manner, both perceptual selectivity and construct accessibility have some relation to selective perception as it has been traditionally addressed in both psychology and communication research. Each of these constructs is discussed in the next section.

A SOCIAL PSYCHOLOGICAL PERSPECTIVE ON PERCEPTION

Social Perception and Construct Accessibility

For the purposes of this discussion, social perception refers to the process of forming impressions of other people or groups. Construct accessibility refers to the ease with which particular constructs, in this case trait concepts, are activated. Bruner and colleagues' work (for a review, see Bruner, 1957) on perception of simple objects (e.g., apple, orange) was applied to perceptions of persons. Bruner (1957) states "given a sensory input with equally good fit to two nonoverlapping categories, the more accessible of the two categories would 'capture' the input" (p. 132). The extrapolation to trait judgments is straightforward. In almost any social situation, a set of circumscribed behaviors can be interpreted in terms of a number of trait concepts. In other words, given only a small amount of information, the motivations for performing a particular behavior are ambiguous. For example, a person sitting on a park bench may be doing so because they have nothing else to do (lazy) or because they just finished work (hardworking). A person who skydives may be considered reckless or adventurous. In both of these cases, lacking any additional diagnostic information, perceptions of the same behavior could be categorized along multiple trait dimensions. According to Bruner, the relative accessibility of applicable trait constructs determines how the behavior is perceived and encoded into an impression of that person. The accessibility of trait concepts in turn may be influenced by external (e.g., communication) and internal (e.g., needs, motivations, attitudes) factors.

Externally Induced Accessibility. The predictions of situational accessibility effects implied by Bruner (1957) were confirmed in a series of studies by Higgins, Rholes, and Jones (1977) and Srull and Wyer (1979, 1980). In both studies, trait concepts were activated, and thus made more accessible, via a priming procedure. The priming procedures varied across the studies. In the Higgins et al. studies, participants were exposed to one of two trait concepts, stubborn or persistent, via an ostensibly unrelated Stroop task. In the Srull and Wyer studies, participants completed a scrambled sentence task to activate either the trait concept of hostile or kind. Later, for both studies, participants read either a description of a situation or list of behaviors that were ambiguous with respect to the trait they might imply. For example, making up ones mind and rarely changing it could be perceived as either stubborn or persistent. Crossing the Atlantic in a sailboat could be perceived as either reckless or adventurous.

The results across all of the studies converged on the same conclusion. When particular trait concepts were made more accessible via priming procedures, participants tended to judge the behaviors of the target person in terms of those primed concepts. Thus, participants in Higgins et al. (1977) had more favorable impressions of a target person when persistent was primed than when stubborn was primed, even though all participants read the exact same trait descriptions. Similar results were found for Srull and Wyer (1979, 1980). Moreover, Srull and Wyer showed that, although these accessibility effects tended to be reduced over time (e.g., 5 min., 1 hr., 24 hrs.), when priming frequency was high, the accessibility effects from priming were observable 24 hours later.

It is important to note the effect of awareness on construct activation, or more specifically, the lack of it, in producing the effects in the priming studies. Situational factors make certain trait concepts more accessible and thus more likely to be used in social perception, but people are usually unaware of this relation. In the Higgins et al. and Srull and Wyer studies, participants were consciously aware of the prime, but they were not aware of its possible influence. This reasoning was confirmed by Bargh and Pietromonaco (1982), who subliminally primed participants with trait concepts prior to an impression formation task. Thus, participants were clearly unaware of the relation between the priming event and the judgment task because they were unaware that the priming event even took place. In addition, Martin (1986) showed that when participants *are* aware that the trait concepts that come to mind may have been influenced by external factors, they may avoid using the trait concepts to form their judgments (see also Lombardi, Higgins, & Bargh, 1987).

One final aspect of the priming studies is important. In both the Higgins et al. and Srull and Wyer studies, the priming of trait categories had an effect only when those trait categories were applicable to the behavior. Thus, in Higgins et al. (1977), the priming of applicable trait concepts (reckless, adventurous) affected interpretations of behaviors such as skydiving (the target person was liked better when the trait concept was more positive than negative) but the priming of inapplicable concepts (neat, disrespectful) had no effect, even though the concepts differ in valence. In the Srull and Wyer studies (1979, 1980), priming of a trait such as hostile affected evaluations of a target person along dimensions applicable to hostility (e.g., unfriendly) but not along trait dimensions unrelated to hostility (e.g., boring). What these findings suggest is that, as Bruner (1957) noted, it is not simply the accessibility per se that guides perception, but also the relative fit between the accessible construct and the stimulus features. Thus, accessible constructs are not necessarily used willy-nilly in the perception of stimuli, but only when the relative fit makes it logical to do so.[1]

Internally Induced Accessibility.   The accessibility of constructs can also result from factors unique to the person. As noted earlier, various temporary or situational need states (e.g., hunger, thirst) may influence how stimulus inputs are categorized. However, individuals may also differ on the extent to which the heightened accessibility of particular constructs persists over time and situations. These individual differences may be the result of internal factors such as personality or from external factors such as frequency of activation as a result of occupations, hobbies, social roles, and so forth. In the latter cases, although the activation of concepts is externally induced, it is internalized over time. When certain constructs exhibit such persistent accessibility, they are said to be "chronically accessible" (Higgins, 1996).

A number of studies have provided evidence of chronic accessibility effects. Higgins, King, and Mavin (1982) first determined participants' level of chronicity for various traits by observing what types of trait concepts they listed first in response to various inquiries. One week later, participants read individually tailored essays (based on their chronicity) that contained trait descriptions, for some of which they were chronic and for some of which they were not chronic. The results showed that both spontaneous impressions of a target person and recall for the behavioral descriptions were more related to the trait concepts for which participants were chronic than for which they were nonchronic. Bargh, Bond, Lombardi, and Tota (1986)

---

[1]However, there is some evidence that very a high degree of accessibility can compensate for a low degree of applicability (Higgins & Brendl, 1995).

extended these findings by looking at the possible additive effects of chronic and experimentally primed accessibility. They determined participants' chronic accessibility from a previous experiment, and then subliminally primed these same participants with traits for which they were found to be chronic. Bargh et al. found that both chronic and primed accessibility had independent, additive effects on impressions of a target person.

It is also worth noting that the effects of chronic accessibility generally persist over time. Higgins et al. (1982) found that the chronic accessibility effects they noted on impression and memory were still detectable two weeks after exposure to the target information. In a different context, Lau (1989) found that chronically accessible constructs can be stable over years and influence the processing of information about a wide variety of politically related people and events (see Higgins, 1996).

Stereotyping. The research just reviewed provides compelling evidence that social perception is a function of the accessibility of applicable trait concepts from memory that are stimulated by the features of a target person. In this research, the primed constructs were traits, and the features of the person were often behaviors or behavioral descriptions. However, it is also possible that other features of a target person other than behavior—for example, length of hair, color of skin, gender—may also activate particular concepts. In fact, this process perfectly describes the nature of stereotyping. Certain concepts, in this case the stereotype itself, may be activated upon mere exposure to the person. Once activated, this stereotype has associated with it a variety of trait concepts that likewise become activated, making those trait concepts more likely to be used in forming an impression of a target person.

Consider the example described earlier of the man sitting on a park bench. Would the consideration of the motive for the behavior (no job vs. just off work), and consequently impressions of the man (lazy, hardworking) differ if the man's skin color was black or white? If the man's hair were long or short? The research on priming suggests that skin color would be likely to activate stereotypical traits associated with race and these traits would be used to form a judgment of the person. Likewise, length of hair (pony tail, shaved head) might also activate stereotypes that would influence judgments of the person.

There is quite a bit of research that validates this reasoning. The prevailing view of stereotype activation is that stereotypes are activated preconsciously as part of the perceptual process. If the stereotype of a person or group is available in memory, it will be activated upon mere exposure to the target person or group (Devine, 1989; Lepore & Brown, 1997; for a review, see Bodenhausen & Macrae, 1998). Because this activation occurs preconsciously, the individual is unaware of both its activation and its use, and this process is often referred to as *implicit stereotyping* (Banaji & Greenwald, 1994). However, this does not mean that stereotyping is necessarily inevitable if the stereotype exists in memory. If people are aware of the stereotype, they may actively resist its application. Devine (1989) showed that even though activation of the stereotype is automatic, low-prejudiced individuals tended to suppress its use to a greater degree than high-prejudiced individuals. In more recent research, Moskowitz, Salomon, and Taylor (2000) provided some qualifications to these findings. They found that some individuals (e.g., those with chronic egalitarian goals) can suppress automatic activation of stereotypes, and this suppression itself occurs automatically, outside of awareness.

Attitude Accessibility. Trait concepts are not the only accessible constructs that can influence perception. Recall from Bruner (1957) that many psychological factors such as expectancies, goals, motivations, values, and attitudes can influence perception. Bruner and colleagues showed that the most important values were not only the most easily recognized, but also tended to be used to interpret ambiguous behaviors (Bruner, 1951; Postman et al., 1948). Attitudes are no different. One of the major findings of attitude research is that attitudes provide an organizing and structuring function for dealing with an ambiguous environment (Eagly & Chaiken, 1993). Gordon Allport (1935, p. 306) presaged the selective perception research to come when he noted that "attitudes determine for each individual what he will see and hear, what he will think and what he will do." Functional theories of attitudes (Katz, 1960; Smith, Bruner, & White, 1956) posited that attitudes simplify an individual's interaction with the world, particularly through their object appraisal function. The attitude provides a useful function in orienting the individual to the attitude object (Fazio, 1989; Roskos-Ewoldsen & Fazio, 1992).

Schemas and Scripts. Schemas and scripts may also be used in the perceptual process (see also Wicks, this volume). Schemas are knowledge structures that represent a set of associations regarding objects or events. They are "organized prior knowledge, abstracted from experience with specific instances" (Fiske & Linville, 1980, p. 543). Scripts are specific types of schemas that refer to procedural knowledge about sequences of events (Abelson, 1976; Eagly & Chaiken, 1993). Schemas and scripts are thus categorization aids that allow perceivers to interpret incoming information in the context of past experience. As such, once activated, they guide expectations with respect to what should and should not happen. Like all constructs, schemas and scripts can vary in their level of accessibility, and thus vary in the probability that they will be activated in any particular situation.

Selective Perception

The previous discussion on construct accessibility, attitudes, and perception provides a useful segue into a subject central to both social psychology and communication, that of selective perception. There are three particular studies that are cited most often in the communication literature concerning selective perception. These are the studies by Cooper and Jahoda (1947), Hastorf and Cantril (1954), and Vidmar and Rokeach (1974). Although these studies are seldom discussed in terms of Bruner's (1957) theory and research (but see Bruner, 1994; Fazio, Roskos-Ewoldsen, & Powell, 1994; Fazio & Towles-Schwen, 1999), as the following discussion suggests, they are very consistent with it.

Hastorf and Cantril's (1954) classic study, "They Saw a Game," investigated the perceptions of spectators at a particularly rough football game between Princeton and Dartmouth, which Princeton ultimately won. Hastorf and Cantril found that spectators' perceptions of the level of, responsibility for, and quantity of dirty play was strongly related to the spectators' attitudinal predispositions. Princeton students thought the Dartmouth team committed many more infractions than did the Dartmouth students, and also thought the Dartmouth team was dirtier and the game less fair than did the Dartmouth students.

Both the Cooper and Jahoda (1947) and Vidmar and Rokeach (1974) studies looked at attempts to use popular communications to change prejudiced attitudes. Cooper and Jahoda found that a cartoon strip that portrayed a prejudiced character ("Mr. Biggott") in a particularly negative light was perceived differently by prejudiced and non-prejudiced readers. They concluded that prejudiced readers avoided psychological conflict by misunderstanding the underlying message. Vidmar and Rokeach found a similar pattern of reactions to the television program *All in the Family*, in which the central character, Archie Bunker, is presented as a "lovable bigot" (p. 36). They found that high- and low-prejudiced viewers liked the show equally well, but for different reasons. Low-prejudiced viewers considered the program a satire about bigotry and saw Archie Bunker as being ridiculed, whereas high-prejudiced saw the program more as an honest depiction and tended to admire Archie more than did low-prejudiced viewers.

Across the three studies, the findings converge on the conclusion that individuals' perceptions are biased toward pre-existing attitudes and beliefs. Although the survey nature of these studies does not allow for any assessment of process (which Cooper and Jahoda themselves noted), the results are consistent with the processes of perceptual vigilance and defense. When exposed to a complex social situation, people will likely interpret actions and events in terms of the constructs that are most accessible in memory. As Postman et al. (1948) demonstrated, important personal values are often the constructs that are most accessible. Thus, in all three of the selective perception studies just described, participants likely interpreted events in terms of their personal values (e.g., Dartmouth good/Princeton bad, values related to racial prejudice), selecting for inclusion instances that fit with those existing values and selecting out those that did not. Moreover, as both Bruner (1957) and Erdelyi (1974) suggest, these processes most likely occur unconsciously. Although Hastorf and Cantril (1954) do not use the cognitive process language of Bruner and Erdelyi, their account of their results falls nicely within that scope:

> Hence the particular occurrences that different people experienced in the football game were a limited series of events from the total matrix of events *potentially* available to them. People experienced those occurrences that reactivated significances they brought to the occasion; they failed to experience those occurrences which did not reactivate past significances (p. 132, emphasis in original).

PERCEPTION AND ENTERTAINMENT MEDIA

The previous discussion of perception and its underlying processes provides the basis for understanding the relation between perception and entertainment media. There are two particular processes that are important in this regard. The first is how individuals perceive entertainment media. Part of this process is represented by what is commonly referred to as the "perceived reality" of the media. For example, as people watch a television program, they receive sensory inputs that must be organized within existing knowledge structures, whether these be simple attitudes and trait concepts, or more elaborate schemas, scripts, and stereotypes. The types of constructs that are activated during the perception process clearly influence both cognitive and affective reactions to the media information, interpretations of the information received, and the integration of the information (and its implications) into existing belief structures.

The second process in the relation between perception and entertainment media that is important is how frequent consumption of entertainment media influences the constructs that play a part in perception. In other words, consumption of entertainment media may influence the internal factors (e.g., attitude accessibility, stereotypes, schemas, and scripts) that play such an important role in the perceptual process. This process falls into the category of media effects, and is exemplified in research on so-called "perceptions of social reality."

Perceiving Entertainment Media

Viewers have certain expectations when they watch an entertainment program. These expectations might be called narrative or story schemas at the general level, or might be specific to the type of program viewed (film, news, sitcom, soap opera, etc.). Interestingly, Bruner himself noted the importance of narratives as an organizing aid to perceptual processing:

> One of the most powerful means we have for making meaning is our narrative capacity: our power to create and to use stories as means for bringing order and sense into experience. Stories are not 'after the fact': we *perceive* stories in progress—we *see*

or *hear* people as heroes, recognize situations as dangerous or benign in terms of an encompassing plot (Bruner, 1994, p. 283, emphasis in original).

Part of what influences our perception and categorization of situations as dangerous, mean-spirited, violent, and so forth are schemas related to program genres. All narratives have shared structures (e.g., plot, event, problem resolution) but certain differences also exist between types of television programs. There are obvious differences between serials (continuing plot and problems from episode to episode) and situation comedies (usually a resolution of a problem in one episode). In addition, we also have schemas for the messages portrayed in different types of programs. Thus, the same act, for instance, pushing or hitting another person in an argument, may be perceived differently given that one knows, for example, that in the situation comedy this act is likely to be related to humor rather than malice, or at the very least, will be resolved happily in the end.

Another interesting aspect of entertainment media, particularly *fictional* entertainment media, is the extent to which viewers let (or do not let) the fictional aspect "get in the way" of narrative processing, and, by extension, let it get in the way of the influence of the narrative processing on the creation and maintenance of beliefs (Green, Garst, & Brock, 2004). I have argued elsewhere that one of the ways in which viewers do not allow the fictional aspect of entertainment media to get in the way is through what Coleridge (1967) called "a willing suspension of disbelief" (cf. Shrum, forthcoming; Shrum, Burroughs, & Rindfleisch, 2004). Although viewers clearly know that a program is fictional, it is simply more enjoyable, more "transporting" (Green & Brock, 2000) to process it as if it is real. What then in turn influences the ease in suspending disbelief is the degree of *perceived reality* of the fictional entertainment program.

Perceived Reality of Entertainment Media[2]

Quite a bit of research has focused on issues related to perceived reality (see Busselle & Greenberg, 2000). Most of this research was concerned with the extent to which perceived reality moderates the effect of media portrayals on attitudes, beliefs, and behaviors. Early research on the relation between media and aggression manipulated perceived reality with techniques such as telling participants the event was real or staged or by telling them the video segment they were to view was from either a movie or a newsreel of an actual event (cf. Feshbach, 1972; Berkowitz & Alioto, 1973; for a review, see Berkowitz, 1984). The general finding was that media portrayals of violence had their greatest effects on violent behavior when the portrayals were considered real. Some studies further suggested that these perceptions of reality have both psychological and physiological effects. Violence perceived as real was more arousing, as evidenced by galvanic skin response (Geen, 1975; Geen & Rakosky, 1973). Violence perceived as fictional appears to allow viewers to distance themselves from the event and it thus may have less of an impact on them (Berkowitz, 1984).

Other research on perceived reality and its relation to media effects has measured the extent to which television is perceived as real by viewers and consequent effects on social perceptions. In particular, perceived reality has been investigated as a potential moderator of cultivation effects. However, this research has not yielded any consistent pattern of results with respect to

---

[2]The main focus of the discussion is on the extent to which fictional portrayals are judged to be real. Nonfiction programming such as news, documentaries, and sports are considered to be judged as real by viewers because viewers know the scenes are either occurring in real time, were recorded, or are based on real events. It should be noted, however, that any of these programs might be perceived as less real if they contain features atypical of conventional news, documentary, or sports programming.

the role of perceived reality (Shrum, forthcoming). Part of the problem may simply be issues of measurement. As Busselle and Greenberg (2000) noted, there has been a lack of clarity and consistency in the way in which the perceived reality has been measured. In addition, given that certain types of programming differ dramatically from others (e.g., news vs. situation comedy),[3] it is unclear what it would mean to measure perceived reality of television in general (Wilson & Busselle, 2004). Just as Bruner (1957) noted that people, events, and situations are interpreted in terms of the most accessible constructs, it is likely that judgments of perceived reality of television in general would be based on the program categories that most readily come to mind (see Busselle, 2004 for evidence of this possibility).

Until recently, little research had focused on the underlying perceptual processes of "perceiving reality" while viewing. That is, how do people make judgments of perceived reality, and what makes one program, scene, or person seem more real than others? Shapiro and his colleagues (Shapiro & Chock, 2003; Shapiro & Fox, 2002) have attempted to address these questions. Their research suggests that *typicality* plays a big role in reality perception. Typicality refers to the extent to which a portrayed event occurs as it would in real life, and, as such, is a function of both plausibility and probability. Shapiro and Chock (2003) found that texts of atypical events were judged as less realistic than texts containing typical events, and Shapiro and Fox (2002) found that atypical events were more easily recalled than typical events. The latter finding suggests that atypical events undergo deeper processing, which improves recall (Craik & Lockhart, 1972). Hall (2003) found six dimensions of perceived reality. In addition to typicality and plausibility, she also identified factuality, emotional involvement, narrative consistency, and perceptual persuasiveness.

Effects of Perceived Reality and Their Underlying Processes. So what effects might perceived reality have? For one, as alluded to earlier, it seems likely that perceived reality would be related to the concept of transportation (Gerrig, 1993; Green & Brock, 2000). Transportation refers to the extent to which viewers are absorbed into the narrative they are processing. Transportation is exemplified by greater attentional focus, imagery development, and emotion. If a program is perceived as atypical, and thus less real (Shapiro & Chock, 1993), it should draw more focus to the atypical events, leaving less resources for focusing on the narrative itself. If so, then it is likely that the narrative would have less persuasive impact.

There is a growing body of evidence supporting both the notion that perceived reality is related to transportation and that transportation is related to persuasion. Wilson and Busselle (2004) found that perceived reality was positively correlated with transportation. Thought-listing data indicated the perceived reality of a program was lowest when participant thoughts were directed away from the program narrative and when they were evaluatively negative. Green and Brock (2000) provided evidence that participants who were more transported while reading a narrative passage found fewer errors in the narrative and had more favorable impressions of the protagonists than did less transported participants.

Green et al. (2004) suggest that these effects are due to processing style. When people perceive a narrative to be fictional, they use this information as a cue to become less critical and more transported. Prentice and Gerrig (1999) suggested that the processing of fact and fiction follow different routes, with fiction being processed less systematically. They also suggested that fiction has the most influence when it is responded to in an experiential rather than rational manner. Zwaan (1994) provided some evidence that this reasoning is related to the processing of factual versus fictional television programming. He found that merely indicating

---

[3] As a tribute to the late Neil Postman, author of *Amusing Ourselves to Death*, it should be noted that the similarities between the news and situation comedies, indeed the similarities between news and *any* other entertainment program category, are increasing rapidly.

that a text passage came from the news or a novel affected how participants interacted with the text and what they got from it. Those who were told the passage was fictional took longer to read the text and remembered more surface features such as verbatims, whereas those who were told the passage was factual recalled more situational information and tended to process at a deeper, rather than surface level.

Entertainment Media Consumption and Social Perception

The previous section addressed how viewers perceive entertainment media and how these perceptions may influence all subsequent stages of information processing. In addition, the previous section discussed how internal qualities that people bring to the viewing situation (e.g., accessible constructs, prior beliefs) guide perceptions and their effects. This section completes the circle. Media consumption, particularly very frequent consumption, can itself influence construct accessibility, which in turn will not only influence perceptions of the media, but also the perception of people and events in everyday life.

Entertainment Media Consumption and Construct Accessibility. The notion that media consumption influences construct accessibility is straightforward. There are a number of factors that can increase construct accessibility. Two internal factors—expectations and motivations—were discussed in earlier sections. However, there are also external factors that can influence accessibility (for reviews, see Higgins & King, 1981; Roskos-Ewoldsen, 1996). Of particular relevance to this discussion are frequency of activation, recency of activation, and vividness of the information. The information can be in the form of simple exemplars or more complex beliefs, attitudes, and values.

Information that has been frequently activated from memory is easily recalled. This is most easily seen in rehearsal effects. In attempting to memorize names, a list of events, or answers to a test, rehearsal is useful in increasing the likelihood that we will be able to recall the information at a later time. The same is true for recency: Information recently activated (and re-stored in memory) is also easily recalled. Finally, the vividness of stored information influences accessibility. More vivid exemplars tend to be more easily recalled that less vivid exemplars. Presumably, the vividness of attributes influences depth of processing, which in turn facilitates recall.

Given these findings, we would expect particular outcomes as a result of entertainment media consumption. Consider television viewing. Television provides a steady stream of formulaic and consistent portrayals of people and events (for a discussion, see Gerbner, Gross, Morgan, Signorielli, & Shanahan, 2002). At the very least, frequent viewing should make these examples more accessible from memory. In addition, frequent activation of evaluative concepts related to the portrayals (i.e., attitudes and values) should also make them more accessible. Moreover, the exemplars portrayed and the evaluative concepts activated during viewing are often vivid or emotional, which in turn also increases construct accessibility.

Increasing the accessibility of constructs activated during television viewing would not be particularly problematic if, in fact, what is portrayed on television reflects reality. However, content analyses make it clear that it is not. The overarching goal of all networks in developing television programs is to induce as many people as possible to tune in. To develop mass appeal, the programs employ some consistent features. For example, in order to entertain and stimulate, they emphasize drama and suspense. One consequence is the frequent use of crime and violence that is over ten times the rate of its real world incidence (Gerbner, Gross, Morgan, & Signorielli, 1986). Television programs also must tell its stories quickly and efficiently: Television time is expensive and viewers' attention spans are short. One technique for telling a story quickly is through the use of stereotypes. A stereotype is a convenient data reduction technique or heuristic (Bodenhausen, Macrae, & Sherman, 1999). As long as viewers are familiar with a stereotype, an abundance of information about a character or situation can be conveyed without resorting to lengthy dialogue. However, as with many stereotypes, characterizations are seldom neutral. Some are positive (hero) and some are negative (criminal). More disturbing are the problems that arise when the pairing of particular stereotypes (e.g., criminal, hero, successful, powerless) and particular attributes (e.g., race, gender, class, age) becomes systematic (e.g., Oliver, 1994).

There is a growing body of evidence that television viewing is positively related to the accessibility of constructs portrayed often on television. Busselle and Shrum (2003) measured the ease of recall of certain exemplars, some of which were frequently portrayed on television (e.g., murder, courtroom trial, highway accident). They found that media examples were more easily recalled for events frequently shown on television but infrequently experienced personally (e.g., trial, murder). Events experienced personally were more easily recalled when the events were encountered often in real life, even when those events were also frequently portrayed on television (e.g., highway accidents, dates). Moreover, the ease of retrieving media examples was related to hours of TV viewing, but only for viewing of television programs in which the events were frequently portrayed and when the direct experience with the events was likely to be low.

Indirect evidence suggesting that television viewing increases accessibility was obtained in studies that measured the speed with which participants constructed their social reality judgments. Greater accessibility was expected to result in faster judgments, and thus heavy viewers were expected to respond faster than light viewers. These expectations were confirmed in a series of studies that varied the type of dependent variables, operationalization of viewing, and control variables (cf. O'Guinn & Shrum, 1997; Shrum, 1996; Shrum & O'Guinn, 1993; Shrum, O'Guinn, Semenik, & Faber, 1991).

In the studies just described, the accessible constructs were exemplars, or some example of a category (e.g., a doctor, a crime). However, external factors such as frequency and recency of activation also influence the accessibility of attitudes (Roskos-Ewoldsen, 1996). Thus, frequent (and recent) viewing should be positively related to the extent to which attitudes related to the messages portrayed on television are activated. Shrum (1999) provided evidence that supports this reasoning in a study of heavy and light soap opera viewers. Study participants were classified as heavy or light soap opera viewers on the basis of a pretest and were recruited for a study two months later (they were unaware of the selection criterion). A content analysis of current soap operas was conducted to identify salient themes, which were determined to be materialism, marital discord, and distrust. Based on this content analysis, participants indicated their attitudes toward owning expensive products, their beliefs that their spouse will cheat on them, and their distrust of people in general and lawyers in particular. Attitude accessibility was operationalized as the speed with which participants indicated their attitudes (measured via computer input). The results showed that even after controlling for attitude extremity, heavy viewers responded faster, and thus indicated more accessible attitudes and beliefs, than light viewers.

Effects of Media-Induced Accessibility on Perceptions. The question now becomes to what extent this heightened accessibility due to exposure to entertainment media influences perceptions about others? By now, the predictions should be clear. More accessible constructs are more likely to be activated in social or decision-making contexts, and are thus more likely to be used as a basis for judgment than less accessible constructs. A number of studies have confirmed this prediction, mostly within the context of testing cultivation theory. Shrum and colleagues (for a review, see Shrum, 2006) have shown that television viewing influences both the magnitude of societal perceptions (% of people involved in a violent crime, % of work force that is lawyers, etc.) and the accessibility of exemplars. More important, this accessibility *mediates* the effect of viewing on societal perceptions. Thus, television viewing influences construct accessibility, which in turn influences the magnitude of societal perceptions.

Attitude accessibility can also have an effect on later stages of information processing. Attitudes that are more accessible from memory are generally stronger, held more confidently, are more persistent and resistant to change, and are more likely to influence behavior than less accessible attitudes (for a review, see Petty & Krosnick, 1995). Thus, even when attitude extremity is the same for two groups, they may differ in attitude accessibility, and hence the probability that they will act on those attitudes. Consider the results from Shrum (1999). In that study, frequency of viewing did not reliably predict attitudes regarding owning expensive products, but it did reliably predict attitude accessibility. Thus, even though the reported attitudes between heavy and light soap opera viewers did not differ, heavy viewers may be more likely to act on those attitudes (e.g., through purchase, judging others in terms of their possessions, etc.) than light viewers.

CONCLUSION

As noted in the introductory paragraph, the exact notion of what perception entails is difficult to grasp. It resembles what Erdelyi calls a "pretheoretic" concept: one that is generally understood in lay language [but] is not a scientific concept that has been formalized (Erdelyi, 1992, 2004). However, even though the precise definition of perception may be elusive, it is nevertheless a fundamental concept in terms of how entertainment media are perceived and how entertainment media shape perceptions. Communication research is just now beginning to take advantage of seminal work in cognitive and social psychology by investigating the interrelations of how media are perceived and how these perceptions influence judgments. Likewise, communication research is also taking advantage of the latest work in construct accessibility to understand how media-induced accessibility frames perceptions of social stimuli. Although it is difficult to predict the future, it seems certain that these perspectives will shed new light on important research questions.

ACKNOWLEDGMENTS

This manuscript benefited greatly from discussions with Bob Wyer, Matt Erdelyi, Tina Lowrey, and Richard Jackson Harris, all of whom contributed to giving me a grasp of the fuzzy concept of perception. Thanks also to my research assistants, Chandra Kalapatapu and Wilson Neto, for a yeoman's job on the literature search under extreme duress. Please direct inquiries to L. J. Shrum, University of Texas at San Antonio, Dept. of Marketing, 6900 N. Loop 1604 W., San Antonio, TX 78249 (lj.shrum@utsa.edu).

REFERENCES

Abelson, R. P. (1976). Script processing in attitude formation and decision making. In J. S. Carroll & J. W. Payne (Eds.), *Cognition and social behavior* (pp. 33–45). Hillsdale, NJ: Lawrence Erlbaum Associates.

Allport, G. W. (1935). Attitudes. In C. Murchison (Ed.), *Handbook of social psychology* (pp. 798-844). Worchester, MA: Clark University Press.

Allport, G. W., & Vernon, P. E. (1931). *A study of values*. Boston: Houghton-Mifflin.

Banaji, M. R., & Greenwald, A. G. (1994). Implicit stereotyping and unconscious prejudice. In M. P. Zanna & J. M. Olson (Eds.), *The psychology of prejudice: The Ontario Symposium* (Vol. 7, pp. 55–76). Hillsdale, NJ: Lawrence Erlbaum Associates.

Bargh, J. A., Bond, R. N., Lombardi, W. J., & Tota, M. E. (1986). The additive nature of chronic and temporary sources of construct accessibility. *Journal of Personality and Social Psychology, 50,* 869–878.

Bargh, J. A., & Pietromonaco, P. (1982). Automatic information processing and social perception: The influence of trait information presented outside of conscious awareness on impression formation. *Journal of Personality and Social Psychology, 43,* 437–449.

Berkowitz, L. (1984). Some effects of thoughts on anti- and prosocial influences of media events: A cognitive-neoassociation analysis. *Psychological Bulletin, 95,* 410–427.

Berkowitz, L., & Alioto, J. (1973). The meaning of an observed event as a determinant of its aggressive consequences. *Journal of Personality and Social Psychology, 28,* 206–217.

Bodenhausen, G. V., & Macrae, C. N. (1998). Stereotype activation and inhibition. In R. S. Wyer, Jr. (Ed.), *Advances in social cognition,* (Vol. XI, pp. 1–52), Mahwah, NJ: Lawrence Erlbaum Associates.

Bodenhausen, G. V., Macrae, C. N., & Sherman, J. W. (1999). On the dialectics of discrimination: Dual processes in social stereotyping. In S. Chaiken & Y. Trope (Eds.), *Dual-process theories in social psychology* (pp. 271–290). New York: Guilford.

Bruner, J. S. (1951). Personality dynamics and the process of perceiving. In R. Blake & G. Ramsey (Eds.), *Perception: An approach to personality*. New York: Ronald.

Bruner, J. S. (1957). On perceptual readiness. *Psychological Review, 64,* 123–152.

Bruner, J. S. (1958). Social psychology and perception. In E. E. Maccoby, T. M. Newcomb, & E. L. Hartley (Eds.), *Readings in social psychology* (3rd ed., pp. 85–94). New York: Holt, Rinehart and Winston.

Bruner, J. S. (1992). Another look at New Look 1. *American Psychologist, 47,* 780–782.

Bruner, J. S. (1994). The view from the heart's eye: A commentary. In P. M. Niedenthal & S. Kitayama (Eds.), *The heart's eye: Emotional influences in perception and attention* (pp. 269–286). New York: Academic Press.

Bruner, J. S., & Postman, L. (1947a). Emotional selectivity in perception and reaction. *Journal of Personality, 16,* 69–77.

Bruner, J. S., & Postman, L. (1947b). Tension and tension release as organizing factors in perception. *Journal of Personality, 15,* 300–308.

Bruner, J. S., & Postman, L. (1949). On the perception of incongruity: A paradigm. *Journal of Personality, 18,* 206–223.

Busselle, R. W. (2004). Television realism measures: The influence of program salience on global judgments. *Communication Research Reports, 20,* 367–375.

Busselle, R. W., & Greenberg, B. S. (2000). The nature of television realism judgments: A reevaluation of their conceptualization and measurement. *Mass Communication and Society, 3,* 249–268.

Busselle, R. W., &, Shrum, L. J. (2003). Media exposure and the accessibility of social information. *Media Psychology, 5,* 255–282.

Coleridge, S. T. (1967). Biographia literaria. In D. Perkins (Ed.), *English romantic writers*. New York: Harcourt, Brace, & World.

Cooper, E., & Jahoda, M. (1947). The evasion of propaganda: How prejudiced people respond to anti-prejudice propaganda. *Journal of Psychology,* 23, 15–25.

Craik, F. I. M., & Lockhart, R. S. (1972). Levels of processing: A framework for memory research. *Journal of Verbal Learning and Verbal Behavior, 11,* 671–684.

Devine, P. G. (1989). Stereotypes and prejudice: Their automatic and controlled components. *Journal of Personality and Social Psychology, 56,* 5–18.

Eagly, A. H., & Chaiken, S. (1993). *The psychology of attitudes*. New York: Harcourt Brace Jovanovich.

Erdelyi, M. H. (1974). A new look at the New Look: Perceptual defense and vigilance. *Psychological Review, 81,* 1–25.

Erdelyi, M. H. (1983). *Psychoanalysis: Freud's cognitive psychology*. New York: W. H. Freeman.

Erdelyi, M. H. (1992). Psychodynamics and the unconscious. *American Psychologist, 47,* 784–787.

Erdelyi, M. H. (2004). Subliminal perception and its cognates: Theory, indeterminacy, and time. *Consciousness and Cognition, 13,* 73–91.

Fazio, R. H. (1989). On the power and functionality of attitudes: The role of attitude accessibility. In A. R. Pratkanis, S. J. Breckler, & A. G. Greenwald (Eds.), *Attitude structure and function* (pp. 153–179). Hillsdale, NJ: Lawrence Erlbaum Associates.

Fazio, R. H., Roskos-Ewoldsen, D. R., & Powell, M. C. (1994). Attitudes, perceptions, and attention. In P. M. Niedenthal & S. Kitayama (Eds.), *The heart's eye: Emotional influences in perception and attention* (pp. 197–216). New York: Academic Press.

Fazio, R. H., & Towles-Schwen, T. (1999). The MODE model of attitude-behavior processes. In S. Chaiken & Y. Trope (Eds.), *Dual-process theories in social psychology* (pp. 97–116). New York: Guilford.

Feshbach, S. (1972). Reality and fantasy in filmed violence. In J. Murray, E. Rubinstein, & G. Comstock (Eds.), *Television and social behavior* (Vol. 2, pp. 318–345). Washington, DC: Department of Health, Education, and Welfare.

Fiske, S. T., & Linville, P. W. (1980). What does the schema concept buy us? *Personality and Social Psychology Bulletin, 6,* 543–557.

Geen, R. (1975). The meaning of observed violence: Real versus fictional violence and effects of aggression and emotional arousal. *Journal of Research in Personality, 9,* 270–281.

Geen, R., & Rakosky, J. (1973). Interpretations of observed violence and their effects of GSR. *Journal of Experimental Research in Personality, 6,* 289–292.

Gerbner, G., Gross, L., Morgan, M., & Signorielli, N. (1986). Living with television: The dynamics of the cultivation process. In J. Bryant & D. Zillmann (Eds.), *Perspectives on media effects* (pp. 17–48). Hillsdale, NJ: Lawrence Erlbaum Associates.

Gerbner, G., Gross, L., Morgan, M., Signorielli, N., & Shanahan, J. (2002). Growing up with television: Cultivation processes. In J. Bryant & D. Zillmann (Eds.), *Media effects: Advances in theory and research* (2nd ed., pp. 43–67). Mahwah, NJ: Lawrence Erlbaum Associates.

Gerrig, R. J. (1993) *Experiencing narrative worlds.* New Haven, CT: Yale University Press.

Green, M. C., & Brock, T. C. (2000). The role of transportation in the persuasiveness of public narratives. *Journal of Personality and Social Psychology, 79,* 701–721.

Green, M. C., Garst, J., & Brock, T. C. (2004). The power of fiction: Determinants and boundaries. In L. J. Shrum (Ed.), *The psychology of entertainment media: Blurring the lines between entertainment and persuasion* (pp. 161–176). Mahwah, NJ: Lawrence Erlbaum Associates.

Greenwald, A. G. (1992). New Look 3: Unconscious cognition reclaimed. *American Psychologist, 47,* 766–779.

Hall, A. (2003). Reading realism: Audiences' evaluations of the reality of media texts. *Journal of Communication, 53,* 624–641.

Hastorf, A. H., & Cantril, H. (1954). They saw a game: A case study. *Journal of Abnormal and Social Psychology, 49,* 129–134.

Higgins, E. T. (1996). Knowledge activation: Accessibility, applicability, and salience. In E. T. Higgins & A. W. Kruglanski (Eds.), *Social psychology: Handbook of basic principles* (pp. 133–168). New York: Guilford Press.

Higgins, E. T., & Brendl, C. M. (1995). Accessibility and applicability: Some "activation rules" influencing judgment. *Journal of Experimental Social Psychology, 31,* 218–243.

Higgins, E. T., & King, G. (1981). Accessibility of social constructs: Information processing consequences of individual and contextual variability. In N. Cantor & J. F. Kihlstrom (Eds.), *Personality, cognition and social interaction* (pp. 69–121). Hillsdale, NJ: Lawrence Erlbaum Associates.

Higgins, E. T., King, G. A., & Mavin, G. H. (1982). Individual construct accessibility and subjective impressions and recall. *Journal of Personality and Social Psychology, 43,* 35–47.

Higgins, E. T., Rholes, W. S., & Jones, C. R. (1977). Category accessibility and impression formation. *Journal of Experimental Social Psychology, 13,* 141–154.

Katz, D. (1960). The functional approach to the study of attitudes. *Public Opinion Quarterly, 24,* 163–204.

Lau, R. R. (1989). Construct accessibility and electoral choice. *Political Behavior, 11,* 5–32.

Lazarus, R. S., Eriksen, C. W., & Fonda, C. P. (1951). Personality dynamics and auditory perceptual recognition. *Journal of Personality, 19,* 471–482.

Lepore, L., & Brown, R. J. (1997). Automatic stereotype activation: Is prejudice inevitable? *Journal of Personality and Social Psychology, 72,* 275–287.

Lombardi, W. J., Higgins, E. T., & Bargh, J. A. (1987). The role of consciousness in priming effects on categorization. *Personality and Social Psychology Bulletin, 13,* 411–429.

Martin, L. L. (1986). Set/reset: The use and disuse of concepts in impression formation. *Journal of Personality and Social Psychology, 51,* 593–504.

Moskowitz, G. B., Salomon, A. R., & Taylor, C. M. (2000). Preconsciously controlling stereotyping: Implicitly activated egalitarian goals prevent activation of stereotypes. *Social Cognition, 18,* 151–177.

O'Guinn, T. C., & Shrum, L. J. (1997). The role of television in the construction of consumer reality. *Journal of Consumer Research, 23,* 278–294.

Oliver, M. B. (1994). Portrayals of crime, race, and aggression in "reality-based" police shows: A content analysis. *Journal of Broadcasting & Electronic Media, 38,* 179–192.

Petty, R. E., & Krosnick, J. A. (Eds.). (1995). *Attitude strength: Antecedents and consequences.* Mahwah, NJ: Lawrence Erlbaum Associates.

Postman, L., Bruner, J. S., & McGinnies, E. (1948). Personal values as selective factors in perception. *Journal of Abnormal and Social Psychology, 43,* 142–154.

Prentice, D. A., & Gerrig, R. J. (1999). Exploring the boundary between fiction and reality. In S. Chaiken & Y. Trope (Eds.), *Dual-process theories in social psychology* (pp. 529–546). New York: Guilford.

Roskos-Ewoldsen, D. R. (1996). Attitude accessibility and persuasion: Review and a transactive model. In B. R. Burleson (Ed.), *Communication yearbook 20* (pp. 185–225). Newbury Park, CA: Sage.

Roskos-Ewoldsen, D. R., & Fazio, R. H. (1992). On the orienting value of attitudes: Attitude accessibility as a determinant of an object's attraction of visual attention. *Journal of Personality and Social Psychology, 63,* 198–211.

Shapiro, M. A., & Chock, T. M. (2003). Psychological processes in perceiving reality. *Media Psychology, 5,* 163 198.

Shapiro, M. A., & Fox, J. R. (2002). The role of typical and atypical events in story memory. *Human Communication Research, 28,* 109–135.

Shrum, L. J. (1996). Psychological processes underlying cultivation effects: Further tests of construct accessibility. *Human Communication Research, 22,* 482–509.

Shrum, L. J. (1999). The relationship of television viewing with attitude strength and extremity: Implications for the cultivation effect. *Media Psychology*, 1, 3–25.

Shrum, L. J. (2006). Cultivation and social cognition. In D. Roskos-Ewoldsen & J. Monahan (Eds.), Communication: Social Cognition Theories and Methods. Mahwah, NJ: Lawrence Erlbaum Associates.

Shrum, L. J., Burroughs, J. E., & Rindfleisch, A. (2004). A process model of consumer cultivation: The role of television is a function of the type of judgment. In L. J. Shrum (Ed.), *The psychology of entertainment media: Blurring the lines between entertainment and persuasion* (pp. 177–191). Mahwah, NJ: Lawrence Erlbaum Associates.

Shrum, L. J., & O'Guinn, T. C. (1993). Processes and effects in the construction of social reality: Construct accessibility as an explanatory variable. *Communication Research, 20,* 436–471.

Shrum, L. J., O'Guinn, T. C., Semenik, R. J., & Faber, R. J. (1991). Processes and effects in the construction of normative consumer beliefs: The role of television. In R. H. Holman & M. R. Solomon (Eds.), *Advances in consumer research* (Vol. 18, pp. 755–763). Provo, UT: Association for Consumer Research.

Smith, M. B., Bruner, J. S., & White, R. W. (1956). *Opinions and personality*. New York: Wiley.

Srull, T. K., & Wyer, R. S. (1979). The role of category accessibility in the interpretation of information about persons: Some determinants and implications. *Journal of Personality and Social Psychology, 37,* 1660-1672.

Srull, T. K., & Wyer, R. S. (1980). Category accessibility and social perception: Some implications for the study of person memory and interpersonal judgment. *Journal of Personality and Social Psychology, 38,* 841–856.

Vidmar, N., & Rokeach, M. (1974). Archie Bunker's bigotry: A study in selective perception and exposure. *Journal of Communication, 24,* 36–47.

Wilson, B., & Busselle, R. W. (2004, May). *Transportation into the narrative and perceptions of media realism*. Paper presented at the meeting of the International Communication Association, New Orleans.

Wyer, R. S., & Srull, T. K. (1989). *Memory and cognition in its social context*. Hillsdale, NJ: Lawrence Erlbaum Associates.

Zwaan, R. A. (1994). Effect of genre expectations on text comprehension. *Journal of Experimental Psychology: Learning, Memory, and Cognition, 20,* 920–933.

5

Comprehension and Memory

Richard Jackson Harris, Elizabeth Tait Cady,
and Tuan Quoc Tran
Kansas State University

Imagine that you are in a theater watching the newest horror movie. As the tension mounts onscreen, suddenly a man in the audience yells "look behind the door!" as the main character searches the house for the killer. Although the man has not seen the movie, he guessed where the killer was hiding based on his previous experience with, comprehension of, and memory for similar horror movies.

As media entertain us, we continually comprehend and remember information, and this process begins early in life. Young children often learn new information such as the letters of the alphabet through songs. Cultures pass traditions and other social knowledge to the next generation using songs, and the memory of the melody assists in remembering the words (Rubin, 1995). Before we can learn or be entertained, we must first comprehend something, which may or may not be exactly the meaning the producer had intended. In order to respond to anything for longer than an immediate reaction, we must remember something of what we had comprehended. Comprehension and memory have a long history of study in psychology and are impossible to separate from one another. In fact, memory may be seen as an inevitable— albeit imperfect—by-product of normal comprehension (Craik & Lockhart, 1972). How we comprehend something has implications for how it is remembered, and what is remembered is in large part a function of what was initially understood.

This chapter reviews the major processes of comprehension and memory as they apply to the consumption of entertainment media. For the most part, we limit the discussion to media traditionally considered entertainment, that is, television entertainment programming, cinema, and popular music. Of course, it is also true that many sorts of "non-entertainment media," such as news, informational articles, and advertising are often consumed as entertainment, and, in fact, the uses and gratifications one has for consuming media may affect how the content is comprehended or remembered (Rubin, 2002). We begin with an overview of research and theory about comprehension and memory from cognitive psychology, then move into particular cognitive processing issues with particular types of media, and, finally, examine a contemporary comprehension model in terms of its application to entertainment.

The Psychology of Memory and Comprehension

During the comprehension process, memory comes into play as incoming perceptual inputs are connected to past knowledge or experience to construct an understanding of the media event. This constructed memory representation then can be used as a reference for interpreting future experience. This continuing interaction of comprehension and memory impacts many media experiences, including memory for media events, remembering whether something we know came from a movie or real life, and constructing worldviews based on media input. We begin this section with a brief general overview of memory, followed by a discussion of how memory contributes to comprehension and how comprehension contributes to memory.

BRIEF OVERVIEW OF MEMORY

Influenced by the computer metaphor beginning in the 1950s, theorists have for much of the last half century conceptualized memory as consisting of different discrete structural stores, with each stage possessing not only different storage capacity and duration, but also employing different types of processing (Atkinson & Shiffrin, 1968). The earliest memory store is referred to as *sensory memory* because at this initial stage information is processed by means of its visual, auditory, or other sensory information (e.g., letter features). Sensory memory is believed to be large in capacity but short-lived (lasting only a couple of seconds) and unstable (Bower, 2000). Much media content that enters our sensory memory is not attended to and thus never proceeds any farther or arrives in consciousness—such as TV or radio in the background to which we are paying no attention.

To prevent decay, information in sensory memory must be attended to and transferred to a limited-capacity short-term memory, or what is currently referred to as *working memory*, which involves the concurrent storage and processing of relevant information while inhibiting or ignoring information irrelevant to the current task (Neath & Surprenant, 2003). In working memory, information is temporarily maintained by a rehearsal mechanism, and information that is directly in the focus of attention is processed primarily on the basis of its visual or verbal code. The resources of working memory are limited, with only finite attentional resources available at any given time (Lang, 2000). Material that is not rehearsed or attended to can decay from working memory within 18 seconds (Peterson & Peterson, 1959).

This storage and processing system allows for the integration of recently presented information with information currently being processed, resulting in the development of an emergent mental model of the current environment, situation, or experiences. Although there are numerous views of working memory, the most recognized and influential is that of Baddeley (1986). In this model, working memory is parsed into three major systems, a *central executive*—assumed to control attention—and two slave systems, a *phonological loop*, responsible for storage and processing of verbal information, and a *visuo-spatial sketch pad*, responsible for storage and processing of visual and spatial information. Although Baddeley's model has been instrumental in accounting for many experimental and neurological phenomena, such as word length effects, phonological suppression, and severe short-term memory deficits in patients with intact long-term memory, a major limitation is that it does not take into consideration the meaningfulness of the stimulus. That is, the model weights different information equivalently, independent of its meaningfulness to the perceiver. According to Neath and Surprenant (2003), stimulus meaningfulness is an important factor in many tasks utilizing working memory, such as the memory span task of holding items in working memory a few seconds and then repeating them back. To address this need, Baddeley (2000) recently postulated an additional system called the *episodic buffer*, as an attempt to alleviate this limitation and other potential shortcomings of the model. The episodic buffer provides a link to long-term memory, an additional storage compartment, and a forum where visual-spatial and verbal information can be integrated. Unfortunately, to date, little experimental research has been conducted to carefully examine the role of the episodic buffer.

In contrast to working memory, information in long-term memory is primarily encoded semantically and is more or less permanently stored, with retrieval failure due to passive memory decay, insufficient retrieval cues, or interference from other material. Although there is no general consensus among memory theorists as to the types of long-term memory, several classifications do exist (Roediger, Marsh, & Lee, 2002). One popular classification is to divide memory into procedural memory ("knowing how") and declarative memory ("knowing that") (Squire, 1987). Procedural memory consists of both motor and cognitive skills that often lack ready verbal descriptions (e.g., how to ride a bike, how to program your VCR), while declarative memory can be further divided into two main categories, semantic memory and episodic memory. Whereas semantic memory can be thought of as a "mental dictionary" of facts about the world (e.g., Herbert Hoover was the 31st President of the United States), episodic memory can be thought of as experience that is tagged with a particular moment in time and place, for instance, the word "sleep" on the list just heard or last summer's vacation to Yosemite National Park (Roediger et al., 2002).

A problem in classifying different kinds of memory arises from the overlapping and multifaceted nature of memories (Roediger et al., 2002). Episodic memory over time can lose its distinctive tag of time and place and become part of semantic memory, such as when you forget the context of when you first learned your multiplication tables, but yet you still know that $8 \times 7 = 56$. Semantic memory can play an influential and sometimes misleading role in episodic memories. For example, former U.S. President Ronald Reagan would sometimes tell movie plots as true stories (e.g., a heroic story about a plane captain who stayed with a wounded man as the plane crashed after being hit by anti-aircraft fire). Although told as a true story, the details were taken from both a movie and a story in a magazine (Loftus & Ketcham, 1994). This confusion of the source of information is called **source amnesia** and is not uncommon.

AUTOBIOGRAPHICAL MEMORY

Autobiographical memory consists of knowledge about events or experiences that have occurred in one's life (Conway, 2001). These memories are most often comprised of episodic memories, such as one's senior prom or the long bike rides in the park with Grandpa, although elements from semantic or procedural memory may also be a part of the memory representation (Roediger et al., 2003).

The structure of autobiographical memory describes three levels of specificity of individual memories: lifetime period, general event, and event-specific knowledge (Conway & Pleydell-Pearce, 2000). The most general level is defined by a period of one's life, such as college years, or living in a certain town. Knowledge at this level includes the start and end points as well as the important people and events of that time. Media events may be integrally connected to a particular lifetime period, as in remembering hearing the song "Ice, Ice, Baby" by Vanilla Ice from your grade-school life, or watching *Titanic* while in middle school. The knowledge base of a certain lifetime period may include an evaluative component, and those attitudes may be used to construct memories at a later time, as when hearing a certain song induces good feelings because of the pleasant lifetime period it evokes, not because of the tune or the lyrics of the song.

The second level of the structure of autobiographical memory is the general event, which could include repeated or single events, or thematic sequences of events. This level of memory is both highly organized and focused on goal-directed and other types of self-discovery behavior. When people try to think of a memory, the general events are usually the easiest to recall (Conway, 2001).

The final level of the structure of autobiographical memory is the event-specific knowledge (ESK), which leads to imagery and other details of the memory itself. As a person recalls a memory, his/her sensory-perceptual experiences during the recalled event lead to vivid memories. Media information may be a part of either the ESK or the context in which it is remembered. Although the presence or absence of the ESK is usually an indicator of the reality of a memory (we do not tend to remember as many supporting details from events that did not occur), at times people do have vivid but false memories of events that never happened (Garry, Manning, Loftus, & Sherman, 1996; Sharman, Manning, & Garry, 2005). In some cases, the memories may be perceptually or temporally clear, but may in fact contain a memory of the cognitions involved in generating that memory. These attributions of reality can be influenced by the motivation, biases, and experiences, as well as metacognitive skills, of the person recalling the event (Mitchell & Johnson, 2000).

Autobiographical memory can be studied by presenting some type of cue and asking for a memory related to that cue. In media research, the cues given relate to autobiographical memories of events, characters, or programs watched at some point in one's life. This method allows research into the effects on children of antisocial media, such as sex or violence (Harrison & Cantor, 1999; Hoekstra, Harris, & Helmick, 1999; Cantor, Mares, & Hyde, 2003), without exposing young participants to it. Although there is an inherent problem of being unable to verify the memories of the media event, the variable of interest is how the memories affect people's perceptions of the media and the world, so the objective accuracy of the memories is of less concern than the participant's reaction to them.

The general method of this research involves asking participants to think of a specific event, when they watched it, and the circumstances or people involved (Harris, Bonds-Raacke, & Cady, 2005). For example, the participant might be instructed to think of the overall experience of watching a frightening or sexual-themed movie in their childhood, or they could be asked to think of one certain character from a specific minority group that they saw on television or in the movies. Once the experience or character has been recalled, the participant rates various aspects of the event or character on several dimensions. For example, the experience of the frightening movie might be rated based on the negative effects (e.g. insomnia) or positive effects (e.g. enjoyment) experienced, while the character might be evaluated in terms of personality dimensions (e.g. likeable, healthy) as well as typicality of the character within his/her minority group. Using the autobiographical memory technique in this way allows research on the perceptions of characters or events seen under normal viewing circumstances, rather than in a short segment viewed in more artificial situations.

Music also plays a large role in autobiographical memory. One study provided adults ages 37–76 with a song title and a segment performed on a piano. Results showed better song recall after hearing the melody and better time estimates of when the song had been popular if they used associated autobiographical memories as cues (Bartlett & Snelus, 1980). Another study (Schulkind, Hennis, & Rubin, 1999) tested both college students and adults ages 66–71. Both groups were more likely to remember and prefer music that had been popular when they were adolescents, although the older adults did not remember the early songs as well, unless the song evoked a strong emotional response (Schulkind et al., 1999).

CONTRIBUTION OF MEMORY
TO COMPREHENSION

Traditionally, information was believed to temporally flow in linear fashion from one memory store to the next; that is, information was first thought to enter sensory memory, then working memory, and, finally, long-term memory (Atkinson & Shiffrin, 1968). More current views (e. g., Cowan, 1995) have argued that encoding both old and new information that share similar features with existing memory concepts uses long-term memory knowledge to create a temporary representation in working memory. In fact, some theorists (e.g., Anderson & Lebiere, 1998; Cowan, 1995; Shiffrin & Schneider, 1977) have reconceptualized working memory as a temporarily activated portion of long-term memory. In this case, incoming information, say, stimulus A, enters sensory memory and is matched with a long-term memory representation to form a temporary mental representation of stimulus A in working memory. While stimulus A is maintained or stored by a rehearsal mechanism in working memory, stimulus B is encoded in sensory memory and is referenced to a different long-term memory. In working memory, stimuli A and B can be linked together to form new associations that can be recorded and stored as a new long-term episodic memory (Cowan, 1995). The view of working memory as an activated portion of long-term memory proposes the following temporal flow of information processing: Incoming information enters sensory memory, is matched and interpreted based on a long-term memory representation, and then is temporarily stored in working memory where different information can form newly established links. Finally, these newly formed associations are stored in long-term memory. Hence, this process explicitly states that long-term memory plays a primary role in interpreting or extracting meaning from incoming information. In other words, long-term memory plays a primary role in comprehension. Applied to media events, this means that our previous experiences with a media event affect how we interpret a current media event. For example, having previously seen a horror film in which a victim was lured into a bedroom and killed by an assassin hiding behind a door, an audience currently watching a horror film will expect that something harmful is about to happen when a movie character walks into an empty bedroom.

CONTRIBUTION OF COMPREHENSION
TO MEMORY

Traditional views of memory (i.e., Atkinson & Shiffrin, 1968) assume that the probability of retrieving information from long-term memory depends on how much that information was rehearsed in working memory; with increased rehearsal comes improved retrieval (Bower, 2000). However, Craik and Lockhart (1972) found that simple repetitive maintenance rehearsal does not necessarily lead to a more durable memory. Instead, retrieval is dependent on how the information is initially processed—the deeper the processing, the more durable the memory. The depth of processing is determined by the degree to which one understands and extracts the sense of the information to form meaningful associations and elaborations with existing knowledge (Bower, 2000). For example, an event that occurs in a TV show may remind you of an event from your own life. To deeply process the show, you would think of aspects relevant to your life and compare your experience with that of the character. Thus, our ability to comprehend materials affects the durability and accessibility of such materials in long-term memory (Craik & Tulving, 1975). If we concentrate very carefully on a TV documentary, we will probably remember the content quite well; if it is only on in the background with our primary attention elsewhere, we probably will not.

How does information from entertainment media come to play such a large role in our comprehension and memory processes? Some findings from communication and cognitive psychology research can help us answer that question.

Exemplification Theory and Cognitive Heuristics

It is a natural part of human thinking that the rich sensory experience of our world must somehow be organized into meaningful categories for interpretation and storage in memory. Indeed, if it were not, sensory input would merely be a disorganized kaleidoscope of stimulation, or at best a set of unorganized exemplars.

Implicit in this categorization process is the fact that specific exemplars or instances of particular categories, classes of events, genres of entertainment, and so forth, will come to represent the entire category, regardless of whether they are, in fact, truly representative. Which exemplars will prevail in mentally defining the category depend on two major factors: 1) frequency; and 2) vividness. The more often an instance occurs, the more representative it will seem. For example, if a disproportionate number of African-American men in movies are criminals or drug dealers, many viewers (especially those with limited life experience with African-American men) will come to see that stereotype as typical of black men.

Secondly, a particularly vivid example is highly memorable and thus is very readily called to mind when thinking of that category. For example, in the 1–2 years after the 1975 blockbuster movie classic *Jaws*, which portrayed numerous shark attacks on swimmers at ocean beaches in New England, coastal beach resorts nationwide reported a significant loss in business and many brave souls who did visit spent little time in the water. The highly vivid fictional attacks from *Jaws*, though extremely rare in real life, were readily remembered and taken to be far more typical beach experiences than was in fact the case. Vivid cases that arouse high levels of emotion (such as a shark attack) are especially memorable, as are vivid cases that are frequently repeated.

Two cognitive heuristics can help describe these processes (Kahneman & Tversky, 1972, 1973; Tversky & Kahneman, 1973). The *representativeness heuristic* says that we evaluate how representative of its class an event is by the extent to which it appears to reflect (a) all the members of that class, and (b) the process by which it was generated. For example, many see a coin-tossing sequence of HHHHHH as much less likely than a sequence of HTTHTH, even though any given sequence of six coin tosses is equally likely as any other. The sequence that has both possible outcomes represented in an order that does not appear systematic will be taken as more representative. Sometimes judging representativeness comes along with ignoring critical base-rate information. For example, if we hear that Robert likes to read French literature and go to wine-tasting parties, we might think he is more likely a philosopher than a truck driver. While those interests appear more representative of the stereotype of philosophers than truck drivers, there are many more truck drivers than philosophers and thus Robert is still more likely to be a truck driver.

The second heuristic, *availability*, posits that we draw conclusions about the frequency or typicality of an event or instance based on how readily we can retrieve examples from memory. Easily retrieved examples are then seen as highly typical, when, in fact, they may not be so. If the first examples of Arabs that come to mind are the villains on TV and film entertainment and terrorists in the news, we will come to think a far larger proportion of Arabs are terrorists than is in fact the case. With over 70% of media characters with mental disorders portrayed as violent, compared to 11% of real people with disorders, no wonder people are scared of people with mental illness (Teplin, 1985). The positive potential of this principle is also considerable; for example, if Will Truman of the popular sitcom *Will and Grace* becomes the prototype for a gay man, the social perception of that group might be improved considerably. Entertainment presents numerous vivid and memorable exemplars of diverse people and situations; when the distribution of these exemplars deviates strongly from the real-world distribution, the risk of viewers having a skewed view of the world markedly increases.

For a good recent discussion of exemplification theory, see Zillmann (2002).

MINORITY GROUP PORTRAYALS

Although the number of African-Americans on U.S. television has greatly increased since the 1960s and now approximates the proportion in the general population, they tend to disproportionately appear in situation comedies or police dramas, and the recent increase in roles for males has not carried over to African-American females (Greenberg, Mastro, & Brand, 2002). In contrast, Hispanic Americans, although even greater in number than African-Americans in the general population, only comprise 2% of prime-time TV characters (Poniewozik, 2001) and are concentrated in humorous, criminal, or police roles. Native Americans used to be almost entirely Plains Indians from TV and movie Westerns and recently have almost entirely disappeared from entertainment media. The small number of characters combined with the stereotypical portrayal of minorities in the media can lead to prejudiced views of these groups, especially in those viewers with limited life experience with members of the group in question (Greenberg, Mastro, & Brand, 2002).

Content analyses investigating the proportion and quality of minority roles on television and film do not tell the whole story of how those portrayals might affect the audience's perception of those minorities. Specifically, a strong character in an immensely popular show or series might "drench" the viewer with an image of the minority that remains strong despite other portrayals the viewer might see, for instance, Bill Cosby's Cliff Huxtable from *The Cosby Show* or Eric McCormack's Will Truman from *Will and Grace*. In this way, certain actors and characters will exert more influence on the perceptions of the group being portrayed than the many other stereotypical portrayals the audience views (Greenberg, 1988).

Sometimes an unusually attractive or respected person or character can greatly influence behavior in positive ways. For example, after a sexy "hunk" actor in a popular Brazilian soap opera in the 1980s played a deaf character, interest in learning sign language soared. Similarly, when NBC news anchor Katie Couric invited *Today* show viewers to watch her colon exam live in 2000, requests for colonoscopies to check for colon cancer rose 20%, almost surely saving numerous lives (Bjerklie, 2003).

CULTIVATION THEORY

Attending to and recalling these types of portrayals can have long-term effects. One theory that describes how knowledge is skewed by comprehending and remembering media experiences is Cultivation Theory, which looks at the way that extensive repeated exposure to entertainment media over time gradually shapes our view of the world and our social reality. See Gerbner, Gross, Morgan, Signorielli, and Shanahan (2002) for an overview of the theory. Cultivation of worldviews happens through a process of construction, whereby viewers learn about the real world from observing the world of television. Memory traces from watching TV are stored relatively automatically (Shapiro, 1991). We then use this stored information to formulate beliefs about the real world (Hawkins & Pingree, 1990; Potter, 1991a, 1991b, 1993; Shrum & Bischak, 2001). The more television one watches, the more one's worldview resembles the world presented on television. For example, people who watch a lot of violent TV believe the world to be a more violent place than it really is (mean world syndrome) (Signorielli,

1990). The cultivated social reality includes many types of knowledge, including gender roles (Morgan & Shanahan, 1995), political attitudes (Morgan, 1989), estimations of crime risk (Shrum, 2001; Shrum & Bischak, 2001), understanding of science and scientists (Potts & Martinez, 1994), attitudes toward the environment (Shanahan & McComas, 1999; Shanahan, Morgan, & Stenbjerre, 1997), adolescent career choices (Morgan & Shanahan, 1995), and effects of prolonged viewing of talk shows (Rössler & Brosius, 2001).

Entertainment as Education

One increasingly popular way to communicate socially positive information to the public is through entertainment media. Over 75 such entertainment-education (E-E) campaigns have been implemented in at least 40 nations worldwide and are especially common in developing countries (Sherry, 2002; Singhal & Rogers, 1999). In many countries, radio and television have long been seen as legitimate tools for development and positive social change, rather than merely vehicles for entertainment. Thus many popular entertainment series are explicitly produced to promote gender equality, adult literacy, sexual responsibility, and family planning.

Sometimes these entertainment programs can be wildly popular and profitable but also lead to considerable knowledge and behavior change. For example, the Tanzanian radio soap opera *Twende na Wakati ("Let's Go with the Times")* reached 55% of the nation's population from 1993–98, with 82% of viewers saying they had changed behavior to reduce the chance of HIV infection (Rogers, Vaughan, Swalehe, Rao, Svenkerud, & Sood, 1999). The South African TV, radio, and public health campaign *Soul City* (Singhal & Rogers, 1999) dealt with HIV prevention, maternal and child health, domestic violence, and alcohol abuse, and it became South Africa's top-rated television show. It elicited discussion and further information-seeking. Before *Soul City*, only 3% agreed with a statement that one's HIV-positive status should be communicated to one's partner, but afterward 75% agreed the partner should be told.

Although no broad-based popular E-E campaigns have been promulgated in the United States, elements of E-E increasingly appear on American entertainment TV. One of the earliest was the designated driver campaign of the late 1980s, where the Harvard School of Public Health worked with 250 NBC writers and producers to incorporate the new idea of a "designated driver" in TV plot lines. By 1994, the designated driver message had appeared on 160 prime-time shows and been the main topic of 25. Two-thirds of the public had noted the mention of designated drivers in TV shows and just over half of young adults reported they had served as a designated driver. By the late 1990s, the drunk-driving fatality rate had fallen by one-third from ten years earlier, in part due to greater use of designated drivers (Rosenzweig, 1999).

Since 1998 the Center for Disease Control has assisted teleplay writers in placing positive health messages in their entertainment scripts for popular TV shows. On *Beverly Hills 90210* Steve bragged about his flawless tan, but his girlfriend noticed a suspicious mole on the back of his neck. Concerned about skin cancer, he later took a megaphone to the beach and shouted about the benefits of using sunscreen. An episode of *ER* used a plot line about morning-after contraception; 6 million of the show's 34 million viewers reported learning about morning-after contraception from watching the show (Rosenzweig, 1999). A Kaiser Family Foundation study found that one third of *ER* viewers reported learning something from the show that had been helpful in making health care decisions in their own families. (Stolberg, 2001). Other embedded health messages have included AIDS awareness themes on soap operas and pro-condom messages on the sitcom *Friends* (Brown & Walsh-Childers, 2002) and the teen-oriented shows *Felicity* and *Dawson's Creek* (Rosenzweig, 1999).

This narrative form of entertainment may be an especially good format for placing persuasive messages (Slater & Rouner, 2002). A good E-E program can increase a viewer/listener's sense of self-efficacy, one's beliefs about one's capabilities to exercise control over events that affect one's life (Bandura, 1997). These beliefs can then lead to prosocial behaviors like using condoms, seeking medical advice, or taking control over one's reproductive health. It can also contribute to a sense of collective efficacy, the belief in joint capabilities to forge individual self-interests into a shared agenda (Bandura, 1995). For example, when viewers of a *Soul City* story on wife-battering gathered at a neighborhood batterer's home and banged their pans in censure, a behavior seen in the series, collective efficacy was achieved (Singhal & Rogers, 2002).

Developmental Issues

The comprehension and memory of information from the media sometimes varies as a function of age, or, more specifically, the level of cognitive development of the viewer.

One area where cognitive development is extremely critical concerns young children's understanding of the difference of programs and commercials. From the adult point of view, television programming is clearly entertainment, while the advertising is something else altogether. However, very young children do not discriminate between commercial and program content and do not understand the persuasive intent of ads; to them television provides a steady uninterrupted stream of entertainment. Although children identify commercials at a very early age, this identification is based on superficial audio and video aspects rather than an understanding of the difference between programs and commercials (Raju & Lonial, 1990). Preschool children have little understanding that commercials are meant to sell. Elementary school children show various intermediate stages of development of the understanding of the purpose of ads (see Martin, 1997, for a meta-analysis). The insertion of video and audio separators between program and commercials has not made this discrimination much easier. Only about a third of 5- to 7-year-olds understand the selling purpose of ads, but almost all do by age 11 (Wilson & Weiss, 1992). Typical explanations of middle elementary children center on the truth (or lack thereof) of the material; not until late elementary school is the distrust based on perceived intent and an understanding of the advertiser's motivation to sell.

A second area where the comprehension of media varies hugely with age is the issue of fear responses to violent entertainment (Cantor, 1996, 1998b, 2002). Different stimuli and events differentially elicit fear responses in viewers of different ages. Distortions of natural forms (e.g., monsters and mutants) are very scary to preschoolers but typically less so to older children. Depictions of dangers and injuries (e.g., attacks, natural disasters) are scarier to upper elementary school children than to preschoolers, in part because the older children are cognitively able to anticipate danger and its possible consequences and thus be fearful *before* the actual event occurs. Only a teen or adult is likely to be very scared of an abstract threat, which may have to be imagined by the viewer from dialogue in the film. The older the child, the more able he or she is to think abstractly and be frightened by seeing situations of endangerment to others.

Consider two brothers watching a movie about a benevolent alien visiting from outer space. The preschooler may be afraid of the fantastic form of the alien, while the older child may be afraid during the scenes where he recognizes that the friendly alien or sympathetic humans may be in potential danger from others. Alternatively, the younger child may not be afraid at all—in fact, he may love the cute little alien—and may not be able to think abstractly enough to understand why his brother is afraid of the potential endangerment to the kind creature. How much each is entertained—and why—will clearly vary as a function of the child's age and how they comprehend fear-inducing stimuli.

Instances of media fear in childhood may often be long-lasting, even traumatic memories. Several studies (Cantor & Oliver, 1996; Harrison & Cantor, 1999; Hoekstra, Harris, & Helmick, 1999) found that practically all young adults were able to readily remember an incident of being extremely scared by a movie as a child or teen. At least the memories, and perhaps some of the effects as well, are long lasting. Some effects reported are general fear/anxiety, specific fears (e.g., fear of swimming after seeing *Jaws*; fear of clowns after seeing *It)*, sleep disturbances, and nightmares.

Sometimes, self-report may not be a completely adequate measure of comprehension or memory, however. For example, Sparks, Pellechia, and Irvine (1999) found that some people reported low levels of negative affect in response to a 25-minute segment from a horror film but high levels of physiological arousal, as measured by skin conductance. Peck (1999, cited in Cantor, 2002) found that women's reports of fear in response to watching scary scenes from *A Nightmare on Elm Street* were more intense than men's, but in some cases men's physiological responses were more intense than women's, if the victim in the scene was male.

Adults of different ages may comprehend media differently. For example, many older women viewers of the hugely successful Indian TV drama *Hum Log (We People)* identified more with the traditional family matriarch character than with her more independent daughters (Brown & Cody, 1991). Similarly, with the 1970s U.S. sitcom *All in the Family*, older and more traditional viewers identified with the bigoted Archie Bunker and found him to be a more positive figure than the producers had envisioned or than the younger audience saw him (Vidmar & Rokeach, 1974). This phenomenon, of some viewers unexpectedly identifying with the negative models, has been observed in several nations and has come to be known as the "Archie Bunker effect."

A MODEL OF COMPREHENSION AND MEMORY FOR ENTERTAINMENT

We now conclude this paper with a proposed application of the comprehension model of Walter Kintsch (1988, 1998) to the cognition of entertainment. Because memory contributes to comprehension, and the degree of comprehension contributes to memory, a picture of an interactive process between comprehension and memory emerges. The process of comprehending or understanding a media event can be described as involving the use of existing knowledge from memory to interpret or extract meaning of perceptual inputs, while simultaneously integrating those inputs to construct a coherent holistic internal mental model of the media event in working memory. This mental model will then be transferred and stored in long-term memory for aiding interpretation of future experiences or events (Kintsch, 1988). According to Kintsch's (1988, 1998) construction-integration (CI) theory, the comprehension process consists of creating three levels of mental representation within two major phases: a construction phase and an integration phase. Both of these take place in working memory, more specifically, in the long-term working memory described in Ericsson and Kintsch (1995) or Baddeley (2000). Although Kintsch's CI theory was originally formulated in the context of text comprehension, the CI theory can be generalized to other contexts such as comprehending media events.

According to Kintsch, understanding something requires building a mental model. This mental model is built up sequentially and in cycles. The three levels of mental representations that act as a unitary whole in the CI model, from lowest to highest, are surface level, textbase level, and situational model. Generalizing to entertainment like television, cinema, and popular music, the construction phase of the CI theory involves the surface and textbase level. The surface level is the actual minimally processed information that the person perceives and encodes from the media. In reading text, this is represented by the exact words or phrases of the text. In media entertainment, this can reflect the actual visual images and sounds that a person encodes from a TV program or listening to music. At the textbase level, information from the surface level is transformed into a propositional representation, which is the smallest component of meaning or thought (Anderson & Bower, 1973; Kintsch, 1974; Bower, 1981). Propositional representation involves a predicate (expressed in the surface forms of verbs, adjectives, or adverbs), argument(s), and time and place that are explicitly mentioned in the media event.

According to Kintsch, during this construction phase, the textbase level activates much information, which is both relevant and irrelevant in memory independent of the current situation. For example, at the textbase level in the movie *The Sixth Sense*, when the lead character Malcolm, played by Bruce Willis, was shot, what we understand about what happens when a person is shot (e.g., serious injury, paralysis, death) comes from seeing previous similar movies involving shootings, which become activated in our memory. In addition, knowledge of prior Bruce Willis characters becomes activated at the textbase level. Not until the second phase (integration) of the CI theory does the mental model becomes context-specific (e.g., de-activating previous Willis roles while maintaining the concept of injury or death) through the use of our knowledge of the genre of the current media event (i.e., *The Sixth Sense* is a thriller). Ideas or concepts that remain activated after the integration phase become embedded into our mental model or situation model in the form of a knowledge network of what we understand of our current environment (i.e., character is shot, he is likely going to die). The mental model is built and lies on the foundation of our interpretation based on the encoded information and our previous knowledge and experiences; hence, our mental model may not always be accurate.

Inaccuracies in our mental model can be due to our misinterpretation or misunderstanding of the perceived event. Sometimes screenwriters intentionally induce such misunderstandings, as in the case of *The Sixth Sense*, where the script purposely leads us to construct an incorrect situation model of the action. Immediately following the scene where Malcolm was shot, we see him sitting on a bench, leading many viewers to interpret and construct a mental model that he recovered from his injury. It is not until near the end of the movie when the movie writers provide the viewers with cues to make them realize that their mental model of the initial shooting scene and its outcome was incorrect (i.e., he did not survive the attack).

In the CI theory, our mental model consists of information that remains activated and embedded within the knowledge net. This information is used to interpret subsequent information about the media event; that is, subsequent information that is consistent with our mental model remains activated and possibly integrated in the knowledge structure, while subsequent information that is not consistent with the mental model tends to be de-activated. This can explain why many watching *The Sixth Sense* were not aware of the subtle cues that the lead character was dead embedded in the movie by the screenwriter but rather continued to accept information consistent with the mental model that he had survived serious injury. Besides being influential in guiding our interpretation of entertainment media, our mental model is also an important construct that affects many other aspects of our entertainment experiences discussed earlier, such as providing a context for use of cognitive heuristics, reality monitoring, cultivating worldviews, and memory for music.

CONCLUSION

This comprehension model applied to media entertainment leads to a better understanding of how some effects of the media may arise. For example, integrating the information about a minority character on one show with information about a similar character on another show may lead to cultivation of an opinion about all members of that particular social group. This process, along with forming exemplars from the media and other processes discussed here, repeats itself many times as people seek out and enjoy media entertainment. In addition, the memories of these media experiences may affect viewers over the course of a lifetime. In this way, comprehending and remembering media events play an important role in our lives.

REFERENCES

Anderson, J. R., & Lebiere, C. (1998). *The atomic components of thought.* Mahwah, NJ: Lawrence Erlbaum Associates.

Anderson, J. R., & Bower, G. H. (1973). *Human associative memory.* Washington: Winston and Sons.

Atkinson, R. C., & Shiffrin, R. M. (1968). Human memory: A proposed system and its control processes. In K. W. Spence & J. T. Spence (Eds.), *The psychology of learning and motivation: Advances in research and theory* (pp. 90–197. New York: Academic Press.

Baddeley, A. D. (1986). *Working memory.* New York: Oxford University Press.

Baddeley, A. D. (2000). The episodic buffer: A new component of working memory? *Trends in Cognitive Sciences, 4,* 417–423.

Bandura, A. (1995). Exercise of personal and collective efficacy. In A. Bandura (Ed.), *Self-efficacy in changing societies* (pp. 1–45). New York: Cambridge University Press.

Bandura, A. (1997). *Self-efficacy: The essence of control.* New York: Freeman.

Bartlett, J. C., & Snelus, P. (1980). Lifespan memory for popular songs. *American Journal of Psychology, 93,* 551–560.

Bower, G. H. (1981). Mood and Memory. *American Psychologist, 36,* 129–148.

Bower, G. H. (2000). A brief history of memory research. In E. Tulving & F. I. M. Craik *(Eds.), The Oxford handbook of memory* (pp. 3–32). New York: Oxford University Press.

Brown, W. J., & Cody, M. J. (1991). Effects of a prosocial television soap opera in promoting women's status. *Human Communication Research, 18,* 114–142.

Bjerklie, D. (2003, July 28). *Time,* p. 73.

Brown, J. D., & Walsh-Childers, K. (2002). Effects of media on personal and public health. In J. Bryant and D. Zillmann (Eds.), *Media effects* (2nd ed.) (pp. 453–488). Mahwah NJ: Lawrence Erlbaum Associates.

Cantor, J. (1996). Television and children's fear. In T.M. Macbeth, (Ed.), *Tuning in to young viewers: Social science perspectives on television* (pp. 87–115). Thousand Oaks CA: Sage.

Cantor, J. (1998a). Ratings for program content: The role of research findings. *The Annals of the American Academy of Political and Social Science, 557,* 54–69.

Cantor, J. (1998b). *"Mommy, I'm scared": How TV and movies frighten children and what we can do to protect them.* San Diego: Harcourt Brace.

Cantor, J. (2002). Fright reactions to mass media. In J. Bryant & D. Zillmann (Eds.), *Media effects* (pp. 287–306). Mahwah NJ: Lawrence Erlbaum Associates.

Cantor, J., & Oliver, M. B. (1996). Developmental differences in responses to horror. In J.B. Weaver and R. Tamborini (Eds.), *Horror films: Current research on audience preferences and reactions* (pp. 63–80). Mahwah NJ: Lawrence Erlbaum Associates.

Cantor, J., Mares, M.-L., & Hyde, J. S. (2003). Autobiographical memories of exposure to sexual media content. *Media Psychology, 5,* 1–31.

Conway, M. A. (2001). Sensory-perceptual episodic memory and its context: Autobiographical memory. In A. Baddeley, J. P. Aggleton, & M. A. Conway (Eds.), *Episodic memory: New directions in research* (pp. 53–70). Oxford: Oxford University Press.

Conway, M. A., & Pleydell-Pearce, C. W. (2000). The construction of autobiographical memories in the self-memory system. *Psychological Review, 107,* 261–288.

Cowan, N. (1995). *Attention and memory: An integrated framework.* New York: Oxford University Press.

Craik, F. I. M., & Lockhart, R. S. (1972). Levels of processing: A framework for memory research. *Journal of Verbal Learning and Verbal Behavior, 11,* 671–684.

Craik, F. I. M., & Tulving, E. (1975). Depth of processing and the retention of words in episodic memory. *Journal of Experimental Psychology: General, 104,* 268–294.

Ericsson, K. A., & Kintsch, W. (1995). Long-term working memory. *Psychological Review, 102,* 211–245.

Garry, M., Manning, C. G., Loftus, E. F., & Sherman, S. J. (1996). Imagination inflation: Imagining a childhood event inflates confidence that it occurred. *Psychonomic Bulletin & Review, 3,* 208–214.

Gerbner, G., Gross, L., Morgan, M., Signorielli, N., & Shanahan, J. (2002). Growing up with television: Cultivation processes. In *Media Effects* (2nd ed.) (pp. 43–68). Mahwah NJ: Lawrence Erlbaum Associates.

Greenberg, B. S. (1988). Some uncommon television images and the drench hypothesis. In S. Oskamp (Ed.), *Television as a social issue* (pp. 88–102). Newbury Park, CA: Sage.

Greenberg, B. S. Mastro, D., & Brand, J. E. (2002). Minorities and the mass media: Television into the 21st century. In J. Bryant & D. Zillmann (Eds.), *Media Effects* (2nd ed., pp. 333–351). Mahwah, NJ: Lawrence Erlbaum Associates.

Harris, R. J., Bonds-Raacke, J. M., & Cady, E. T. (2005). What we remember from television and movies: Using autobiographical memory to study media. In R. Walker & D. J. Herrmann (Eds.), *Cognitive technology: Essays on the transformation of thought and society* (pp. 130–148). Jefferson NC: McFarland Publishers.

Harrison, K., & Cantor, J. (1999). Tales from the screen: Enduring fright reactions to scary media. *Media Psychology, 1*, 97–116.

Hawkins, R. P., & Pingree, S. (1990). Divergent psychological processes in constructing social reality from mass media content. In N. Signorielli & M. Morgan (Eds.), *Cultivation analysis* (pp. 35–50). Newbury Park, CA: Sage.

Hoekstra, S. J., Harris, R. J., & Helmick, A. L. (1999). Autobiographical memories about the experience of seeing frightening movies in childhood. *Media Psychology, 1*, 117–140.

Kahneman, D., & Tversky, A. (1972). Subjective probability: A judgment of representativeness. *Cognitive Psychology, 3*, 430–454.

Kahneman, D., & Tversky, A. (1973). On the psychology of prediction. *Psychological Review, 80*, 237–251.

Kintsch, W. (1974). *The representation of meaning of memory.* Hillsdale NJ: Lawrence Erlbaum Associates.

Kintsch, W. (1988). The use of knowledge in discourse processing: A construction-integration model. *Psychological Review, 95*, 163–182.

Kintsch, W. (1998). *Comprehension: A paradigm for cognition.* New York: Cambridge University Press.

Lang, A. (2000). The limited capacity model of mediated message processing. *Journal of Communication, 50*, 46–70.

Loftus, E. F., & Ketcham, K. (1994). *The myth of repressed memory.* New York: St. Martin's Press.

Martin, M. C. (1997). Children's understanding of the intent of advertising: A meta-analysis. *Journal of Public Policy and Marketing, 16*, 205–216.

Mitchell, K. J., & Johnson, M. K. (2000). Source monitoring: Attributing mental experiences. In E. Tulving & F. I. M. Craik (Eds.), *The Oxford handbook of memory* (pp. 179–195). New York: Oxford University Press.

Morgan, M. (1989). Television and democracy. In I. Angus & S. Jhally (Eds.), *Cultural politics in contemporary America* (pp. 240–253). New York: Routledge.

Morgan, M., & Shanahan, J. (1991). Television and the cultivation of political attitudes in Argentina. *Journal of Communication, 41*(1), 88–103.

Morgan, M., & Shanahan, J. (1992). Comparative cultivation analysis: Television and adolescents in Argentina and Taiwan. In F. Korzenny & S. Ting-Toomey (Eds.), *Mass media effects across cultures* (pp. 173–197). Newbury Park, CA: Sage.

Morgan, M., & Shanahan, J. (1995). *Democracy tango: Television, adolescents, and authoritarian tensions in Argentina.* Cresskill NJ: Hampton Press.

Neath, I., & Surprenant, A. M. (2003). *Human memory: An introduction to research, data, and theory* (2nd Ed.). Belmont, CA: Wadsworth.

Peterson, L. R., & Peterson, M. J. (1959). Short-term retention of individual items. *Journal of Experimental Psychology, 58*, 193–198.

Poniewozik, J. (2001, May 28). What's wrong with this picture? *Time*, pp. 80–82.

Potter, W. J. (1991a). Examining cultivation from a psychological perspective: Component subprocesses. *Communication Research, 18*, 77–102.

Potter, W. J. (1991b). The relationships between first- and second-order measures of cultivation. *Human Communication Research, 18*, 92–113.

Potter, W. J. (1993). Cultivation theory and research: A conceptual critique. *Human Communication Research, 19*, 564–601.

Potts, R., & Martinez, I. (1994). Television viewing and children's beliefs about scientists. *Journal of Applied Developmental Psychology, 15*, 287–300.

Raju, P. S., & Lonial, S. C. (1990). Advertising to children: Findings and implications. *Current Issues and Research in Advertising, 12*, 231–274.

Roediger, H. L., Marsh, E. J., & Lee, S. C. (2002). Varieties of memory. In D.L. Medin & H. Pashler (Eds.), *Stevens' Handbook of Experimental Psychology, Third Edition, Volume 2: Memory and Cognitive Processes* (pp. 1–41). New York: John Wiley & Sons.

Rogers, E. M., Vaughan, P. W., Swalehe, R. M. A., Rao, N., Svenkerud, P., & Sood, S. (1999). Effects of an entertainment-education radio soap opera on family planning behavior in Tanzania. *Studies in Family Planning, 30*(3), 193–211.

Rosenzweig, J. (1999). Can TV improve us? *The American Prospect, 45*, 58–63.

Rössler, P., & Brosius, H.-B. (2001). Do talk shows cultivate adolescents' view of the world? A prolonged-exposure experiment. *Journal of Communication, 51*(1), 143–163.

Rubin, A. M. (2002). The uses and gratifications perspective of media effects. In *Media Effects* (2nd ed.) (pp. 525–548). Mahwah NJ: Lawrence Erlbaum Associates.

Rubin, D. C. (1995). *Memory in oral traditions: The cognitive psychology of epic, ballads, and counting-out rhymes.* New York, NY: Oxford University Press.

Schulkind, M. D., Hennis, L. K., & Rubin, D. C. (1999). Music, emotion, and autobiographical memory: They're playing your song. *Memory & Cognition, 27*, 948–955.

Shanahan, J., & McComas, K. (1999). *Nature stories.* Cresskill NJ: Hampton Press.

Shanahan, J., Morgan, M., & Stenbjerre, M. (1997). Green or brown? Television's cultivation of environmental concern. *Journal of Broadcasting & Electronic Media, 41*(3), 305–323.

Shapiro, M. A. (1991). Memory and decision processes in the construction of social reality. *Communication Research, 18,* 3–24.

Sharman, S. J., Manning, C. G., & Garry, M. (2005). Explain this: Explaining childhood events inflates confidence for those events. *Applied Cognitive Psychology, 19,* 67–74.

Sherry, J. L. (2002). Media saturation and entertainment-education. *Communication Theory, 12,* 206–224.

Shiffrin, R. M., & Schneider, W. (1977). Controlled and automatic human information processing: II. Perceptual learning, automatic attending, and a general theory. *Psychological Review, 84,* 127–190.

Shrum, L. J. (2001). Processing strategy moderates the cultivation effect. *Human Communication Research, 27,* 94–120.

Shrum, L. J., & Bischak, V. D. (2001). Mainstreaming, resonance, and impersonal impact: Testing moderators of the cultivation effect for estimates of crime risk. *Human Communication Research, 27,* 187–215.

Signorielli, N. (1990). Television's mean and dangerous world: A continuation of the Cultural Indicators perspective. In N. Signorielli & M. Morgan (Eds.), *Cultivation analysis: New directions in media effects research* (pp. 85–106). Newbury Park, CA: Sage.

Singhal, A., & Rogers, E. M. (1999). *Entertainment-Education: A communication strategy for social change.* Mahwah NJ: Lawrence Erlbaum Associates.

Singhal, A., & Rogers, E. M. (2002). A theoretical agenda for entertainment-education. *Communication Theory, 12,* 117–135.

Slater, M. D., & Rouner, D. (2002). Entertainment-education and elaboration likelihood: Understanding the process of narrative persuasion. *Communication Theory, 12,* 173–191.

Sparks, G. G., Pellechia, M., & Irvine, C. (1999). The repressive coping style and fright reactions to mass media. *Communication Research, 26,* 176–192.

Squire, L. R. (1987). *Memory and brain.* New York: Oxford University Press.

Stolberg, S. G. (2001). *C.D.C. injects dramas with health messages.* www.nytimes.com/2001/06/26/health/26CDC.html

Teplin, L. A. (1985). The criminality of the mentally ill: A dangerous misconception. *American Journal of Psychiatry, 142,* 593–599.

Tversky, A., & Kahneman, D. (1973). Availability: A heuristic for judging frequency and probability. *Cognitive Psychology, 5,* 207—232.

Vidmar, N., & Rokeach, M. (1974). Archie Bunker's bigotry: A study in selective perception and exposure. *Journal of Communication, 24*(1), 35–47.

Wilson, B. J., & Weiss, A. J. (1992). Developmental differences in children's reactions to a toy advertisement linked to a toy-based cartoon. *Journal of Broadcasting & Electronic Media, 36,* 371–394.

Yalch, R. F. (1991). Memory in a jingle jungle: Music as a mnemonic device in communicating advertising slogans. *Journal of Applied Psychology, 76,* 268–275.

Zillmann, D. (2002). Exemplification theory of media research. In J. Bryant and D. Zillmann (Eds.), *Media effects* (2nd ed.) (pp. 19–42). Mahwah NJ: Lawrence Erlbaum Associates.

6

Media Information Processing

Robert H. Wicks
University of Arkansas

INTRODUCTION

Processing media information is a fundamental part of the enjoyment of media. Information processing theory considers the ways in which individuals cope with all of the sensory information they encounter. It explains the relationship between diverse and disparate ideas about how individuals sift through the flood of incoming information, how they determine whether to attend to the information, how they consider it in the context of stored knowledge, and, finally, how they commit the information to memory.

Information processing theory is concerned with the relationship between theories of information selection, attention, encoding, schema activation, information retrieval, and information storage in memory. It also involves mechanisms that enable individuals to filter out or ignore irrelevant information. As such, it is more of an amalgam of theories that explain how people perceive symbols, images, and sounds, and then convert them into mental representations. The process is complete when these representations are capable of being reproduced in memory in a similar or altered form.

Earlier chapters in this volume explored issues such as motivation, selective exposure, attention, perception, comprehension, and memory, which serve as the foundation for information processing theory. This chapter connects these concepts into a coherent model. To do so, the concepts relevant to message framing, memory, and schematic processing and reception processes are considered.

The foundation for information processing theory can be traced to social psychology (Lachman, Lachman & Butterfield, 1979). However, media effects scholars (e.g., Berger & Chaffee, 1989) quickly recognized the potential for such theory in a media context. Psychologists explain that information processing theory attempts to explicate the sequence of events that occur between a stimulus and a response by considering four basic steps in the process (Fiske & Taylor, 1991): (1) understanding the meaning of information; (2) searching for information on the topic; (3) verifying the answer, and; (4) stating the answer. Potter (2004) suggests that media information processing involves: (1) message filtering and encoding; (2) the task of matching information with stored knowledge, and; (3) the eventual construction of meaning.

Several exemplary communication reports on media information processing (e.g., Basil, 1994; Lang, 2000; Potter, 2004) have explicated the processes described above by explaining how human beings, with their limited processing skills, are quite adept at sifting through sensory stimuli and making determinations about where to invest cognitive effort in processing information. These reports suggest that people are quite capable of shifting processing resources as needed to cope with the flood of sensory information to which they are exposed.

This chapter differs from these reports in a number of important ways. First, information processing theory is evaluated in the context of message framing. How producers frame a message or construct it in order to elicit a particular cognitive or affective response is important with respect to understanding why and how audience members process that message. Thus, while most of this chapter focuses on how people receive and process messages, it also focuses on the importance of the source and type of stimuli. Therefore, this chapter considers the dynamic interaction between the *source,* the *message*, and the *receiver*.

Second, while information processing theory has often been used to explain how people interpret news content, it is also relevant to the entertainment experience for a variety of reasons. As the lines between news and entertainment continue to blur, some media observers have asserted that the vast majority of news is, in fact, entertainment (sometimes called *infotainment*). Sensational murder cases or the bizarre behavior of celebrities often crowd out news that may be truly relevant to audience members (e.g., economics, health, or politics). Thus, efforts by message producers to stimulate processing among audience members are relevant both to serious news and news programming that increasingly emulates entertainment.

Finally, information processing theory has become increasingly important in the entertainment realm. Network censors in the early days of television prohibited the use of words like *pregnant*. Married couples (e.g., Rob and Laura Petrie in the Dick Van Dyke Show) slept in twin beds. Comedian George Carlin confronted the wrath of the FCC for uttering seven dirty words in a comedy routine on radio. In short, the regulated broadcast media that continued into the 1980's assured that sexual themes and graphic violence were kept to a minimum. But the explosion of media outlets, including the emergence of networks distributed by cable and satellite and the remarkable proliferation of the Internet, have forced broadcast television networks and radio stations to experiment with programs that are designed to hold the attention of the audience members. Hence, the advent of programs like *Fear Factor* and *Rush Limbaugh* and the proliferation of reality programming now compete for the attention of audience through the use of stunning techniques that would have been off limits years ago. These program genres warrant consideration in the context of information processing theory.

THE DEVELOPMENT OF INFORMATION PROCESSING THEORY

Information processing theory is based on the processes of accepting or ignoring information, matching information against stored knowledge and constructing meaning. Once a media message has successfully engaged an individual, a search takes place in which the person attempts to match information with stored knowledge. This is achieved through a search within the memory. Linkages are made between the symbols, words, sounds, or images that enter the senses, and the information stored in knowledge structures known as schemas, (Anderson, 1990).

McQuail (1994) asserts that the focus on trying to understand how people process information and construct meaning represents the fourth major paradigm shift relevant to information processing theory. The first paradigm from the early 1900s through the 1930s assumed that media have strong or *powerful effects* on shaping attitudes, opinions, and beliefs (e.g., Lasswell, 1927; 1948). Effective propaganda campaigns during World War I demonstrated that media could shift public opinion.

The second paradigm, called the *limited effects* phase, suggested that when media do have powerful effects, it is a departure from the norm because media primarily reinforce beliefs. Klapper (1960) contended that media effects are mitigated by many variables including: (1) *message content*; (2) the manner in which communicators *construct messages*; and (3) the *knowledge, attitudes, beliefs* and *predisposition's* an individual held prior to exposure.

The third paradigm began in the 1970s with an emphasis on a *search for stronger effects* (Noelle-Neumann, 1984) along with a focus on how media may cognitively influence individuals (Beniger & Gusek, 1995). The 1970s also gave rise to the *agenda-setting* research tradition (McCombs & Shaw, 1972), the *uses and gratifications* approach (Rubin, 2002), and *cultivation theory* (Gerbner & Gross, 1976). Each of these approaches argued, from differing perspectives, that media must have more than minimal effects.

By the middle of the 1980s, rapidly expanding information processing research in the field of psychology prompted Gardner (1985) to suggest that a *cognitive revolution* was underway. Metallinos (1999) explains: "Cognition is the process by which the classified and intensified bits of information are decoded, interpreted, and turned into holistic units. Cognition is synonymous with comprehension, recognition, understanding, interpreting" (p. 433).

This cognitive revolution profoundly stimulated mass communication research in the domains of agenda setting, persuasion, the spiral of silence, uses and gratifications, and cultivation theory. It also accelerated collaborative research between communication scholars and psychologists leading to significant cross-pollination between the allied fields (e.g., Bryant & Zillmann, 2002; Reeves & Anderson, 1991; Reeves & Thorson, 1986). All of this helped to advance information processing research as scholars working from different perspectives studied how audience members use media messages to construct meaning.

While McQuail (1994) may be correct in asserting that the focus on information processing and meaning construction represents the most current paradigm in communication scholarship, its roots as a communication theory can be traced to the early part of the last century. As early as 1922, Walter Lippmann wrote that people are all captives of the "pictures in their heads" (Lippmann, 1922, p.3). Lippmann believed human beings tend to process information in the context of previously stored knowledge for the purpose of *reinforcing* beliefs and *stereotypes* foreshadowing research published 40 years later by Joseph Klapper (1960). Hence, information processing theory as an early construct can be traced to the early part of the last century. But, in the 1980s, the central focus of these ideas had shifted to asking *what people do with media information* rather than *how media messages influence people*.

MESSAGE CONTENT, STRUCTURE, AND FRAMING

To understand *what people do with media information*, we must first consider issues such as message content and structure and framing techniques used to attract audience members. The main objective of media companies is to produce content to attract audiences and make profits through advertising. Therefore, attention-getting devices are employed to attract and engage the audience members (Bickham, Wright, & Huston, 2001). Audience members filter out information that is perceived as dull, irrelevant or unimportant (Potter, 2004). Therefore, news and entertainment programmers and journalists "frame" or explain news stories or issues in ways to resonate with the audience to retain their attention.

Entman (1993) defined framing as selecting "some aspects of a perceived reality" to "make them more salient in a communicating text . . . to promote a particular problem definition, causal interpretation, moral evaluation, and/or treatment recommendation for the item described." (p. 52). Reese (2001) defined frames as organizing principles that are shared and remain persistent over time in a society, working symbolically to structure the social world. Entman (1993) and Scheufele (1999) said frames simplify complex issues for the audience, selecting and calling attention to particular aspects of an issue, thus consciously or unconsciously diverting attention from other aspects of that issue. McCombs and Ghanem (2001) argued that although the mass media do not tell audiences what to think, they emphasize certain aspects of a topic, influencing how audiences think about that topic.

Entman (1993) asserts salience (or taking a piece of information and highlighting it to make it more memorable to the audience) is central to framing. People are more likely to attend to, consider, and remember salient information. Kim, Scheufele and Shanahan (2002) said that covering certain aspects of an issue more prominently apparently increased the salience of those same aspects among the audience.

Iyengar (1991) argued that framing provides context and cues, such as using a particular word or language to help the audience understand where a journalist is going with a story. For example, the Watergate incident was initially framed as a partisan issue in the 1972 presidential race. It was first called the Watergate caper, then the Watergate scandal, and finally presented as widespread corruption in the Nixon White House (Lang & Lang, 1983). Media frames are the themes or salient viewpoints regarding topics or issues that are presented in mass media news stories, which can influence how the audience comes to understand that topic or issues (Entman, 1991; Gamson & Modigliani, 1987; Price, Tewksbury, & Powers, 1997; Scheufele, 1999).

Theorists began to consider media frames as early as the 1960s, when it became clear that television was having an impact on society, culture, and individuals (e.g., Burke, 1968; Gitlin, 1980; Goffman, 1974; McLuhan, 1964; Schramm, 1971). Framing research is now advancing simultaneously across disciplines such as communication (D'Angelo, 2002; Entman, 2003; Entman & Rojecki, 1993; Neuman, Just, & Crigler, 1992; Price & Tewksbury, 1997; Scheufele, 2004), sociology (Gamson, 1988; 1992), political science (Norris, Kern, & Just, 2003), journalism (Pan & Kosicki, 1993; 2001; Reese, Gandy, & Grant, 2001), and political psychology (Iyengar, 1991).

In the context of news information, media framing begins when journalists and editors decide what news content to present. Then, mechanisms are employed that place the information in a field of meaning. These mechanisms may be structural in terms of production styles, or they may reflect efforts, either conscious or unconscious, on the part of the creator to convey meaning (Crigler, Just, & Neuman, 1994; Lang, Zhou, Schwartz, Bolls, & Potter, 2000). Entman (1993) asserts that the news message framing process has four components: (a) identify a problem; (b) assess the cause of an event and assign responsibility; (c) consider the issue or problem in the context of legal, ethical, or moralistic principles; and (d) identify and recommend solutions to the problem.

Reese (2001) has extended Entman's (1993) model, suggesting that message frames also represent a system that can exert influence and power over public opinion. He asserts, "The power to frame depends on access to resources, a store of knowledge, and strategic alliances" (p. 20). Taken together, Reese (2001) and Entman (1993) suggest the way a given piece of information is presented or framed can create differences in the attitudes, opinions, and beliefs of audience members. Finally, frames may be manipulated to influence public opinion if proper resources are available.

Information processing begins when people are exposed to stimuli in the form of images, symbols, or sounds that enter the sensory organs. However, exposure alone does not guarantee that a message will be processed. For processing to occur, the perceiver must invest cognitive energy in the form of attention, meaning she or he must be consciously aware of the message (Cowan, 1995). So, although a person cannot attend to a message without exposure, mere exposure does not guarantee attention. To attract and maintain the attention of audience members, media message producers employ framing or other techniques.

Information suppliers use structural characteristics to engage audience members and initiate processing. The style of music and pacing, shot selection, editing techniques, special effects, and narrative sequencing may lead viewers to become more or less involved with the information (Lang, Geiger, Strickwerda, & Sumner, 1993). The interaction between structure and content can also produce effects that may engage viewers (Geiger & Newhagen, 1993). Political advertisements produced on behalf of a candidate might include a hopeful message illustrated with colorful patriotic images and inspiring music. An ad produced to attack the record on crime of an incumbent might feature gray tones along with images of illicit drugs and police activity placed over a somber narrative describing the ineffectiveness of the administration in power (Kern & Wicks, 1994; Wicks & Kern, 1993). Hence, the interaction between message content and structure may influence the degree to which people attend to, process and store media content.

Media messages also contain contextual cues to help people understand information. Iyengar (1991) noted that news messages supply information as either *episodic* or *thematic* frames. Episodic framing is event-oriented coverage of breaking news stories while thematic framing is coverage that provides background and perspective on public issues. Iyengar (1991) explained:

> *"The episodic news frames take the form of a case study or an event-oriented report and depicts public issues in terms of concrete instances (for example, the plight of the homeless person or a teenage drug user, the bombing of an airliner or an attempted murder). The thematic frame, by contrast, places public issues in some more general or abstract context and takes the form of a timeout, or backgrounder, report directed at general outcomes or conditions. Examples of thematic coverage include reports on changes in government welfare expenditures, congressional debates over the funding of employment training programs, the social or political grievances of groups undertaking terrorist activities, and the backlog in the criminal justice process" (p. 14).*

Finally, frames can be created either by individuals observing news coverage or by the message producers themselves (reporters, editors, producers, etc). Various presidential administrations have taken advantage of the *Cold War Frame* that developed after World War II. The framing of "good" versus "evil" can produce rally effects that strengthen support for presidents and other elected officials (Norris, Kern, & Just, 2003). The administration of George W. Bush initially framed the 2003 invasion of Iraq as a defensive measure to protect U.S. interests in the *War on Terror.* But as it became clear that Iraq no longer possessed weapons of mass destruction, the rhetoric changed to freeing an oppressed people and planting the seeds of democracy in the Middle East.

THE ACTIVE AUDIENCE

Encoding and Filtering Processes

Later in this chapter we consider the concept of individual framing or the process by which media frames interact with frames held by audience members. To understand this process, we must first consider the audience as active participants in the mediated communication process. The concept of the audience playing an active role in information processing can be traced to the period of the limited effects paradigm noted earlier (Bauer, 1964). Audience activity denotes emotional and intellectual engagement with the message (Biocca, 1988). It implies that media can alter the mood, disposition, and even the physiology of the viewer (Zillmann, 1991). Horror and erotic films, for example, provide frightening or voyeuristic experiences that may affect a person physiologically, causing perspiration or an increase in heart rate (Cantor, 1991; Tamborini, 1991; Weaver, 1991). Information processing models begin with two assumptions. People are information processors because they are capable of using stimuli to build upon existing knowledge that can alter or produce attitudes, opinions, and beliefs. The second assumption is that people have a limited and fixed pool of mental resources. You may be able to think about several things simultaneously, but a threshold exists at which an individual is incapable of dealing with more input (Basil, 1994; Graber, 1988; Lang, 2000).

The encoding task is critical to understanding information processing. The first process occurs when sensory receptors such as eyes, ears, the nose, the mouth, or skin encounter information. This information gathered by the senses enters a sensory store (Zechmeister & Nyberg, 1982). Mediated messages should automatically enter the sensory store, but only a fraction of the information they contain makes it into the short-term or active working memory. The first step in the encoding process involves separating which bits of information will be processed and which will be filtered out. Both automatic (unintentional) and controlled (intentional) processes influence the information selection process (Schneider & Shiffrin, 1977; Shiffrin & Schneider, 1977). Therefore, encoding involves transforming messages into mental representations. Exposure and attention are considered complex processes in which the bits of information in a mediated message must engage the sensory receptors and enter the sensory store, where only a fraction of them are selected (as a result of automatic and controlled processing) and transformed into mental representations in a person's working memory (Lang, 2000, p. 49).

Audience members consciously and unconsciously decide what media messages to process and which to filter out (Zillmann & Bryant, 1985). *Controlled information processing* occurs when people consciously decide to pay attention to a message. Although message producers may not think in terms of stimulating controlled processing in a psychological sense, their main objective is to produce programs that engage the audience and hold their attention. Thus, controlled processing implies an intense focus on media messages rooted in a motivation to learn or interact in some cognitive or affective way with the message.

Automatic information processes are unintentional, involuntary, effortless, autonomous, and outside awareness (Schneider & Shiffrin, 1977; Shiffrin & Schneider, 1977). Such processing does not require a specific goal and is relatively effortless. Leaving the television turned to a news program while tending to chores or engaging in other activities is an example of automatic processing. The viewer may be monitoring the news without exerting much cognitive effort. Much of the filtering task that takes place does not require the use of significant cognitive resources because the demands on cognitive capacity would be overwhelming. Thus, the vast numbers of messages that are encountered on a daily basis are automatically filtered out to provide cognitive capacity for messages that are deemed interesting, important or useful.

Before proceeding, it is important to note that automatic and controlled processing represent points at the end of a continuum rather than a dichotomy (Shiffrin, 1988). Although viewing television may appear to be a relatively passive and mindless exercise, research shows that processing the information and images takes significant training and skills (Davis, 1990; Davis & Robinson, 1989; Graber, 1988; Robinson & Davis, 1990; Robinson & Levy, 1986). Arousing content increases the resources needed to process information, shifting attention more in the direction of controlled processing (Lang, Potter, & Bolls, 1999). This may explain the rapid proliferation of reality programming on television and the success of radio personalities such as Howard Stern or Rush Limbaugh. The programmers and personalities present content that is designed to attract and hold the attention of the audience using often bizarre gimmicks or controversial commentary.

VARIABLES INFLUENCING INFORMATION PROCESSING

Once information processing is initiated, it is important to understand the variables that will assure that processing continues. Maintaining attention and processing information is dependent upon a number of variables including *comprehension, emotions, perceived gratifications*, and *level of involvement*.

Comprehension

The debate as to whether reading text or watching television produce differences in information processing strategies has been ongoing. Noble (1983) suggested that television viewing is a more passive activity because it tends to tap the emotions more than cognitions. But the processes of simultaneously encoding visual and audio channels that may or may not reinforce themes presents viewers with a formidable processing challenge (Singer, 1980).

Some theorists speculate that people must create mental images when reading newspaper articles while television and films provide the images to viewers. As a result, human beings expend cognitive energy to interpret text. However Hoijer (1989) has argued that the process of constructing mental images is not necessarily confined to the print medium. Television audience members also create mental representations that are interpretations and impressions rather than photocopies. The process requires drawing upon schemas of interpretation that are stored in memory. Schemas are discussed later in this chapter, but generally they represent stored and organized knowledge that help people make sense of incoming information. Furthermore, interpretation may be influenced by emotions that are tapped during the viewing experience (Zillmann, 1983). This process of making sense and attaching meaning to incoming stimuli is what is meant by comprehension.

Emotion or Mood Adjustment

Research on mood adjustment or management in connection with media exposure suggests audiences actively use information to alter their emotional state or to elevate mood. A televised football game may provide a vicarious thrill, producing euphoria and even outright laughter when a favorite team is beating the opposition (Zillmann, 1983; 1985). Depressed women experiencing symptoms of premenstrual syndrome have been found to gravitate toward comedy over serious dramatic programs or game shows (Meadowcroft & Zillmann, 1987). Although viewers may not consciously decide to adjust their mood through the selection of media programs, psychological mechanisms encourage them to actively select fare to alter their mental states. In sum, people use certain media and program genres as a means of emotional release (Hearn, 1989).

Selective Attention and Media Gratifications

Audience members also selectively attend to messages that provide certain gratifications (Blumler & Katz, 1974; Rosengren, Wenner, & Palmgreen, 1985). Compelling or frightening news events, such as the attacks of September 11, 2001 or the second war with Iraq, may cause people to selectively attend to information. When audience members are engaged with media messages, even subtle features of televised content can have an important impact on the meaning people construct from the message. For example, Newhagen (1994a; 1994b) experimentally studied the impact of censorship disclaimers (i.e., text noting that censors cleared the news items) during the 1991 Persian Gulf War. He reported that although many participants had difficulty remembering they had seen the disclaimers, they interpreted the news stories quite differently from participants who had not seen reports with the disclaimers. The participants exposed to the disclaimers were much less likely to trust the credibility of the messages. So once members of the audience attend to a message, a good deal more of the content may be processed than just the main theme of the message.

Involvement

Messages perceived as involving or important to audience members also lead to processing. Perse (1990) examined the audience activity in the context of local television news by testing the relationships among: (1) strength of news viewing motivation and involvement intensity; (2) type of news viewing motivation and involvement orientation; and (3) cognitive and emotional involvement. She found that audience utilitarian (i.e., news that was perceived as helpful or useful) news viewing was associated with higher cognitive involvement and feelings of anger. Diversionary viewing was used to relax or escape, producing feelings of happiness on the part of the viewers. She found a link between cognitive and affective involvement with the news.

In sum, the literature does not suggest that people are always active processors of media content. Rather, factors such as framing techniques, message structure (i.e., use of graphics, color, pacing, music, etc.), message content (i.e., program type or genre) and level of involvement with the message can stimulate active processing (Burnkrant & Sawyer, 1983).

MATCHING INFORMATION WITH STORED KNOWLEDGE

Memory

Thus far, we have considered that messages may cause audience members to pay attention and initiate the task of processing. But to understand how information is processed, we must turn to the basic model of memory. Although disagreements exist concerning how memory operates, certain principles are widely accepted. First, stimuli must enter the sensory organs, such as eyes and ears (Broadbent, 1958; Craik & Lockhart, 1972; Kellermann, 1985). The region of the brain labeled as the sensory register (Atkinson & Shiffrin, 1968a; 1968b), the sensory store (Wyer & Srull, 1980; 1981), or the sensory buffer (Hastie & Carlston, 1980) interprets external stimuli such as sights, sounds, or smells.

Memory models generally agree that the overall memory system in the brain contains both a short-term memory (STM) and a long-term memory (LTM) subsystem (Anderson, 1983; Guillund & Shiffrin, 1984; Murdock, 1982). All of the environmental stimuli to which we are exposed comprise the repertoire of information in the STM. The STM is believed to have limited capacity because memories within it decay quite rapidly (Baddeley, 1976). The LTM is capable of storing information for indefinite periods of time or even for a lifetime. A region of the STM known as the working memory (WM) is where current thinking is believed to occur.

Communication-based memory models such as multiple resource theory (e.g., Basil, 1994) or the limited capacity model (e.g., Lang, Newhagen, & Reeves, 1996; Lang, 2000) argue that distinct mental tasks are performed simultaneously, as opposed to earlier models that presumed sequential processing (e.g., Craik & Lockhart, 1972). Humans are theorized to possess limited information processing resources, but they can shift these resources as needed to assist in processing information. The requirements of the processing task dictate whether resources are concentrated on encoding, decoding, or retrieving stored information. Audience members shift resources as needed to attend to and process media messages.

Working and Short-Term Memory

The WM is a subsystem of the STM and is the portion that is actively engaged. When we are currently thinking about something, we are using the WM (Baddeley, 1986; Gathercole & Baddeley, 1993). Psychologists once thought that people could (metaphorically speaking) maintain about seven chunks of information in an active state at a given time (e.g., Bower, 1972; 1977; Ehrlich & Johnson-Laird, 1982; Miller, 1956; Simon, 1974). A chunk is considered to be a finite set of information such as words or images that decay in 20 to 30 seconds. However, this conceptualization of memory fails to acknowledge that people are capable of monitoring many environmental stimuli simultaneously (Anderson, 1990; Basil, 1994; Rumelhart & McClelland, 1986).

Driving a car is an automatic process, so people routinely talk on the phone, listen to news on the radio, or drink coffee on the way to work. People are quite adept at filtering out irrelevant information such as street name signs if the route being followed is habitual. The issue is not a lack of capacity, but rather the inability to maintain a great amount of information in a high state of activation. Some studies have begun to suggest correlations between cell phone usage and automobile accidents, indicating that driving resource skills may be impaired, leading to less attention to the road. But a ball rolling into the street (with the likelihood of a child not far behind) can cause an individual to shift driving resources into a high state of activation in the WM (Baddeley, 1986).

Hence, a surprising, interesting, exciting, or frightening television news story can draw our attention away from the automatic (i.e., monitoring mode) to the controlled (i.e., active mode) in the WM. Such was the case for many when fiery images of the burning World Trade Center, or the stunning images of the bombing of Baghdad, suddenly appeared on their television screens.

Long-Term Memory

The long-term memory (LTM) subsystem contains the *semantic long-term memory* (SLTM) and the *episodic long-term memory* (ELTM) (Rumelhart, Lindsay, & Norman, 1972; Tulving, 1972). The SLTM stores general knowledge that has been accumulated through repeated exposure to information or situations. The ELTM contains information about specific events or episodes. In short, the ELTM involves using information to address specific issues. ELTM is constantly changing to accommodate new events and information. The SLTM remains more stable because it pertains to general events, principles, or ideas.

Humans tend to draw on episodic memory shortly after exposure to new stimuli (Bower, Black & Turner, 1979). But, as time passes, memory processes become more abstract and constructive, as inferential associations and linkages are made between new and stored information. This is basically the process of integrating new information into *schemas* (to be discussed shortly), which represent general information or knowledge rather than memories about specific incidents.

Reconstructing knowledge implies interpreting new information in the context of stored knowledge. Reconstruction may result in conclusions or interpretations that are incorrect because individuals may not remember where they initially received the facts (Brosius, 1993).

People are quite adept at processing information but far less skillful at remembering where the information originated in the first place. One reason that negative political advertising is effective is that people misremember where they encountered the unfavorable information about a candidate. They sometimes remember that they read it in newspaper articles or learned about it during television news broadcasts rather than in a campaign commercial (Biocca, 1991; Jamieson, 1992; Kern & Wicks, 1994).

Studies have also shown that attention can, but does not necessarily, lead to recall (Grimes & Meadowcroft, 1995; Kahneman, 1973). Rather, thinking about information (i.e., rehearsing) in the working memory may lead to the development of stronger and more durable memories that are stored in the LTM. Strong memories are those that are well-learned and relatively easy to retrieve from the LTM because they have been regularly rehearsed (Baddeley, 1986).

Associative Networks

Theorists believe that concepts in the working memory can activate a set of related concepts and associations stored within the long-term memory. The principles of such activation were well-established in a variety of theoretical perspectives including *spreading activation theory* by Collins and Loftus (1975), *the associative network model* of social memory by Hastie, (1986) and the *connectionist network model* of knowledge by Rumelhart and McClelland (1986). These theories assume that processing one news item in WM may lead to accessing other memories of related news items. Therefore, when the media supplies cues in the form of audio or visual stimuli, audience members strive to make sense of this new information by searching for associated or related information. This process may add new content to a schema, which will refine the frame through which interpretation of information occurs.

Theories of associative memory posit that when memory is in use, related memories are more easily activated (Eysenck, 1993). In the search of associative memory (SAM) model introduced by Raaijmakers and Shiffrin (1980; 1981), memory is conceptualized as a mental apartment complex in which related concepts are stored hierarchically and in close proximity. Thus, the concept of dog resides within the animal region, and the concept of television resides within the media region. Concentration on a particular domain or region will tend to retrieve the contents of that domain rather than other unrelated concepts. For example, it is more likely that mentioning a dog will retrieve ideas associated with animals like cats rather than ideas associated with television or newspapers. These mental processes make internal cuing possible.

Internal cueing may result in accessing memories (Tulving & Pearlstone, 1966; Tulving, 1974). This type of cueing represents a kind of networking that ultimately leads to the development of associations between related ideas and concepts. An individual may watch but fail to recall many of the details of a televised news story about the murder of a convenience store clerk in a Los Angeles neighborhood. However, in the course of running errands, he or she may consciously or unconsciously avoid driving through that neighborhood. By inference, the individual may believe that it is likely that the incidence of rape, burglary, and assault is also high in that neighborhood. Stored information led this individual to infer it would be wise to avoid the neighborhood because it had been linked to the crime schema (i.e., murders, burglaries, assaults, and rapes).

In sum, theories of associative memory began to appear in the 1970s and 1980s (see Brewer & Nakamura, 1984; Hastie, 1981; Markus & Zajonc, 1985; Murdock, 1982 for historical accounts). These theories suggest that incrementing (i.e., the act of recalling an item) makes it more likely an item will be recalled in the future. Forgetting occurs because the process of retrieving stored information is difficult and grows even more difficult as time passes and new memories are added (Raaijmakers & Shiffrin, 1980; 1981). This is consistent with general information processing theory, which posits that moving items between the long-term and working memory strengthens the memory trace, thereby leading to greater access of stored information (Anderson, 1976; 1983; 1990).

Schemas

Schemas are an important component of memory theory, theories of associative memory, and overall information processing theory. Schema theory dates to 1932, when Frederick A. Bartlett began experimenting to find out how well people remember figures and stories. He believed that people organize information in logical associated clusters to make understanding easier and more efficient. A schema is defined as a cognitive structure that includes knowledge about a concept, person, or event (Fiske & Taylor, 1991; Hastie, 1981; Rumelhart, 1984; Wicks, 1992; 2001). Schematic thinking enables people to use information quickly in ordinary human interactions and it assists in making judgments. Schemas also serve interpretive and inferential functions because they guide people in decision-making and help them to surmise from an incomplete set of facts.

Schemas are believed to perform four primary functions (Fiske & Taylor, 1991; Fiske & Linville, 1980).

1. Schemas guide the process of noticing and storing information so that it may be retrieved from memory at a later time.
2. Schemas help organize and evaluate new information. They enable people to store related information together, making it unnecessary to establish a new conceptual domain for each new piece of information.
3. Schemas serve an inference function by helping people fill in the gaps when information is incomplete.
4. Schemas enable people to solve problems by using information about similar scenarios to assess new information. This helps people decide how to act in certain situations.

Media information is capable of summoning one or more schemas from the long-term memory. New information is analyzed in the context of these schemas in the working memory. When new information is added to a schema, the structure becomes more complex, which makes it more able to interact with other related schemas. For example, repeated reporting on the Israeli versus Palestinian conflict over a prolonged period of time reduces the likelihood that the *peace in Israel* schema, as well as the *peace in the Middle East* schema, are obtainable goals. Well-developed schemas are more difficult to alter than simple schema, because more cognitive effort was used creating them.

Images, symbols, and sounds that stimulate our senses cause us to think about new information within the context of stored schemas (Bobrow & Norman, 1975). Media messages that attract attention may cause controlled processing. This leads to retrieving schemas from the LTM to make sense of the new information. If we access information stored in these schemas on a fairly regular basis, the schemas become more durable, organized, and easier to access in the future. Thus, the fundamental reason that human beings are able to process information quickly has to do with the organization of memory. This may explain why studies have shown that discrete news stories may be forgotten quickly, but when cued, people can retrieve information presented in the stories that were added to a relevant schema.

Schema theory maintains that schemas are strengthened by the addition of new, related knowledge. In addition, associations between related schemata will be strengthened as individuals evaluate information over time. Consistent with the connectionist models, retrieval of information becomes easier as a consequence of the deepened memory trace. In other words, the more one rehearses (i.e., thinks about recently encountered information), the more likely one can retrieve previously stored knowledge.

In sum, understanding how we interact with media involves the use of cognitive schemas, which facilitate the interpretation of new information. Integration of information takes place in WM and the act of rehearsing information strengthens memories as mental pathways between the LTM and WM are deepened, producing a durable trace (Anderson, 1990; Baddeley, 1976). Memories that were stored for quite some time may be summoned if the proper cues are encountered or *primed* (Raaijmakers & Shiffrin, 1980; 1981; Roskos-Ewoldson, Roskos-Ewoldson, & Carpenter, 2002).

CONSTRUCTING MEANING

Individual Framing

As noted early in this chapter, it is necessary to understand the dynamic interaction between the *source,* the *message,* and *receiver* to fully comprehend the information processing theory. In the first part of this chapter, we considered how media message producers frame information to stimulate attention to messages. We considered how media messages contain contextual cues supplied by professional communicators to help people understand information (Reese, Gandy & Grant, 2001). We then considered the steps that occur in processing information. Now we consider how people construct meaning using *individual frames* to evaluate and interpret information based on shared conceptual constructs.

People interpret information in the context of their life experiences and by drawing upon schemas. Many of these experiences are stored in memory and drawn upon to make sense of new information. Individual framing focuses on how the receiver interprets messages based upon individual predispositions, attitudes, and beliefs. This is an area where some convergence appears to be taking place between reception theorists (Hall, 1980) critical and cultural theorists and social scientists (D'Angelo, 2002; Scheufele, 1999). Individual frames are both cognitive representations of an individual's memory, and devices that are communicated through public discourse (Pan & Kosicki, 1993; 2001). Central to the audience framing concept is the idea that people are more receptive to information that is consistent with previously held beliefs and attitudes. People selectively attend to messages or avoid ones they find unpleasant. Politicians may use popular frames like "support our troops" or "hold the line on taxes" that are then amplified by news organizations. These salient messages may resonate with individual frames, thus attracting and holding the attention of audience members.

Some scholars suggest that individual framing involves a second level of agenda-setting theory (Golan & Wanta, 2001) because journalists make attributes of issues salient through the selection and repetition process. Gamson (1992) also notes that the different story frames that newsmakers use to construct their messages provide citizens with a basic tool kit of ideas they apply to thinking and talking about politics and public policy issues. Hence, frames connect news media messages to cognitive elements such as thoughts, goals, motivations, and feelings and attitudes held by individuals (Shah, Watts, & Fan, 2002; Price, Tewksbury, & Powers, 1997).

Individual framing effects occur when the media frame interacts with attitudes, opinions, beliefs, and schemas within the viewer, which may activate particular elements of messages over others (e.g., Iyengar, 1991; Pan & Kosicki, 1993). The cognitive elements that are activated from media coverage of an event are more likely to influence viewers' interpretations, evaluations, and judgments with respect to that event (Gamson, 1992; Iyengar, 1991; Lau, Smith, & Fiske, 1991; Neuman et al., 1992; Zaller, 1992).

Individual frames may thus be viewed as an abstract principle, tool, or "schemata" of interpretation that works through media texts to structure social meaning. The media actively sets the frames of reference that readers or viewers use to interpret and discuss public events. At the same time, people's information processing and interpretation are influenced by pre-existing meaning structures, or schemas (Wicks, 1992). When audience members receive media frames, each member makes a conscious or subconscious choice to accept, deny, or mold that frame to fit an existing schema. In other words, each audience member decides whether to incorporate the media frame into the individual frame.

INFORMATION PROCESSING AND THE ENTERTAINMENT EXPERIENCE

In the chapters to follow, experts from communication, marketing, and psychology will explore many entertainment genres, the accompanying reaction processes, and the applications of psychological entertainment theory. The editors will synthesize these chapters into an integrated model of the psychology of entertainment. This model of entertainment theory will incorporate a wide range of theories including the role of fantasy and imagination, empathy, involvement, the role of excitation and mood, the search for fear and fright, and how cognitions and emotions interact to influence audience members. In so doing, they will illustrate that while understanding information processing depends upon certain steps, we continue to learn more about how the mind interacts with different types of media entertainment content.

Understanding information processing theory in the context of current and developing entertainment genres is important for a variety of reasons. The myriad of entertainment sources to which we have access will continue to grow. Competition for the attention of audience members will prompt producers to introduce new attention-getting techniques. It is likely that programmers will continue to test the limits of sexual, violent, and reality programming. The competitive news environment will most likely continue the trend of attracting audiences with graphic, outrageous, and even salacious news reports. Audience members should recognize that they are the targets of media and should understand how and why messages are developed and processed as they are. In so doing, the audience may become literate consumers of media. Scholars and students of media should understand how and why they attend to messages, and how these messages may eventually occupy space in memory.

Media literacy with respect to news information processing is especially important. A recent content analysis of the days leading up to the fall of Baghdad revealed that CNN and Al-Jazeera framed much of their news reporting in the context of explaining why the war was an important turning point in Middle East and global history, and analyzed the potential impact. By contrast, Fox News tended to rely on a *Cold War* type of frame in which the United States (the remaining superpower) representing good was compelled to destroy evil (the regime of Saddam Hussein) to assure global peace and stability (Wicks & Wicks, 2004). With important decisions to be made regarding the future course of the nation, students and scholars of media should understand how they process information and develop belief systems. And the audience should understand how, and ultimately why, messages are produced to inform, persuade, and entertain.

CONCLUSION

In sum, a myriad factors contribute to the ways in which we come to know about the world in which we live. Content of television news stories and entertainment programs influence how and why people construe information as they do (Epstein, 1973; Gans, 1979; Tuchman, 1978).

Media outlets frame messages and attempt to persuade people to adopt a perspective, purchase products and pay attention. With respect to news content, journalistic conventions and biases held by professional communicators can contribute to the message framing process.

As media consumers, we must be aware that interpreting media messages is a two-way street. The attitudes, beliefs, opinions, and predispositions we hold inevitably contribute to the construction of social reality (Berger & Luckmann, 1966). Interpreting media involves complex processes in which the messages as produced interact with each individual receiver. As a result, no two people interpret a media message in precisely the same way, although many people may construct quite similar meanings (Pinker, 1997). In certain instances, strongly held attitudes or beliefs cause different people to attach very different meanings to the same message. This process is normal and healthy in terms of constructing meaning. However, to be media literate, we must understand why and how we interpret and misinterpret media information.

REFERENCES

Anderson, J. R. (1976). *Language, memory and thought.* Hillsdale, NJ: Lawrence Erlbaum Associates.

Anderson, J. R. (1983). *The architecture of cognition.* Cambridge, MA: Harvard University Press.

Anderson, J. R. (1990). *Cognitive psychology and its implications* (3rd ed.). New York: W. H. Freeman and Company.

Atkinson, R. C., & Shiffrin, R. M. (1968a). Human memory: A proposed system and its control processes. In K. W. Spence & J. T. Spence (Eds.), *Advances in the psychology of learning and motivation: Research and theory* (Vol. 2). New York: Academic Press.

Atkinson, R. C., & Shiffrin, R. M. (1968b). The control of short-term memory. *Scientific American, 225,* 82–90.

Baddeley, A. (1976). *The psychology of memory.* New York: Basic Books.

Baddeley, A. (1986). *Working memory.* New York: Oxford University Press.

Bartlett, F. C. (1932). *Remembering: A study in experimental and social psychology.* London: Cambridge University Press.

Basil, M. D. (1994). Multiple resource theory 1: Application to television viewing. *Communication Research, 21,* 177–207.

Bauer, R. (1964). The obstinate audience. *American Psychologist, 19,* 319–328.

Beniger, J. R., & Gusek, J. A. (1995). The cognitive revolution in public opinion and communication research. In T. L. Glasser & C. T. Salmon (Eds.), *Public opinion and the communication of consent* (pp. 217–248). New York: Guilford.

Berger, P. L., & Luckmann, T. (1966). *The social construction of reality.* Garden City, NY: Doubleday.

Berger, C. R., & Chaffee, S. H. (1989). *Handbook of communication science.* Newbury Park, CA: Sage.

Bickham, D. S., Wright, J. C., & Huston, A. C. (2001). Attention, comprehension and the educational influences of television. In D. G. Singer & J. L. Singer (Eds.), *Handbook of children and media* (pp. 101–119). Thousand Oaks, CA: Sage.

Biocca, F. (1988). Opposing conceptions of the audience: The active and passive hemispheres of mass communication theory. In James Anderson, (Ed.), *Communication Yearbook 11* (pp. 51–80). Beverly Hills: Sage.

Biocca, F. (1991). Viewers' mental models of political commercials: Towards a theory of the semantic processing of television. In F. Biocca (Ed.), *Television and political advertising: Psychological processes* (Vol. 1, pp. 27–89). Hillsdale, NJ: Lawrence Erlbaum Associates.

Blumler, J., & Katz, E. (1974). *The uses of mass communication.* Newport Beach: Sage.

Bobrow, D. G., & Norman, D. A. (1975). Some principles of memory schemata. In D. G. Bobrow & A. G. Collins (Eds.), *Representation and understanding: Studies in cognitive science* (pp. 131–150). New York: Academic Press.

Bower, G. H. (1972). A selective review of organizational factors in memory. In E. Tulving & W. Donaldson (Eds.), *Organization of memory.* New York: Academic.

Bower, G. H. (1977). *Human Memory.* New York: Academic Press.

Bower, G. H., Black, J. B., & Turner, T. J. (1979). Scripts in memory for text. *Cognitive Psychology, 11,* 177–220.

Brewer, W. F., & Nakamura, G. V. (1984). The nature and functions of schemas. In R. S. Wyer, Jr., & T. K. Srull (Eds.), *Handbook of Social Cognition: Vol. 1* (pp. 119–160). Hillsdale, NJ: Lawrence Erlbaum Associates.

Broadbent, D. E. (1958). *Perception and communication.* London: Pergamon Press.

Brosius, H. B., (1993). The effects of emotional pictures on television news. *Communication Research, 20,* 105–124.

Bryant, J., & Zillmann, D. (2002). *Media effects: Advances in theory and research.* Mahwah, NJ: Lawrence Erlbaum Associates.

Burke, J. (1968). *Language as symbolic action: Essays on life, literature, and method.* Berkeley: University of California Press.

Burnkrant, R. E., & Sawyer, A. G. (1983). Effects of involvement and message content on information-processing intensity. In R. J. Harris (Ed.), *Information processing research in advertising.* Hillsdale, NJ: Lawrence Erlbaum Associates.

Cantor, J. (1991). Fright responses to mass media productions. In J. Bryant & D. Zillmann, (Eds), *Responding to the screen: Reception and reaction processes* (pp. 169–198). Hillsdale, NJ: Lawrence Erlbaum Associates.

Collins, W. A., & Loftus, E. (1975). A spreading activation theory of semantic processing. *Psychological Review, 82*(6), 407–448.

Cowan, N. (1995). *Attention and memory: An integrated framework.* New York: Oxford University Press.

Craik, F. I. M., & Lockhart, R. S. (1972). Levels of processing: A framework for memory research. *Journal of Verbal Learning and Verbal Behavior, 11,* 671–676.

Crigler, A. N., Just, M., & Neuman, W. R. (1994). Interpreting visual audio messages in television news. *Journal of Communication, 44*(4), 132–149.

D'Angelo, P. (2002). News as a multiparadigmatic research program: A response to Entman. *Journal of Communication, 43*(4), 870–888.

Davis, D. K. (1990) News and politics. In D. L. Swanson, & D. Nimmo (Eds.), *New Directions in political communication.* Newbury Park, CA: Sage.

Davis, D. K., & Robinson, J. P. (1989). Newsflow and Democratic society in an age of electronic media. In G. Comstock (Ed.), *Public communiation and behavior,* Vol. 3. New York: Academic Press.

Ehrlich, K., & Johnson-Laird, P. N. (1982). Backward updating of mental models during continuous reading of narratives. *Journal of Experimental Psychology: Learning, Memory, and Cognition, 21,* 296–306.

Entman, R. (1991). Framing U.S. coverage of international news: Contrasts in narratives of the KAL and Iran air incidents. *Journal of Communication, 41*(4), 6–27.

Entman, R. (1993). Framing: Toward clarification of a fractured paradigm. *Journal of Communication, 43*(4), 51–58.

Entman, R. M. (2003). Cascading activation: Contesting the White House's frame after 9/11. *Political Communication, 20,* 415–432.

Entman, R., & Rojecki, A., (1993). Freezing out the public: Elite and media framing of the U.S. anti-nuclear movement. *Political Communication 10*(2), 151–167.

Epstein, E. J. (1973). *News from nowhere.* New York: Random House.

Evans, W. (1990). The interpretive turn in media research. *Critical Studies in Mass Communication, 7*(2), 145–168.

Eysenck, M. W. (1993). *Principles of cognitive psychology.* Hillsdale, NJ: Lawrence Erlbaum Associates.

Fiske, S. T., & Linville, P. T. (1980). What does the schema concept buy us? *Personality and Social Psychology Bulletin, 6,* 543–557.

Fiske, S. T., & S. E. Taylor (1991). *Social cognition* (2nd ed.). New York: McGraw Hill.

Gamson, W. A. (1988). The 1987 distinguished lecture: A constructionist approach to mass media and public opinion. *Symbolic Interaction, 11*(2), 161–174.

Gamson, W. A. (1992). *Talking politics.* Cambridge, England: Cambridge University Press.

Gamson, W. A., & Modigliani, A. (1987). The changing culture of affirmative action. In R. Braungart, & M. Braungart (Eds.), *Research in political sociology 3* (pp. 137–177). New Haven, CT: Yale University Press.

Gans, H. (1979). *Deciding what's news.* New York: Pantheon Books.

Gardner, H. (1985). *The mind's new science.* New York: Basic Books.

Gathercole, S. E., & Baddeley, A. D. (1993). *Working memory and language.* Hillsdale, NJ: Lawrence Lawrence Erlbaum Associates.

Geiger, S., & Newhagen, J. (1993). Revealing the black box: Information processing and media effects. *Journal of Communication, 43*(4), 42–50.

Gerbner, G., & Gross, L. P. (1976). Living with television: The violence profile. *Journal of Communication, 26*(2), 172–199.

Gitlin, T. (1980). *The whole world is watching: Mass media in the making and unmaking of the New Left.* Berkeley, CA: University of California Press.

Goffman, E. (1974). *Frame analysis: An essay on the organization of experience.* New York: Harper and Row.

Golan, G., & Wanta, W. (2001). Second-level agenda setting in the New Hampshire Primary: A comparison of coverage in three newspapers and public perceptions of candidates. *Journalism and Mass Communication Quarterly, 78,* 247–59.

Graber, D. A. (1988). *Processing the news: How people tame the information tide* (2nd ed.). New York: Longman.

Grimes, T., & Meadowcroft, J. (1995). Attention to television and some methods for its measurement. In B. Burleson (Ed.), *Communication Yearbook 18.* Thousand Oaks, CA: Sage.

Guillund, G., & Shiffrin, R. C. (1984). A retrieval model for both recognition and recall. *Psychological Review, 91,* 1–67.

Hall, S. (1980). Encoding and decoding in the television discourse. In S. Hall, (Ed.), *Culture, media, language.* London: Hutchinson.

Hastie, R. (1981). Schematic principles in human memory. In E. T. Higgins, C. P. Herman, & M. P. Zanna (Eds.), *Social cognition: The Ontario Symposium: Vol. 1.* (pp. 39–88). Hillsdale, NJ: Lawrence Lawrence Erlbaum Associates.

Hastie, R. (1986). A primer of information processing theory for the political scientist. In R. R. Lau, & D. O. Sears (Eds.), *Political cognition: The 19th annual Carnegie symposium on cognition* (pp. 11–40). Hillsdale, NJ: Lawrence Erlbaum Associates.

Hastie, R., & Carlston, D. (1980). Theoretical issues in person memory. In R. Hastie, T. M. Ostrom, E. B. Ebbesen, R. S. Wyer, D. Hamilton, & D. E. Carlston, (Eds.), *Person memory: The cognitive basis of social perception.* Hillsdale, NJ: Lawrence Erlbaum Associates.

Hearn, G. (1989). Active and passive conception of the television audience: Effects of a change in viewing routine. *Human Relations, 42,* 857–875. Hillsdale, NJ: Lawrence Erlbaum Associates.

Hoijer, B. (1989). Television-evoked thoughts and their relation to comprehension. *Communication Research 16*(2), 179–203.

Iyengar, S. (1991). *Is anyone responsible? How television frames political issues.* Chicago: University of Chicago Press.

Jameson, K. H. (1992). *Dirty politics: Deception, distraction and democracy.* New York: Oxford University Press.

Kahneman, D. (1973). *Attention and effort.* Englewood Cliffs, NJ: Prentice-Hall.

Kellermann, K. (1985). Memory processes in media effects. *Communication Research, 12*(1), 83–131.

Kern, M., & Wicks, R. H. (1994). Television news and the advertising-driven "new" mass media election: A more significant role in 1992. R. Denton (Ed.), *The 1992 Presidential campaign: A communication perspective.* Westport, CT: Praeger.

Kim, S. H., Scheufele, D. A., & Shanahan, J. (2002). Think about it this way: Attribute agenda-setting function of the press and the public's evaluation of a local issue. *Journalism and Mass Communication Quarterly, 79*(1), 7–25.

Klapper, J. T. (1960). *The effects of mass communication.* New York: Free Press.

Lachman, R., Lachman, J. L., & Butterfield, E. C. (1979). *Cognitive psychology and information processing: An introduction.* Hillsdale, NJ: Lawrence Erlbaum Associates.

Lang, A. (2000). The limited capacity model of mediated information processing *Journal of Communication, 50*(1), 46–70).

Lang, A., Geiger, S., Strickwerda, M., & Sumner, J. (1993). The effects of related and unrelated cuts on television viewers; attention, processing capacity, and memory. *Communication Research, 20,* 4–29.

Lang, A., Newhagen, J., & Reeves, B. (1996). Negative video as structure: Emotion, attention, capacity and memory. *Journal of Broadcasting and Electronic Media, 40*(4), 460–478.

Lang, A., Potter, R. F., & Bolls, P. D. (1999). Something for nothing: Is visual encoding automatic? *Media Psychology, 1,* 145–163.

Lang, K., & Lang, G. E., (1983). The unique perspective of television and its effect: A pilot study. In W. Schramm (Ed.), *Mass Communications, 2nd Edition* (544–560). Urbana: University of Illinois Press.

Lang, Z., Zhou, S., Schwartz, N., Bolls, P. D., & Potter, R. F. (2000). The effects of edits on arousal, attention, and memory for television messages: When an edit is an edit can an edit be too much? *Journal of Broadcasting and Electronic Media, 44,* 94–109.

Lasswell, H. D. (1927). *Propaganda technique in the World War.* New York: Alfred A. Knopf.

Lasswell, H. D. (1948). The structure and function of communication in society. In L. Bryson (Ed.), *The communication of ideas* (pp. 37–51). New York: Harper.

Lau, R. R., Smith, R. A., & Fiske, S. T. (1991). Political beliefs, policy interpretations, and political persuasion. *J. Polit, 53,* 644–75.

Lippmann, W. (1922). *Public opinion.* New York: Macmillan.

Markus, H., & Zajonc, R. B. (1985). The cognitive perspective in social psychology. In G. Lindzey, & E. Aronson, (Eds.), *The handbook of social psychology:* (3rd ed.) (Vol. 1, pp. 137–230). New York: Random House.

McCombs, M. E., & Ghanem, S. I. (2001). The convergence of agenda-setting and framing. In S. D. Reese, O. H. Gandy, Jr., & A. E. Grant (Eds.), *Framing public life: Perspectives on media and our understanding of the social world* (pp. 7–31). Mahwah, NJ: Lawrence Erlbaum Associates.

McCombs, M. E., & Shaw, D. L. (1972). The agenda-setting function of mass media. *Public Opinion Quarterly, 36,* 176–187.

McLuhan, M. (1964). *Understanding media.* New York: American Library.

McQuail, D. (1994). *Mass communication theory: An introduction* (3rd ed.). Thousand Oaks, CA: Sage.

Meadowcroft, J. M., & Zillmann, D. (1987). Women's comedy preference during the menstrual cycle. *Communication Research, 14,* 204–218.

Metallinos, N. (1999). The transformation of biological precepts into mental concept in recognizing visual images. *Journal of Broadcasting and Electronic Media, 43,* 432–442.

Miller, G. A. (1956). The magical number seven, plus or minus two: Some limits on our capacity for processing information. *Psychological Review, 63,* 81–97.

Murdock, B. B. Jr., (1982). A theory for the storage and retrieval of item and associative information. *Psychological Review, 89,* 609–626.

Neuman, W. R., Just, M. R., & Crigler, A. N. (1992). *Common knowledge: News and the construction of political meaning.* Chicago: University of Chicago Press.

Newhagen, J. E. (1994a). Effects of censorship disclaimers in Persian Gulf War television news on negative thought elaboration. *Communication Research, 21*(2), 232–248.

Newhagen, J. E. (1994b). Effects of televised government censorship disclaimers on memory and thought elaboration during the Gulf War. *Journal of Broadcasting & Electronic Media, 38*(3), 339–352.

Noble, G. (1983). Social learning from everyday television. In M. J. Howe (Ed.), *Learning from television: psychological and educational research.* New York: Academic Press.

Noelle-Neumann, E. (1984). *The spiral of silence—Our social skin.* Chicago: University of Chicago Press.

Norris, P., Kern, M., & Just, M. (2003). *Framing terrorism: The news media, the government and the public.* New York: Routledge.

Pan, Z., & Kosicki, G. M. (1993). Framing analysis: An approach to news discourse. *Political Communication, 10,* 55–75.

Pan, Z., & Kosicki, G. M. (2001). Framing as a strategic action in public deliberation. In S. D. Reese, O. H. Gandy, Jr., and A. E. Grant (Eds.), *Framing public life: Perspectives on media and our understanding of the social world* (pp. 35–65). Mahwah, NJ: Lawrence Erlbaum Associates.

Perse, E. M. (1990). Involvement with local television news: Cognitive and emotional dimensions. *Human Communication Research, 16*(4), 556–81.

Pinker, S. (1997). *How the mind works.* New York: W. W. Norton.

Potter, W. J. (2004). *Theory of media literacy: A cognitive approach.* Thousand Oaks, CA: Sage.

Price, V., & Tewksbury, D. (1997). News values and public opinion: A theoretical account of media priming and framing. In G. Barnett & F. J. Bolster (Eds.), *Progress in communication sciences* (pp. 173–212). Greenwich, CT: Ablex.

Price, V., Tewksbury, D., & Powers, E. (1997). Switching trains of thought: The impact of news frames on readers' cognitive responses. *Communication Research,* 24, 481–506.

Raaijmakers, J. G. W., & Shiffrin, R. M (1980). SAM: A theory of probabilistic search of associative memory. In G. H. Bower, (Ed.), *The psychology of learning and motivation: Vol. 14* (pp. 207–262). New York: Academic Press.

Raaijmakers, J. G. W., & Shiffrin, R. M. (1981). Search of associative memory. *Psychological Review, 88,* 93–134.

Reese, S. D. (2001). Prologue—Framing public life: A bridging model for media research. In S. D. Reese, O. H. Gandy, Jr., & A. E. Grant (Eds.), *Framing public life: Perspectives on media and our understanding of the social world* (pp. 7–31). Mahwah, NJ: Lawrence Erlbaum Associates.

Reese, S. D., Gandy, O. H., Jr., & Grant, A. (2001). *Framing public life: Perspectives on media and our understanding of the social world.* Mahwah, NJ: Lawrence Lawrence Erlbaum Associates.

Reeves, B., & Anderson, D. R., (1991). Media studies and psychology. *Communication Research, 18*(5), 597–600.

Reeves, B, & Thorson, E. (1986). Watching television: Experiments on the viewing process. *Communication Research, 13,* 343–361.

Robinson, J. P., & Davis, D. K. (1990). Television news and the informed public: An information processing approach. *Journal of Communication, 40*(3), 106–119.

Robinson, J., & Levy, M. (1986). *The main source: Learning from television news.* Beverly Hills, CA: Sage.

Rosengren, K. E., Wenner, L. E., & Palmgreen, P. (1985). *Media gratification's research: Current perspectives.* Beverly Hills, CA: Sage.

Roskos-Ewoldson, D. R., Roskos-Ewoldson, B., & Carpenter, F. R. D. (2002). Media priming: A synthesis. In D. Zillmann, & J. Bryant (Eds.), *Media effects: Advances in theory and research* (pp. 97–120). Mahwah, NJ: Lawrence Erlbaum Associates.

Rubin, A. M. (2002). The uses-and-gratifications perspective of media effects. In J. Bryant, & D. Zillmann (Eds.), *Media effects: Advances in theory and research (2nd ed.).* Hillsdale, NJ: Lawrence Erlbaum Associates.

Rumelhart, D. E. (1984). Schemata and the cognitive system. In R. S. Wyer, Jr., & T. K. Srull (Eds.), *Handbook of social cognition* (Vol. 1, pp. 161–188). Hillsdale, NJ: Lawrence Erlbaum Associates.

Rumelhart, D. E., Lindsay, P. H. & Norman, D. A. (1972). A process for long-term memory. In E. Tulving, & W. Donaldson (Eds.), *Organization of memory.* New York: Academic Press.

Rumelhart, D. E., & McClelland, J. L. (1986). *Parallel distributed processing* (Vol. 1). Cambridge: MIT Press.

Scheufele, B. T. (2004, May). Making frames: Testing a model of news production, Paper presented at the meeting of the *International Communication Association,* New Orleans, LA.

Scheufele, D. A. (1999). Framing as a theory of media effects. *Journal of Communication, 49,* 102–122.

Schneider, W., & Shiffrin, R. M. (1977). Controlled and automatic human information processing: 1. Detection, search, and attention. *Psychological Review, 84,* 1–66.

Schramm, W., (1971). The nature of communication between humans. In W. Schramm, & D. Roberts (Eds.), *The process and effects of mass communication* (Rev. ed., pp. 3–53), Urbana: University of Illinois Press.

Shah, D. V., Watts, M. D., & Fan, D. V. (2002). News framing and cueing of issue regimes: Explaining Clinton's public approval in spite of scandal. *Public Opinion Quarterly, 66,* 339–370

Shiffrin, R. M. (1988). Attention. In R. A. Atkinson, R. J. Herrnstein, G. Lindzey, & R. D. Luce (Eds.), *Stevens' handbook of experimental psychology:* Vol. 2, *Learning and cognition* (pp. 739–811). New York: John Wiley.

Shiffrin, R. S., & Schneider, W. (1977). Controlled and automatic human information processing: II. Perceptual learning, automatic attending and a general theory. *Psychological Review, 84,* 127–190.

Simon, H. A. (1974). How big is a chunk? *Science, 183,* 482–488.

Singer, J. L. (1980). The power and limitations of television: A cognitive-affective analysis. In P. H. Tannenbaum, & R. Ables (Eds.), *The entertainment functions of television* (pp. 31–65). Hillsdale, NJ: Lawrence Erlbaum Associates.

Tamborini, R. (1991). Responding to horror: Determinants of exposure and appeal. In Bryant J., & Zillmann, D. (Eds.) *Responding to the screen: Reception and reaction processes* (pp. 305–328). Hillsdale, NJ: Lawrence Erlbaum Associates.

Tuchman, G. (1978). *Making news: A study in the construction of reality.* New York: Free Press.

Tulving, E. (1972). Episodic and semantic memory. In E. Tulving & W. Donaldson (Eds.), *Organization of memory.* New York: Academic Press.

Tulving, E. (1974). Cue-dependent forgetting. *American Scientist, 62,* 74–82.

Tulving, E., & Pearlstone, Z. (1966). Availability versus accessibility of information in memory for words. *Journal of Verbal Learning and Verbal Behavior, 87,* 1–8.

Weaver, J. (1991). Responding to erotica. Perceptual processes and dispositional implications. In Bryant J., & Zillmann, D., (Eds.) *Responding to the screen: Reception and reaction processes* (pp. 329–354). Hillsdale, NJ: Lawrence Erlbaum Associates.

Wicks, R. H. (1992). Schema theory and measurement in mass communication research: Theoretical and methodological issues in news information processing. In S. Deetz (Ed.), *Communication Yearbook 15* (pp. 115–145). Beverly Hills, CA: Sage.

Wicks, R. H. (2001). *Understanding audiences.* Mahwah, NJ: Lawrence Erlbaum Associates.

Wicks, R. H., & Kern, M. (1993). Cautious optimism: A new proactive role for local television news departments in local election coverage? *American Behavioral Scientist, 37*(2), 262–271.

Wicks, R. H., & Wicks, J. L., (2004). Televised coverage of the War in Iraq on Al Jazeera, CNN and Fox News. Presented at the annual meeting of the *International Communication Association,* New Orleans, LA, and May 2004.

Wyer, R. S., & Srull, T. K., (1980). The processing of social stimulus material: A conceptual integration. In R. Hastie, T. M. Ostrom, E. B. Ebbesen, R. S. Wyer, D. Hamilton, & D. E. Carlston, (Eds.), *Person memory: The cognitive basis of social perception.* Hillsdale, NJ: Lawrence Erlbaum Associates.

Wyer, R. S., & Srull, T. K. (1981). Category accessibility: Some theoretical and empirical issues concerning the processing of social stimulus information. In E. T. Higgins, C. P. Herman, & P. Zanna (Eds.), *Social cognition: The Ontario Symposium* (Vol. 1, pp. 161–197). Hillsdale, NJ: Lawrence Erlbaum Associates.

Zaller, J. (1992). *The nature of origins of mass opinion.* New York: Cambridge University Press.

Zechmeister, E. B., & Nyborg, S. E. (1982). *Human memory: An introduction to theory and research.* Monterey, CA: Brooks/Cole.

Zillmann, D., & Bryant, J. (1985). Affect, mood and emotion as determinants of selective exposure. In D. Zillmann, & J. Bryant (Eds.), *Selective exposure to communication* (pp. 157–190). Hillsdale, NJ: Lawrence Erlbaum Associates.

Zillmann, D. (1983). Transfer of excitation in emotional behavior. In J. T. Cacioppo, & R. E. Petty (Eds.), *Social psychophysiology: A sourcebook* (pp. 215–240). New York: Guilford.

Zillmann, D. (1991). Television viewing and physiological arousal. In J. Bryant J., & D. Zillmann, (Eds.) *Responding to the screen: Reception and reaction processes* (pp. 103–134). Hillsdale, NJ: Lawrence Erlbaum Associates.

PART II

REACTION PROCESSES

7

Fantasy and Imagination

Patti M. Valkenburg and Jochen Peter
University of Amsterdam

The past three decades have witnessed a considerable increase in empirical research into the origins, contents, and effects of people's fantasy and imagination. What exactly is meant by fantasy and imagination, however, often remains unclear. Many authors leave the concepts undefined, and the definitions that have been given are not uniform. Moreover, the terms are often used without distinction, suggesting that they capture one and the same experience. This tendency to equate fantasy and imagination is found in both everyday life and academic circles.

Of course, fantasy and imagination overlap to some extent. Both activities require the generation of thoughts and in both activities associative thinking plays a role. However, there are at least two differences between the two activities. First, fantasy (including mental processes, such as daydreaming, internal dialogue, and mindwandering) usually takes place separate from the context from which the fantasy emerged. Fantasizing or daydreaming is a state of consciousness characterized by a "shift of attention away from an ongoing physical or mental task or from a perceptual response to external stimulation towards a response to some internal stimulus" (Singer, 1966, p. 3). According to this definition, one cannot be fantasizing and simultaneously be involved in a physical or mental task (Knowles, 1985). Imagination, by contrast, does not necessarily take place apart from an external context. Some authors even believe that perception or sensation of an external stimulus is an essential part of the imaginative process (e.g., Sartre, 1948). According to Singer (1999, p. 13) imagination is a "form of human thought characterized by the ability of the individual to reproduce images or concepts originally derived from the basic senses but now reflected in one's consciousness."

A second difference between fantasy and imagination lies in the degree of goal directedness. Although fantasies can sometimes be evoked deliberately, fantasizing is typically a free-floating mental activity (Klinger, 1990). Imagination, on the other hand, is characterized more often by goal directedness. According to Lewin (1986, p. 51), typical examples of imagination are efforts to visualize the appearance of a monster described in a book, to 'see' a friend's face or to 'hear' her voice when she is not around, or to give an accurate account of a movie just seen.

Changing Theories of Fantasy and Imagination

Fantasy and imagination have long been understood as a primitive and maladaptive aspect of human consciousness. Freud (1908, 1962) viewed fantasy as an attempted solution to a deprivation state or an underlying conflict. Moreover, he assumed that fantasy and imagination undermine people's conscious rational thinking. Piaget, too, saw fantasy and imagination as an infantile form of thinking that will be given up during development in favor of rational and realistic thinking (Harris, 2000).

In the past decades, however, the adaptive function of fantasy and imagination have progressively been recognized (e.g., Klinger, 1990; Singer, 1999). Current theories agree that fantasy and imagination serve several important functions. Fantasy and imagination may enhance self-knowledge and self-understanding by helping individuals to clarify their thoughts and to stay in touch with their needs and feelings. They can promote decision making by allowing one to spell out the anticipated consequences of one's choices. Finally, they can regulate moods and emotions, and relieve tension—for example, by allowing individuals to relive the positive or negative emotions associated with previous experiences (Klinger, 1990).

Contemporary cognitive psychologists agree that human thought consists of two information processing modes that each have their own adaptive value (see Epstein, 1994, for a review). Jerome Bruner (1986), for example, assumes that human thought can be ordered along two complementary dimensions, one paradigmatic and one narrative. The paradigmatic dimension involves logical and verbal thinking and its object is to test for empirical truth. Paradigmatic thinking seeks good argument, tight analysis, and falsifiable empirical discovery. The narrative dimension, in contrast, entails storylike, imagistic thinking, and its object is not truth but 'versimilitude' or 'lifelikeness' (Bruner, 1986, p. 11). Narrative thinking seeks "good stories, gripping drama, and believable (though not necessarily true) historical accounts" (p. 13). Fantasy and imagination play an essential role in narrative thinking.

Another cognitive theory that acknowledges the functional importance of associative or imaginative thinking is Epstein's (1994) cognitive-experiential self theory. Like Bruner, Epstein posits a distinction between two systems of information processing, rational and experiential thought. Rational thought is intentional, verbal, abstract, logical, and analytical. Experiential thought is automatic, effortless, concrete, and affect-driven, and it encodes reality in concrete images and narratives.

Epstein's theory extends earlier dual-mode information processing theories by attaching importance to the role of emotions in the experiential system. This emphasis on emotion as an aspect of information processing dovetails with philosophical theories of imagination, which documented much earlier than psychological theories that cognitions, imagination, and emotions are closely connected. In his phenomenological essay "The psychology of imagination," Sartre (1948) argued that the human imagination consists of two layers: a primary or constituent layer, which involves the formation of mental images in response to external stimuli, and a secondary layer, which comprises the affective reactions (e.g., love, hate) and motor reflexes (e.g., nausea, papillary dilation) to these mental images.

In summary, fantasy and imagination are progressively considered as essential characteristics of human information processing. From birth on, we are engaged in the construction of mental representations of environmental experiences and events. Once such representations are encoded for storage in the brain, they may be retrieved in rational thinking, for example during logic reasoning. However, they also reappear in imaginative activities, where they may provide a rich source for creative products or simply for self-entertainment (Singer, 1999).

In this chapter, we review the literature on the relationship between exposure to media entertainment, and fantasy and imagination. We distinguish between three successive phases in which fantasy and/or imagination are related to the entertainment experience, namely before, during, and after exposure. Our review of the literature will be organized in three sections that correspond to these three phases. As will become clear later, most empirical research has focused on the function of fantasy and/or imagination before and after exposure to entertainment, whereas their role during exposure has received far less attention. As a result, our reasoning on the role of fantasy and imagination during exposure will be more speculative than our discussion about their role before and after exposure.

FANTASY AND IMAGINATION BEFORE EXPOSURE TO ENTERTAINMENT

The role of fantasy and imagination before exposure to entertainment lies in their potential to influence one's selective exposure to media entertainment. Although we believe that both fantasy and imagination may affect people's selective exposure to certain types of media entertainment, previous pre-exposure research has dealt only with *fantasy* as a predictor of selective exposure to entertainment, and not with imagination. Three hypotheses have been proposed of how certain types of fantasy may cause changes in people's exposure to media entertainment: the escapism hypothesis, the thematic correspondence hypothesis, and the thematic compensation hypothesis.

Escapism Hypothesis

According to the escapism hypothesis, exposure to media entertainment is stimulated by an overproduction of unpleasant fantasies. We have identified two versions of the escapism hypothesis, the thought-blocking hypothesis and the boredom-avoidance hypothesis. The thought-blocking hypothesis argues that individuals suffering from many unpleasant fantasies watch more entertainment in order to drive away these unpleasant thoughts. The boredom-avoidance hypothesis argues that individuals suffering from a fantasy style called "poor attentional control" spend more time watching entertainment. Individuals with poor attentional control are easily bored and distracted, and hence experience a great deal of fantasies, mindwandering, and drifting thoughts.

Both versions of the escapism hypothesis have been investigated only in correlational research, in which the causal direction of the relationships could not be established. Consistent with the thought-blocking hypothesis, people with an unpleasant fantasy style watched more television (McIlwraith, 1998; McIlwraith & Schallow, 1983). It is unknown whether these people also watched more entertainment because only measures of general viewing were employed in studies investigating this hypothesis.

In line with the boredom-avoidance hypothesis, people suffering from poor attentional control watched more television in general and more entertainment programs in particular (Schallow & McIlwraith, 1986). News and informational programs were watched less frequently by these people. This latter finding is consistent with Schramm, Lyle, and Parker's (1961) proposition that news and informational programs are less likely to fulfill an escapist function than entertainment programs.

Thematic Correspondence Hypothesis

The thematic-correspondence hypothesis argues that the themes people fantasize about directly influence the types of entertainment they prefer to view. It assumes, for example, that people select more violent or heroic entertainment contents if they have more aggressive or heroic fantasies. This hypothesis received support in a series of correlational studies (e.g., Huesmann & Eron, 1986). It is, however, an open question whether fantasy is the causal factor in this relationship, because watching violent entertainment may also stimulate people to fantasize more about such themes. An experiment by Feshbach and Singer (1971) and a causal-correlational study by Valkenburg and Van der Voort (1995) suggest that violent entertainment is the cause and violent fantasy the effect. However, because the evidence for causality is still scarce, it would be premature to conclude that the thematic correspondence hypothesis has been proved wrong. The fantasy–entertainment relationship may be reciprocal: Certain types of entertainment could stimulate corresponding fantasy themes, which in turn could stimulate interest in watching these types of entertainment (for a review, see Valkenburg & van der Voort, 1994).

Thematic Compensation Hypothesis

The thematic-compensation hypothesis proposes that people select entertainment themes that reflect those types of fantasies that they cannot produce themselves. For example, individuals who are unable to produce arousing sexual fantasies may turn to erotica or pornography. This hypothesis is consistent with Freud's (1908, 1962) assumption that one's motivation for fantasies are unsatisfied wishes.

The thematic-correspondence hypothesis presumes a negative relationship between fantasizing about specific contents and the viewing of corresponding program contents. However, the research to date has only shown null and positive relationships between exposure to specific media contents and corresponding fantasies. As discussed above, frequent viewing of aggressive media content goes together with more fantasies about aggressive and heroic themes. Frequent watching of erotic content is related to fantasizing about similar themes (Leitenberg & Henning, 1995). These findings suggest that the thematic compensation hypothesis may not be tenable.

THE ROLE OF FANTASY AND IMAGINATION
DURING EXPOSURE TO ENTERTAINMENT

Fantasy and imagination during exposure to entertainment may influence cognitive and emotional involvement. It is important to note that whereas we assume *imagination* to exert an important influence on involvement in entertainment, we do not believe that *fantasy* can exert such an influence. In the opening section of this chapter, we defined fantasy as a shift of attention away from an ongoing physical or mental activity toward a response to some internal stimulus (Singer, 1966). According to this definition, fantasizing is mutually exclusive to any other ongoing task or activity, and, therefore, cannot be combined with cognitive and emotional involvement in media entertainment.

In this section, we focus on the function of imagination during exposure to *fictional* media entertainment (i.e., film, drama, and other narratives). Unfortunately, however, research dealing with the entertainment experience has entirely ignored the role of imagination during exposure. In most books on entertainment published in the past decade, the entry 'imagination' does not even appear in the index. This lack of attention complicates our review because we cannot draw on existing hypotheses and research. As a result, we can only use cross-disciplinary knowledge that we have taken from, for example, emotion theory, films studies, and information-processing theories.

The lack of research on the function of imagination during exposure to entertainment is remarkable because, in our view, imagination is an essential aspect of all cognitive and emotional processes related to the entertainment experience. If one wants to explain why we enjoy fictional entertainment and why we respond emotionally to fictional characters and events, one is necessarily led to examining the relationships between fiction and imagination, imagination and emotion, and emotion and fiction.

Fiction, Imagination, and Emotion

One of the still most puzzling questions for entertainment and emotion researchers is why we experience joy and other emotions in response to fictional entertainment, that is, to imaginary worlds that portray characters and events that have never existed and certainly are not existing now. Most traditional emotion theories assume that emotions can only be evoked by stimuli and events that are appraised as real (Frijda, 1988, Lazarus, 1991). According to Nico Frijda's Law of apparent reality, "emotions are triggered by events appraised as real, and their intensity corresponds to the degree to which this is the case" (1988, p. 352). This emotion law rules out the possibility that people can experience emotions while being exposed to imaginary events in fiction.

However, we are all familiar with experiencing strong emotional involvement with a movie or a novel. These emotional responses to art and fiction are referred to as "aesthetic" or "imagined" emotions (Boruah, 1988; Scruton, 1974). Frijda's Law of apparent reality was therefore seriously criticized for disregarding aesthetic emotions (e.g., Walters, 1989). In a follow-up article, Frijda (1989) had to concede that aesthetic emotions are real emotions and that they are an ordinary phenomenon. According to Frijda (1989), viewers experience aesthetic emotions because they regard the events in films as true events in an imaginary world. Viewers do not perceive the occurrence of these events as unreal; they just discount any proof in the film that points to it being unreal.

Harris (2000) gives a useful extension to Frijda's (1989) explanation. Harris believes that fictional entertainment can be consumed in two ways. First, in a default mode, whereby viewers do not employ their knowledge of the reality status of the movie to suppress their emotions. In this default mode, viewers are emotionally touched by movies, not because they constantly think that the movie is real, but because they do not include their knowledge of the reality status of the movie in their appraisal.

In the second way in which viewers consume fictional entertainment, they do use their knowledge of the reality status of the movie. This can occur spontaneously, for example when the protagonists act unconvincingly. However, viewers also do this resolutely, for example when they see a shocking scene such as a mutilation. In that case, they deliberately question the reality status of the film to protect themselves against the emotions elicited. "It is only make-believe" is one such protective statement.

The two ways in which viewers consume fiction may find their roots in the two systems of information processing distinguished by Epstein (1994). When consuming entertainment in the non-default mode—when emotional involvement is minimized—viewers rely on their rational system of information processing: They use logic and reason to come to a reality appraisal of the fictional world. When viewers consume fiction in the default mode, that is, escorted by emotions, they rely on their experiential system of information processing. The experiential system involves rapid, automatic processing, is pleasure-oriented, emotionally driven, and characterized by a primacy of affective reactions.

The experiential mode of information processing may also underlie emotional involvement in fictional entertainment. According to Epstein (1994), the constructs of the world that are represented in the experiential system are called schemata. These schemata, which are all linked to other schemata, can represent thoughts (realistic or imaginary), emotions, and behavioral tendencies. If an external stimulus or event (e.g., watching a movie) activates one or more of these schemata, it can simultaneously trigger many other related schemata in an individual's brain. Not all of these schemata are necessarily related to the external stimulus or event. The spreading-activation principle (Collins & Loftus, 1975), implies that thoughts, fantasies, and emotions can all automatically and preconsiously be activated by each other, and, as a result, can all simultaneously occur while watching fictional entertainment.

Fictional Entertainment and the Role of Imagination

Epstein's experiential mode of information processing offers a plausible explanation of why we experience joy and other emotions in response to imaginary events presented in fictional entertainment. In this section, we focus on how imagination can influence different psychological processes while being exposed to fictional entertainment. As discussed, we assume that imagination is an important part of all emotional processes related to the fictional entertainment experience. We will illustrate this assumption by means of three entertainment processes: immersion, empathy, and parasocial interaction.

Immersion.   Most people have had the sensation of being lost in a book or movie. This phenomenon of being immersed in an imaginary world has been covered in the literature by several concepts. In empirical literature research, it is known as *transportation* (Green & Brock, 2002), in film theory as the *Diegetic effect* (Burch, 1979), and in research on virtual realities as *presence* (Slater, Usoh, & Steed, 1994). Although there are some differences in the processes that these concepts describe, they share some important mechanisms. In all three concepts, the media consumer's mental system becomes focused on the events occurring in an imaginary space, while the real-life world is temporarily suppressed. Second, in all three concepts, the media consumer witnesses the events occurring in the imaginary space, and these events drive his or her emotional system (Polichak & Gerrig, 2002; Tan, 1996). Finally, all three concepts derive their force from our capacity for imagination.

Polichak and Gerrig (2002) focused on types of participatory responses while watching drama. Because several types of their participatory responses fall under our definition of imagination, their taxonomy is relevant to our aim. They identified 5 types of participatory responses that may result from being involved in drama.

1. *Inferences:* These responses are used to fill in the gaps in the scenes that are not visible, but causally or temporally implied.
2. *'As if" responses:* These are the immediate responses that a viewer experiences when observing the scene as a participant. For example, when someone is killed in a detective movie, the viewer may generate a mental list of potential perpetrators and evaluate the likeliness of each of them.
3. *Problem-solving responses:* Viewers often strategically gather evidence from the story that will enable them to more confidently predict the outcomes of the story, especially the outcomes they favor.
4. *Replotting responses:* These are similar to problem-solving responses, but are retrospective. If, for example, a story develops differently than was expected, the viewer may feel discomfort and may start to replot the story to reduce this feeling of discomfort.
5. *Evaluatory responses:* These responses reflect the viewers evaluation of the general or specific events or messages in the film story. For example, when viewers have seen an aggressive boxing movie, they may question whether boxing is an appropriate branch of sports.

Polichak and Gerrig's study demonstrates that viewers bring a wealth of imaginary and participatory responses to the entertainment experience. The number and types of these responses depend on several variables, including the nature of the fictional story, viewers' capacity for imagination, their beliefs, and earlier experiences, and the motives with which they selectively expose themselves to the particular type of entertainment.

Empathy.   Imagination is a necessary condition for empathetic responses to fictional characters. According to Boruah (1988), empathy consists of two dimensions, an affective one and an imaginative one. Geared to the entertainment experience, the affective dimension refers to the vicarious affective response to a protagonist's situation. However, the emotions of the viewer and the protagonist do not necessarily have to be identical. In a scene in which an aggressive alien is lurking in the shadows, waiting to attack an innocent and unsuspecting child, the emotions of viewer and child do not agree, even though in such cases empathetic reactions are still common. This phenomenon, whereby empathy is felt without actually seeing the emotions of the protagonist, is called anticipatory empathy (Stotland, 1969).

The imaginative dimension of empathy consists of the viewer's cognitive representation of the thoughts and emotions of the protagonist. By adopting the protagonist's psychological point of view, the viewer imagines the protagonist's thoughts, feelings, and mental states. In the example of anticipatory empathy, the viewer imagines the implications of the alien's attack for the child. Such participatory or anticipatory responses are manifestations of one's imagination. According to Boruah (1988), empathy involves a confluence of emotion and imagination in one mental state. Without imagination, it is impossible to empathize with protagonists, and without empathy for protagonists, it is impossible to be involved in entertainment.

Parasocial Interaction.   Imagination is probably most essential in parasocial relationships (Horton & Wohl, 1956), the one-sided imaginative relationships that some media consumers develop with media characters or celebrities. John Caughey (1984) discussed many cases in which people fantasized that they were involved with a celebrity. A well-known example in his book is the story of John Hinckley, who attempted to assassinate President Ronald Reagan in 1981. Hinckley had seen the movie *Taxi Driver* and become infatuated with Iris, a teenage prostitute played by Jodie Foster. Hinckley wrote love letters to her and began to imagine that he was her lover. He eventually started to believe that he could win her by killing president Reagan. The Hinckley case reflects an extreme example of dysfunctional imagination. More common are parasocial relationships in which individuals imagine talking to or interacting with media characters. Such individuals are often lonely and isolated from people in their real-life environment (Caughey, 1984; Honeycutt & Cantrill, 2001)

THE ROLE OF FANTASY AND IMAGINATION
AFTER EXPOSURE TO ENTERTAINMENT

This final section deals with the question of how media entertainment can influence people's fantasy and imagination after exposure. The effects hypotheses in this section identify media entertainment as the cause of changes in the viewer's fantasy and imagination. We discuss the stimulation hypothesis, and four types of reduction hypotheses: the visualization hypothesis, the rapid-pacing hypothesis, the passivity hypothesis, and the arousal hypothesis.

Stimulation Hypothesis

According to the stimulation hypothesis, media entertainment influences people's fantasy and imagination through the content of the programs watched. As for fantasy, this hypothesis assumes that viewers who frequently consume certain types of media entertainment tend to fantasize more frequently about themes that correspond to that content. In regard to imagination, it argues that media entertainment provides viewers with a rich source of ideas from which they can draw when engaged in imaginative tasks, such as drawing and story telling, with the result that the quality or quantity of their imaginative products is enhanced.

Evidence for Fantasy. Environmental stimuli can evoke fantasies, especially when these stimuli correspond to our current concerns (Klinger, 1990). Because media entertainment forms an important part of people's everyday environment, it is likely that fantasies are evoked not only by real-life events but also by exposure to media entertainment. It is no surprise, therefore, that the results of most studies on the entertainment–fantasy relationship are consistent with the stimulation hypothesis (see Valkenburg & van der Voort, 1994 for a review). However, as noted in the section on the thematic correspondence hypothesis, none of the correlational studies on the fantasy–entertainment relationships permit a conclusive causal interpretation. It is possible that both the stimulation hypothesis and the thematic-correspondence hypothesis are valid explanations of the positive relationships between television entertainment and fantasy themes.

Evidence for Imagination. Television entertainment may enrich the viewer's repertoire of ideas. A series of media comparison experiments have demonstrated that children who have just seen a film story then incorporate elements from the film in their story completions or drawings (e.g., Greenfield & Beagles-Roos, 1988). However, there is as yet little evidence that the quality or quantity of their imaginative products is improved through exposure to these film stories. Actually, none of the studies into the influence of media on imagination demonstrated positive relationships between media exposure and imagination. Therefore, there is little indication that the stimulation hypothesis holds true when it comes to imagination (Valkenburg & van der Voort, 1994).

Reduction Hypotheses

All reduction hypotheses on fantasy and imagination focus on the possible negative effects of *audiovisual* entertainment. There is no researcher who believes that books or auditory stories hinder fantasy or imagination. The reduction hypotheses are all based on the idea that audiovisual entertainment has a number of structural characteristics (e.g., a rapid pace, ready-made images) that hinder the development of fantasy and imagination.

Most reduction hypotheses have been forwarded for both fantasy and imagination. However, as discussed above, in the case of fantasy there is no evidence of negative fantasy–entertainment relationships. In fact, so far only positive effects of media exposure on people's fantasy have been reported. Because in the case of fantasy the validity of the reduction hypotheses is entirely lacking, our review of the reduction hypotheses will focus on imagination.

Visualization Hypothesis

In this reduction hypothesis, the visual nature of audiovisual entertainment is held responsible for a negative effect on imagination. Unlike verbal entertainment, audiovisual entertainment presents the viewer with ready-made visual images, and thus leaves media consumer little room for forming his or her own visual images. It is assumed that when engaged in imagination, it is hard to dissociate oneself from the images supplied by audiovisual entertainment, with the result that one has difficulty generating novel ideas.

The visualization hypothesis has been investigated only with children. In these studies, referred to as media comparison studies, children were presented with a story or a problem.

The text of the story or problem was usually kept the same, whereas the presentation modality (i.e., print and/or audio versus audiovisual) was varied.

It is obvious that consumers of audiovisual entertainment generally have to produce fewer visualizations than those of auditory or print media. After all, in print or audio modalities, readers and listeners are required to convert verbal information into their own visual images. However, this does not necessarily mean that readers or listeners produce more imaginative thoughts while being exposed to a print or audio story. Gerrig and Prentice (1996) observed that some forms of imaginative thinking, such as thoughts about characters or reflections on the broader implication of the story, are more often evoked by an audiovisual story rather than a printed story. An explanation for these differences is, according to the authors, that audiovisual entertainment is often more engaging than verbal entertainment.

The assumption that viewers have difficulty dissociating themselves from audiovisual images during thinking has never been investigated in the media comparison experiments. Therefore, it is still an open question whether the visual images are responsible for the presupposed reduction effect. A media comparison experiment by Kerns (1981) demonstrated that a silent film, which contained only visual information, elicited more imaginative responses than verbal media did. The author attributed her finding to the fact that silent films are more ambiguous than verbal media and leave more room for one's own interpretations. This finding contradicts the assumption of the visualization hypothesis that the ready-made visual images are responsible for the presupposed difficulty to dissociate oneself from the stimulus materials.

The final assumption of the visualization hypothesis, that audiovisual stories result in less imagination in comparison to verbal stories, has received mixed support in the media comparison experiments. When an audiovisual story was compared to a written story, one study found a small negative effect for the audiovisual story (Kerns, 1981), whereas another study did not find significant differences (Meline, 1976). When compared to an auditory story, the audiovisual story led to slightly fewer imaginative (i.e., stimulus-free) responses, but not for children younger than 7 (Valkenburg & Beentjes, 1997), and not when the stimulus stories were difficult to comprehend (Greenfield & Beagles-Roos, 1988) or unrelated to the imaginative assignment (Runco & Pezdek, 1984). Finally, when a visual imaginative task was used, the audiovisual story encouraged children to produce less conventional drawings than the audio story (Vibbert & Meringoff, 1981).

In summary, although the visualization hypothesis is still a popular hypothesis in both academic and popular literature, none of the assumptions of this hypothesis have received enough evidence to establish its validity.

Rapid-Pacing Hypothesis

In this hypothesis, the reductive effect of entertainment is attributed to its rapid pace and continuous movement. The rapid pace of entertainment allows the viewer little time to process the information and reflect on the program content. As a result, audiovisual entertainment encourages cognitive overload, and a nonreflective style of thinking. Because reflective thinking is a prerequisite for imagination, imagination is hindered.

The validity of the rapid-pacing hypothesis has never been directly investigated. Although it is unknown whether the mechanisms proposed by the rapid-pacing hypothesis indeed underlie a reductive effect of television on imagination, we may examine whether the available evidence suggests that these mechanisms operate at all. First, rapidly paced programs leave viewers less room for reflection on program content than slowly paced programs. Until now, however, there are no indications that a rapid program pace leads to cognitive overload, impulsive thinking, and shortened attention spans (Anderson, Levin, & Lorch, 1977). Zillmann (1982) even suggests that fast-paced educational programs result in superior attention and information acquisition. Because there is no evidence of ill effects from fast-paced programs on viewers' cognitive style, it is not likely that a rapid program pace is a potential cause of reductive entertainment effects on imagination.

Passivity Hypothesis

This reduction hypotheses particularly focuses on television. Television is seen as an "easy" medium, requiring little mental effort (Salomon, 1984). With a minimum of mental effort, the viewer consumes fantasies produced by others. This leads to a passive "let you entertain me" attitude that undermines the willingness to use one's own imagination.

The validity of the passivity hypothesis has also never been directly investigated. Some studies however have examined some of the assumptions on which the passivity hypothesis is based. The passivity hypothesis first assumes that the processing of television information requires little mental effort, and that this low level of mental effort elicited during television viewing leads to a tendency to invest little mental effort during other activities. It also assumes that viewers' imagination is undermined, because they consume fantasies produced by others.

Although several studies have shown that viewers are cognitively far from passive while watching television or drama (Collins, 1982; Polichak & Gerrig, 2002), there is some evidence that television viewing requires less mental effort than reading does (Salomon, 1984). It has, however, never been investigated whether television viewing leads to a general tendency to invest little mental effort.

Of course, television viewers consume fantasies produced by others, as proposed in the passivity hypothesis, but there is little reason to assume that this reduces imagination. People who read a story, listen to a story, or watch a play also consume fantasies produced by others. Nevertheless, it has never been argued that verbal stories or theater hinder the development of imagination. Therefore, there is little reason to assume that television's reductive effect on imaginative play and creativity is caused by a television-induced passive "let-you-entertain-me" attitude.

Arousal Hypothesis

Like the rapid-pacing hypothesis, this hypothesis assumes that television promotes hyperactive and impulsive behavior. However, hyperactivity is not seen as a result of the rapid pace of television programs, but is attributed to the arousing quality of action-oriented and violent programs. This arousing quality is assumed to foster a physically active and impulsive behavior orientation in viewers, which in turn disturbs the sequential thought and planning necessary for imagination.

Although television viewing appears to be generally associated with relaxation, violent programs can produce intense arousal (Zillmann, 1991). In addition, there is evidence that the frequency with which children watch violent and/or action-oriented programs is positively related to restlessness in a waiting room (Singer, Singer, & Rapaczynski, 1984a) and impulsivity at school (Anderson & McGuire, 1978).

Because research does indicate that violent programs can induce impulsive behavior, it is no surprise that some television-imagination effect studies have demonstrated that watching violent programs can adversely affect imagination in children (e.g., Singer, Singer, & Rapaczynski, 1984b). However, although these studies established that violent programs can hinder imagination, they failed to investigate whether it was the arousal provoked by television violence that was responsible for the reductions in imagination.

Summary and Conclusion

In this chapter, we clarified that fantasy and/or imagination play an important role before, during, and/or after the entertainment experience. Fantasy and imagination before the entertainment experience may guide the media consumer's selective exposure. Two out of three hypotheses forwarded in that section seem to be valid. The escapism hypothesis, which states that people's fantasy lives predispose them to watch more television and media entertainment, was supported by correlational studies that related certain fantasy styles to media exposure. The thematic-correspondence hypothesis, which suggests that certain fantasy themes predispose media consumers to choose similar entertainment contents, also received support. However, the thematic-compensation hypothesis, which assumes that people choose entertainment themes that are opposite to their fantasies, appeared to be invalid.

We argued that emotional processes during the entertainment experience are largely founded on imagination. Fantasy cannot play a role during the entertainment experience, because we defined fantasy as an activity that is mutually exclusive with any ongoing physical or mental task. Although empirical research on imagination during the entertainment experience is still lacking, we provided a first conceptualization of the role of imagination during exposure. Future research should formulate and test some assumptions on the specific role of imagination during exposure. These hypotheses can in part be inspired by those that pertain to the pre- and post-entertainment setting. It would be interesting to test whether the assumptions that some reduction hypotheses make about information processing during exposure are valid. The rapid-pacing hypothesis, for example, assumes that audiovisual entertainment precludes reflection while watching. It is worthwhile to compare the kind of reflections that people make while watching different types of entertainment in different modalities.

The role of fantasy and imagination after the entertainment experience is rather complex. In the case of fantasy, a stimulation effect seems most plausible, whereas in the case of imagination, a slight reduction effect seems most plausible. However, the existing research is too scant to allow us to single out which of the four reduction hypotheses presented in this chapter is most plausible. Unfortunately, empirical research on the media–imagination/fantasy relationship has usually not been guided by explicit theoretical models. Most studies have examined the relationships between media exposure and fantasy or imagination as an input–output process, without specifying the underlying mechanisms of the presumed relationships. Future media research on the role of imagination and/or fantasy should derive from more sophisticated theoretical models and pay closer attention to the question of how and why imagination and fantasy may affect the entertainment experience and vice versa.

REFERENCES

Anderson, D. R., Levin, S. R., & Lorch, P. E. (1977). The effects of TV program pacing on the behavior of preschool children. *AV Communication Review, 25,* 159–166.

Anderson, C., & McGuire, T. (1978). The effect of TV viewing on the educational performance of elementary school children. *The Alberta Journal of Educational Research, 24,* 156–163.

Boruah, B. H. (1988). *Fiction and emotion: A study in aesthetics and the philosophy of mind.* Oxford: Clarendon Press.

Bruner, J. S. (1986). *Actual minds, possible worlds.* Cambridge, MA: Harvard University Press.

Burch, N. (1979). *To the distant observer.* Berkeley: University of California Press.

Caughey, J. L. (1984). *Imaginary social worlds.* Lincoln, NE: University of Nebraska Press.

Collins, A. M., & Loftus, E. F. (1975): A spreading-activation theory of semantic processing. *Psychological Review, 82,* 407–428.

Collins, W. A. (1982). Cognitive processing in television viewing. In D. Pearl, L. Bouthilet, & J. Lazar (Eds.), *Television and behavior: Ten years of scientific progress and implications for the eighties* (pp. 9–23). Washington, DC: U.S. Government Printing office.

Epstein, S. (1994). Integration of the cognitive and the psychodynamic unconsciousness. *American Psychologist, 49*, 709–724.

Feshbach, S., & Singer, R. D. (1971). *Television and aggression: An experimental field study.* San Francisco, CA: Jossey-Bass.

Freud, S. (1908, 1962). Creative writers and daydreaming. In J. Strachey (Ed.), *The standard edition of the complete psychological works of Sigmund Freud.* Vol IX. London: Hogarth.

Frijda, N. H. (1988). The laws of emotion. *American Psychologist, 43*, 349–358.

Frijda, N. H. (1989). Aesthetic emotions and reality. *American Psychologist, 44*, 1546–1547.

Gerrig, R. J., & Prentice, D. A. (1996). Notes on audience response. In D. Bordwell & N. Caroll (Eds.), *Post-Theory: Reconstructing film studies* (pp. 388–403). Madison, WI: University of Wisconsin Press.

Green, M. C., & Brock, T. C. (2002). In the mind's eye: Transportation-imagery model of narrative persuasion. In M. C. Green, J. J. Strange, & T. C. Brock (Eds.), *Narrative impact: Social and cognitive foundations* (pp. 315–343). Mahway, NJ: Lawrence Erlbaum Associates.

Greenfield, P. M., & Beagles-Roos, J. (1988). Radio vs. television: Their cognitive impact on children of different socioeconomic and ethnic groups. *Journal of Communication, 38*(2), 71–92.

Harris, P. L. (2000). *Understanding children's worlds: The work of the imagination.* Oxford, UK: Blackwell.

Honeycutt, J. M., & Cantrill, J. G. (2001). Cognition, communication, and romantic relationships. Mahwah, NJ: Lawrence Erlbaum Associates.

Horton, D., & Wohl, R. R. (1956). Mass communication and para-social interaction: Observations of intimacy at a distance. *Psychiatry, 19*, 215–229.

Huesmann, L. R., & Eron, L. D. (1986). The development of aggression in American children as a consequence of television violence viewing. In L. R. Huesmann & L. D. Eron (Eds.), *Television and the aggressive child: A cross-national comparison* (pp. 45–80). Hillsdale, NJ: Lawrence Erlbaum Associates.

Kerns, T. Y. (1981). Television: A bisensory bombardment that stifles children's creativity. *Phi Delta Kappa, 62*, 456–457.

Klinger, E. (1990). *Daydreaming: Using waking fantasy and imagery for self-knowledge and creativity.* Los Angeles, CA: Tarcher.

Knowles, R. T. (1985). Fantasy and imagination. *Studies in Formative Spirituality, 6*, 53–63.

Lazarus, R. S. (1991). Progress on a cognitive-motivational-relational theory of emotion. *American Psychologist, 46*, 819–834.

Leitenberg, H., & Henning, K. (1995). Sexual fantasy. *Psychological Bulletin, 117*, 469–496.

Lewin, I. (1986). A three dimensional model for the classification of cognitive processes. *Imagination, Cognition and Personality, 6*, 43–54.

McIlwraith, R. D. (1998). "I'm addicted to television": The personality, imagination, and TV watching patterns of self-identified TV addicts. *Journal of Broadcasting, 42*, 371–386.

McIlwraith, R. D., & Schallow, J. R. (1983). Adult fantasy life and patterns of media use. *Journal of Communication, 33*(1), 79–91.

Meline, C. W. (1976). Does the medium matter? *Journal of Communication, 26*(3), 81–89.

Polichak, J. W., & Gerrig, R. J. (2002). Get up and Win! In M. C. Green, J. J. Strange, & T. C. Brock (Eds.), *Narrative impact: Social and cognitive foundations* (pp. 71–97). Mahwah, NJ: Lawrence Erlbaum Associates.

Runco, M. A., & Pezdek, K. (1984). The effect of television and radio on children's creativity. *Human Communication Research, 11*, 109–120.

Salomon, G. (1984). Television is "easy" and print is "tough": The differential investment of mental effort as a function of perceptions and attributions. *Journal of Educational Psychology, 76*, 647–658.

Sartre, J. P. (1948). *The psychology of imagination.* London: Routledge.

Schallow, J. R., & McIlwraith, R. D. (1986). Is television viewing really bad for your imagination? Content and process of TV viewing and imaginal styles. *Imagination, Cognition and Personality, 6*, 25–42.

Schramm, W., Lyle, J., & Parker, E. (1961). *Television in the lives of our children.* Stanford, CA: Stanford University Press.

Scruton, R. (1974). *Art and Imagination.* London: Methuen & Co.

Singer, J. L. (1966). *Daydreaming.* New York: Random House.

Singer, J. L. (1999). Imagination. In M. Runco, & S. R. Pritzker (Eds.) *Encyclopedia of Creativity*, Vol. 2, (pp. 13–25). New York: Academic Press.

Singer, J. L., Singer, D. G., & Rapaczynski, W. S. (1984a). Family patterns and television viewing as predictors of children's beliefs and aggression. *Journal of Communication, 34*(2), 73–89.

Singer, J. L., Singer, D. G., & Rapaczynski, W. S. (1984b). Children's imagination as predicted by family patterns and television viewing: A longitudinal study. *Genetic Psychology Monographs, 110*, 43–69.

Slater, M., Usoh, M., & Steed, A. (1994). Depth of presence in virtual environments. *Presence: Teleoperators and Virtual Environments, 3*, 130–144.

Stotland, E. (1969). Exploratory investigations of empathy. In L. Berkowitz (Ed.), *Advances in experimental social psychology* (pp. 271–314). New York: Academic Press.

Tan, E. (1996). *Emotion and the structure of narrative film: Film as an emotion machine.* Mahwah, NJ: Lawrence Erlbaum Associates.

Valkenburg, P. M., & Beentjes, J. W. J. (1997). Children's creative imagination in response to radio and TV stories. *Journal of Communication, 47,* 21–38.

Valkenburg, P. M., & van der Voort, T. H. A. (1994). Influence of TV on daydreaming and creative imagination: A review of research. *Psychological Bulletin, 116,* 316–339.

Valkenburg, P. M., & van der Voort, T. H. A. (1995). The influence of television on children's daydreaming styles: A one-year panel study. *Communication Research, 22,* 267–287.

Walters, K. S. (1989). The law of apparent reality and aesthetic emotions. *American Psychologist, 44,* 1545–1546.

Vibbert, M. M., & Meringoff, L. K. (1981). *Children's production and application of story imagery: A cross-medium investigation* (Technical Report No. 23). Cambridge, MA: Project Zero, Harvard University. (ERIC Document Reproduction Service No. ED 210 682.)

Zillmann, D. (1982). Television viewing and arousal. In D. Pearl, L. Bouthilet, & J. Lazar (Eds.), Television and behavior: Ten years of scientific progress and implications for the eighties. (Vol. 2.) Washington, DC: US Government Printing Office.

Zillmann, D. (1991). Television viewing and psychological arousal. In J. Bryant, & D. Zillmann (Eds.), *Responding to the screen: Reception and reaction processes* (pp. 103–133). Hillsdale, NJ: Lawrence Erlbaum Associates.

8

Attribution and Entertainment: It's Not Who Dunnit, It's Why

Nancy Rhodes and James C. Hamilton
University of Alabama

The typical mystery novel begins when an unsuspecting average person, going about his or her daily life, discovers a dead body. That person or some other unlikely hero (e.g., Miss Marple) then attempts to piece together the circumstances of the death. Inevitably, it is discovered that there were many people in the deceased's life who had ample motive to commit murder. The tension builds as the amateur sleuth follows mistaken leads to a person who the reader suspects to be innocent. Then, in the engrossing climax, the sleuth suddenly figures out who really murdered the victim, and confronts the perpetrator just before becoming victim number two.

In the unfolding plot of a mystery novel, what engages the reader's interest is the piecing together of the motives for murder; specifically, exploring the question of why the perpetrator killed the victim. The intrigue is in trying to understand why, given what we have learned about the character of the perpetrator, this individual chose to take the life of another human. Is it that the victim was such a horrible person that he or she deserved to die and the murderer just beat everybody else to it? Is it that the murderer is an evil person? Is it perhaps that there was something unique about the relationship between the murderer and the victim that drove the murderer to the dastardly deed?

One of the earliest and most profound accomplishments of social psychology was the scientific appreciation of the role that causal thinking plays in the regulation of human action and emotion. Collectively referred to as attribution theory, this early work and subsequent advances suggest that our experience of the world, other people, and ourselves is based upon our causal explanations for actions and events. For instance, negative emotions do not result directly from failure or loss, but from the belief that the failure or loss reflects poorly on our abilities or personal attributes (Carver, DeGregorio, & Gillis, 1980; J. Greenberg, Pyszczynski, & Solomon, 1982). Children's aggressive reactions to accidental harm or insult caused by other children are not direct responses to those slights or injuries, but are mediated by attributions of hostile intent (Dodge & Newman, 1981; Nasby, Hayden, & DePaulo, 1980).

Attribution research is ubiquitous. If you access a computerized index of scientific publications, like PsychInfo, and search for "attribution" (or "attributional") in the title you will get nearly 10,000 results. In addition to research on the basic psychology of causal thinking, attribution theory has been applied to understanding a dizzying array of social phenomena. However, a search for attribution articles whose titles also contain "entertain," "amuse," or "enjoy," or the noun forms of these words, produced just one result, a paper on maintaining exercise in weight loss programs (Hodgins & Fuller, 2001). In other words, the role of causal thinking in the experience of mediated entertainment is uncharted territory.

In this chapter we will explore some of the ways that attribution processes might affect the enjoyment people derive from various sources of entertainment. We will discuss a small number of research areas that relate to the role of attribution in entertainment experiences. We will then put forth a new hypothesis about the role that attributional thinking might play in the enjoyment of fictional drama. First, however, we will review basic attribution theory principles, with a particular focus on the ones that are relevant to our analysis of attribution and entertainment.

DEVELOPMENTS IN ATTRIBUTION THEORY

In modern social psychology, Heider (1958) is generally credited with the notion that ordinary people function in their daily lives as naïve scientists in attempting to understand their social worlds. Generally speaking, attribution theory refers to the prediction of whether an observer will attribute an actor's behavior to internal or external causes. Recall the mystery stories described in the introduction. If the audience assumes that the perpetrator of the crimes is an evil person, the audience is making an internal attribution—the murderous behavior is caused by something internal to the prepetrator. In contrast, if the audience makes the assumption that the victim's repeatedly horrible behavior toward the perpetrator drove the murderer to commit the crime, then the audience is making an external attribution. Specifically, forces external to the perpetrator caused the behavior. In this case, the audience might also make an internal attribution about the victim; the victim may have deserved to be murdered. This is a special example of attributional thinking, which we will discuss later, that is the foundation for the just-world hypothesis.

Overall, it appears that people have a preference for attributing the behaviors of others to internal causes. Dating back to Heider's initial descriptions, and especially continuing with Jones' and Davis' (1965) concept of correspondent inferences, it has long been understood that there is an overall bias toward making internal attributions for observed behavior. Assuming that the behaviors of other people are caused by stable dispositions is enormously adaptive: it helps the observer to predict, and perhaps control, the future behavior of other people. Without assuming stability, the social world would appear to be a bewildering sea of random occurrences.

Early on, this preference for internal attributions was labeled the fundamental attribution error (Ross, 1977). Specifically, when asked to make an attribution for the behavior of another person or actor, most people are biased to explain the behavior in terms of the actor's personality. Think about the last time you were driving, and another driver cut you off. Did you explain the breach of driving etiquette by focusing on situation characteristics such as that the driver must be in a hurry, or having a bad day? Typically not. Under those circumstances, most Americans would make disparaging remarks about the driver's bad character, that is, they would make an internal attribution.

A notable exception to the fundamental attribution error has been termed the actor-observer effect (Jones & Nisbett, 1972). Interestingly, the tendency to favor internal attributions is more prevalent when an observer is trying to understand the behavior of someone else. When someone is explaining his or her own behavior, i.e., making an attribution from the perspective of the actor, attributions are more likely to emphasize the situation rather than the disposition. Thinking back to the example of the driver, if you have ever become aware that you have made a similar driving faux pas, you are much more likely to say that it was due to the situation: the traffic was bad, the lighting was poor, and you were late for an appointment. The internal attribution that came so quickly when you were explaining the other driver's behavior doesn't occur to you in this case: You are unlikely to decide that you, in fact, are a jerk.

Various explanations have been offered for the fundamental attribution error and the actor-observer effect. Perhaps the most prevalent explanation focuses on the perceptual salience of the actor and the situation or their point of view (Ross & Nisbett, 1991). When an actor is functioning in a particular situation, his or her attention is usually focused outwardly, on the features of the situation that are relevant to what he or she is trying to accomplish in that situation. However, from the perspective of the observer of an individual functioning in such a situation, the person being observed becomes the perceptual focus; attention is drawn to the person almost as one attends to the actions of the star of a TV show. Thus, attributions that are made are consistent with the attentional focus: As the actor in the situation, one perceives the environment as most salient, and attributes the causes of actions to that environment. In contrast, for the observer of someone in a given situation, the individual performing the action is most salient, and the causes for behavior are attributed to characteristics of the actor.

The perceptual focus explanation for dispositional attributions stems from what Gilbert and Malone (1995) described as the "invisibility" of the situation. The situational constraints that guide behavior are often unclear to observers. Indeed, the media can sometimes exaggerate the propensity to make an internal attribution by making a situation even less salient than it is in reality. Consider, for example, the behavior of Howard Dean, a candidate for the Democratic nomination for president in the spring of 2004. After the Iowa caucuses, in which Dean had not done as well as hoped, he gave what appeared to be a somewhat hysterical concession speech which culminated with a "primal scream" that was played over and over on various media outlets, and contributed to the demise of his candidacy. Clearly, the news cameras that focused tightly on the candidate and the directional microphones that picked up only Dean's voice contributed to a variety of internal attributions. Many commentators concluded that he was emotionally unstable, and certainly lacked "presidential" stature. Interestingly, contrasting the TV footage with other video shot from the floor of the rally leads to a vastly different interpretation. Video shot by an amateur in the hall presented a small image of the candidate partially occluded by a foreground of rowdy supporters. The sound on the floor was quite loud, with a great deal of yelling and chanting going on; in fact, it was difficult to hear what Dean was saying over the sounds of the crowd. What had appeared to be overly emotional and completely inappropriate ranting on TV news reports appears to be fully consistent with the emotional tenor of the situation in the amateur video. Clearly, the salience of the situation can affect the attributions that are made.

An additional explanation for actor/observer effects is that individuals choose to make attributions that show themselves in the most positive light. These self-serving attributions help to protect self-esteem. Attributions for success and failure experiences generally follow self-serving predictions (McFarland & Ross, 1982; Miller & Ross, 1975; Weiner, Russell, & Lerman, 1978). Success experiences are attributed to internal causes, whereas failure experiences are attributed to external causes. This type of attribution is thought to be one of the mechanisms through which procrastinating behavior is reinforced. The procrastination sets up a self-serving attribution that can be applied regardless of the outcome. Specifically, if you procrastinate on a project and get a bad grade on it, it is easy to explain away the failure because, of course, you did not have sufficient time to do a good job on the project. In contrast, if you manage to do well on the project in spite of your procrastination, you can bask in the glory of possessing such brilliance that you excelled *in spite of* having had so little time for the project.

A preference for internal attributions has been shown to be prevalent in a wide range of studies (Ross & Nisbett, 1991). However, it seems that this bias may not be as "fundamental" as had originally been thought. Cross-cultural research has demonstrated that the bias toward internal attributions may be a function of the culture in which a person was raised. In particular, it appears that people who are raised in Western, individualist cultures are more prone to the fundamental attribution error than those raised in eastern, collectivist cultures (Miller, 1984). Research findings have demonstrated that the attributional process functions somewhat differently in eastern cultures, with people from those cultures making situation attributions for others' behavior more frequently than people from western cultures (Knowles et al., 2001; Lee, Hallahan, Herzog, 1996; Morris & Peng, 1994).

Recent research in attributional processes has focused on the possibility that these findings can best be explained by a dual-process account. Dual process theories typically describe an initial, memory-based, heuristic judgment in response to a stimulus that can be later modified by a more careful, systematic consideration of the specific features of the case at hand (Chaiken & Trope, 1999; Smith & DeCoster, 2000; Strack & Deutsch, 2004). In general, dual process theories propose that under most conditions, the initial judgments that are made on the basis of heuristic processing are good enough. Only when people have the motivation and ability to think further about their judgments will they take the time and put forth the effort to consider additional aspects of their judgments and revise them. In attributional thinking, when there is an important reason to consider an attribution further, the correspondence bias is greatly reduced and attributions more accurately reflect the constraints of the situation. Research supports this dual theory process: Findings indicate that there is indeed an initial tendency toward an internal, dispositional attribution that is later corrected by more careful consideration of the situation when there is ample reason for people to do so (Gilbert, Pelham, & Krull, 1988). Thus, the fundamental attribution error may seem fundamental because for the most part, it represents a "good enough" understanding of why an individual behaved in a particular way.

Given what we know, then, about how attributional thinking functions in everyday situations with social stimuli, we turn to a consideration of how these attributional processes function in the processing of entertainment media.

ATTRIBUTION AND ENTERTAINMENT

There are two ways to conceptualize the possible role of attributional thinking in entertainment. The first of these concerns the ways that attributional thinking is related to the act of participating in entertainment experiences. The second concerns the ways that attributional thinking is related to the degree of entertainment that people derive within the entertainment experience. Put more plainly, the first issue concerns attributional processes that lead us to go to the movies, open a book, or sit down in front of the television; the second has to do with the attributional process that affect our enjoyment of the story. We organize our discussion around these two related issues.

Participating in Entertainment Experiences
The Third-Person Effect and Obstacles to Media Entertainment

The degree to which an activity is judged to be entertaining and enjoyable is determined not only by the pleasures it provides, but also by the perceived costs of the experience. Over the past 20 years, social scientists have determined that some popular and enjoyable media experiences contribute to serious social problems. Most notably, researchers have found reliable associations between media portrayals of violence and violent behavior among viewers (Anderson & Bushman, 2002; Bushman & Anderson, 2001). Similarly, pornographic depictions of degrading treatment of women have been found to decrease empathy toward rape victims (Sharp & Joslyn, 2001). In juxtaposition to the significant social costs of these sorts of media experiences are well-publicized media trends toward more graphic violence, especially in film and video games, and the proliferation of pornography, especially in images and video available over the internet. Assuming the average media consumer does not support or condone murder or sexual assault, why would people enjoy media experiences that contribute to violence or the degradation of women?

One answer to this question is provided by a social psychological effect called the third-person effect. The third-person effect is the robust and highly replicable finding that persons believe they are less strongly influenced by media effects than are other people (Hoffner et al., 2001; Scharrer, 2002). This sort of cognitive activity bears similarities to other well-documented social psychological phenomena. Notably, the third-person effect appears to be a specific example of a general tendency toward unrealistic optimism (Weinstein, 1980, 1987). In dozens of domains, Weinstein and others have shown that people believe that they are less likely than others to experience accidents, illnesses, or injuries. For officials charged with promoting preventive health behaviors or those trying to change the television viewing habits of children, this cognitive bias presents a unique challenge. Persons must not only be convinced of the general connection between particular behavior patterns and undesired consequences, they must be persuaded that they, as individuals, are not immune from those influences.

For the purposes of this chapter, the third-person effect is distinguished from the general phenomenon of unrealistic optimism because it is more explicitly related to causal attributions. The generic unrealistic optimism paradigm is concerned with our estimates of the likelihood that bad things will happen to us. In contrast, the third-person effect specifically concerns our estimates of the likelihood that various media experiences will cause us to engage in bad behavior. In this sense the study of the third-person effect is informed by research on control beliefs. Research on control beliefs has consistently shown that people underestimate the degree to which their behaviors and outcomes are determined by external causal factors that are outside of their control (Thompson, Armstrong, & Thomas, 1998).

As described above, the actor-observer bias in attributions suggests that people are more likely to acknowledge the situational (external) influences on their own behavior than to acknowledge situational influences on the behavior of others. They are especially likely to attribute undesired or unflattering outcomes to external causes, that is, to make self-serving attributions. These biases might predict that individuals would overestimate the likelihood of being influenced by a situational factor such as media violence. Although the actor-observer bias might seem to be at odds with the third-person effect, they are easily reconciled by differentiating between explaining causes of past behavior and prediction of future behavior. The actor-observer effect and the self-serving bias are observed in explanations of past, undesired, behavioral outcomes. The third-person effect concerns the role of media influence as a cause of future behavior. The two effects are nicely reconciled by self-enhancement theories of self-attribution. With regard to failures or misdeeds that have already occurred, it is most flattering to explain them as a response to mitigating situational factors. In the case of predicting future behavior, it is most flattering to portray oneself as having the strength of character to be unaffected by media influences to which others are susceptible (David, Liu, & Myser, 2004). A recent study by Goerke and colleagues found precisely this pattern of causal thinking about past and future failures (Goerke et al., 2004).

Convenient Fiction: Entertainment as a Product of Insufficient Justification

One always tends to overpraise a long book, because one has got through it.

—E. M. Forster (E. M. Forster, 1936, p. 142)

One plausible answer to the question of why people enjoy media entertainment is that they conclude that they must. Estimates from the United States and Canada suggest that each day the average television owner spends approximately 3 to 5 hours watching television (Inc, 2004). Apart from work and sleep, there is no other single activity that occupies as much time in the lives of people in contemporary western societies (Kubey et al., 2004). Most people who live in these modern societies report that the multiple and competing demands of work and family leave little spare time for activities such as exercise or community involvement. In this context, the practice of spending several hours a day watching television begs not only for a scientific explanation, but also a personal explanation. In other words, how do people explain the time they spend watching television?

The uses and gratifications literature has provided some insights into viewers' understanding of why they watch television. In general, people report that watching television serves instrumental functions, e.g., to learn about the world, manage mood, or ritual functions, for example, habit or routine (Rubin, 1984). The specific functions that have been identified vary along an intuitive social desirability dimension. For example, 70% of respondents in one study (Rubin, 1984) reported that they watch television to learn about people and events, which most people would agree is a worthwhile way to spend one's time. However, large numbers also admitted they watch as an escape, or because television is inexpensive, or out of habit—all somewhat less noble reasons.

The social psychological theories of cognitive dissonance and self-perception were developed to explain the causal influence of behaviors on attitudes. Cognitive dissonance theory posits that when we act in a way that is inconsistent with our beliefs or attitudes, we experience psychic discomfort and restore our sense of equanimity by forming attitudes that are consistent with our behaviors (Festinger & Carlsmith, 1959). Self-perception theory asserts that when our attention is called to a particular behavior, we generate a self-attribution to explain the behavior (Fazio, Zanna, & Cooper, 1977). In either case, attitudes do not change if there is an obvious situational attribution for our behavior. In most cases, our decision to watch television is uncoerced; no one makes us watch for hours at a time. Nor would most of us endorse the idea that we have no better way to spend several hours each weeknight. Thus, television watching, when it is brought to our attention, may cause us to generate legitimate, but post-hoc, reasons for our viewing behavior.

Many studies in the uses and gratifications approach rely exclusively on viewers' self-reports about the reasons they watch television. Only a small proportion of studies collect additional data that could validate viewers' self-reports. In the absence of validating information it is difficult to determine the degree to which the uses and gratifications that viewers supply to explain their television watching are simply their own constructions, or accurate descriptions of real psychological processes. An admittedly incomplete review of the literature on uses and gratification shows good support for the idea that television viewing serves as an escape and a distraction (Moskalenko & Heine, 2003), which some audiences admit. It is harder to find support for some of the more flattering explanations for television watching, for instance those having to do with learning or mental stimulation. For example, Henning and colleagues showed that higher levels of need for cognition, that is, the tendency toward precocious thinking, was inversely correlated with television viewing (Henning & Vorderer, 2001). In a more troubling study, Kaye and colleagues found that the more news media audience members consumed, the greater their *illusion* of being well informed (Kaye & Johnson, 2002).

Enjoying the Entertainment Experience

The prototypic entertainment experience is watching a film. The film runs; we sit and watch. Key ingredients of the prototype are an actor or entity that actively entertains, and an audience of relatively passive recipients. Another important aspect of the entertainment prototype is a sense that the experience is both for itself and for the moment. The things we regard as entertainment are not expected to serve any other purpose than giving us pleasure at that moment. Just as entertainment expects little of the entertained, it also says little about the entertained.[1] We do not speak of people as being relatively good or bad at being entertained. Being entertained does not require any special skills or talents, and no improvement is expected over time. As soon as an entertaining pastime begins to take on these properties, it ceases to be considered entertainment. The definition of entertainment as a preoccupation that requires little of us, means little about us, and has little enduring value is certainly not a very flattering one.

The view of entertainment as a mere distraction may be responsible for the minimal attention that the study of entertainment has received from social scientists. In contrast to whole disciplines that have emerged to understand the artistic, cultural, and technical aspects of literature, films, and television, little attention has been paid to the psychological experience of being entertained by those media. Little is known about what distinguishes distractions that are highly entertaining from those that are boring or annoying. However, the past 10 years has seen an increase in empirical research aimed at understanding media entertainment. To date, this research has focused primarily on the affective consequences of media, for example, the use of entertainment to regulate mood (Zillmann, 1991; Zillmann & Bryant, 1985). Much less has been written about the cognitive processes that might mediate the mood regulating effects of media entertainment, or about the cognitive consequences of media entertainment.

Disposition Theory

Disposition theory provides one account of the conditions under which media exposure is experienced as enjoyable (Raney, 2003; 2004; Raney & Bryant, 2002). Specifically, this theory describes a process by which audiences develop emotional alliances with characters they encounter in fictional media. According to the theory, audiences make moral judgments about the characters in movies, television shows, and the like. These moral judgments form the basis of the audience's emotional reactions to the characters: Those characters who behave in accord with the audience's moral code will be evaluated positively; those who behave contrary to the audience's moral code will be evaluated negatively. These evaluations then trigger the audience to experience an emotional affiliation toward the characters. Because of these emotional ties, audiences develop expectations for the outcomes of the drama, such that they wish for good things to happen to the characters they like, and for bad things to happen to characters they dislike. Enjoyment of the program or movie results when the outcomes are as expected.

Although disposition theory has not been explicitly linked to attribution theory, we believe that attributions may play an important role in the experience of enjoyment as laid out in disposition theory. Specifically, we believe that the fundamental attribution error, or the propensity to view someone's actions as caused by an enduring feature of their personality, is at the heart

---

[1]It is possible that media interest and expertise could serve as an important self-definition for some people.

of the judgments that underlie an audience's affective disposition toward a character. In effect, this is a demonstration that judgments of mediated characters are made in the same way that judgments are made of real people one encounters in real life.

Interestingly, recent extensions of this model (Raney, 2004) acknowledge that, consistent with dual process considerations, audiences of television programs and movies may not always have the motivation and cognitive resources to make well-reasoned moral judgments of characters they encounter in mediated fiction. For example, audiences' goals may be to escape reality for a time and transport themselves into an alternate reality (Green, Brock, & Kaufman, 2004a). This motivation would not be conducive to engaging in a high level of processing about the causes of the characters' behavior. Furthermore, there is evidence that judgments about the goodness or badness of characters are made rather quickly, and that these judgments often rely on stereotypes or beliefs about the typical unfolding of a plot in fictional media (Green, Garst, & Brock, 2004b). These are precisely the types of conditions under which cognitive shortcuts, such as the fundamental attribution error, are most likely to be used.

One heuristic that seems particularly relevant is the belief in a just world (e.g., Hafer & Begue, 2005). This is an example of a shortcut in heuristic thinking that permits optimism in the face of tragedy. Specifically, when confronted with a person who has encountered misfortune, it is typical for an observer to engage in attributional thinking so as to place the blame for the misfortune on the victim of the misfortune. This is protective, because the observer is able to distance him- or herself from the misfortune. For example, assuming that the homeless people one encounters on the streets of a large city are homeless because of their own faults allows one to discount the likelihood of personally being victimized by poverty and homelessness.

In summary, according to disposition theory, the judgments audiences make about characters' morality lead to expectations for the outcomes of the drama. Satisfaction of one's expectations is related to enjoyment of the movie or program. Attributions for the causes of the characters' actions are a likely mechanism through which moral judgments are made, and, in fact, the fundamental attribution error and belief in a just world are likely heuristics that allow audiences to arrive quickly and effortlessly at a satisfactory judgment of a fictional character. By implication, when the protagonist with whom the audience identifies experiences a bad outcome, it is displeasing. This displeasure is likely related to the contradiction of our belief in a just world.

A New Thesis on the Role of Attributional Thinking in the Entertainment Experience

Most psychological theories of the appeal of entertainment, such as disposition theory, have focused on the value of entertainment for mood regulation. Certainly mood regulation is an important function of entertainment, but we suspect that careful research will eventually reveal that the cognitive processes involved in entertainment experiences are complex, and that entertainment experiences affect our thought processes and products. We believe that entertainment experiences are like children's play. Although play appears at first glance to be a meaningless pastime, psychologists continue to discover important ways in which play contributes to the cognitive and social development of children. We suspect that media entertainment, especially fictional literature, and film and television drama, will turn out to have similar significance. Specifically, we believe that these forms of entertainment contribute to the development and exercise of cognitive skills related to formulating and testing hypotheses about the causes and consequences of social behavior. Our thesis, put simply, is that fictional drama allows the audience to practice thinking about the complex blend of dispositional and situational factors that influence complex human social behavior.

There are three categories of observations upon which we base our thesis. The first concerns parallels between historical changes in the nature of human identity and historical changes in fictional literature. The second concerns the degree to which modern formulas for effective fiction writing involve techniques that provoke attributional thinking. Finally, we discuss the gender differences in entertainment preferences that are consistent with our thesis.

Historical Trends

The underlying premise of attribution theories is that understanding the causes of others' behavior serves important personal or social functions. If fictional literature serves to develop or refine attributional thinking, it stands to reason that historical changes in the difficulty or importance of these functions ought to be paralleled by changes in the nature of fictional literature and drama.

Baumeister has offered a provocative argument which posits that conceptions of the self—how and what people think about themselves—have changed dramatically across the past sixteen centuries (Baumeister, 1986; 1987). He argues that in the medieval period, identity was unproblematic. The medieval person's place in the world and in Christian cosmology was fixed at birth and unlikely to change throughout his or her lifetime. From a practical standpoint of navigating one's relations with others, little knowledge of others, as individuals, was required beyond knowing their place in the social order. As a result of changes in social and economic structures and changes in Christian views of morality and salvation, identity became more complex and more dependent upon individual self-determination. One's occupation, one's place in the social hierarchy, and one's Christian salvation all came to depend on personal decisions. Identity became a function of difficult choices one could make, as well as living up to various externally and internally defined standards of self-evaluation. As these changes occurred, social relations became commensurately more complicated, with the success of social relations becoming more dependent on knowing others as unique individuals.

For our purposes, Baumeister's most interesting observations relate to the historical emergence of concerns about deception and self-knowledge in the seventeenth century, and the decline of public life that occurred in the Victorian era. In the seventeenth and early eighteenth century, Calvinists believed that some people were destined for salvation and others were not. They wondered a lot about who was whom and eventually came to believe that a persons's actions or affectations may provide subtle clues. This belief eventually led to a fear that engaging in pious acts might be a way to fool oneself or others that one was among "the elect." This mentality gave rise to the general idea that a person's outward actions and appearances could not be taken for an accurate indication of his or her true inner nature. In other words, it became apparent that effective social functioning became dependent upon accurate explanations of others' behavior, and accurate understanding of their motives.

Baumeister suggests that these concerns over sincerity and deception reappeared in the Victorian era, but with an added twist. For the Victorians, the fear arose that subtle aspects of an actor's appearance or behavior could reveal to observers information about the actor's true self about which even the actor was unaware. Among other things, this belief caused Victorians to retreat into private life in order to escape the penetrating scrutiny of strangers. This change had the effect of limiting the number and variety of people with whom Victorians had close personal contact. One way to sum up the Victorian predicament is that understanding people's unique character and motivations became simultaneously more important—because people might not be who they appeared to be—and more difficult—because people closed themselves off from public observation. The state of contemporary western societies is largely unchanged in these respects.

According to Baumeister, at roughly the same time that self-hood and identity started to become problematic, the nature of fictional characters began to change. Writers began to focus on the individual experience of characters and less frequently used characters as a vehicle for addressing universal themes (e.g., allegorical stories). In addition, Baumeister points out that novels in the last half of the nineteenth and the early twentieth century reflected the Victorian preoccupation with deception and the complexities of human motives. Novels such as the popular Sherlock Holmes stories explicitly reflect the theme of the astute observer of small details who is able to uncover important hidden truths about people. Moreover, the modern novel came to rely on realistic characterizations. That is, they were stories about average people. At a time when the inner lives of others had become so important to understand and when intimate knowledge of others had simultaneously become so restricted, the novel provided a uniquely intimate way of accessing the private lives of others.

The following quote by E.M. Forster (Forster, 1974, p. 54), an early twentieth-century novelist, shows explicit recognition of the contrast between the intimacy that could be achieved with real people and the intimacy that could be achieved with fictional characters.

> In daily life we never understand each other, neither complete clairvoyance nor complete confessional exists. We know each other approximately, by external signs, and these serve well enough as a basis for society and even for intimacy. *But people in a novel can be understood completely by the reader*, if the novelist wishes; their inner as well as their outer life can be exposed. And this is why they often seem more definite than characters in history, or even our own friends; we have been told all about them that can be told; even if they are imperfect or unreal they do not contain any secrets, whereas our friends do and must, mutual secrecy being one of the conditions of life upon this globe.

Another development in literature of the late nineteenth century was the use of different narrative points of view. Prior to the seventeenth century, story narration was limited to a description of events from the perspective of an observer with no special insights into the thoughts and emotions of the characters. With the development of the novel in the eighteenth century, authors began to use a point of view called third-person omniscient; the narrator not only described the actions of the characters, he or she also described the characters' thoughts and feels about the unfolding events of the plot. This device is relevant to attributional thinking in two ways. First it allows the author to quickly and directly shape the reader's view of a character's personality. This sets up a situation in which plot events can be created which conform to or violate the reader's expectations of the characters. Second, it allows the author to regulate the reader's frame of reference: The author can either place the reader in the position of the character, or situate the reader outside of the character looking in. As our discussion of the actor-observer effect suggests, these alterations in perspective should lead to changes in the relative strength of person and situational attributions.

In sum, the evolution of the modern novel parallels historical developments in the difficulty and complexity of understanding human social behavior. Although these parallels do not provide proof that attributional thinking is exercised in the processes of reading novels, the specific features that have come to characterize the modern novel reflect the key ingredients of attributions.

Characteristics of "Good" Fiction

Advice on novel writing and scriptwriting can be found in many sources and amounts to expert opinion on the key ingredients of enjoyable fiction. There is no clear way to establish the expertise of those who offer such advice. However, it appears that there is nearly universal agreement about the importance of three elements of modern fiction writing; character, plot, and point-of-view. Advice related to these elements of fictional stories is consistent with the idea that novels and movies are enjoyable to the extent that they engage the reader in a process of attributional thinking: Characterization and point-of-view define realistic characters whose thoughts, feelings, and intentions are at least partly available to the audience for making dispositional inferences; plots place these characters in various moral and interpersonal dilemmas. The essence of novel reading is predicting how the characters will resolve these dilemmas and revising inferences about them when these predictions fail. In short, the recipe for a good novel is almost identical to the recipe for provoking attributional thinking.

Characterization.   The advice of modern novelists confirms the importance of realistic characters, and often speaks explicitly to the importance of elucidating the motives of story characters. This quote from Ayn Rand (Rand & Boeckmann, 2000) make the point explicitly.

> "Characterization is the presentation of the nature of the people in a story. Characterization is really the presentation of *motives*. We understand a person if we understand what makes him act the way he does." (pp. 59–60)

Forster (1972) makes a point in his explication of flat versus round characters. A quote by literary critic Robert DiYanni (1998, p. 60) directly asserts that characteriztion is the process of leading the reader to draw inferences about characters.

> To make inferences about characters, we look for connections, for links and clues to their function and significance in the story. In analyzing a character or character's relationships, we relate one act, one speech, one physical detail to another until we understand the character.

There is very little psychological research about the specific ingredients of fictional stories that make them enjoyable to audiences. One study of responses to fictional stories examined the relations between the readers' perceptions of the typicality of situations depicted, the realism of the story, and their interest and enjoyment (Shapiro & Chock, 2003). The results suggest that fictional drama depicting typical people in typical situations was regarded as most realistic, and that the more realistic a fictional drama is, the more interesting and enjoyable it is. It is important to note that these patterns did not apply to other genres of fiction. For example, comedies were rated as most enjoyable when they were judged as unrealistic. This pattern is consistent with our hypothesis that fictional drama might serve a uniquely useful purpose related to the interpersonal lives of audience members. Another quote by DiYanni (1998, p. 60) emphasizes the importance of realistic characters:

> We should approach fictional characters with the same concerns with which we approach people. We need to be alert for how we are to take them, for what we are to make of them, and we need to see how they may reflect our own experience. We need to observe their actions, to listen to what they say and how they say it, to notice how they relate to other characters and how other characters respond to them, especially to what they say about each other.

Plot Twists and Near Misses.   If a determinant of good fiction is the ability of a story to provoke causal thinking, it should be possible to identify literary devices that have this effect. Moreover, it should be possible to show that the enjoyment experienced by consumers of fictional drama is related to the use of these devices. Two commonly used literary devices seem to fit the bill—plot twists and near misses.

Plot twists can be operationalized as story events that make it difficult to predict how a main character will behave, or behaviors that violate expectations of a character that the author has fostered in the consumer. Inner conflicts between loyalty and greed, love and power, and so on, place the reader in a position of predicting what the character will choose. Unexpected or surprising behavior prompts the reader to reconcile the characters actions with their understanding of the character's basic dispositions. This quote from Forster supports the combined importance of realistic characters acting in ways that challenge the reader's expectation (Forster, 1974).

> The test of a round character is whether it is capable of surprising in a convincing way. If it never surprises, it is flat. If it does not convince, it is flat pretending to be round (p. 231).

Surprisingly few social psychological studies have explicitly addressed situational factors that lead people to engage in effortful attributional thinking. However, an early study by Pyszczynski and Greenberg (1981) showed that participants engaged in active attempts to understand the causes of a target's behavior when the target behaved in an unexpected manner. Other studies have replicated this finding (Gendolla & Koller, 2001; Hammer & Ruscher, 1997). It is interesting to note that in none of these studies did the participants expect future interactions with the target. Instead, the participants' interest in explaining the unexpected behavior appears to have been more academic than practical. This observation of attributional thinking "for its own sake" provides crucia support for our thesis.

Near misses are story events that create tension or irony by directing the reader's attention to the non-occurrence of an event that could have significantly changed the course of the story. The protypical example is the situation in which the protagonist is almost able to reach a weapon to defend him or herself from a villain, but ultimately is unable to reach it. Suspense and adventure stories are replete with near-miss incidents. Romance stories also contain near-miss incidents. A typical example would be a moment in which one character is about to reveal his or her love for another, but is thwarted by some sort of interruption. The viewer or reader is left to imagine what would have happened had the words been said.

Social psychologists refer to cognitive responses to these near-miss events as counterfactual thinking (Roese, 1997). Research on counterfactual thinking shows that counterfactuals are spontaneously generated in response to unexpected events (Sanna & Turley, 1996). It also appears that counterfactual thinking is affected by the ease with which alternative outcomes could be generated. For example, sportscasters utter more spontaneous counterfactual remarks (e.g., "if only ... ") during close games, and more at the end of games than during lopsided victories or losses. Although researchers who explore the complex details of counterfactual thinking are not exactly sure how counterfactual reasoning and causal attributions are related, it is clear that counterfactual thinking involves thinking about the antecedent causes of events and about the causal relations between events and their consequences (Spellman & Mandel, 1999).

There is evidence that writers and directors are conscious of the entertainment value of counterfactual reasoning. There are dozens of movies that are based almost entirely on a series of near misses, missteps, miscues, and mistakes. The 1985 Martin Scorsese film, *After Hours*, is a striking example. The entire premise of the holiday classic, *It's A Wonderful Life*, is that the lives of everyone in a small town would have been different had the main character, George, never lived. These films appear to be designed to engage the viewer to think about alternative story lines by drawing the audience's attention to plot elements that powerfully evoke counterfactual reasoning.

There is no disputing the frequent use of such devices in both film and literature. The question is how do these devices contribute to the audience's experience of entertainment? One obvious possibility is that these moments create tension or suspense. They increase autonomic arousal and, in doing so, probably heighten the audience's emotional response to the story by creating swings between hope and disappointment, fear and relief. The possibility that other psychological processes are involved is suggested by the fact that counterfactual devices do not always create tension. For every near-miss incident, there are two stories to be told: The one that the author or scriptwriters decide upon, and the one that each audience member creates as they image what would have transpired had the protagonist reached the weapon, or had the shy school boy had enough time to proclaim his love for the homecoming queen.

Women and Fictional Drama

Our final piece of circumstantial evidence for the idea that fiction stimulates attributional thinking has to do with the prominent role played by women as consumers of fictional drama in print, film, and serial television. In a recent discussion of British reading habits, Bloom (2002) notes that throughout the entire twentieth century, women have accounted for three-quarters of all book sales and library loans. A survey conducted for the National Endowment for the Arts found that among U.S. citizens in 2002, 55% of women reported having read a book in the preceding 12 months, compared to just 37% of men. And though women read more than men in general, the predominance of women reader, is particularly impressive among audiences of romance and detective novels. A survey conducted by the Romance Writers of America found that women account for 91% of audiences of romance novels.

Interestingly, there are complementary developmental trends in reading habits. Bloom notes that whereas reading rates among boys and girls are comparable up to age 6 or so, the differential reading rates observed for adult men and women are fully apparent by age 11 or 12. Moreover, the adult pattern of differences between men and women with regard to preference for fictional romance is observed by age 13.

Similar patterns of gender differences can be observed in television viewing and preferences for film and television genres. According to one report, women watch more television than men and are five times more likely to report watching television drama. The audience for daytime serial dramas (i.e., "soap operas") is 91% female. These trends are also seen in film choices. Although men and women are more closely matched in the number of films they watch, genre preferences match those reported for books and television: Women prefer movies that focus on romance and relationships, whereas men prefer action and adventure themes.

So, across media, there is a consistent pattern indicating that women are more interested than men in fictional drama. There have been a few studies that help elucidate the uses and gratification that underlie women's preferences for relationship-focused fiction. In the preceding section we stressed the idea that attributional thinking is provoked by unexpected events. Thus, if the attraction of romantic drama or fiction is that it stimulates attributional thinking, one would expect these genres to be rife with interpersonal strife, conflict, and moral dilemmas. Carveth and Alexander (1985) found exactly that in a content analysis of daytime television dramas. They report that the frequency of these life challenges as reflected in daytime television dramas far exceeded their rates of occurrence in the general population. Consistent with Forster's formula, the soaps represent ordinary (i.e., realistic) characters in extraordinary situations.

When women are asked directly why they watch daytime television dramas, their answers are surprisingly consistent with idea that they are learning something about the social world through viewing these programs. In a study of soaps and audience motivation, Greenberg and Woods (Greenberg & Woods, 1999) reviewed the evidence on this point: Compesi (1980) notes that one of the reasons female subjects cited for watching these programs was social utility/advice; Greenberg, Neuendorf, Burkel-Rothfuss, and Henderson (1982) reports that the women in their study viewed soaps as a source of information to help them deal with their own life problems; Barbow's (1987) participants actually cited the complexity of the characters and the issue of predicting their behavior in their responses to the question of why they watch soap operas. Furthermore, women's style of interacting with both books and television, compared with men's style, is to become more involved with the characters and to see them as real people (Bleich, 1978). All of these findings converge on the idea that for some audience members, daytime television drama is a vehicle for understanding the workings of the social world.

A wealth of research has addressed the prominent role of women in developing and maintaining social networks, especially family networks. Although the debate over whether sex differences are innate or learned is far from settled (Travis, 2003; Wood & Eagly, 2002), research has established that women, in general, are more focused on interpersonal relationships than men (Hoyenga & Hoyenga, 1993). Research in gender roles has shown that women possess more traits related to emotional expression and caring for others than men, whereas men tend to exhibit more traits related to social dominance than women (Deaux, 1985; Eagly & Steffen, 1984; Eagly & Wood, 1991; Spence & Helmreich, 1980). Whether this is a cause of or effect of gender distributions in social roles, women serve different functions in groups than men do. For example, even in task-focused groups, women are more concerned than men with promoting harmonious interactions in the group, whereas men are more focused on establishing dominance through success on the task (Eagly & Karau, 1991; Hyde, 1990; Wood & Rhodes, 1991). Furthermore, women's greater emphasis on interpersonal concerns is related to the finding that women's self-concepts are more likely to be closely tied to their evaluations of the quality of their relationships than are men's (Worell, 1988).

From what we know about the relative emphasis that women place on communal values and the maintenance of family and social networks, it is reasonable to guess that their conspicuous consumption of dramatic fiction might be somehow related to these values and social roles. Although the accuracy of self-reports about uses and gratifications is probably not perfect, female consumers of these media experiences generally endorse the idea that there is some instrumental value to watching daytime television dramas—and we assume that the same sorts of uses and gratifications apply to consumption of romantic fiction in books and film.

A Research Program for Understanding the Role of Attributions in Entertainment

Given the paucity of research on cognitive processes in the experience of mediated entertainment, it is worth considering the sorts of research questions that need to be asked and some possible approaches to answering them. Our theory that reading or viewing fictional stories serves as a kind of cognitive exercise suggests several testable hypotheses. One category of related hypotheses concerns individual differences associated with the consumption of fictional drama. First, if the consumption of fiction is related to the sorts of social cohesiveness values that are typically associated with female gender roles, it should be possible to show that those who consume more fictional drama show higher levels of affiliative concerns. Second, if the consumption of fiction serves to help the consumer better understand and predict the behavior of others, it should be the case that people who consume more dramatic fiction are better at making judgments of the causes of others' behaviors. These tests might utilize paradigms that examine the degree to which attributions are responsive to distinctiveness and consistency information.

A second category of related hypotheses would involve tests of the degree to which specific story features provoke attributional thinking. For example, our argument would predict that the use of omniscient point of view would stimulate more attributional thinking than a narration that is limited to acts and events. Our hypothesis also suggests that stories are enjoyable to the extent that they involve realistic characters in extraordinary situations. This idea could be tested in research in which independent ratings of the verisimilitude of the characters and the presence of relevant and challenging situations were predictive of enjoyment of the story.

Implications

If it turns out that the consumption of fictional drama is in fact related to the exercise of attributional thinking, how might this information be used? One implication of such a discovery would be that literature might become a useful tool for helping people who suffer from psychological impairments related to poor social skills. This approach might be particularly helpful for those whose social skill deficits lay primarily in the area of receptive communication deficits; that is, those who have difficulty understanding the underlying meaning of others' behaviors. For these individuals, assigned reading of carefully chosen novels might serve as a cost-effective adjuvant to traditional individual psychotherapy approaches. The assignment of novels might also be a useful tool for group therapy, with the story characters' actions and motives serving as the basis for learning how to understand the behavior of people in the group members' lives.

Another possible implication of our hypothesis concerns the art of novel writing. At the most general level, understanding the cognitive processes that affect the enjoyment of mediated entertainment should provide guidance for writers who seek to improve the entertainment value of their work. If the foregoing analysis is correct, it appears that writers are already using an implicit or intuitive understanding of the principles of attribution theory. However, a more explicit and technical application of these principles as guides to fiction writing might further enhance the quality of the readers' entertainment experiences. For example, findings related to the actor-observer bias could be used to train new writers how to use point of view to manipulate the sorts of attributions readers will make for the actions of their characters.

Our hypothesis might also have implications for emerging forms of mediated entertainment, such as video games and interactive fiction. These media experiences do not often rely on the sorts of rich characterizations that distinguish fictional drama. If one component of entertainment is the opportunity to exercise attributional thinking, the value of these interactive experiences might be enhanced if performance required understanding motivationally complex characters, or, at least, the creation of motivationally complex characters. If our hypothesis is correct, adding elements of characterization to video games and interactive fiction would improve the overall enjoyment provided by these activities, but might also be a particularly effective way of increasing the participation of women in these activities.

REFERENCES

Anderson, C. A., & Bushman, B. J. (2002). The effects of media violence on society. *Science* (*295*, p. 2377): American Association for the Advancement of Science.

Barbow, A. S. (1987). Student motives for watching soap operas. *Journal of Broadcasting and Electronic Media, 31*, 309–321.

Baumeister, R. F. (1986). *Identity: Cultural change and the struggle for self*. New York: Oxford University Press.

Baumeister, R. F. (1987). How the self became a problem: A psychological review of historical research. *Journal of Personality and Social Psychology, 52*(1), 163.

Bleich, D. (1978). *Subjective criticism*. Baltimore: The Johns Hopkins University Press.

Bloom, C. (2002). *Bestsellers: Popular fiction since 1900*. Houndmills, Basingstoke, Hampshire; New York: Palgrave Macmillan.

Bushman, B. J., & Anderson, C. A. (2001). Media violence and the American public: Scientific facts versus media misinformation. *American Psychologist, 56*(6/7), 477.

Carveth, R., & Alexander, A. (1985). Soap opera viewing motivations and the cultivation and the cultivation process. *Journal of Broadcasting and Electronic Media, 29,* 259–273.

Carver, C., DeGregorio, E., & Gillis, R. (1980). Field-study evidence of an ego-defensive bias in attribution among two categories of observers. *Personality & Social Psychology Bulletin, 6*(1), 44.

Chaiken, S., & Trope, Y. (1999). *Dual-process theories in social psychology.* New York: Guilford.

Compesi, R. J. (1980). Gratifications of daytime TV serial viewers. *Journalism Quarterly, 57,* 155–158.

David, P., Liu, K., & Myser, M. (2004). Methodological artifact or persistent bias? Testing the robustness of the third-person and reverse third-person effects for alcohol messages. *Communication Research, 31*(2), 206.

DiYanni, R. (1998). *Literature: Reading Fiction, Poetry, Drama, and the Essay.* Fourth Edition. Boston: McGraw Hill, p. 60.

Deaux, K. (1985). Sex and gender. *Annual Review of Psychology, 36,* 49–81.

Dodge, K., & Newman, J. (1981). Biased decision-making processes in aggressive boys. *Journal of Abnormal Psychology, 90*(4), 375.

Eagly, A. H., & Karau, S. J. (1991). Gender and the emergence of leaders: A meta-analysis. *Journal of Personality and Social Psychology, 60,* 685–710.

Eagly, A. H., & Steffen, V. J. (1984). Gender stereotypes stem from the distribution of women and men into social roles. *Journal of Personality and Social Psychology, 46,* 735–754.

Eagly, A. H., & Wood, W. (1991). Explaining sex differences in social behavior: A meta-analytic perspective. *Personality and Social Psychology Bulletin, 17,* 306–315.

Fazio, R. H., Zanna, M. P., & Cooper, J. (1977). Dissonance and self-perception: An integrative view of each theory's proper domain of application. *Journal of Experimental Social Psychology, 13*(5), 464.

Festinger, L., & Carlsmith, J. M. (1959). Cognitive consequences of forced compliance. *Journal of Abnormal & Social Psychology, 58,* 203.

Forster, E. M. (1936). *Abinger harvest.* Harcourt, Brace & Co.: Orlando, Florida, pp. 141–148.

Forster, E. M. (1967). Flat and round characters. In P. Stevick (Ed.), *The Theory of the Novel* (pp. 223–231). New York: Free Press.

Forster, E. M. (1974). *Aspects of the novel and related writings* (Vol. 12). London: Edward Arnold.

Gendolla, G. H. E., & Koller, M. (2001). Surprise and motivation of causal search: How are they affected by outcome valence and importance? *Motivation And Emotion, 25*(4), 327–349.

Gilbert, D. T., & Malone, P. S. (1995). The correspondence bias. *Psychological Bulletin, 117*(1), 21–38.

Gilbert, D. T., Pelham, B. W., & Krull, D. S. (1988). On cognitive busyness: When person perceivers meet persons perceived. *Journal of Personality and Social Psychology, 54,* 73–740.

Goerke, M., Möller, J., Schulz-Hardt, S., Napiersky, U., & Frey, D. (2004). "It's not my fault, but only I can change it": Counterfactual and prefactual thoughts of managers. *Journal of Applied Psychology, 89*(2), 279.

Green, M. C., Brock, T. C., & Kaufman, G. F. (2004a). Understanding media enjoyment: The role of transportation into narrative worlds. *Communication Theory, 14,* 311–327.

Green, M. C., Garst, J., & Brock, T. C. (2004b). The power of fiction: Determinants and boundaries. In L. J. Shrum (Ed.), *The psychology of entertainment media* (pp. 161–176). Mahwah, NJ: Lawrence Erlbaum Associates.

Greenberg, B., & Woods, M. (1999). The soaps: Their sex, gratifications, and outcomes. *Journal of Sex Research, 36*(3), 250–257.

Greenberg, B. S., Neuendorf, K., Buerkel-Rothfuss, N., & Henderson, L. (1982). The soaps: What's on and who cares? *Journal of Broadcasting, 26,* 519–535.

Greenberg, J., Pyszczynski, T., & Solomon, S. (1982). The self-serving attributional bias: Beyond self-presentation. *Journal of Experimental Social Psychology, 18*(1), 56.

Hafer, C. L., & Begue, L. (2005). Experimental research on just-world theory: Problems, developments, and future challenges. *Psychological Bulletin, 131,* 128–167.

Hammer, E. D., & Ruscher, J. B. (1997). Conversing dyads explain the unexpected: Narrative and situational explanations for unexpected outcomes. *British Journal Of Social Psychology, 36,* 347–359.

Heider, F. (1958). *The psychology of interpersonal relations.* New York, NY: Wiley.

Henning, B., & Vorderer, P. (2001). Psychological escapism: Predicting the amount of television viewing by need for cognition. *Journal of Communication, 51*(1), 100.

Hodgins, M., & Fuller, R. (2001). "Come for the weight loss, stay for the enjoyment"—Exploring attributions for initiating and maintaining exercise. *Irish Journal of Psychology, 22*(2), 38–50.

Hoffner, C., Plotkin, R., Buchanan, M., Anderson, J., Kamigaki, S., Hubbs, L. (2001). The third-person effect in perceptions of the influence of television violence. *Journal of Communication, 51*(2), 283–299.

Hoyenga, K. B., & Hoyenga, K. T. (1993). *Gender-related differences: Origins and outcomes.* Boston, MA: Allyn and Bacon.

Hyde, J. S. (1990). Meta-analysis and the psychology of gender differences. *Signs: Journal of Women in Culture and Society, 16*, 55–73.

Inc, W. A. E. G. (2004). *Average U.S. television viewing time, October 2002* (No. 00841382): World Almanac Education Group Inc.

Jones, E. E., & Davis, K. E. (1965). A theory of correspondent inferences: From acts to dispositions. In L. Berkowitz (Ed.), *Advances in Experimental Social Psychology (Vol. 2)*. New York: Academic Press.

Jones, E. E., & Nisbett, R. E. (1972). The actor and the observer: Divergent perceptions of the causes of behavior. In E. E. Jones, D. E. Kanouse, H. H. Kelley, R. E. Nisbett, S. Valins & B. Weiner (Eds.), *Attribution: Perceiving the causes of behavior* (pp. 79–94). Morristown, NJ: General Learning Press.

Kaye, B., & Johnson, T. (2002). Online and in the know: Uses and gratifications of the web for political information. *Journal of Broadcasting & Electronic Media, 46*(1), 54–71.

Knowles, E. D., Morris, M. W., Chiu, C., & Hong, Y. (2001). Culture and the process of person perception: Evidence for automaticity among East Asians in correcting for situational influences on behavior. *Personality and Social Psychology Bulletin, 27*, 1344–1356.

Kubey, R., Csikszentmihalyi, M., R, K., & M, C. (2004). Television addiction is no mere metaphor. *Scientific American 14*, pp. 48.

Lee, F., Hallahan, M., & Herzog, T. (1996). Explaining real-life events: How culture and domain shape attributions. *Personality and Social Psychology Bulletin, 22*, 732–741.

McFarland, C., & Ross, M. (1982). The impact of causal attributions on affective reactions to success and failure. *Journal of Personality and Social Psychology, 43*, 937–946.

Miller, D. T., & Ross, M. (1975). Self-serving biases in attribution of causality: Fact or fiction? *Psychological Bulletin, 82*, 213–225.

Miller, J. G. (1984). Culture and the development of everyday social explanation. *Journal of Personality and Social Psychology, 46*, 961–978.

Morris, M. W., & Peng, K. (1994). Culture and cause: American and Chinese attributions for social and physical events. *Journal of Personality and Social Psychology* (67), 949–971.

Moskalenko, S., & Heine, S. J. (2003). Watching your troubles away: Television viewing as a stimulus for subjective self-awareness. *Personality & Social Psychology Bulletin, 29*(1), 76.

Nasby, W., Hayden, B., & DePaulo, B. (1980). Attributional bias among aggressive boys to interpret unambiguous social stimuli as displays of hostility. *Journal of Abnormal Psychology, 89*(3), 459.

Pyszczynski, T., & Greenberg, J. (1981). role of disconfirmed expectancies in the instigation of attributional processing. *Journal of Personality & Social Psychology, 40*(1), 31–38.

Rand, A., & Boeckmann, T. (2000). *The art of fiction: A guide for writers and readers*. New York: Plume.

Raney, A. A. (2003). Disposition-based theories of enjoyment. In J. Bryant, D. R. Roskos-Ewoldsen, & J. Cantor (Eds.), *Communication and emotion: Essays in honor of Dolf Zillmann* (pp. 61–84). Mahwah, NJ: Lawrence Erlbaum Associates.

Raney, A. A. (2004). Expanding disposition theory: Reconsidering character liking, moral evaluations, and enjoyment. *Communication Theory, 14*, 348–369.

Raney, A. A., & Bryant, J. (2002). Moral judgment and crime drama: An integrated theory of enjoyment. *Journal of Communication, 52*, 402–415.

Roese, N. J. (1997). Counterfactual thinking. *Psychological Bulletin, 121*(1), 133.

Ross, L. (1977). The intuitive psychologist and his shortcommings: Distortions in the attribution process. *Advances in Experimental Social Psychology (Vol. 10)*. New York: Academic Press.

Ross, L., & Nisbett, R. E. (1991). *The person and the situation: Perspectives of social psychology*. New York: McGraw-Hill.

Rubin, A. M. (1984). Ritualized and instrumental television viewing. *Journal of Communication, 34*(3), 67.

Sanna, L., & Turley, K. (1996). Antecedents to spontaneous counterfactual thinking: Effects of expectancy violation and outcome valence. *Personality & Social Psychology Bulletin, 22*(9), 906–919.

Scharrer, E. (2002). Third-person perception and television violence, *Communication Research* (Vol. 29, pp. 681): Sage Publications Inc.

Shapiro, M. A., & Chock, T. M. (2003). Psychological processes in perceiving reality. *Media Psychology, 5*(2), 163.

Sharp, E., & Joslyn, M. (2001). Individual and contextual effects on attributions about pornography. *Journal of Politics, 63*(2), 501–519.

Smith, E. R., & DeCoster, J. (2000). Dual process models in social and cognitive psychology: Conceptual integration and links to underlying memory systems. *Personality and Social Psychology Review, 4*, 108–131.

Spellman, B. A., & Mandel, D. R. (1999). When possibility informs reality: Counterfactual thinking as a cue to causality. *Current Directions in Psychological Science, 8*(4), 120.

Spence, J. T., & Helmreich, R. L. (1980). Masculine instrumentality and feminine expressiveness: Their relationships with sex role attitudes and behaviors. *Psychology of Women Quarterly, 5*, 147–163.

Strack, F., & Deutsch, R. (2004). Reflective and impulsive determinants of social behavior. *Personality and Social Psychology Review, 8,* 220–247.

Thompson, S. C., Armstrong, W., & Thomas, C. (1998). Illusions of control, underestimations, and accuracy: A control heuristic explanation. *Psychological Bulletin, 123*(2), 143.

Travis, C. B. (2003). Talking evolution and selling difference. In C. B. Travis (Ed.), *Evolution, gender, and rape.* Cambridge, MA: MIT Press.

Weiner, B., Russell, D., & Lerman, D. (1978). Affective consequences of causal ascriptions. In J. H. Harvey, W. J. Ickes & R. F. Kidd (Eds.), *New directions in attribution research* (Vol. 2). Hillsdale, NJ: Lawrence Erlbaum Associates.

Weinstein, N. D. (1980). Unrealistic optimism about future life events. *Journal of Personality & Social Psychology, 39*(5), 806.

Weinstein, N. D. (1987). Unrealistic optimism about susceptibility to health problems: Conclusions from a community-wide sample. *Journal of Behavioral Medicine, 10*(5), 481.

Wood, W., & Eagly, A. H. (2002). A cross-cultural analysis of the behavior of women and men: Implications for the origins of sex differences. *Psychological Bulletin, 128,* 699–727.

Wood, W., & Rhodes, N. (1991). Sex differences in interaction style in task groups. In C. L. Ridgeway (Ed.), *Gender, interaction, and equality.* New York, NY: Springer-Verlag.

Worell, J. (1988). Women's satisfaction in close relationships. *Clinical Psychology Review, 8,* 477–498.

Zillmann, D. (1991). Television viewing and physiological arousal. In J. Bryant & D. Zillmann (Eds.), *Responding to the screen: Reception and reaction processes* (pp. 103–133). Hillsdale, NJ: Lawrence Erlbaum Associates.

Zillmann, D., & Bryant, J. (1985). Affect, mood, and emotion as determinants of selective exposure. In D. Zillmann & J. Bryant (Eds.), *Selective exposure to communication* (pp. 157–190). Hillsdale, NJ: Lawrence Erlbaum Associates.

9

The Psychology
of Disposition-Based
Theories of Media Enjoyment

Arthur A. Raney
Florida State University

As the other chapters in this volume attest, the enjoyment of media entertainment is indeed a complex phenomenon. Seeking to understand "why we like what we like" is a goal shared by social scientists and media content providers alike. In this quest, one thing has become abundantly clear: Enjoyment—perhaps not in all, but surely in many ways—is an individual-level phenomenon, with personality traits and subjective evaluations playing key roles. One leading explanation of the media-enjoyment process centers on how individuals evaluate and form affiliations with media characters and how enjoyment is impacted by what happens with and to those characters. Collectively these explanations have been referred to in several ways: disposition theory, disposition theories, affective disposition theory, and disposition-based theories.

The first such theory was developed by Zillmann and Cantor (1972) to describe how people appreciate jokes involving the disparagement of a person or group. The principals of the so-called disposition theory of humor were later applied to the appreciation of drama and sports, yielding the disposition theories of drama (Zillmann & Cantor, 1976) and sports spectatorship (Zillmann, Bryant, & Sapolsky, 1989) respectively. Since then, entertainment scholars have applied the key concepts of the theories to examine fright-inducing entertainment (Hoffner & Cantor, 1991a; Oliver, 1993), action films (King, 2000), reality-based programming (Oliver, 1996), crime-based fiction (Raney & Bryant, 2002), and news programming (Zillmann, Taylor, & Lewis, 1998). Differences between these media contents dictate subtle differences in the application of the theories, rendering efforts to develop a general disposition theory of media content problematic. However, the process by which enjoyment is derived through dispositional affiliations and subsequent anticipatory emotions with media characters is quite similar regardless of the media content.

With this in mind, the purpose of this chapter is to explore the psychological components of media enjoyment as predicted by the various disposition theories. To accomplish this goal, a summary of each disposition-based theory of enjoyment is offered first. We will then explore the commonalities between these theories, with an emphasis on the psychological processes of disposition formation, disposition maintenance, empathic reactivity, and enjoyment. Suggestions for future research and theory development will be integrated into this discussion.

DISPOSITION THEORY OF HUMOR

Until the early 1970s, our understanding of humor appreciation relied heavily upon superiority theories (cf. Hobbes, 1976); and as a result, early works focused almost entirely upon humorous situations containing the debasement or disparagement of another. Several researchers attempted to explain how and why people find such situations humorous (e.g., Priest, 1966; La Fave, 1972; Wolff, Smith, & Murray, 1934). In short, the literature concluded that we can appropriately find humor in situations containing a disparagement as long as the slighted party belongs to or can be identified with groups other than those to whom we are affiliated. That is, we can laugh when a joke makes fun of the New York Yankees as long as we are not a Yankee fan. However, simple dichotomous conceptualizations of group membership proved troublesome. For instance, early humor studies only considered whether the person hearing (or reading) the joke liked the group or person who served as the butt of the joke. Such a perspective failed to take into account potential *dislike* for parties involved in the humorous situation. Using the previous example, a joke that pokes fun at the New York Yankees might not be enjoyed by a Yankee fan, but might be exceedingly enjoyed by a Boston Red Sox supporter.

To address this limitation, Zillmann and Cantor (1976) proposed that when we witness a joke containing a disparagement, we form affective dispositions toward the joke's characters on a continuum of affect ranging from extreme negative through a neutral point of indifference to extreme positive. This proposed continuum of affect embraces the potential for the witness to hold either a positive or negative affiliation with the disparaged party or group, as well as acknowledges varying degrees (as opposed to mere absence or presence) of affect that might be felt toward the offended and/or offending party. As the reader will note, the "continuum of affective dispositions" is a key element in all disposition-based theories of media enjoyment.

A second limitation to early group-membership approaches to humor involved the implied cognitive effort needed to comprehend and enjoy a disparaging joke. Using the Yankee-joke example, a witness must presumably identify the character in the joke as a Yankee player or fan and then mentally compare that group membership with his/her own allegiances (in specific relation to the Yankees). Zillmann and Cantor (1972) noted that such a process seems unlikely. The scholars argued that witnesses tend to respond quite quickly to (or experience "gut reactions" toward) characters in a joke, that they do not require the cognitively taxing job described above. Instead, the researchers contended that when we encounter a humorous situation, we first identify the roles and activities of the characters and then (1) react with empathy toward characters whose roles and activities we associate with positive experiences, (2) react with counterempathy toward characters whose roles and activities we associate with negative experiences, or (3) both. So, empathy governs how we react to characters in a humorous situation containing disparagement: Empathic reactions lead us to align with those who are more experientially close to us, while counterempathic reactions lead us to align against those who are more experientially distant.

Finally, Zillmann and Cantor also argued that our responses to humorous situations vary across time because our emotional state plays a significant role in those responses. This is a departure from previous humor theories that operationalized affiliations as manifestations of personality and thereby presumably stable and consistent. By suggesting that affective dispositions are more contingent upon emotional states than personality traits, the researchers recognized and accommodated the dynamics of an individual's affect.

Similarly, by accommodating affective dynamics, the researchers help explain how we can find humor in a joke that offers little or no description of its characters, like "Two women walk into a bar." In such jokes, of course, one party offends or provokes the other, with the latter regaining the upper hand in a novel and humorous fashion. According to previous theories, our humorous response to this example would depend upon our ability to classify and evaluate the two women based on cues in the joke work (cf. Freud, 1960). The problem is that the joke provides no such cues; as a result, we have no way of determining which woman to align with and which to align against. However, Zillmann and Cantor argued that we can find the joke humorous because we can react *in situ* with negative emotions toward the character who provokes and with mirthful emotions toward the retaliator (presumably because of justice considerations). The disparaging actor gets her comeuppance, and we get a nice laugh.

To summarize the new direction of their thinking, Zillmann and Cantor offered a *disposition theory of humor* that characterized a humorous response to a disparaging joke as a function of affective dispositions situationally formed toward the characters in the joke. When a humorous situation or joke containing disparagement is encountered, the witness emotionally reacts to the characters along a continuum of affect from extremely positive through indifference to extremely negative. The strength and valence of those dispositions are governed by empathic reactions to the characters. Dispositions are more positive for characters with whom the witness shares relevant, positive experiences and more negative for those with whom they share relevant, negative experiences. Humor appreciation will be higher the more negative the disposition held toward the disparaged character, and/or the more positive the disposition held toward the character responsible for the disparagement. Humor appreciation will be lower the more positive the disposition held toward the disparaged character, and/or the more negative the disposition held toward the character responsible for the disparagement. Ultimately, Zillmann and Cantor (1976; p. 101) argued, "Appreciation should be maximal when our friends humiliate our enemies, and minimal when our enemies manage to get the upper hand over our friends."

While a large number of studies have found support for the disposition theory of humor, we will limit our discussion to two seminal pieces. As mentioned, Zillmann and Cantor (1972) lay the groundwork for the formal development of the theory. In the study, the researchers sought to initially question de-facto reference group classifications as a predictor of humorous responses to jokes containing a disparagement. The authors contended that dispositions formed toward characters are contingent upon perceived experiential similarities (that will be both transitory and dynamic) rather than readily-identifiable social groups. Jokes involving superiors and subordinates (i.e., fathers-sons, professors-students, and employers-employees) were used because such a distinction fell outside the social-group classifications previously identified. Because of the existing power dynamics and subsequent feelings of resentment between the two, it was hypothesized that individuals who are more experientially close to the group getting the upper hand in the joke will enjoy the joke more (with the converse also holding true).

University students and working professionals were given the experimental materials. The students were expected to be more experientially close to the subordinates in the jokes, to better like those characters, and to respond with more humor to jokes in which the subordinate gets the upper hand. Similarly, professionals were expected to be more experientially similar to the superiors, to better like those characters, and to respond with more humor to jokes in which the superior gets the upper hand. The findings from the study support these hypotheses, and, in turn, the supremacy of affective dispositions to social-group classifications as predictors of humor appreciation.

Zillmann, Bryant, and Cantor (1974) lent further support to disposition theory. Political cartoons featuring physical harm on either incumbent President Richard Nixon or his Democratic opponent for the 1972 presidency, George McGovern, were evaluated for humor by college students. Previous research (e.g., Priest, 1966) suggested that mere preference for one candidate over the other was enough to predict one's humor response to such political cartoons. However, the researchers, in an attempt to validate the importance of affective dispositions, considered candidate preference alone as an insufficient criterion of affiliation. The researchers gathered candidate appreciation data as well (i.e., to what extent do you agree with the policies of the candidates?"). Those participants who preferred one candidate but did not particularly like him were discarded from the initial analysis. As hypothesized, the cartoon assaults on the disliked candidates elicited significantly higher levels of humor appreciation than the same assaults on the favored candidates. However, as reported in Zillmann and Cantor (1976), a reexamination of the discarded data indicated that simple candidate preference—without the accompanying affective disposition as measured by the appreciation measures—was insufficient in replicating the findings. Therefore, the disposition theory of humor's reliance on affective dispositions as a predictor of humor appreciation has been met with much support in the humor literature. Suffice it to say at this point, the introduction of the disposition theory of humor changed the direction of humor research, as well as the study of all entertainment appreciation.

DISPOSITION THEORY OF DRAMA

The principals of the disposition theory of humor were subsequently applied to the appreciation of drama (Zillmann & Cantor, 1976). The resulting disposition theory of drama posits that enjoyment of media content is a function of a viewer's affective disposition toward characters and the outcomes experienced by those characters in the unfolding narrative. Simply stated, it predicts that enjoyment will increase when liked characters experience positive outcomes and/or when disliked characters experience negative ones. Conversely, enjoyment will suffer when liked characters experience negative outcomes and/or disliked characters experience positive ones.

As with the disposition theory of humor, the feelings that viewers hold toward characters are of supreme importance to enjoyment. Consistent with the discussion above, the disposition theory of drama contends that viewers form alliances with characters on a continuum of affect from extremely positive through indifference to extremely negative. Thus, as drama viewers, we like and cheer for certain characters, while despising and rooting against others. However, our human nature requires that our selection of favored and unfavored characters not be capricious; our emotional side-taking must be morally justified. With the disposition theory of humor, the presence of joke work presumably offers the requisite moral amnesty to quickly form affective dispositions; that is, the joke work offers us an excuse to violate existing social sanctions against finding humor in the misfortune of others (Zillmann & Cantor, 1976). No such equivalent is present in drama to give us the guilt-free opportunity to enjoy the misfortune—and benefaction for that matter—of others.

Instead, viewers must be vigilant in monitoring the morality of characters, continually rendering verdicts about the rightness or wrongness of character's actions (Zillmann, 2000). For instance, if a guilty party is punished in a drama, then the situation can be enjoyed without guilt because it is morally appropriate to punish guilty parties (Zillmann & Bryant, 1975). Through moral scrutiny of the actions and motivations of dramatic characters, viewers are able to justify their emotional side-taking. Specifically, we form more positive dispositions with characters whose actions and motivations we judge to be proper or morally correct, while we form more negative dispositions toward characters whose actions and motivations we judge as improper or morally incorrect. The strength of these affective dispositions, again, ranges on a continuum of affect, and because of our constant moral monitoring, are subject to change as the dramatic accounts progress. This intertwining of affective dispositions and moral judgment permits and governs our emotional involvement in the drama.

This is possible because once characters are liked, we are able to empathize with their plights and hope for their triumph. Conversely, once characters are hated, we are unable to empathize with them and are free to wish for their downfall In fact, it is logical to assume that the stronger the positive feelings, the stronger our empathic reaction. Likewise, the stronger the negative affect, the stronger the negative or counterempathic reaction. Ultimately, enjoyment increases in proportion to our dispositions as the outcomes we wish for are portrayed. Enjoyment suffers in proportion to the dispositions held if the outcomes we wish for are not portrayed. The key is the disposition: The lack of a positive or negative feeling toward the character (i.e., indifference) does not trigger an emotional response to the drama. No emotion, no enjoyment.

As one might then imagine, a key factor in determining emotional reactions to characters is empathy, which has been repeatedly identified as a key mechanism for enjoyment of drama (Hoffmann, 1987; Zillmann, 1991, 1994, 2000). Researchers consistently report that viewers with varying individual levels of empathy differ in their reactions to media characters. For example, Raney (2002) demonstrated that persons with higher levels of empathy were more likely to sympathize with the victim of a media crime and were, in turn, more likely to enjoy a presentation where the crime is avenged. But empathy is of interest only *after* dispositions are formed. As Zillmann noted, "affective dispositions toward persons or their personas virtually control empathy... Empathy seems to be governed by such morally derived dispositions" (1994, pp. 44–45).

Others have attempted to analyze the process of moral judgment that presumably leads to dispositions in the first place. It has been noted that the strength (and perhaps even valence) of dispositions will vary between viewers because of an individual's unique moral composition. Most developmental perspectives assume that people vary in the manner and sophistication with which they approach moral reasoning (cf. Kohlberg, 1981; Rest, 1979). Furthermore, moral reasoning is thought to be governed and influenced by a complex constellation of factors. As a result, people think about and respond to characters differently, with differing results. Therefore, because dispositions are based on the moral evaluation of a character's actions and motives, and because we tend to differ in the way that we make moral evaluations, dispositions toward characters should vary between individual viewers. In turn, viewers should not be surprised when they like certain *characters* (and therefore different dramatic presentations) more (or less) and for different reasons than their friends and vice versa. It then follows that—because enjoyment is dependent upon character liking—viewers also enjoy certain dramatic *programs* more (or less) than their friends. In an attempt to further examine how exactly disposition are formed toward media characters, a few scholars have sought to identify psychological factors that govern crucial moral judgments.

Oliver (1996) found that authoritarianism was associated with greater liking of reality-based crime dramas; others (Raney, 2005; Raney & Bryant, 2002) isolated attitudes about vigilantism and punitive punishment as predictors of crime-based drama enjoyment. In fact, Raney and Bryant (2002) offered a model to describe the process of disposition formation in both affective and cognitive terms. The researchers found that both affective and cognitive variables were predictive of moral judgments within crime-punishment entertainment, with the outcome of those moral judgments (termed deservedness and victim sympathy respectively) predicting overall enjoyment.

To summarize, according to the disposition theory of drama, affiliations are developed as a viewer monitors the actions and motivations of characters, continually evaluating the moral appropriateness of those actions and motivations through their subjectively held moral lens.

In response, viewers either like or dislike characters (to varying degrees). The valence and intensity of those affective dispositions lead the viewers to develop anticipations about the unfolding outcomes associated with those characters. For liked characters, success is hoped and failure is feared. For hated characters, failure is hoped and success is feared. Enjoyment is a product of those anticipations in relation to the actual outcomes portrayed. Thus, highly liked characters generate intensely hoped-for outcomes that when met result in relief, pleasure, and enjoyment. However, when these hoped-for outcomes are not observed or when feared-for outcomes are observed, enjoyment suffers. This is the basic disposition theory of drama formula.

Support for the disposition theory of drama is abundant (Hoffner & Cantor, 1991a; Oliver, 1993, 1996; Raney, in press, 2002; Raney & Bryant, 2002; Zillmann & Cantor, 1977). In one of the earliest studies, Zillmann and Bryant (1975) established the role of moral judgment in the formation of affective dispositions in drama. In the study, children at varying levels of moral development viewed one of three versions of a fairy tale in which a good king had an opportunity to punish a rival king who had planned to banish him. The three versions differed on the severity of punishment handed out by the good king: under punishment (the rival king was forgiven), equitable punishment (the rival king received the banishment planned for the good king), or excessive punishment (the rival king received a public beaten and life imprisonment). Children at later stages of moral development were expected to freely enjoy the equitable-retribution version, but not the others that violated sanctions on fairness. In contrast, children at an earlier stage of development were expected to be unable to make such moral judgment-based distinctions and thus enjoy the condition in which the rival king was punished most severely. These predictions were supported in full. As a result, the important role of moral judgment in drama appreciation was first identified.

More recently, Raney (2002) further examined the role of moral judgment in the formation and maintenance of dispositions toward characters in drama. The author predicted that different crimes would elicit different moral judgments about those crimes, the punishments for those crimes, and the individuals committing those crimes, all of which would then impact enjoyment. Participants viewed one of two clips from a crime-punishment movie and then rated their enjoyment of the clip. In one version of the video, the crimes portrayed were a rape, physical abuses, and property damage; in the second version, the crimes were non-sexual physical abuses and property damage. All other aspects of the two versions were identical, including the punishment of the perpetrator.

As predicted, the fictional crime generated at least two moral judgments: sympathy toward the victim and an evaluation of fairness (or deservedness) about the punishment delivered. Furthermore, in the no-rape condition, the results of those two moral judgments predicted overall enjoyment of the film clip, with viewers who more sympathized with the victim of the crime (which was ultimately avenged) reporting higher levels of enjoyment. Furthermore, viewers who thought the perpetrator deserved a less severe punishment reported higher levels of enjoyment. Additionally, the moral judgment of sympathy toward the victim was predicted by a viewer's level of empathy and social-justice attitudes (as discussed above).

In the rape condition, however, only the moral judgment of sympathy toward the victim predicted enjoyment; the viewer's level of empathy predicted victim sympathy. The moral judgment of punishment deservedness was neither predicted by the various personality factors nor did it predict enjoyment. The author noted that the heinous nature of the rape apparently eliminated all variance on the moral judgment of deservedness; in other words, all viewers found the crime repulsive on a visceral level and the criminal equally deserving of his fate. As a result, differences between viewers on the more cognitive factors (i.e., attitudes toward vigilantism and punitive punishment) were rendered inconsequential. These findings offered additional support to disposition theory in general and to several of the key components of the theory including the role of moral judgment in dispositional formation and enjoyment.

DISPOSITION THEORY OF SPORTS SPECTATORSHIP

The disposition theory of sports spectatorship (for a comprehensive summary see Raney, 2003; see also Zillmann, Bryant, & Sapolsky, 1989; Zillmann & Paulus, 1993) was developed by applying the disposition-based principals to sporting events. The theory holds that fanship allegiance with a team or player forms along the now-familiar continuum of affect from intense liking through indifference to intense disliking. Enjoyment from viewing a sports event is a function of the outcome of the event in relation to the strength and valence of the dispositions held toward its competitors. Specifically, enjoyment is thought to increase the more the viewer likes the winning team and/or dislikes the losing team. Conversely, enjoyment should decrease the more the winning team is disliked by the viewer and/or the more the losing team is favored. Some researchers have therefore suggested that maximum enjoyment from sports viewing should be experienced when an intensely liked team defeats an intensely disliked team. Conversely, maximum disappointment or "negative enjoyment" should be experienced when a loved team is defeated by a hated one (Zillmann & Paulus, 1993).

Several studies have found support for the disposition theory of sports spectatorship across a variety of contests. Zillmann, Bryant, and Sapolsky (1989) detail several of these studies, two of which are noted below. In the first, the researchers examined how dispositions held toward two National Football League (NFL) teams might impact enjoyment of viewing a televised game between the two teams. Team fanship was measured for both teams resulting in each participant being categorized as having a positive, neutral, or negative disposition toward each team. Participants then viewed a live-broadcasted contest between the two teams. The researchers measured enjoyment of every play in the game, as well as the overall contest. It was predicted that those with a positive disposition toward the winning team would enjoy the game more than those who had a negative disposition toward the team and more than those who held a positive disposition toward the losing team. In fact, it was predicted that enjoyment of the contest would be the greatest for those viewers who really liked the winning team and really disliked the losing team. It was also predicted that those viewers who liked the losing team and disliked the winning team would enjoy the game the least. All of these expectations were observed in the study.

A second study—involving an Olympic basketball game—reported by Zillmann and his colleagues found additional support for the disposition theory of sports spectatorship. College students viewed a portion of the 1976 men's gold medal game between the United States and Yugoslavia. Respondents once again rated their enjoyment of each play in the game. As expected, the American students reported more enjoyment of plays in which the U.S. scored and less enjoyment when Yugoslavia scored. The researchers found additional support for the theory by isolating responses for plays that involved U.S. players who has previously played at the university where the research was conducted. Respondents reported the higher enjoyment of scoring plays involving those two players than for any other U.S. players. The researchers concluded, although positive dispositions were held for the U.S. team in general, that the bonds held with the two former students were the strongest among players on the U.S. squad; as a result, enjoyment was highest on plays in which they scored. Additional support for the disposition theory of sports spectatorship has been found in studies of professional tennis (Zillmann, Bryant, & Sapolsky, 1989) and high-school basketball competition between an all-White and an all-Black team (Sapolsky, 1980).

PSYCHOLOGICAL FACTORS ASSOCIATED WITH DISPOSITION-BASED THEORIES

After reviewing the various expressions of disposition theory, we find several principles and features that are common across them all. In the remaining pages, six such principles and related psychological factors and issues will be examined.

1. Disposition-Based Theories Are Concerned With the Enjoyment or Appreciation of Media Content.   While the theories cannot ultimately predict whether an individual will like or dislike a specific character or story, they can serve as useful aides in understanding the *process* through which people enjoy these things.

As noted above, what exactly enjoyment *is* has yet to be fully determined. The earliest disposition literature utilizes the term *appreciation*—not *enjoyment*—when referring to one's hedonic response to media content. Only in the last decade or so has the term *enjoyment* started being used. Perhaps we need to determine if these two terms are indeed synonymous. At a minimum, we need to better understand what enjoyment really is. One definition—the pleasure experienced from consuming media entertainment (Raney, 2003)—does little to help our understanding of what leads to this pleasure, how and why it differs between individuals, and how it differs between media content. People use the term *enjoyed* to refer to their experiences with listening to classical music, viewing horror movies, and playing the latest video game. The experiences surely differ from one another in many ways; only a few researchers have started exploring these differences in relation to enjoyment (Carpentier, Yu, Butner, Chen, Hong, Park, & Bryant, 2001; Oliver & Raney, 2005).

Vorderer, Klimmt, and Ritterfeld (2004) most recently discussed the complexity of enjoyment, noting its central role in media entertainment experience. The researchers offered a conceptual model that embraces this complexity and identifies several requisite psychological (e.g., suspension of disbelief, empathy, interest) and motive (e.g., mood management, escapism) conditions necessary for the experience. The researchers acknowledged the predictive power of disposition theories in explaining a piece of the enjoyment puzzle (in specific situations as described above), but also pointed to the limitations of any one (extant) theory to explain, describe, or predict all instances of media entertainment. This author completely agrees with such an assertion.

Additional research is desperately needed to examine how the principles and phenomena described by disposition theories interact with and complement the psychological and motive conditions of the enjoyment experience described by Vorderer and his colleagues. For example, uses and gratifications researchers have for years contended that people regularly turn to media content in an attempt to escape the stress that they encounter in their daily lives (e.g., Herzog, 1940; Katz & Foulkes, 1962; McQuail, Blumler, & Brown, 1972). In fact, we find the so-called escapism motivation expressed in relation to many media contents (cf. Henning & Vorderer, 2001). To date, however, no work has attempted to investigate how the principles of disposition theory might influence escapism. How do affective dispositions formed toward characters help with this escape? Does exercising moral judgment in a passive environment constitute a part of escapism? Or does vigorous moral reasoning about characters actually detract from escapism? Similar questions should be asked about the relationship between affective disposition and parasocial interaction, presence, suspension of disbelief and mood, and how all these relationships influence or create enjoyment in various media contexts. Exploring these differences should prove to be extremely worthwhile; disposition-based theories will certainly play a large role in those explorations.

2. Disposition-Based Theories Are Concerned With Emotional Responses to Media Content. Affect is at the heart of disposition theories, with empathy identified as the chief mechanism guiding emotional responses to media characters and their plights. More specifically, individual differences in empathy influence the extent to and manner in which viewers respond to media characters. Additionally, researchers also acknowledge that cognition—in particular, moral judgment—plays a tremendous role in the disposition-formation process (especially with drama). In fact, as will be noted below, certain cognitive structures associated with moral judgment may precede and coexist with affective responses (Raney, 2004).

One word of caution is offered: Although the affect-cognition distinction can be easily discussed in theory, because of the interdependence between the two, such hard distinctions are of course difficult (if not impossible) to measure in reality. But, let it not be misunderstood, the greatest amount of variance explained in the enjoyment processes (according to disposition theories) is that explained by a viewer's emotional reactions to characters and to the outcomes experienced by those characters. While communication scientists continue to investigate the role of other psychological functions and processes involved in the entertainment experience, the primacy of emotions remains.

3. Disposition-Based Theories Contend That Media Enjoyment Starts With and Is Driven by the Viewer's Feelings About Characters. As discussed prior, a great deal of the disposition research seeks to identify the psychological factors influencing these feelings, such as empathy and social justice attitudes. However, these variables surely represent the proverbial tip of the disposition-formation iceberg. For instance, it is reasonable to assume that various content features (e.g., camera movements, sound effects, music) would influence the formation and maintenance of dispositions toward characters. One's viewing environment, age, or moral maturity would also likely impact character liking.

Likewise, prior exposure to an actor may influence the formation of dispositions toward characters subsequently portrayed by that actor. For instance, information about the private lives of entertainers could reasonably influence viewer evaluations of the characters they portray. Knowledge of an actor's off-screen infidelities could may influence her believability as (and thus the viewer's liking of) a sympathetic victim of love's cruelties on screen. Similarly, viewers used to seeing an actor play a certain type of character (e.g., a villain) might later find it difficult to like the same actor as another type of character (e.g., a hero) subsequently. In each of these cases, the extent to which dispositional affiliations are affected, then enjoyment should likewise be affected. To date, the disposition literature has largely ignored these prior attitudes; they have relied solely on the dispositions formed as a result of specific content.

Recently, some have sought to better understand how dispositional affiliations are formed and how prior knowledge impacts those formations. Although not specifically addressing the cases stated above, Raney (2004) argued that schema theories might provide such understanding (Brewer, 1987; Fiske & Kinder, 1981; Wyer & Gordon, 1984). In particular, the author noted that story schemas (cf. Rumelhart, 1980; Mandler, 1984) often assist viewers in making determinations of character liking immediately when the character is introduced into the narrative. As the reader will recall, the theories contend that disposition affiliations result from a viewer's moral monitoring of the behaviors and motivations of characters. However, the author argues that with many narratives such moral judgment is unnecessary, especially as a viewer seeks to determine an initial dispositional valence toward a character. Through repeated exposure to media entertainment, viewers learn how stories are constructed, how various actions are related, and how themes are repeated, among other things. Over time, a viewer develops various schema structures that become activated during subsequent exposures to related media texts. These structures then guide expectations about, and interpretation of, the ongoing narrative and, importantly for our purposes, the characters involved.

Thus, viewers arguably do not encounter most narratives empty handed, or empty minded, as the case may be. Existing story schema—developed over time through encountering a variety of different narratives—are activated that enable viewers to quickly understand various elements of, and to immediately form expectations about, the narrative accordingly. For example, viewers of drama are generally able to classify characters as "good" or "bad" almost instantaneously. Therefore, the initial determination of dispositional valence for a character may require few or no moral considerations at all. Existing schemas help guide initial character interpretation. As a result, viewers can typically identify almost immediately which characters we should—and will—presumably like and dislike.

The implications of this new line of thinking have yet to be explored; in fact, the assertions have yet to be empirically examined. However, this integration of schema theories with disposition theories may prove useful to entertainment scholars as they further seek to understand the psychological features associated with the complex disposition-formation process.

4. Disposition-Based Theories Contend That Affiliations Toward Characters Are Formed and Maintained on a Continuum From Extreme Positive Through Indifference to Extreme Negative Affect. One major advantage offered by the disposition theory of humor over previous humor theories is the emphasis on dispositional intensity or magnitude. Conceptualizing affiliations as simply dichotomous proved ineffective in predicting enjoyment. Furthermore, the continuum of affect embraced both ambivalent (or neutral) and negative (or disliking) responses to media characters, which proved crucial in our understanding of both dramatic and sports appreciation.

Although this conceptualization of the actual affiliations is extremely useful to entertainment researchers, further work is needed to help us understand how the dynamics of these affiliations operate during the entertainment experience. Raney (2004) addressed possible limitations to the ways that the continuum of affect might operate in practice. In discussing the possible ways that preexisting story schemas might influence the initial dispositional judgments of characters, the author asserted that schemas might similarly limit the range of affective response that would be experienced for any one character. Specifically, if schemas allow viewers to immediately classify characters as, for instance, good or evil, then it is reasonable to assume that those characters would subsequently be evaluated through that interpretive lens. Or stated another way, while the range of a viewer's affective reaction to a character can be theorized to exist along a continuum of affect from extremely positive through indifference to extremely negative, story schemas provide moral-judgmental shortcuts that theoretically limit the application of the full affective continuum to characters. Consequently, viewers should only need to evaluate stereotypically good characters (e.g., police officers, doctors) as good and stereotypically bad characters (e.g., drug dealers, gangsters) as bad. Moral judgment, therefore, is theoretically abridged as the standard of moral comparison is restricted to one end of the continuum.

As a result, Raney (2004) postulates that viewers form affiliations with certain characters on an affective continuum that extends only from extremely positive to indifference, based on the expectations that they are the protagonist (and thereby morally acceptable). The opposite is the case for antagonists. In other words, the author contends that schemas prejudice viewers to evaluate certain characters in certain and predictable ways. It should be noted that the application of these new ideas is limited to fairly standard narrative structure (i.e., the type of story formulas that are often used in mainstream, Hollywood-style offerings). Further, these claims have yet to be tested empirically. However, the ideas do potentially offer new understandings and directions for disposition research and our understanding of the media enjoyment process.

5. Because Disposition-Based Theories Rely Upon the Evaluation of Conflict Outcomes Between Characters, Justice Consideration Are a Necessary Component of the Theories. As was previously stated, disposition theorists have repeated shown that dispositions—both in terms of valence and intensity—are influenced by a viewer's moral judgment of the characters actions and motivations. Furthermore, because characters (especially in drama) are typically placed in situations involving conflict, then it is apparent that enjoyment of the situation will be bound to an appraisal of the conflict's resolution. The more a viewer favors the side that emerges victorious, the more she/he should enjoy the situation. The less the side that succeeds is favored, the less the situation should be enjoyed. Justification for such moral side-taking is found in the judgment leading to the initial disposition formation. Not only do we enjoy characters succeeding because we like them, but we like them (and thus enjoy their success) because we think that they ought to win. Liked characters are thought to be morally proper and justified in their victory over others; our initial judgment of their behaviors and motivations gives us justification for our thinking. Moral judgments of justice, equity, and propriety are, thus, essential to any disposition-based theory of enjoyment; the extant empirical research in the area supports these claims.

Raney (2004) offered an additional perspective on the role of justice in and moral judgment of enjoyment in light of the proposed role of schema theories in some entertainment experiences. First, he suggested that the initial formation of an affective disposition toward a character may, at times, actually precede specific moral evaluations of the character; this proposal was introduced above. It seems reasonable to expect that various story schema assist viewers in readily identifying the proper affective valence to hold toward many characters. That is, viewers often know who the good and bad characters are as soon as (or even before) they show up on screen. As a result, the initial dispositional valence and the accompanying range of affective response possible for a character may be set without considerations of justice coming into play. These initial valences then govern expectations about those characters, such that protagonists are expected to act in a virtuous manner, while antagonists and villains are expected to act in reprehensible ways. So, in some situations, dispositions may precede moral judgment and justice considerations.

Furthermore, Raney (2004) argued that because viewers expect liked characters to do good things and disliked characters to do bad things, those expectations should lead viewers to interpret character actions and motivations in line with the established dispositional valences rather than to morally scrutinize each action and motivation. Various attitude-maintenance strategies—such as in-group favoritism (Brewer, 1979; Levine & Campbell, 1972; Sherif, Harvey, White, Hood, & Sherif, 1961), selective perception (Billig & Tajfel, 1973; Rabbie & Horwitz, 1969; Tajfel, 1970), group attribution error (Allison & Messick, 1985; Taylor & Doria, 1981), and moral disengagement (Bandura, 1986; 1999)—that are used to justify or condemn the actions of others in reality would seemingly also be activated during entertainment consumption.

So, at times, it might be more proper to say that viewers read or interpret the actions of liked characters as morally proper for the sake of maintaining and defending their positive attitudes about those characters, rather than morally evaluating them for their appropriateness. The same is doubtless the case for disliked characters, as well. The author further contends that, in doing so, the viewer extends the latitude of moral propriety for liked characters, such that liked characters are given additional moral leeway so that the viewer can continue liking them unfettered. A similar process is thought to possibly operate in the case of disliked characters, where negative feelings toward those characters are perpetuated regardless of evidence that might place them in a more favorable light. In fact, it seems reasonable that our willingness to excuse or defend expectancy-inconsistent actions or motivations increases with the strength of our dispositional intensity toward the character.

As stated before, these proposals have yet to be empirically tested. However, at least in theory, they point to the complex relationships between a viewer's individually held attitudes, schema development, moral judgment and justification, and enjoyment that still offer ripe fields of intellectual harvest.

6. Disposition-Based Theories Further Acknowledge and Rely Upon the Differences Between Individuals in Terms of Emotional Responsiveness, Personal Experiences, Basal Morality, and Countless Other Psychological and Social-Psychological Factors. As stated repeatedly above, the role of these psychological factors in the process of enjoyment is well established. Furthermore, disposition theories (perhaps inherently) acknowledge that humans are constantly changing creatures. As a result, our responses to media content should likewise be dynamic and often unpredictable across time. For instance, it seems reasonable that mood (mood theory, see Zillmann, 1991), ability or willingness to attend to the narrative (elaboration likelihood model of persuasion, see Petty, Priester & Briñol, 2002), preexisting attitudes about genre (selective exposure, see Zillmann & Bryant, 1985), motivation for viewing (uses and gratifications, see Rubin, 2002), and other factors would also influence the enjoyment process as well. To date, little work has attempted to integrate these well-established traditions in media studies with disposition theories. Such work is encouraged as entertainment scholars seek to unlock the unfolding mysteries of enjoyment.

CONCLUDING THOUGHTS

The goal of this chapter has been to introduce the reader to the various psychological factors and processes involved in the media enjoyment experience explained by disposition theories. The process of disposition formation is similar across various media genre, involving both state and trait characteristics of the viewer utilized in the perceiving and evaluating of media characters. To date the essential characteristics that have been identified include empathy and various attitudes influencing moral judgment. Certainly others are yet to be identified. Furthermore, additional user inputs seem reasonably associated with the disposition-formation and–maintenance process such as existing character and story schemas, prior attitudes about various archetypical and stereotypical characters, mood, and motivation for viewing; all of which most certainly influence enjoyment. Finally, because enjoyment is experienced as a psychological (and physiological) phenomenon, it seems reasonable that each individual media entertainment experience itself influences all future ones. Disposition theories help to explain a small but significant part of these experiences.

REFERENCES

Allison, S. T., & Messick, D. M. (1985). The group attribution error. *Journal of Experimental Social Psychology, 21*, 563–579.

Bandura, A. (1986). *Social foundations of thought and action: A social cognitive theory.* Englewood Cliffs, NJ: Prentice-Hall.

Bandura, A. (1999). Moral disengagement in the perpetuation of inhumanities. *Personality and Social Psychology Review* [Special Issue on Evil and Violence], *3*, 193–209.

Billig, M., & Tajfel, H. (1973). Social categorization and similarity in intergroup behaviour. *European Journal of Social Psychology, 3*, 27–52.

Brewer, M. B. (1979). In-group bias in the minimal intergroup situation: A cognitive-motivational analysis. *Psychological Bulletin, 86*, 307–324.

Brewer, W. F. (1987). Schemas versus mental models in human memory. In P. Morris (Ed.), *Modeling cognitions* (pp. 187–197). London: Wiley.

Carpentier, F. D., Yu, H., Butner, B., Chen, L., Hong, S., Park, D., & Bryant, J. (2001, April). *Dimensions of the entertainment experience: Factors in the enjoyment of action, comedy, and horror films.* Presented at the annual meeting of the Broadcast Education Association, Las Vegas, NV.

Fiske, S. T., & Kinder, D. R. (1981). Involvement, expertise, and schema use: Evidence from political cognition. In N. Cantor & J. F. Kihlstron (Eds.), *Personality, cognition, and social interaction* (pp. 171–190).

Freud, S. (1960). *Jokes and their relation to the unconscious.* (James Strachey, Ed. and trans). New York: W. W. Norton & Company. (Original work published 1905).

Henning, B., & Vorderer, P. (2001). Psychological escapism: Predicting the amount of television viewing by need for cognition. *Journal of Communication, 51*, 100–120.

Herzog, H. (1940). Professor quiz: A gratification study. In P. Lazarsfeld (Ed.), *Radio and the printed page* (pp. 64–93). New York: Duell, Sloan, & Pearce.

Hobbes, T. (1976). *Leviathan.* (John Gordon Davis, Ed.). New York: Dutton. (Original work published 1651).

Hoffmann, M. L. (1987). The contribution of empathy to justice and moral judgment. In N. Eisenberg & J. Strayer (Eds.), *Empathy and its development* (pp. 47–80). Cambridge: Cambridge University Press.

Hoffner, C., & Cantor, J. (1991). Factors affecting children's enjoyment of a frightening film sequence. *Communication Monographs, 58*(1), 41–62.

King, C. M. (2000). Effects of humorous heroes and villains in violent action films. *Journal of Communication, 50*(1): 5–24.

Katz, E., & Foulkes, D. (1962). On the use of mass media for escape: Clarification of a concept. *Public Opinion Quarterly, 26*, 377–388.

Kohlberg, L. (1981). *Essays on moral development.* San Francisco: Harper & Row.

La Fave, L. (1972). Humor judgments as a function of reference groups and identification classes. In J. H. Goldstein & P. E. McGhee (Eds.), *The psychology of humor* (pp. 195–210). New York: Academic Press.

Levine, R. A., & Campbell, D. T. (1972). *Ethnocentrism: Theories of conflict, ethnic attitudes, and group behavior.* New York: John Wiley & Sons.

Mandler, J. M. (1984). *Stories, scripts, and scenes: Aspects of schema theory.* Hillsdale, NJ: Erlbaum.

McQuail, D., Blumler, J. G., & Brown, J. R. (1972). The television audience: A revised perspective. In D. McQuail (Ed.), *Sociology of mass communications* (pp. 135–165). Middlesex, England: Penguin.

Oliver, M. B. (1993). Adolescents' enjoyment of graphic horror: Effects of attitudes and portrayals of victim. *Communication Research, 20*(1), 30–50.

Oliver, M. B. (1996). Influences of authoritarianism and portrayals of race on Caucasian viewers' responses to reality-based crime dramas. *Communication Reports, 9*(2), 141–150.

Oliver, M. B., & Raney, A. A. (2005). *Exploring the multi-dimensionality of media enjoyment.* Manuscript in preparation.

Petty, R. E., Priester, J. R., Briñol, P. (2002). Mass media attitudes change: Implications of the elaboration likelihood model of persuasion. In J. Bryant, & D. Zillmann (Eds.), *Media effects: Advances in theory and research* (2nd ed., pp. 155–198). Mahwah, NJ: Erlbaum.

Priest, R. F. (1966). Election jokes: The effects of reference group membership. *Psychological Reports, 18*, 600–602.

Rabbie, J. M., & Horowitz, M. (1969). Arousal of ingroup-outgroup bias by a chance win or loss. *Journal of Personality and Social Psychology, 13*, 269–277.

Raney, A. A. (2005). Punishing media criminals and moral judgment: The impact on enjoyment. *Media Psychology, 7*(2), 145–163.

Raney, A. A. (2002). Moral judgment as a predictor of enjoyment of crime drama. *Media Psychology, 4*, 305–322.

Raney, A. A. (2003). Disposition-based theories of enjoyment. In J. Bryant, J. Cantor, & D. Roskos-Ewoldsen (Eds.), *Communication and emotions: Essays in honor of Dolf Zillmann* (pp. 61–84). Mahwah, NJ: Erlbaum.

Raney, A. A. (2004). Expanding disposition theory: Reconsidering character liking, moral evaluations, and enjoyment. *Communication Theory, 14*(4), 348–369.

Raney, A. A., & Bryant, J. (2002). Moral judgment and crime drama: An integrated theory of enjoyment. *Journal of Communication, 52*, 402–415.

Rest, J. (1979). *Development in judging moral issues.* Minneapolis, MN: University of Minnesota Press.

Rubin, A. M. (2002). The uses-and-gratifications perspective in media effects. In J. Bryant, & D. Zillmann (Eds.), *Media effects: Advances in theory and research* (2nd ed., pp. 525–548). Mahwah, NJ: Erlbaum.

Rumelhart, D. E. (1980). Schemata: The building block of cognition. In R. J. Spiro, B. C. Bruce, & W. F. Brewer (Eds.), *Theoretical issues in reading comprehension: Perspectives in cognitive psychology, linguistics, artificial intelligence, and education* (pp. 33–58). Hillsdale, NJ: Erlbaum.

Sapolsky, B. S. (1980). The effect of spectator disposition and suspense on the enjoyment of sport contests. *International Journal of Sport Psychology, 11*(1), 1–10.

Sherif, M., Harvey, O. J., White, B. J., Hood, W. R., & Sherif, C. W. (1961). *Intergroup conflict and cooperation: The Robbers Cave experiment.* Norman, OK: University Book Exchange.

Tajfel, H. (1970). Experiments in intergroup discrimination. *Scientific American, 223,* 96–102.

Taylor, D. M., & Doria, J. R. (1981). Self-serving and group-serving bias in attribution. *The Journal of Social Psychology, 113,* 201–211.

Wolff, H. A., Smith, C. E., & Murray, H. A. (1934). The psychology of humor. *Journal of Abnormal and Social Psychology, 28,* 341–365.

Wyer, R. S., & Gordon, S. E. (1984). The cognitive representation of social information. In R. S. Wyer & Srull, T. K. (Eds.), *Handbook of social cognition* (Vol. 2, pp. 73–150). Hillsdale, NJ: Lawrence Erlbaum Associates.

Vorderer, P., Klimmt, C., & Ritterfeld, U. (2004). Enjoyment: At the heart of media entertainment. *Communication Theory, 14*(4), 388–408.

Zillmann, D. (1991). Empathy: Affect from bearing witness to the emotion of others. In J. Bryant and D. Zillmann (Eds.), *Responding to the screen: Reception and reaction processes* (pp. 135–167), Hillsdale, NJ: Erlbaum.

Zillmann, D. (1994). Mechanisms of emotional involvement with drama. *Poetics, 23,* 33–51.

Zillmann, D. (2000). Basal morality in drama appreciation. In I. Bondebjerg (Ed.), *Moving images, culture, and the mind* (pp. 53–63). Luton: University of Luton Press.

Zillmann, D., & Bryant, J. (1975). Viewer's moral sanction of retribution in the appreciation of dramatic presentations. *Journal of Experimental Social Psychology, 11,* 572–582.

Zillmann, D., & Bryant, J. (1985). *Selective exposure to communication.* Hillsdale, NJ: Lawrence Erlbaum Associates.

Zillmann, D., Bryant, J., & Cantor, J. (1974). Brutality of assault in political cartoons affecting humor appreciation. *Journal of Research in Personality, 7,* 334–345.

Zillmann, D., Bryant, J., and Sapolsky, B. S. (1989). Enjoyment from sports spectatorship. In. J. H. Goldstein (Ed.), *Sports, games, and play: Social and psychological viewpoints* (2nd ed, pp. 241–278). Hillsdale, NJ: Lawrence Erlbaum Associates.

Zillmann, D., & Cantor, J. (1972). Directionality of transitory dominance as a communication variable affecting humor appreciation. *Journal of Personality and Social Psychology, 24,* 191–198.

Zillmann, D., & Cantor, J. (1976). A disposition theory of humor and mirth. In T. Chapman & H. Foot (Eds.), *Humor and laughter: Theory, research, and application* (pp. 93–115). London: Wiley.

Zillmann, D., & Cantor, J. (1977). Affective responses to the emotions of a protagonist. *Journal of Experimental Social Psychology, 13,* 155–165.

Zillmann, D., & Paulus, P. B. (1993). Spectators: Reactions to sports events and effects on athletic performance. In R. N. Singer, M. Murphey, & L. K. Tennant (Eds.), *Handbook on research in sport psychology* (pp. 600–619). New York: Macmillan.

Zillmann, D., Taylor, K., & Lewis, K. (1998). News as nonfiction theater: How dispositions toward the public cast of characters affect reactions. *Journal of Broadcasting and Electronic Media, 42*(2), 153–169.

10

Empathy: Affective Reactivity to Others' Emotional Experiences

Dolf Zillmann
University of Alabama

Empathy is often thought of as an affective state that is mediated by an ability of persons to place themselves, mostly deliberately, but on occasion spontaneously, into observed others' emotional experiences. The resultant affections are construed as 'feeling with' or 'feeling for' the persons whose emotions were witnessed. Such feelings are also deemed 'vicarious' emotions. Whatever the construal, the affections in question are usually thought to foster a deeper understanding of the observed persons' lot and, conditions allowing, inspire supportive actions.

The indicated empathic ability tends to be treated as an enduring disposition or trait. People are presumed to have it to varying degrees; that is, to have this trait in moderation or abundance, or to be deficient in it. No doubt, empathic ability differs greatly among people and should be treated as an individual-difference variable of consequence. As such, empathic ability is certainly deserving of attention, in particular with regard to its constituent and developmental determinants. Often enough, however, the focus on empathy as a traitlike ability has detracted from the vast variation in empathic reactivity across transitional experiential states, especially across interpersonal dispositional circumstances. It is common knowledge that people can show great empathic concerns in some situations and fail to show any in others. Theories of empathy, then, must not only address the inter-individual variation in responsiveness, but must explain the intra-individual vicissitudes of empathic sensitivity. Of these vicissitudes, the situational conditions under which even acutely manifest empathic sensitivity is entirely lost, and emotions between observers and the observed change from compatible to extreme discordance, constitute an impasse in most theories and call for elaboration and elucidation.

Furthermore, theoretical approaches to the phenomenon of empathy must come to terms with the fact that empathic reactivity is not restricted to social situations that are immediately witnessed. They must provide satisfactory accounts of why people empathize with persons and personlike entities that exist only in descriptions or in images. In particular, they must address the fact that actually existing persons, when presented in formats such as the news, can evoke empathic reactions of intensities comparable to those triggered by events in their immediate social environment. Even more important, these approaches must explain why we respond empathically to media displays of the emotions of fictional characters, when it is clearly understood that these characters are figments of their creators' imagination. The crux is that storytelling of any kind is a principal forum of empathic reactivity. It would appear, in fact, that empathic engagement is what fuels interest in tales—fictional or otherwise—and that all impetus for attending revelations of the characters' fate would be lost if we were not lovingly disposed to care for them, or, for that matter, if we were not disposed by disdain to wish harm upon them. The empathy concept can thus be considered pivotal to any interest in, and likely any gratification from, storytelling via the media of communication.

In this chapter, the merits of various conceptual approaches to the phenomenon of empathy are explored. The principal theories of empathy are outlined, and an integrative theoretical model of empathy is presented. It is shown that this model incorporates and integrates much established theory. The presentation of theory is followed by a discussion of pertinent research findings. Finally, the integrative model's implications for affective development are projected. Special consideration is given to the changing ecology of empathic experience, and focus is on the new communication technology and its enormous capacity for replacing immediate, affect-producing social exchanges with mediated events that abstract, simulate, and represent such exchanges.

CONCEPTUALIZATIONS OF EMPATHY

Empathy has meant different things to different scholars, both in philosophy (e.g., Scheler, 1913; Smith, 1759/1971; Stein, 1970) and in psychology (e.g., Berger, 1962; Hoffman, 1977, 1984, 1987; Stotland, 1969). It has been construed, for instance, as the ability to perceive accurately the emotions of others (e.g., Borke, 1971; Ickes, 1997; Tagiuri, 1969), the proficiency of putting oneself into another person's lot (e.g., Dymond, 1949, 1950; Katz, 1963; Mead, 1934), the skill of understanding the affective experiences of others (e.g., Cline & Richards, 1960; Davis et al., 1987; Truax, 1961), the deliberate sharing of particular emotions with others (e.g., Aronfreed, 1968; Feshbach, 1978; Lipps, 1907), hedonic concordance of affect in a model and an observer (e.g., Berger, 1962; Stotland, 1969; Stotland et al., 1978), affinity in the autonomic response patterns associated with the model's and observer's affective behavior (e.g., Berger, 1962; Hygge, 1976a, 1976b; Tomes, 1964), the conscious or unconscious assimilation of another ego through a process called identification (e.g., Fenichel, 1954; Freud, 1921/1950, 1933/1964), the mental entering into another person or thing that results in fused consciousness (e.g., Hart, 1999; Lipps, 1903, 1906; Worringer 1908/1959), instinctlike affect propagation and primitive action-inspiring emotional contagion (e.g., McDougall, 1908; Trevarthen, 1984), and the instigation to act so as to relieve distress in others (e.g., Mehrabian & Epstein, 1972; Stotland et al., 1978).

This diversity in the specification of what is to be considered empathy may give the impression that different investigators have addressed different phenomena, and that at least some of the discrepant specifications are irreconcilable (Duan & Hill, 1996; Thornton & Thornton, 1995), and that the concept suffers from imprecision to a degree that totally compromises its usefulness (Levy, 1997). Such an impression is overly pessimistic, however, as there is sufficient commonality in the definitional approaches to consider them delineations of one particular, albeit multifaceted, behavioral phenomenon.

The impression of incompatible specifications seems created by selective attention to different facets of empathy. Some definitions have focused on specific mechanisms of empathic behavior as well as on limited sets of manifestations of the behavior in question. Others have concentrated on behavioral implications, such as the motivation to render help to fellow beings.

Yet others have emphasized the utility of empathic processes presumed to be involved in interpersonal sensitivity and diagnostic skills. Nonetheless, with the exception of Lipps' (1903, 1906) and Worringer's (1908/1959) proposals concerning aesthetic experience, all definitional approaches seem to address a process by which persons respond emotionally to the emotional experiences of others, a process that tends to yield a considerable degree of affinity between witnessed emotional experiences and the witnesses' emotional reactions to them.

For our purposes, this descriptive account of empathy is insufficient, however. It is too restrictive in that it limits the empathy concept to affective responses to the expression of emotions by others. Some time ago, Adam Smith (1759/1971) observed that the anticipation of a model's emotional reaction alone could induce the affective response that actual witnessing of the reaction would produce. "When we see a stroke aimed and just ready to fall upon the leg or arm of another person," he wrote, "we naturally shrink and draw back our own leg or our own arm [p. 3]." Stotland (1969) reviewed research that corroborated the existence of such 'anticipatory empathic reactions' and felt compelled to include these responses under the empathy heading. He defined empathy as "an observer's reacting emotionally because he perceives that another is experiencing *or is about to experience* an emotion [p. 272; italics added]." Emotional expression being manifest or impending, the definition still limits empathy to others' expressions. This restriction is also unacceptable because affect concordance between a model and an observer is frequently evident when the observer is merely exposed to information that seemingly precipitates the model's facial and bodily expressions of emotion. Characteristically, the observer responds to both (a) *the circumstances that produce the model's emotional reaction*; and (b) *the expressive elements of that reaction*. In recognition of this fact, Hoffman (1978) defined empathy as "a largely involuntary, vicarious response to affective cues from another person *or from his situation* [p. 227; italics added]." Aronfreed (1968), on the other hand, thought it necessary to keep the two potential sources of affect conceptually separated and suggested that the empathy construct should be restricted to affective reactions induced by exposure to others' emotional expressions. He further suggested that affect in response to witnessing the conditions that produce emotional reactions in others, or to learning about them indirectly, should be termed *vicarious reactions*.

There seems to be merit in both approaches. First, it is most important to recognize the joint operation of (a) information about the apparent causes of a model's affect; and (b) the model's expression of affect. This joint operation may be considered a 'natural confounding,' that is, an ecologically undeniable concurrence. The separation of the confounded elements would appear to create a degree of ambiguity that prevents meaningful affective reactions. Or meaningful reactions occur only after the observer 'infers' deleted events. For instance, the facial expression of discomfort, in and of itself, might well produce an appreciable impact on an observer. This might occur because of facial mimicry and resultant afferent feedback. It cannot be ruled out, however, that the respondent guesses a cause for the discomfort, and that an empathylike affective reaction comes about only thereafter. It could be argued that respondents make sense of a model's expressions in terms of their own affective experiences with specific stimulus situations, and then respond on the basis of these experiences. Exposure to facially expressed affect, then, via presumptions about their stimulation, might liberate readily accessible affective memories, and these memories might foster affect in the observer.

Regardless of the specific mediation of emotional reactions to a model's expressions of affect, such expressions are frequently too vague to allow meaningful empathic reactions to a fellow being's affect. For example, tears are shed on joyous occasions as well as in states of misery, laughter can accompany despair as well as gaiety, and smiles do not necessarily signal a state of well-being. It is common observation, however, that empathy is seldom erroneously elicited by 'inappropriate' affective expressions. It should be the rare exception, for example, that people respond grievously upon seeing others cry at what they know to be a happy reunion, even if they should shed some tears themselves. This attests to the fact that information about the instigation of affect in displayers is of overriding significance in the determination of the observer's potentially empathic experiential state. Consequently, information about a model's affective responding may be considered incomplete, in general, and the role of information about the *causal circumstances* of the model's affective responding will have to be acknowledged if an ecologically valid comprehension of empathy and its function in human affairs is to be achieved.

It should be clear that the argument is reversible. Causal circumstances may be ambiguous, leaving it unclear how particular persons might respond to them. If a person's response is not exhibited, an observer may well anticipate, accurately or erroneously, a specific reaction and its expressive manifestations. The anticipation of particular reactions may then create feelings of empathy.

Second, the conceptual separation of causal and expressive elements in a model's affect is nonetheless useful, if not imperative, because it assures attention to the relative contributions that these components make to empathic reactions. It is conceivable that the contribution of the witnessed expression of affect is substantial in some circumstances and insignificant in others. The two components also seem sequentially dependent. Prior knowledge of the causal circumstances of a model's display of affect should promote pronounced empathic reactions. Exposure to similar displays whose causation is unclear during exposure should produce comparatively subdued reactions. But, more importantly, the conceptual separation leads to focusing attention on types of causal circumstances that precipitate a model's affect, as well as on the relationship between the model's and the observer's responsiveness to the causal conditions.

Presentation and representation of causal conditions are extremely variable. Affect-inducing circumstances may be provided verbally. They can be reported in a roundabout or in a precise fashion, and they can be dramatically embellished to different degrees. In our daily lives, all emotional happenings that we did not witness directly are subject to the indicated variation. Novels, with their partiality to the display of emotions, are by definition limited to verbal accounts. In cases of direct witness to emotional events, the causal circumstances tend to be audiovisually defined. They may be manifest in vivid action and, most importantly, in events that, in and of themselves, are capable of inducing emotional reactions in an onlooker. The same holds true for audiovisual representations of such events, that is, of events known to abound in movies and television programs of any kind.

With regard to the type of presentation, it seems likely that the affective impact on an observer increases with the fidelity of the portrayal of the circumstances that foster emotions in a model. A witnessed person's facial and bodily expression of fear and panic, for instance, may be the apparent result of his entrapment in a house on fire. The affective reaction of a respondent who is exposed to the entire scenario may derive, at least in part, from the model's expression of fear. However, it also may result, in part, from the verbal or audiovisual presentation of information about the causal circumstances, that is, about the blaze engulfing the model. One might be inclined to believe that the more closely any representation of such inferno mimics the actual stimulus conditions, the greater its affective impact. This reasoning suggests that audiovisual representations (or in semiotic terms, iconic representations) tend to generate more affect than alternative forms. It is nonetheless conceivable that verbal or other non-iconic representations can be similarly powerful in eliciting affect because they instigate the individual to imagine the circumstances in terms of experiences that proved arousing in the individual's past. This likely involvement of stimulus-bound affective experience applies, of course, to all modes of representation. For instance, individuals who directly experienced fire as a threat to personal welfare should respond more strongly than those lacking such experience with infernolike situations. This should hold true whether fire is presented in vivid images or presented verbally, because affective memory should be revived in either case.

Our account entails a formidable dilemma for the conceptualization of empathy. First of all, negative affect in response to witnessing a model confronted with life-threatening conditions is likely to come partly from exposure to an affect-inducing stimulus, such as a home ablaze. Individuals are unlikely to know which part of their affective reaction is the result of this kind of exposure and which derives from responding to the model's despair. In fact, individuals are highly inefficient in separating contributions to affect, if they are at all capable of making such a separation (Zillmann, 1978, 1983). Individuals will, as a rule, construe their reactions as resulting *in toto* from the most plausible, immediately present inducing conditions. Under empathy-evoking conditions, the model's obtrusive behavior is a likely candidate. If focused on, individuals may erroneously construe their reactions as entirely empathic, although these reactions may largely derive from exposure to ulterior affect-inducing stimuli. To complicate matters further, it is conceivable that the non-empathic response component of an affective reaction, such as affect from exposure to a particular threatening condition, ultimately enhances the empathic reaction because it amplifies the model's endangerment. In the inferno example, affect triggered by the sight of a home ablaze should facilitate the perception of peril for the entrapped persons and thereby intensify feelings of empathic distress about their dilemma. The same considerations apply, of course, to positive affect in empathic reactions. Observers might respond strongly to the rewards that a model attains and, in focusing on the model's expression of contentment and satisfaction, come to construe their responses as purely empathic.

Finally, misreadings of contributions to affect in empathic experiences are likely to arise in situations where the respondents truly share, to some degree, the model's satisfaction or endangerment. This is to say that a model's benefaction may also benefit the observer, and likewise, a model's endangerment may also constitute a threat to the observer. Parents, for instance, might construe their feelings in response to seeing their son or daughter rejoice upon receiving a college diploma as empathy, although much of their excitedness is likely to stem from self-gratification and the anticipation thereof (i.e. from celebrating their own accomplishment). Similarly, empathic grief for a brother or sister who has been stricken by a hereditary disease is likely to be fueled by the fear that the disease might eventually victimize the respondents themselves.

Given these conceptual difficulties, and given that the specifiable types of stimuli and conditions that contribute to affective reactions elude measurement and probably will do so in the foreseeable future, it appears appropriate to conceive of empathy as an affective reaction that the reacting individual deems produced by happenings to another person and/or by this person's expressive and behavioral responses to them. Empathy is thus viewed as a feeling state brought on by the observation of a fellow being in a specific situation. The feeling state is particular, however, as not all affective reactions would and could be considered empathy. Clearly, affect that is hedonically opposite to that observed in another person is unlikely to be construed as 'feeling with' or 'feeling for' somebody; that is, as empathic. The only reactions that qualify are hedonically compatible and concordant. If someone's pain is an observer's joy, or someone's joy is an observer's pain, the construct of empathy as subjective experience does not apply. Efforts to retain it for these conditions (e.g., by labeling the discordant affective experience *negative empathy* or *counter-empathy*) may be considered to invite confusion because they can be construed as suggesting that counter-empathy is a form of empathic experience, when in fact it refers to the absence of such an experience or to a response contingent on its absence. Experientially, then, counter-empathy is not an empathic reaction. However, the concept of counter-empathy can be granted descriptive utility in that it unambiguously specifies an affective reaction opposite in hedonic valence to empathy.

The involvement of attributional processes in the conceptualization of empathy renders non-subjective specifications of hedonic concordance or affect affinity unnecessary. It creates some problems of its own, however. The reliance on causal attribution, no matter how implicit and rudimentary this process is presumed to be, presupposes conceptual and linguistic skills that are lacking in linguistically immature children. As a result, the conceptualization seems wanting and useless for the assessment of empathy in such children. But the situation can be remedied. First, the empathic response in pre-attributional children can be considered incomplete because the conscious component of the possible feeling state is absent. Second, the problem can be resolved at the operational level by having a third party establish (a) that the child focuses on events that come a model's way as well as on the model's expressive and behavioral response to them; and (b) that the resulting affect is hedonically concordant with that exhibited by the model. It would seem preferable, however, to consider such concordant affective reactions just that—rather than full-fledged empathic experiences.

Before formalizing these considerations in a definition of empathy, some clarifications are in order. First, in the cited definitions, as well as in the preceding conceptual efforts, it has been assumed that affect is a response that is associated with a notable increase in arousal. Those who have measured the arousal component of affect (Cacioppo & Tassinary, 1990; Grings & Dawson, 1978; Wagner & Manstead, 1989) have invariably operationalized it as a sympathetic reaction; that is, as an increase of sympathetic excitation in the automatic nervous system. We shall follow this tradition in conceptualizing affect as a reaction associated with an appreciable increase in sympathetic activity—but, without a commitment to a specification of what, exactly, would qualify as an appreciable increase.

Second, in our consideration of empathy and affect we have avoided the term 'vicarious reaction.' Vicarious, if taken to mean 'instead of' or 'in place of,' projects a mechanism of empathy that entails the assumption of some sort of ego fusion or confusion in respondents. Confronted with a model's situation and behavior, observers are apparently viewed as imagining themselves in the model's place and then to feel and respond accordingly. As it cannot possibly be considered established that all affective reactions to the emotions of others or to their apparent causes are brought about by the respondents' placing themselves into others' stead, the use of the term 'vicarious' seems careless and unfortunate. The definitional use of the term amounts to a foregone conclusion about the mechanics of empathy. In this connection, Aronfreed's (1968) suggestion to consider as vicarious any affect that is produced by exposure to the circumstances that eventually lead to a model's emotions seems particularly misleading. How can a response to affect-inducing circumstances (e.g., a house on fire) be vicarious? How can it be in someone's stead, especially when models are only scarcely defined or there are none at all?

Third, we have suggested that empathy be considered an experiential state in which observers attribute model-concordant affect to exposure to the model. It was recognized that this global attribution is likely to entail the misattribution of some elements of affective reactivity. The excitatory component of affect, in particular, may aggregate response elements from stimuli salient to self rather than to the model's welfare. In case an empathic response contains such elements, or even is primarily composed of them, one might be inclined to label the affective reaction 'pseudo-empathy.' This possible characterization is based in part on the conceptual distinction that Aronfreed (1968) introduced, and it thus maintains the distinction. The characterization seems of questionable utility, however, because it is difficult to operationalize and, more importantly, the likely misconceptions about contributions to affect in observers' attributions do not detract from the experiential reality of their empathic responses.

Empathy, then, may be defined as responding (a) to information about circumstances presumed to cause emotions in other individuals and/or; (b) to other individuals' facial and bodily expression of their emotions and/or; (c) to other individuals' overt actions presumed to be instigated by their emotions; this responding being associated with; (d) a noticeable increase of excitation and; (e) an appraisal of the entire experience as feeling with, or feeling for, the other individuals.

THEORIES OF EMPATHY

Theories of empathy can be grouped into two main categories: (1) theories that consider the behavioral functions of empathic reactivity, mostly within a bio-evolutionary context; and (2) theories that focus on the cognitive and excitatory mediation of empathic reactivity, largely disregarding the particular objectives that ultimately may be served by such reactivity. Theories of the second group can be further divided into subcategories: (a) theories positing that empathic reactivity is mediated by innate, often reflexive processes; (b) theories positing that empathic reactivity is acquired through learning without necessitating explicit and deliberate cognitive operations; and (c) theories positing that empathic reactivity is the direct result of explicit and deliberate cognitive operations. We will briefly discuss the theories in all these categories.

Evolutionary Approaches

Numerous scholars have stressed the adaptive significance of reflexive expressions that, within a species, trigger reflexive actions in others and thereby effect the social coordination of behavior (e.g., Darwin, 1872; Eibl-Eibesfeldt, 1970; Hinde, 1970; Lockard, 1980; MacLean, 1958, 1967, 1990; Plutchik, 1980). The survival of a species is obviously well served by some individuals' presumably involuntary expressive signaling of danger that readies others for fight or flight. Such signaling also may alert others to vital opportunities, essentially the attainment of food, shelter, and sexual access. In a safe environment, this signaling may further promote mutual appeasement and social bonding between individuals and groups.

It is generally assumed that intraspecific signaling of this kind is innate. Buck and Ginsburg (1997), who consider such signaling and responding a genetically fixed empathic process, speak of a primordial conveyance of "social knowledge" that "is based upon spontaneous communication systems involving innate displays and preattunements" (p. 37). In even stronger anthropomorphic language they contend that, "if the sender produces a display and the receiver attends to the display, the receiver comes to know directly its meaning as a kind of inherited knowledge" (p. 17). This kind of knowledge conveyance is viewed as "the root of empathy" (p. 22) and "the bootstrap by which simple life forms were able to attain higher levels of organization" (Buck & Ginsburg, 1991, p. 166). Empathy in humans is ultimately understood in these terms, although the possibility of extending the range of empathic responding via learning is acknowledged.

Also probing the phylogenetic roots of empathy, Plutchik (1980, 1987) looked for patterns in animal behavior that could be construed as homologous with empathy in humans. He surveyed, among other things, schooling behavior in fish, flocking and mobbing behavior in birds, and herding behavior in numerous mammalian species. Moreover, he examined the specificity of alarm calls in a variety of species. His analysis led him to conclude that unlearned, innate signal-response connections abound. Using the concept of allelomimetic behavior as introduced by Scott (1959, 1969), he found ample evidence for the adaptive coordination of fight and flight reactions in groups of animals. The genetically fixed expression of apparent fear, for instance, elicits genetically fixed parallel emotive reactions. Emotive contagion of this kind is primarily intraspecific, but may be interspecific in related species. It also became clear, however, that such innate signal-response relationships serve many functions other than fostering reactions similar to those of the signaling animal. Threat displays within a species, for example, prompt unlearned reactions that obviously serve the welfare of threatening animals but not of the animals threatened by the display. Threatened animals may well counter-threaten at first, thus responding in parallel fashion. The ultimate response is nonparallel, however, as threatened animals are likely to resort to fight or flight. Behaviors akin to empathy, namely those that are motivationally parallel between signal provider and signal receiver, apparently define only a subset of innate communicative behavior.

Notwithstanding this restriction, there can be no doubt that genetically fixed allelomimetic behavior comprises an essential and seemingly universal part of the behavioral repertoire of a multitude of animal species. Its adaptive utility also cannot be questioned. Given that, it would seem to be justified to consider such behavior the evolutionary precursor of empathy in humans.

The issue is not, however, that empathy has a primitive precursor in allelomimetic behavior, but whether such behavior shares the experiential qualities of human empathy. Do we need to assume, as Buck and Ginsburg (1997) apparently do, that responding to innately coded signals entails an understanding of the behavior's "meaning" and reveals something about the signal sender's affective state or motives? Would it not be more parsimonious to accept that genetically fixed signals trigger genetically fixed responses and that this is all we know about it? It would seem to be prudent to consider the elicitation of parallel emotive behavior *contagious*, as it is only the contagion of behavior that is secured by observation.

Along these lines, Plutchik (1987) considers empathy a state that third parties have to infer on the basis of observation or that an empathizing person has to disclose. This position agrees with our conceptualization. Part *c* of our definition stipulates the inferential aspect of empathy. In addition, however, part *e* stipulates that individuals must comprehend their reaction to the emotions of others as an empathic one. Especially this latter part of the definition, as it makes empathy contingent upon an introspective appraisal, compels us to construe the elicitation of parallel emotive behavior by genetically fixed connections, as well as by learned signal-response connections, as allelomimetic. We consider the process fostering such behavior signal-mediated behavioral contagion or pseudo-empathy rather than empathy proper.

Empathy as Reflexive Reactivity

Theories of this kind are implicit in the analysis of behavior across species whose survival hinges on coordinated action, primarily for fight and flight. They are, consequently, part and parcel of bio-evolutionary theorizing. But they are also used to explain empathic reactivity in a broader spectrum of human behavior.

Reflexive mediation of empathy has been proposed by McDougall (1908) and Lipps (1907), among others. McDougall held that an innate response disposition, which he labeled 'primitive passive sympathy,' simply compels observers to experience the emotions of others. He feared, in fact, that individuals might get so caught up in empathizing with the joys and miseries of others, especially with the latter, that this innate disposition would prove maladaptive; and he developed theoretical amendments to show how debilitating empathy is prevented and sanity maintained (e.g., McDougall, 1922). Lipps' related proposal focused more strongly on the expressive elements of emotional experience. Specifically, he posited that observers, in giving way to innate dispositions, mimic the observed party's postural and gestural expressions and that afferent feedback from this motor mimicry liberates empathic affect because it connects to the observer's affective experiences that are associated with the expressions in question. The self's experience of pain, for instance, is associated with afferent feedback from a specific facial expression, and the elicitation of a similar expression via mimicry, especially the afferent feedback from this expression, is thought to produce the experience of pain to some degree. The second step of Lipps' theory of empathy (i.e. the assumed capacity of emotional expression, especially facial expression, to produce affective states) is an integral part of several more recent theories of emotion (e.g., Izard, 1977; Levenson, Ekman, & Friesen, 1990; Tomkins, 1962, 1963; Zajonc, 1980).

Evidence for motor and facial mimicry comes from studies with both children and adults. For instance, infants cry reflexively to the cries of other infants, rather than to similarly noxious cries of another kind (e.g., Sagi & Hoffman, 1976; Simner, 1971). In the so-called smiling response, they smile at smiling faces (e.g., Spitz & Wolf, 1946; Washburn, 1929). Children mimic a variety of facial expressions (e.g., Hamilton, 1972). Adults show increased lip movement and a higher frequency of eye blinking upon observing stuttering or a high eye-blink rate, respectively, in others (e.g., Berger & Hadley, 1975; Bernal & Berger, 1976). Even the frequency of individuals' yawning is known to vary with that of yawning models (Cialdini & McPeek, 1974).

The proposal that afferent feedback from the facial muscles fosters expression-specific affect has received some support from research in which subjects were made to express emotions that were either consistent or inconsistent with concurrent external stimulation (Laird, 1974). Replication of these supportive findings proved difficult, however (Tourangeau & Ellsworth, 1979). In related research, the suppression of the expressive component of affect was found to reduce both the intensity of felt affect and the excitatory concomitant of that affect (Lanzetta, Cartwright-Smith, & Kleck, 1976). On the other hand, exaggeration of facial expression was observed to amplify affect-linked excitatory reactions (Vaughan & Lanzetta, 1980). Voluntary facial expressions were also found to instigate moderate emotion-related autonomic reactions (Levenson et al., 1990). Taken together, the findings suggest that afferent feedback from the expression of affect is capable of modifying affective behavior. However, given the apparent failure to replicate some of the pivotal research demonstrations (cf. Tourangeau & Ellsworth, 1979), the findings also indicate that the affect-controlling power of afferent feedback is more limited than projected by the proposals under consideration (cf. Buck, 1980).

Theories of reflexive empathic reactivity obviously rely on rather basal mechanisms of behavior elicitation. Cognitive elaboration of situational circumstances and reflection focusing on intentional or experiential aspects of elicited behavior are not invoked as necessary components of empathic behavior. However, it is generally held that post-reactive awareness of the quasi-concurrent, parallel reaction with another's reaction tends to be construed as feeling for or with the other whose behavior was witnessed. The involvement of cognition may thus be deemed secondary, but it nonetheless ensures that the parallel reactivity is experienced as empathic (see our definition of empathy at the end of *Conceptualizations of Empathy*).

Empathy as Acquired Reactivity

Theories of this type, meaning proposals that project empathic reactivity as acquired in a more or less mechanical fashion, have been implied in the reasoning of many investigators. The presumed mechanics of the acquisition process have been most clearly articulated by Aronfreed (1970) and Humphrey (1922).

Aronfreed suggested that empathic reactions are acquired in socially parallel affective experiences. Specifically, in the concurrent, externally induced experience of pain or pleasure in a model and an observer, the model's expression of pain or pleasure becomes associated with these responses. With repeated experiences of this kind, the model's expression gradually assumes the power to elicit sensations akin to pain and pleasure. Two children, for instance, may be punished for a jointly committed misdeed and observe each other's expressions of distress during and after the ordeal. Upon witnessing similar expressions in others, they should come to feel a touch of their own distress reaction that is associated with these expressions. The behavioral conditions under which empathic responses are acquired need not be entirely parallel, however. Many investigators have pointed to the mother-infant caretaker relationship as a most critical one in which empathy is likely to develop (e.g., Hoffman, 1973; Sullivan, 1940). The mother's expression of contentment is predominantly linked to the child's benefaction and the experience of well-being. Likewise, the mother's expression of distress is mostly connected with the child's experience of distress. A model's expression of negative and positive affect is thus consistently paired with an observer's affect of the same kind, despite the asymmetry in the behavioral situation.

Humphrey's theorizing proves more inclusive yet. He started on the premise that the elicitation of affect is mostly mediated by stimuli from so-called distance receptors and that, because of this circumstance, affect-inducing stimuli for self and for others are often very similar. For instance, the visual experience of cutting one's finger is essentially the same as that of seeing another person cutting his or her finger. Analogously, a child's perception of his or her own crying is much the same as that of the crying of another child. Surely, the sensations associated with these expressions are private and subjective. But Humphrey insisted that visual, auditory, and olfactory access to self is principally no different than that to others. In making his point on distance reception he wrote: "... my own body and that of my neighbour are on a par, not identical but similar [1922, p. 115]." Given such perceptual similarity between self and other, he then proposed that the individual, upon exposure to stimuli that are highly similar to those that were associated with positive or negative affective reactions in the past, will experience this affect again, at least to some degree. Using the classical-conditioning paradigm, Humphrey argued that stimuli that were consistently paired with or followed by affective reactions would assume the power to elicit these reactions. Empathy-like responses may thus be expected after a stimulus-sensation linkage has been established and the individual is exposed to the critical stimulus condition in the behavior of others, especially in their expressive behavior. Humphrey went beyond classical conditioning, however, when he proposed that empathic reactions could be induced by exposure to events that, as percepts, have not been linked with affective reactions before. He suggested that in the absence of sufficient stimulus similarity, a *complex* integrating related percepts and sensations would be activated, and that this complex would be capable of mediating empathic responses. For instance, a person may never have seen him- or herself stand on the edge of a precipice, especially not on one that is in the process of breaking off. But upon witnessing another stand there, confronted with disaster, the onlooker is bound to empathize in virtually sensing the model's loss of balance. According to Humphrey, the empathic reaction is the result of stimulus components of a complex such as 'soil slipping beneath' that is activated by the perception of the model's dilemma. Should the observer lack affective experience with slipping soil, the concept of 'losing balance' could be invoked as the empathy mediator. Essentially, then, any percept is viewed as capable of activating experientially pertinent complexes. If complexes integrating immediately related experiences do not exist, those that integrate more remotely related experiences are called upon. Because Humphrey did not focus attention solely on the expressive component of affect, his reasoning, especially his expansion of the conditioning paradigm, resulted in an empathy model that is sensitive to both causal circumstances and delayed expressive behavior. Perhaps most importantly, however, it created a model that is sensitive to affective experience and affective memory generally.

The proposal that initially neutral stimuli can, through conditioning, assume the power of eliciting empathic reactions, especially their excitatory concomitants, has received considerable support (e.g., Berger, 1962; Craig & Lowery, 1969). In such acquisition of empathic reactivity, information about the experiential quality of the model's reaction (regardless of its immediately perceptible facial or bodily expression) appears to be more important than information about the experience-inducing conditions per se (e.g., Hygge, 1976a, 1976b). Most of the research done in this area has focused on the affect-eliciting properties of facial expression, however. Generally speaking, facial expressiveness, autonomic activity, and self-reported intensity of affect tend to be positively correlated (e.g., Zuckerman, Klorman, Larrance, & Spiegel, 1981). As the likely consequence of this state of affairs, conditioning of empathic reactions to facial expressions proved to be comparatively easy under conditions of face-affect congruity, and rather difficult under conditions of face-affect incongruity (e.g., Lanzetta & Orr, 1981; Orr & Lanzetta, 1980). An investigation by Englis, Vaughan, and Lanzetta (1982) concentrated on the affective consequences of persons' conditioning history of facial expressions. Symmetry between a model's facially expressed affect and an observer's contingent affective experience was found to produce particularly strong empathic reactions to facially expressed affect after discontinuance of the contingent stimulus. Asymmetry, in contrast, produced anti-empathic reactions (i.e. discordant affect) or indifference. It has further been shown that the excitatory component of empathic reactions that were established under conditions of face-affect symmetry are more resistant to extinction than are reactions established under other circumstances (e.g., Öhman & Dimberg, 1978). Empathic responding thus seems readily acquired and maintained under some conditions, but not under others. Learning is apparently partial to ecologically concurrent response components (cf. Seligman, 1970).

The involvement of cognition in acquired empathic reactivity may again be considered secondary, as it becomes operative only in the aftermath of immediate affective responding. Mediation of such responding via the basic learning mechanisms does not necessitate cognitive elaboration, but post-reactive awareness of parallel reactivity by self and other is once more considered to provide the reactivity's experiential quality that renders it empathy.

Empathy as Cognitively Mediated Reactivity

Finally, theories in an explicitly cognitive vein have been pioneered by Smith (1759/1971) and, among others, by Stotland (1969). In these theories, cognition that accomplishes, or that seem to accomplish, the imaginary placement of an observer into the observed and his or her experiential state, whether deliberate or instigated by environmental stimuli, are assigned the key role in the elicitation of empathy.

Smith, in his classic theory of moral sentiments, anticipated much of contemporary cognitive psychology, or, more accurately, its application to affect and empathy. "By the imagination," he wrote, "we place ourselves in his [i.e. the observed person's] situation, we conceive ourselves enduring all the same torments, we enter as it were into his body and become in some measure him, and thence form some idea of his sensations, and even feel something which, though weaker in degree, is not altogether unlike them [pp. 2–3]." Despite early criticism of this view as an over-intellectualization of the empathy process (Humphrey, 1922), the suggested mechanism (i.e. empathy as more-or-less deliberate and conscious place-taking) is probably the one that is best known and most widely adopted (cf. Katz, 1963; Mead, 1934; Rogers, 1967).

The work of Stotland and his coworkers (e.g., Mathews & Stotland, 1973; Stotland, 1969; Stotland, et al., 1978) can be considered to have firmly established that imagination indeed does produce and enhance empathy, both the subjective experience and its physiological accompaniments. In particular, the instruction to imagine oneself in an affect-exhibiting person's place has been shown to foster empathic reactions of greater intensity than observing this person without such instruction. Pronounced affective reactions were often observed to occur with considerable latency after the onset of exposure to others in distressing situations, and this circumstance has been interpreted as showing that empathy does not so much result from exposure to external stimuli, even when favorable perceptual conditions exist (i.e. an empathy-directed cognitive set), as from the cognitive operations involved in the observer's imagination of the observed person's experiential situation. It is conceivable that imaginative efforts that take off on the observation of others' emotions revive related images from the observer's own past, and that this ideational activity triggers affective responses that are then construed as empathy. If so, empathic reactivity would not so much be the result of 'putting oneself into the place of another' as it would reflect the effect of affective memories of experiences similar to those witnessed in another. It would be the causal attribution of these memory-generated affective reactions to the imagination of taking an emotional other's place, implicit and erroneous as this attribution may be, that would justify considering the reaction one of empathy. It should be noticed that the discussed cognitive mediation of empathic reactivity are very similar to those in Humphrey's proposals. Granted vast differences in terminology and conceptual specifics, empathic responding is thought greatly influenced, if not entirely mediated by, the activation of observation-related affective experiences in memory. It is therefore subjectively bound, which renders the 'taking of another's place' illusory yet necessary for the experience of feeling with or feeling for another being.

In these theories, then, cognitive elaboration is the primary empathy-mediating process. The very intent or desire to emphasize with another is often the starting point of empathic reactivity. The volition to feel for or with another is thought to activate affective memories pertaining to the broadly conceived circumstances that instigate the other's expressed or inferred experiential state. Such cognitive focus on the other's situation, especially because it entails a memory scan for related personal experiences, is presumed to encompass conscious reflection of affective reactivity. If this reactivity is volitionally inspired, it is bound to foster awareness and ultimately empathic experience. However, the reactivity also may be triggered by the incidental encounter of others in emotional situations, and thus be spontaneous rather than deliberate. It is expected to foster awareness of parallel feelings nonetheless, thereby qualifying it as empathy.

Toward an Integrated Theory of Empathy

If we label empathic or empathylike reactions that are controlled by built-in dispositions *reflexive*, those that are acquired through some form of conditioning *learned*, and those that are mediated by purposeful, comparatively complex cognitive operations *deliberate*, it can be said that there is some supportive evidence for reflexive, learned, and deliberate empathy. At the same time it becomes clear, however, that none of the individual mechanisms can provide a satisfying explanation for all empathy phenomena. It also becomes clear that the basic mechanisms of empathy tend to be confounded in most empathic experiences of interest. Reflexive empathic reactions, for instance, make for an incomplete empathic experience, unless some form of appraisal is involved (Hoffman, 1978, 1987). Any appraisal is likely to entail components of deliberate empathy, possibly of learned empathy. Analogously, learned empathy is likely to entail components of reflexive empathy and to foster an appraisal that may have a deliberate empathy-enhancing component. Deliberate empathy, finally, might exploit reflexive and learned reactions to ideational activity that is purposely produced.

The limitations of individual mechanisms of empathy are thus obtrusively evident. It is difficult to see, for instance, how facial mimicry could be explained as the result of time-consuming, deliberate cognitive operations. It is equally difficult to see how any reference to complex, deliberate cognitive activity could explain the elicitation of learned excitatory reactions that accompany empathy and that critically influence the depth of the experience. Moreover, reflexive and learned mechanics, as long as stimulation is restricted to immediately present environment stimuli, are at a loss in explaining empathy that is brought on deliberately through imaginative processes. It would seem imperative, therefore, to integrate the basic paradigms into a unified theory that more fully accounts for the diversity of empathy phenomena. Such an integration is attempted in the text to follow. A paradigm is proposed that not only brings the principal mechanisms together, but that considers and delineates modes of interaction for these mechanisms.

THREE-FACTOR THEORY OF EMPATHY

The theory to be specified is an application of the three-factor theory of emotion (Zillmann, 1978, 1979, 1983, 1984, 1996, 2003) to empathy phenomena. Briefly, three-factor theory projects emotional behavior as the result of the interaction of three behavior-controlling forces: the dispositional, the excitatory, and the experiential component.

The *dispositional* component is conceived of as a response-guiding mechanism. Immediate motor reactions to emotion-inducing stimuli are assumed to be largely under stimulus and reinforcement control. In emotional behavior, then, the immediate skeletal-motor reactions are viewed as responses to stimuli that are made without mediation by elaborate cognitive information processing, such as complex appraisals of circumstances; that is, they are viewed as responses made without the substantial latency that is characteristic of complex cognitive mediation.

The *excitatory* component is conceived of as a response-energizing mechanism. Excitatory reactions, analogous to skeletal-motor reactions, are assumed to be reflexive, but also under stimulus and reinforcement control, again without the necessary involvement of complex cognitive mediation. Excitation is operationalized as heightened activity in the sympathetic nervous system, primarily, that prepares the organism for the temporary engagement in vigorous action such as needed for fight or flight. This excitatory preparation is largely independent of cognitive mediation and, in fact, eludes volitional cognitive intervention to a high degree. It is not assumed, however, that the preparedness for vigorous action has appetitive properties in the sense that it motivates specific goal-directed behaviors.

The *experiential* component, finally, is conceived of as the conscious experience of the skeletal-motor and/or the excitatory reaction to emotion-inducing stimuli. It is assumed that exteroceptive and/or interoceptive information about many facets of an immediate emotional reaction reaches awareness, and that this awareness fosters an appraisal of the response-eliciting circumstances. It is assumed that individuals continually monitor their emotional behavior, and furthermore, that they, by applying pertinent criteria to the monitoring process, determine the utility and appropriateness of emotional reactions and actions. Actions that are deemed inappropriate are volitionally inhibited and discontinued. Incipient, unfolding reactions are similarly appraised and, if need be, inhibited or, to the extent that this can be accomplished by deliberate intervention, terminated. Immediate emotional reactions, when deemed appropriate, are continued and may be redirected, via cognitive mediation, to better achieve desirable ends. The experiential component of emotion is thus viewed as a corrective capable of altering the course of emotional behavior and experience. It might be considered a cognitive means of control that can modify and override, to some degree, the operation of the more archaic mechanics stipulated in the dispositional and excitatory components of emotional behavior.

Applied to empathy, the three components can be specified as follows.

1. Reflexive and learned skeletal-motor reactions that are elicited by exposure to another person's manifest or impending emotional behavior along with information about its causal circumstances constitute the *dispositional component of empathy*. Motor mimicry, especially mimicry involving the facial muscles, comprises a multitude of specific reflexive responses. Many facial and gross motor responses are likely to be learned, however. Especially in confrontation with acutely dangerous situations, individuals tend to acquire personally specific, immediate coping reactions.

Reflexive and learned skeletal-motor reactions elicited by exposure to representations of high iconicity (i.e. photography, cinematography) that exhibit another person's manifest or impending emotional behavior are also subsumed in the dispositional component of empathy. So are such reactions that are elicited by exposure to non-iconic representations (i.e., signs with arbitrarily defined stimulus-referent relationships; practically speaking, almost all linguistic signs and sign aggregates) that specify another person's manifest or impending emotional behavior. It is assumed that the control exerted by iconic versus non-iconic representations over skeletal-motor reactions is different for reflexive responses, but similar for learned ones. Specifically, iconic representations are considered capable of eliciting reflexive reactions because of great stimulus similarity with the represented events. Non-iconic representations cannot have this capacity, as responses to arbitrarily established signs would have to be learned. Learning makes iconicity immaterial. Potentially, all stimulus conditions can acquire the power to elicit particular responses. Additionally, it is assumed that iconic and non-iconic elements of the representation of emotional events are interconnected in memory, forming complex networks that integrate pertinent percepts and operations (Kieras, 1978; Kintsch, 1974; Lang, 1979, 1984; Pylyshyn, 1973; see also Damasio, 1994; LeDoux & Phelps, 2000; Zillmann, 2003, this volume, Chapter 13 *Dramaturgy for Emotions from Fictional Narration*). External stimuli in one representational mode may thus activate stored representational information in another mode and thereby extend their impact.

2.   Excitatory reactions resulting from exposure to another person's manifest or impending emotional behavior along with information about its causal circumstances that are jointly elicited with reflexive and learned motor responses, or that are reflexive and learned but independently elicited, constitute the *excitatory component of empathy*.

Reflexive and learned excitatory reactions elicited by exposure to iconic or non-iconic representations that exhibit another person's manifest or impending emotional behavior are also subsumed in the excitatory component of empathy. All considerations concerning the semiotic modality of representation that were developed in connection with the dispositional component apply equally to the excitatory component.

3.   In the *experiential component of empathy*, three subcomponents can be distinguished: processes that serve (a) the experience proper; (b) the correction and redirection of affective reactions; and (c) the generation of affective reactions.

(a) An affective reaction elicited by exposure to another person's manifest or impending emotional behavior along with information about its causal circumstances or to any kind of representation thereof constitutes empathy only to the extent that observers are cognizant of their reaction and appraise it as feeling with or feeling for the person observed. It is assumed that such appraisal presupposes hedonic concordance between the observed and observing persons' responses; that is, the observers' affective reaction that is hedonically opposite to that apparent in the observed person is unlikely to be construed by the observers as an empathic reaction.

An affective reaction that is elicited by exposure to the conditions specified above and that a third party perceives as being similar to and hedonically concordant with that displayed by the observed person can be considered an inferred empathic reaction.

As an affective experience, a complete empathic reaction is comprised of dispositional, excitatory, and experiential response elements.

(b) An affective reaction elicited by exposure to another person's manifest or impending emotional behavior or to any kind of representation thereof is monitored for appropriateness. Respondents assess their reaction in terms of social and moral judgment; that is, they employ their knowledge of the prevailing contingencies of social approval and reproach, as well as internalized moral standards of conduct, in determining the appropriateness of the reaction. If the reaction is deemed appropriate, it is allowed to unfold. If it is deemed inappropriate, it is inhibited and redirected so as to conform with, or at least be less in violation of, accepted rules of social conduct and moral precepts.

It is assumed that affective reactions are initially comprised of reflexive and learned response elements. Because of neural mediation, skeletal-motor responses, in particular responses in the facial muscles, follow the onset of stimulation quasi-instantaneously. In contrast, owing to the involvement of systemic hormonal processes, the excitatory reaction develops with appreciable latency. Cognizance of an affective reaction (i.e. of a response that has both a dispositional and an excitatory component) thus can manifest itself only after some delay. It is further assumed that motor and facial muscles are subject to volitional control, whereas excitatory reactivity largely eludes such control (Zillmann, 1979, 1983, 1988).

Based on these assumptions, it is proposed that incipient reactions of empathy that individuals deem inappropriate take the following characteristic course: Upon stimulation, skeletal-motor reactions as well as facial responses materialize quasi-instantaneously. The excitatory response unfolds with some latency. Individuals attain proprioceptive and, to some degree, exteroceptive information from both their muscular and excitatory reactivity. The reception of proprioceptive and exteroceptive feedback instigates appraisal processes. If these processes render the affective reaction inappropriate, the ongoing motor activity and its facial accompaniment are inhibited and redirected into an affective reaction that better conforms to prevailing rules of conduct. As the inhibition of the jointly elicited excitatory reaction cannot be similarly accomplished, at least not immediately, excitation will enter into and thereby influence the affect arrived at through cognitive intervention and correction (Zillmann, 1978, 1983). This intervention in, and correction and redirection of, ongoing affective behavior has been termed *cognitive override*. Most importantly, such override of affective reactivity is considered capable of converting the hedonic valence of reactions. Whenever the indicated hedonic conversion occurs, initially empathylike affective reactions—because they cannot be construed as feeling with—not only lose their empathic quality in becoming non-empathic, but attain an experiential quality opposite to empathy in becoming anti- or counter-empathic.

Affective dispositions, manifest in degrees of liking and disliking of beings and objects, that are held toward persons observed undergoing emotional experiences, entail a readiness for the cognitive sanctioning (in case of liking) or overruling (in case of disliking) of affective reactivity. Positive affective dispositions are conducive to empathy. Negative affective dispositions, in contrast, demand the hedonic conversion of affect, mostly because the continuance of empathylike responding would be noxiously experienced whereas anti-empathic responding should be pleasurable. Corrective readiness mediates the rapid adjustment of facial expressiveness to converted, altered affective responding. Cognizance of socially sanctioned display rules for affective experiences should assist the indicated mediation (Saarni, 1982; Saarni & von Salisch, 1993). However, such cognizance also may demand the expression of affect independent of, and even opposite to, affective dispositions; that is, display rules may prescribe, for example, the dissimulation of the expression of sadness in experiences of joy and the expression of joy in experiences of sadness.

(c) Affective reactions that are construable as empathy can be deliberately elicited by the imagination of situations that are closely related to those confronting an observed party and that have produced intense affective reactions in the observer's past. The process need not be deliberate, however. Imagination of this kind may be instigated by exposure to conditions conducive to the elicitation of empathic reactions, and affect generated by ideational stimuli may complement and enhance affect in response to external stimulation.

It is assumed that the imagination of situations to which motor and excitatory reactions have been conditioned is capable of eliciting these reactions (Lang, 1979, 1984). However, owing to proficient volitional control of motor responses, apparently inappropriate gestural, postural, and gross motor responses are, as a rule, effectively inhibited. Furthermore, owing to habituation (Grings & Dawson, 1978; Zillmann, 1984), excitatory responses are likely to be of diminished strength. These qualifications notwithstanding, ideational representations of stimulus conditions to which excitatory reactions have been conditioned are capable of eliciting these reactions to some degree (Keltner & Ekman, 2000; Levenson et al., 1990; Schwartz, 1977). To the extent that these representations can be controlled volitionally, the excitatory concomitants of affect can be generated deliberately. Empathy through role-taking can thus be viewed as mediated through the partial revival of related past experiences.

It should be recognized that the possibility of generating affective reactions and empathy through the ideational representations of potent stimuli also has implications for the inhibition of empathic reactions. If empathy is deemed inappropriate, individuals can avoid ideation of affectively potent representations, or they can practice ideation of unrelated representations to stifle and prevent unwanted ideation.

Dispositional Override of Empathy

As a part of the three-factor approach, the concept of dispositional override is the most significant point of departure from alternative models and seems in need of further explication. The concept projects, essentially, that an empathic experience entails both archaic and, phylogenetically speaking, more recently developed elements. Comparatively primitive processes, such as motor mimicry and involuntary excitatory responses, are viewed as being superseded by complex cognitive processes, with these latter processes taking control of the experience in determining its final status. The concept is reminiscent of McDougall's (1908, 1922) position discussed earlier. As will be recalled, he projected that 'primitive passive sympathy,' when it produces suffering that the individual deems unwarranted, may be converted to mirth. This conversion was thought to serve the maintenance of emotional health. The position taken here is far broader, however, and it is independent of considerations concerning adaptation and emotional adjustment. Regardless of the experiential quality of immediate affective reactions, cognitive monitoring should effect their inhibition or hedonic conversion if this brings the responses into better agreement with socially and/or individually sanctioned emotional behavior. It is expected, then, that inappropriate initial affect, when positive, can become nil or negative, and, when negative, can become nil or positive.

Because of the proposed sequence of events, the three-factor model can make sense of mixed affective reactions that are left unexplained by alternative views. If, for example, a student witnesses from a private corner her oppressive, resented professor mount a bicycle, lose balance, fall to the ground, get up, and hobble around in apparent pain, she may well cringe upon seeing him fall, but quickly come to chuckle and grin, if not to laugh out loud. The dominant affective experience being one of delight, the student is unlikely to construe her response as empathy, despite the initial cringe. Had the student been in a public situation, chuckling and laughing would probably have been drowned in their incipient stages. The deliberate curtailment of expression should have been of little consequence for the experience of delight, however. This is because it is not feedback from the expression of affect, but the cognitive appraisal of the circumstances, that is expected to function as the primary determinant of affective experience. The example can readily be altered to illustrate dispositional preparedness. If the student did not merely feel resentment toward her professor, but acutely feared and hated him, virtually hoping for some misfortune to come his way, seeing the bike sway might foster expectations of delight. If so, witnessing the accident might not produce initial cringing, but unqualified euphoria from the beginning. On the other hand, even a lot of spite might not assure continued delight. Should it become apparent, for example, that the accident crippled the professor, it would be unlikely that the student goes on rejoicing. Appraisal would make continued delight an unacceptable response. To the extent that thoughts of pity are evoked, the experience might actually come to be construed as empathy. Such arrival at empathy would not preclude that the student had awareness of the fact that her initial reaction was of a different kind.

Clearly, the hedonically opposite sequence of events (i.e. the case not considered by McDougall) is equally conceivable. Students who tenaciously competed for a particular award, for example, may well facially exhibit a moment of joy as they respond to the winner's display of triumph and happiness. Such mimicry is likely to be overpowered, however, by expressive responses deriving from disappointment and envy, from feelings of having suffered an injustice, or from contempt for the winner and the system. The important point is that, although such complex and mixed affective reactions are likely to contain rudiments of empathy, the affective experience as a whole is not likely to be construed as empathy because of the dominance of cognitions alien to the concept of feeling with and feeling for.

Development of Empathy and Dispositional Override

As discussed earlier (see *Evolutionary approaches)*, motor mimicry constitutes an archaic response form with considerable adaptive utility. The limits of this adaptive quality are equally obvious, however. Motor mimicry would be quite maladaptive, for instance, if members of a species that witness debilitating illness in others were to take that as a cue for emulating subdued behaviors, thereby increasing their chances to fall victim to predation. At the human level, as McDougall (1922) has so aptly articulated, such 'primitive sympathy' would also be of questionable value. Co-suffering, in and of itself, has no utility. Only to the extent that it promotes behavior directed at lessening the model's misery does it become adaptive (Eisenberg, 2000; Hoffman, 1978; Stotland et al., 1978). In order to accomplish the behavioral plasticity needed for adaptive intraspecific interaction, mechanisms more complex than those governing motor mimicry had to evolve. The mechanics of stimulus and reinforcement control can be viewed as the post-reflexive stage that provided much needed plasticity. And the human capacity for the arbitrary handling of information, the greater independence from immediately present stimuli in particular, can be viewed as having afforded us yet greater behavioral plasticity. Phylogenetically, then, the progression of control in empathylike behaviors is from reflexive through learning and, in humans, from learning through cognitive mediation.

This progression is also evident in individual development. Hoffman (1978) and others (e.g., Eisenberg, 2000) have reviewed the pertinent research on empathy and found it to be consistent with such a developmental ordering. Empathylike, behaviorally contagious responses are initially fixed. They then become varied but remain largely involuntary. Finally, cognitive modification of affect comes into being, and deliberate empathy becomes possible.

Although the developmental age at which the capacity for the dispositional overriding of empathic reactions materializes has not been determined, it may be assumed that it is closely tied to the emergence of monitoring affect of self and of moral judgment in its basal forms. As soon as individuals recognize prevailing contingencies of reinforcement (in the sense of responding to them properly), they are likely to be disturbed at seeing these contingencies waived for others (Zillmann, 1979, 2000). Witnessing the benefaction of an undeserving party is annoying and perturbs moral sentiment. Witnessing the aversive treatment of a party considered deserving of such treatment, in contrast, is in line with basal morality and consequently may be applauded and enjoyed. At the very least, seeing those who attained gratifications by violating established rules of social conduct, or simply by brutalizing others, be duly punished for their transgressions, does not call for empathic co-suffering. Onlookers seem morally entitled to the callousness that is manifest in responding non-empathically, even euphorically, to seeing aversive treatments applied to villainous parties and to seeing the impact of these treatments (Zillmann, 1980; Zillmann & Cantor, 1976).

It can only be speculated that, in the transition from reflexive and stimulus-reinforcement control of empathy to the predominantly cognitive mediation of affect, children experience considerable affective conflict: empathic inclinations, on the one hand, and non-empathic emotions such as spiteful envy (in case of a model's undeserved benefaction and joy) and satisfying pleasure (in case of a model's deserved and justified punishment and pain), on the other. Possibly ambivalent affect should soon give way, however, to unequivocal affective reactions that are mediated by considerations of appropriateness and justice. To the extent that such cognitively controlled affect produces anti-empathic reactions, every reaction can be considered a learning trial in the formation of response dispositions for discordant affect. Discordant affect eventually becomes associated with witnessing parties undeserving of good fortunes and deserving of misfortunes. In short, discordant affect is consistently connected with villainous persons; that is, with persons who, for assorted reasons, are disliked, resented, or acutely despised by the respondent. A negative affective disposition toward a model thus becomes predictive of anti-empathy. Disliking comes to signal that empathy is unnecessary and, in fact, inappropriate. Because discordant affect is consistently sanctioned under these dispositional circumstances, it should become the characteristic response mode. Negative affect should create a readiness for counter-empathic reactions; and the reactions, once initiated, should become mechanical in the sense that the individual need not engage in explicit judgmental deliberation concerning the circumstances. The involvement of moral judgment in dispositional override and, especially, the concept of moral sanction in this context are further developed in this volume's Chapter 13, *Dramaturgy for Emotions from Fictional Narration.*

Affective Experience and Empathy

As discordant affect can become a highly mechanical response (one that takes its cue from dislike and resentment that at first might be inspired by apprehension and fear, and that later grows on disapproval and condemnation of the model's actions), so can empathy, via role-taking. Although the initial stages of the mental exercise of placing oneself into another person's situation are entirely deliberate, affect is likely to be evoked by the recall of related affective experiences of self. Once such recall occurs and the deliberate empathizer senses responding in an excited fashion, the arousal state itself may function as a cue to recalling further emotional experiences (Clark, 1982; LeDoux & Phelps, 2000; Leight & Ellis, 1981). It can be argued, in fact, that the initial 'make-believe' effort in deliberate empathy is incapable of eliciting appreciable affect unless it hooks into related affective experiences of self (Scheler, 1913). Deliberate empathy, in this view, is akin to system acting; that is, to a procedure in which actors and actresses use volition to access affective memory in order to trigger the emotional experience they seek to display (cf. Konijn, 2000). Such reasoning makes it very clear that role-play empathy relies on affective memory in the observer, and that it cannot be more than mere pretense if the observer does not have related experiences that can be revived. This dependence of empathy on pertinent experience has been emphasized by many investigators (e.g., Allport, 1924; Murphy, 1937, 1983; Sapolsky & Zillmann, 1978). Scheler (1913) used it to show that empathy can never be a true sharing of feelings, but is at best a close approximation of affect in observer and observed, because observers are only capable of responding in terms of their own, private and unique affective experience.

Disapprobation, Disliking, and Discordant Affect

The observation of a model's overt behavior, along with intentions inferred on the basis of this behavior, is expected to take the following course (see also Figure 3 of Chapter 13, *Dramaturgy for Emotions from Fictional Narration,* this volume).

If a model's behavior and apparent intentions are approved and deemed commendable, a disposition of liking and caring is formed. This disposition of amity fosters, in turn, the anticipatory duality of welcoming, even hoping for, the benefaction of the model while, at the same time, rejecting, even fearing, the victimization of the model. As either eventuality materializes, the observer will respond *empathically* when witnessing the model's good or bad fortune, responding with *concordant* affect to both the benefited, gratified model as well as to the victimized, suffering model. In this case, the model's joy is the observer's pleasure; and analogously, the model's agony is the observer's anguish.

If a model's behavior and apparent intentions are deplored and call for condemnation, a disposition of disliking and detesting is formed. This disposition of enmity fosters, in turn, the anticipatory duality of rejecting, even fearing, the benefaction of the model while, at the same time, welcoming and even hoping for the victimization of the model. As either eventuality materializes, the observer will respond *anti-empathically* when witnessing the model's good or bad fortune, responding with *discordant* affect to both the benefited, gratified model as well as to the victimized, suffering model. In this case, the model's joy is the observer's anguish, and analogously, the model's agony is the observer's pleasure.

These parallel yet diametrically opposite propositions can be applied recursively. By feeding later evaluations of actions, apparent intentions, and circumstances back to the beginning or to early stages of the process chain, initially developed dispositions can be appreciably modified or substantially altered. The repeated dispositional adjustment creates a dynamic system that accommodates complex characters with, morally speaking, inconsistent behavioral histories. This adjustment provides a means of dealing with characters of any kind; for example, with those who turn from good to evil or from exceedingly bad to kind, caring, and simply wonderful, not to mention those who are stable conglomerates of good and evil traits.

Zillmann and Cantor (1977) conducted an investigation that strongly supports the proposed chains of events. Films were especially produced for child audiences. Specifically, second- and third-grade boys and girls were exposed to different versions of a film, and the children's affective reactivity to the film was assessed in facial expressions and in structured interviews that ascertained perceptions of, and dispositions toward, the film's protagonist. The film versions featured a peer protagonist who was presented either as a most pleasant and helpful character or as an obnoxious and hostile one. The different character development was accomplished by depicting the protagonist as kind and supportive in interacting with his peers, pet, and younger brother versus as rude and mean in the same interactions. The findings showed that, in accord with the manipulation, the character was perceived either as benevolent or as malevolent. They showed, furthermore, that respondents came to like the 'nice' protagonist and to dislike the 'bad' one. These findings corroborate the proposal that a person's behavior is judged in terms of approval and disapproval, and that affective dispositions toward a person are formed on the basis of this approval and disapproval. The film versions, finally, were given different endings. In a happy-ending version, the protagonist was seen receiving a new bicycle. He expressed euphoria in response to the gift, jumping for joy with a happy smile on his face. In the alternative ending, tragedy came about as the protagonist jumped on his old bike, and starting off down the street, lost balance and crashed. His dysphoria was bodily and facially evident as he cringed in pain. Affective responses to these concluding events, as measured through interview, showed empathy to both the euphoric and the dysphoric outcome only under conditions of positive affective dispositions toward the protagonist. Under conditions of negative affective dispositions, empathy was not only absent, but discordant affect was observed. The transverse interaction between affective disposition and affective response is displayed in Figure 10.1. This data pattern was obtained for both girls and boys.

Apparently, the respondents did pass judgment on how good or bad the protagonist's behavior was, and accordingly, they were prepared to accept the protagonist's good or bad fortune only when it could be deemed fair and deserved. Such judgment made it inappropriate to enjoy witnessing the benefaction of the malevolent and, hence, undeserving character. It seems to have resulted in acute annoyance with the outcome that could only be perceived as utterly unfair and unjust. Such judgment made it very appropriate, however, to enjoy witnessing the suffering of the malevolent and, hence, deserving character.

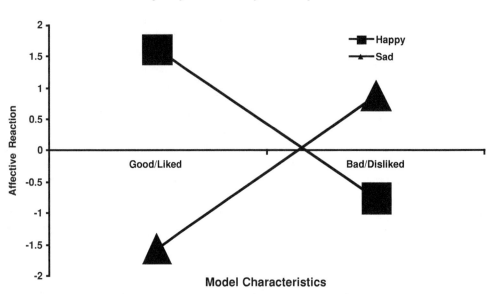

FIG. 10.1.   Empathic and counter-empathic reactions of normally developed children to the expression of happiness versus sadness by a model perceived to be either good or bad. The children exhibited empathy only when the model was good and liked. Disapproval of action fostered disliking, and disliking produced discordant affect in response to the bad model's emotions (adapted from Zillmann & Cantor, 1977).

The analysis of respondents' facial expressions, interestingly, failed to exhibit a close correspondence with verbally expressed empathy and discordant affect. This finding can be interpreted as evidence against the view that facial responses serve as the primary determinant of feeling states. The influence of cognitive operations apparently dominated that of facial expressions in the production of concordant and discordant affect. The analysis leaves it unclear, however, to what extent the low correlation between reported affect and facial expression may have resulted from inconsistent and contradictory facial responses in the conditions producing discordant affect. It is conceivable that the children, when exposed to the malevolent character's bicycle accident, initially cringed reflexively, but then quickly brought their face in line with their cognitive appraisal of the events before them. It is also conceivable that such a correction was made in case of the malevolent character's surprise benefaction. The children may have responded positively at first, if only momentarily, and then corrected their reaction. The possible reversion of facially expressed affect is entirely consistent with the proposed three-factor model of empathy. According to this model, faces can follow cognitive appraisals as much as cognitive appraisals can follow faces. But the findings do not help resolve this issue, as facial responses were assessed across the entirety of the concluding events. When the need for a more detailed analysis became apparent, records had been destroyed already to assure confidentiality.

The moral-judgment mediation of empathy and, especially, of discordant affect is more clearly apparent in comparing the responses of children who have developed equity judgment with responses from children who have not. Wilson, Cantor, Gordon, and Zillmann (1986) used the films employed by Zillmann and Cantor (1977) in an investigation with mentally challenged children who were judged incapable of equity judgment. Consistent with data from emotionally disturbed children, reported by Feldman, White, and Lobato (1982), the challenged children proved capable of perceiving the character correctly as 'nice' or 'bad' (although, similar to children at the lowest levels of moral development, their perception was partly a function of events for which the character could not be held accountable; for instance, the nice protagonist was perceived as being nicer when he received a gift than when he suffered a misfortune). The challenged children thus seem to have managed the approbation versus disapprobation of a model's action, at least in the sense of correctly classifying and labeling persons in accord with social sanction versus condemnation. However, these children gave no evidence of executing assessments of deservingness, nor of forming a disposition consistent with such assessments. They seemed not to hope for particular outcomes, nor fear others. In particular, they failed to arrive at disliking and its consequence of empathic deficiency and callousness. The absence of this moral mediation led them to respond empathically, regardless of particular dispositional circumstances. As can be seen from Figure 10.2, they responded euphorically when the bad character was benefited, and they responded dysphorically when this character suffered a misfortune. These findings make it clear that discordant affect in response to witnessing a resented agent's euphoria or dysphoria is indeed mediated by considerations of deservingness (i.e. by moral judgment).

The moral-judgment mediation of discordant affect has been further established by another investigation with children (Zillmann & Bryant, 1975). An audiovisually presented fairy tale was manipulated to create different punitive treatments. A thoroughly good prince combated a thoroughly bad prince. In their struggle, the bad prince got the upper hand and condemned the good prince to exile in an undesirable part of the kingdom. The good prince eventually returned to power, however, and now was in a position to punish his tormentor. This punishment was

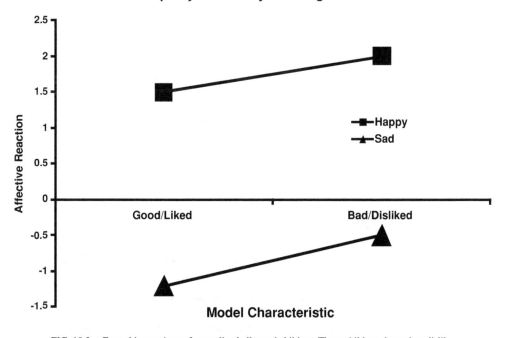

FIG. 10.2. Empathic reactions of mentally challenged children. These children showed no disliking for a badly behaving model and, hence, did not develop a disposition to respond with discordant affect to the model's happiness or sadness. The children responded empathically throughout, irrespective of whether or not outcomes for a model could be deemed deserved (adapted from Wilson, Cantor, Gordon, & Zillmann, 1986).

either equitable (i.e. the good prince applied the punishment that the bad prince had planned for him), too mild (i.e. the good prince proved forgiving and did not retaliate), or too severe (i.e. the good prince's retaliation was far more brutal than the bad prince's initial misdeed). As punishment was justified, normal children were expected to enjoy its application (i.e., to approve it and show callousness in discordant, anti-empathic reactivity). Dependent on the level of moral development, different approvals may be expected, however. Children matured to the level of equitable retribution (Kohlberg, 1964; Piaget, 1948) should approve equitable punishment, but neither under- nor over-retaliation. In these cases of inequitable retribution, their sense of justice should be left disturbed, and this disturbance should hamper and reduce the euphoric reaction to witnessing the application of punishment. Children at the level of expiatory retribution, in contrast, should freely enjoy witnessing any amount of punishment. In fact, as they tend to infer the magnitude of a transgression from the severity of punishment, it can be expected that enjoyment will increase with the severity of punishment. The implicit assessment is that the more severe the punishment, the greater must have been the violation of unquestioned rules of conduct that the punishment corrected.

Figure 13.3 shows data that fully corroborate the predictions concerning the joy that is liberated by the infliction of aversion upon a party apparently deserving of such treatment. The findings give further support to the process-chain schema of empathic and anti-empathic reactivity detailed at the beginning of this section (see also Fig. 13.3 of the Chapter 13, *Dramaturgy for Emotions from Fictional Narration,* this volume). The schema stipulates that

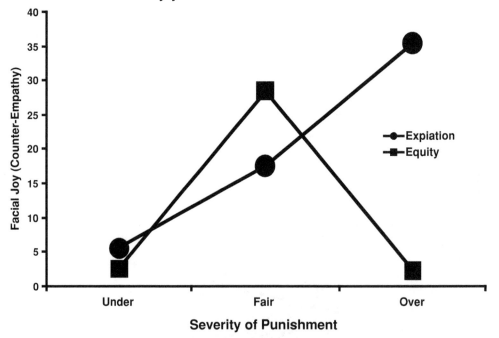

FIG. 10.3.   Enjoyment of punitive action against a transgressor as a function of moral-judgment development. For children at the developmental level of expiatory retribution (represented by 4-year-olds), the enjoyment of punishment increased with its severity. Children at the developmental level of equitable retribution (represented by 7- and 8-year-olds) enjoyed equitable punishment more than inequitable punishment, regardless of the direction of inequity (adapted from Zillmann & Bryant, 1975).

the post-affect approval or disapproval of model experiences is capable of feeding back into affective dispositions toward observed persons. The assessment is expected to be in moral terms and mediate liking and disliking of punished and punishing parties. This mediation was in fact observed. It conformed entirely to the respondent's system of moral judgment. Children at the level of equitable retribution exhibited intense dislike for the bad prince when he was not duly punished. When his punishment was overly severe, in contrast, disliking became negligible. It was as though they felt pity for him. These children liked the good prince only when he had been fair in applying punishment. His failure to be punitive, but especially his being unnecessarily brutal, prompted a considerable loss in liking. Children at the level of expiatory retribution responded very differently. They maintained a strong disliking for the bad prince. In fact, they disliked him the most when he had been unduly brutalized. They also liked the good prince the most when he had been overly punitive. This bears out Piaget (1948) who commented that for this developmental level the severity of punishment correlates with the perception of justice: "The sterner it is, the juster" (p. 199).

These and other findings concerning discordant, anti-empathic affect (e.g., Zillmann, 1980; Zillmann & Knobloch, 2001) leave little doubt that moral considerations play a significant part in justifying, allowing, and motivating discordant affective reactions to the emotional experiences of others.

EMPATHY IN THE INFORMATION AGE

The new communication technology, in particular its seemingly unlimited capacity for recording, storing, recreating, manipulating, and transporting complex audiovisual events with superlative fidelity, is commonly thought to have a profound influence on the public's emotions, related affective experiences, and empathic sensitivities or deficiencies thereof. Projections of such influence have been greatly divergent, however. But notwithstanding the particular direction of any presumed influence, its mediation has remained rather unclear. Research into the effects of the communication media has largely neglected the specifics of hypothesized affect-enhancing and affect-diminishing processes. It has failed, moreover, to establish compellingly that effects on the development of empathy and on empathic sensitivity have significant social consequences.

The Changing Media Landscape

Clearly, communication technology has enormously altered the conditions for affect development and, perhaps more importantly, for affect maintenance. Compared to exposure to others' affects and emotions in pre-technological times, both children and adults in contemporary society are inundated with portrayals of others' affective behaviors and experiences. In pre-technological society, exposure to the emotions of others was limited to the immediate social environment and to storytelling. The frequency of exposure undoubtedly increased with the broadening availability of printed material, but portrayals of other's emotions remained limited to verbal reporting, occasionally embellished with drawings and later with a photograph or two. As far as the portrayal of others' emotions is concerned, the invention of the motion picture ushered in a new era of exposure magnitude. It became a daily experience to witness a broad range of human emotions expressed by a vast cast of humans and humanlike characters. The movies, initially delivered in theaters and later via television and computers, reached essentially all members of technological society, the so-called mass-media audience. More-or-less all members of society were, with increasing frequency, exposed to the emotional experiences of others. Some of these experiences they may have witnessed prior, though infrequently, but others may have been nonexistent in their immediate social environment. Adults who had never witnessed, for example, the agony of a severely injured soldier, the despair of a rape victim, the furor of a sadistic murderer, or the exultation of a record-setting athlete, now could witness such emotions. Analogously, children are now exposed to persons in extreme fear, anger, rage, hatred, or joy about a multitude of conditions that rarely, if ever, materialize in the family or peer context. As entertainment thrives on the exhibition of human emotions (Tannenbaum, 1980; Zillmann, 1980; Zillmann & Vorderer, 2000), there can be no doubt that in modern society children and adults alike are massively exposed to others' acute emotional experiences. News programs and documentaries are similarly partial to the exhibition of emotion-laden happenings, and further escalate the indicated exposure situation (Haskins, 1984; Zillmann, 1998).

Implications for Empathic Reactivity

With the inception of the new communication technology, then, the sheer magnitude of exposure to audiovisual representations of a vast array of others' emotional experiences has risen in leaps and bounds. Less obvious is the similarly dramatic and potentially significant concurrent change in the portrayal of these emotional experiences themselves. Whereas individuals in their own social environment tend to catch only glimpses of someone else's affective responses, the visual and audiovisual media present these reactions most graphically. The news is obviously partial to the portrayal of extreme emotions, especially of exceedingly aversive ones. In fiction, moreover, the portrayal of emotions is not only detailed and graphic, but tends to be elaborated and exaggerated with cinematographic techniques such as slow motion and numerous other reality-enhancing special effects. Respondents get to see and hear at close range, for example, a couple jumping for joy upon learning that they won a multi-million dollar jackpot, a hysterically screaming woman after being informed that her two sons drowned, a government official as he commits suicide by putting a gun to his head and pulling the trigger, or a little girl who slowly drowns in a mudslide. Close-ups can be extreme to the point where the respondent must feel being physically in front of the head, say, of the girl drowning in the mud, seeing the slush enter her mouth and cover her nose and eyes as she succumbs to the mire. To the extent that the facial expression of emotions and also the vocalization of these emotions are critically involved in the elicitation of empathy (as well as in that of anti-empathy), such 'supernormal' portrayal of human affective reactivity should prove capable of altering the reactivity that was established in non-artificial social environments.

First, it is conceivable that at early developmental stages the supernormal facial exhibition of a model's emotion fosters facial mimicry of great fidelity. If so, the child's empathic responding to a variety of others' emotional experiences could be facilitated. The expression of extreme emotions, in particular, is likely to foster long-lasting connections between affective displays and associated affective responses because the displays are likely to leave nearly indelible traces in emotional memory (LeDoux, 1996; McGaugh, 1992). Second, and regardless of the role that facial mimicry may play in the development of affective responding, frequent exposure to extremely detailed portrayals of others' affective reactions, as it informs about the circumstances under which such reactions are performed and seem appropriate, should further the acquisition of emotional display rules (cf. Saarni, 1982, 1999; Saarni & von Salisch, 1993) as well as competence in their usage. Under the assumption that media portrayals of the prevailing display rules are predominantly accurate, children (and adults) should become acquainted with a multitude of such rules and learn to enact them in novel situations; that is, in situations for which their limited social environment may not have prepared them.

Another principal difference between exposure to others' affect in pre-technological and technological societies concerns the semiotic modality of representations. In earlier times, representations were predominantly non-iconic, if not exclusively so. The poor stimulus relationship between sign and referent (i.e. between designating and designated stimuli) made empathic and counter-empathic reactions dependent on the generation of ideational representations that, in turn, depended on experience by self with similar affect-inducing conditions. In short, to come from verbal reports of others' emotions, even when aided by some grimacing and gesticulating by the reporter, to intense concordant or discordant affect relied on considerable imaginative activity. Today's iconic representations, because of their extreme reproductive fidelity, seem to render such activity superfluous. Little, if anything, is to be transformed. It remains unclear, however, whether the greater or the lesser reliance on imaginative processes fosters stronger empathic or counter-empathic reactions. It is conceivable that the conversion of non-iconic signs to an iconic ideational representation of others' emotions, because it draws heavily from related experiences of self, is particularly conducive to affective responsiveness. But it is equally possible that high-fidelity iconic representations also activate related affective memory structures. Such iconic representations may function as particularly potent affect inducers because, in the affect they elicit, they combine affect from memory revival with unmitigated affect from exposure to the iconic portrayals themselves.

Yet another fundamental difference in the portrayal of others' emotions in pre-technological and technological societies concerns the pace of presentations. Whereas affect from the direct perception of others' emotional experiences is usually not temporally curtailed (i.e., persons can focus on their experience as long as they please) and verbal representations tend to be paced by a speaker or are self-paced by a reader so as to permit the complete unfolding of affective responses, in the media of communication the episodic chunks of iconic representations follow one another without interruption. Specifically, the story moves on before affective reactions to an episode featuring the emotions of others can run their course.

In fast-paced, action-packed adventure films, for example, children may have seen a much-loved protagonist get caught, tormented, and killed by evil antagonists, along with the tearful grieving of the hero's friends. All this may have happened in about seven seconds. Without delay, the story might have continued with the prolonged suspenseful pursuit of the perpetrators and their eventual capture and punishment. Similarly, adults may react to a news program featuring an emotion-laden interview with a women who had just learned that her husband got killed in a coal mine explosion, the interview being immediately followed by reports about a bus-driver strike, record trading on Wall Street, and a commercial in which a female model attest to a cologne's libido-enhancing effect on men. Such unrelated informative or entertaining chunks of presentations are often arbitrarily placed into a sequence, apparently with little, if any, regard for the audience's emotional reactions to the various chunks.

Whatever the reasons for the practiced message pacing may be, the pace that characterizes contemporary audiovisual storytelling and reporting is likely to produce affective confusion with considerable regularity in both children and adults. Consideration of excitatory activity as the primary intensity-determining component of affective reactions leads to the projection that, owing to the latency of excitatory responses, many of these responses materialize too slowly and manifest themselves too late for the 'affective' reactions to which they belong. More importantly, owing to the slow dissipation of excitatory activity, it leads to the projection that excitation elicited by a particular episode or report will come to intensify the subsequently presented episode or report (Zillmann, 1978, 1984; see also Chapter 13, *Dramaturgy for Emotions from Fictional Narration*, this volume). A temporally curtailed sadness reaction to the misfortunes of a liked character (i.e., empathic distress), for instance, will produce increased amusement with subsequent humorous happenings (e.g., Cantor, Bryant, & Zillmann, 1974); and temporally curtailed reactions of empathic distress to nonfictitious events are likely to enhance the appeal of subsequently advertised products (e.g., Mattes & Cantor, 1982), among other things.

The rapid pacing of affect-inducing events in modern media presentations, then, appears to minimize affective and, especially, empathic reactions (a) by preventing the complete unfolding of the response because distracting, competing information is provided too soon after the elicitation of affect; and (b) by creating affective confusion through the intensification (by excitation transfer) of subsequent, potentially non-empathic affect. Empathic reactions, therefore, are often deprived of their inherent intensity, whereas subsequent reactions are often artificially intensified. On occasion, however, empathic responding may also be intensified by excitation from incidental exposure to preceding unrelated arousing material (e.g., Zillmann, Mody, & Cantor, 1974.). To the extent that the fast-paced presentation of chunked episodes impairs reflection on the respondents' own affective reactivity, and moreover that such reflection is vital to empathic experience, the fast-paced exposition of affect-inducing contents may be considered another factor in the development of impoverished empathic responding (Singer, 1980; Singer & Singer, 1990).

Finally, the sheer magnitude of exposure to others' emotions must be expected to diminish the intensity of empathic reactivity. The main reason for the waning of this reactivity is the habituation of excitatory responding associated with affective experiences (Grings & Dawson, 1978; Henry, 1986; Zillmann, 1996). Diminished affective reactivity has been most clearly demonstrated for the prolonged consumption of erotica (Howard, Reifler, & Liptzin, 1971; Mann et al., 1974; Zillmann, 1989, 1991). It is also obtrusively evident in the consumption of violent action and horror films by adolescents (Sparks & Sparks, 2000; Tamborini, 2003; Zillmann & Gibson, 1996; Zillmann & Weaver, 1996). Although direct research demonstrations of the loss of empathic reactivity from excitatory habituation caused by frequent exposure to others' emotions is lacking, there is ample reason, then, to expect that the indicated process of affect diminution extends to empathy.

Regarding the possibility of an enhancement of empathic reactivity by media presentations, the research on affective dispositions and their implications for empathy and counter-empathy points to moral distinctness and clarity in the portrayal of characters in any kind of drama— dramatic events in the news included. This applies especially to drama for children at the lower levels of moral development. For these children, the actions of characters unambiguously define the characters as either good or bad, loveable or hateful, friends or foes, or simply as heroes or villains. Unqualified liking constitutes the optimal condition for the development of empathic reactivity; unqualified disliking for the development of callousness necessary for counter-empathic responses. Indifference or ambivalence toward a person constitutes the condition under which neither concordant nor discordant affect is likely. Should affective reactivity occur, it will be of negligible intensity. This condition is consequently not conducive to the development of empathic sensitivity in children. Drama in which dispositions toward protagonists and antagonists are uncomplicated and clear-cut emerges as a genre that is well-suited for the teaching of such sensitivity. It might prove essential for the maintenance of empathic and counter-empathic sensitivity in adults as well, this because it avoids the complicating involvement of bad heroes and good villains, of monsters with a heart, and of heroes with a tragic flaw. In short, it avoids the dispositional circumstances that make respondents not care one way or another, and that eventually may foster affective indifference and insensitivity. It may also be expected that, in the interest of retaining enjoyable affective reactivity to drama, the depiction of increasingly potent emotion-laden events will be necessary. It would appear that the creators of media entertainment treat a decline in the affective sensitivities in various segments of their audience as a fact of life. Their beliefs are manifest in the use of ever-more graphic portrayals of increasingly extreme emotional happenings. This escalation is apparent, for instance, in the depiction of threats, dangers, and violence, as well as in other highly arousing behaviors such as erotica (Anscombe, 1987; Bryant & Zillmann, 2001; Zillmann & Bryant, 1986), and constitutes a spiral toward the gratuitous depiction of utterly excessive and often exaggerated emotionality; that is, a spiral of displays likely to foster empathic callousness with others' more common and less intense emotional experiences.

Clearly, most of these projections are derivations from theoretical proposals. The empirical exploration of the impact of the media of communication on the development of affective reactivity in children, on the affective sensitivity of adults, on the maintenance of empathic sensitivity, and on the habituation of affect and the formation of affective insensitivity and callousness is in its incipient stages. The truism that 'more research is needed' applies to the study of media effects on affect and emotions much more than to any other aspect of media influence.

REFERENCES

Allport, F. (1924). *Social psychology*. New York: Houghton Mifflin.

Anscombe, R. (1987, May 4). Stranger than fiction. *Newsweek*, pp. 8–9.

Aronfreed, J. (1968). *Conduct and conscience: The socialization of internalized control over behavior*. New York: Academic Press.

Aronfreed, J. (1970). The socialization of altruistic and sympathetic behavior: Some theoretical and experimental analyses. In J. Macaulay & L. Berkowitz (Eds.), *Altruism and helping behavior* (pp. 103–126). New York: Academic Press.

Berger, S. M. (1962). Conditioning through vicarious instigation. *Psychological Review, 29,* 450–466.

Berger, S. M., & Hadley, S. W. (1975). Some effects of a model's performance on observer electromyographic activity. *American Journal of Psychology, 88,* 263–276.

Bernal, G., & Berger, S. M. (1976). Vicarious eyelid conditioning. *Journal of Personality and Social Psychology, 34,* 62–68.

Borke, H. (1971). Interpersonal perception of young children: Egocentrism or empathy? *Developmental Psychology, 5,* 263–269.

Bryant, J., & Zillmann, D. (2001). Pornography: Models of effects on sexual deviancy. In C. D. Bryant (Ed.), *Encyclopedia of criminology and deviant behavior* (pp. 241–244). Philadelphia, PA: Brunner-Routledge.

Buck, R. (1980). Nonverbal behavior and the theory of emotion: The facial feedback hypothesis. *Journal of Personality and Social Psychology, 38,* 811–824.

Buck, R., & Ginsburg, B. (1991). Emotional communication and altruism: The communicative gene hypothesis. In M. Clark (Ed.), *Review of personality and social psychology* (Vol. 12, pp. 149–75). Newbury Park, CA: Sage.

Buck, R., & Ginsburg, B. (1997). Communicative genes and the evolution of empathy. In W. Ickes (Ed.), *Empathic accuracy* (pp. 17–43). New York: Guilford.

Cacioppo, J. T., & Tassinary, L. G. (Eds.). (1990). *Principles of psychophysiology: Physical, social, and inferential elements*. Cambridge: Cambridge University Press.

Cantor, J. R., Bryant, J., & Zillmann, D. (1974). Enhancement of humor appreciation by transferred excitation. *Journal of Personality and Social Psychology, 30,* 812–821.

Cialdini, R. B., & McPeek, R. W. (1974, May). *Yawning, yielding, and yearning to yawn*. Paper presented at the meeting of the Midwest Psychological Association, Chicago.

Clark, M. S. (1982). A role for arousal in the link between feeling states, judgments, and behavior. In M. S. Clark & S. T. Fiske (Eds.), *Affect and cognition: The seventeenth annual Carnegie Symposium on Cognition* (pp. 263–289). Hillsdale, NJ: Erlbaum.

Cline, V. B., & Richards, J. M., Jr. (1960). Accuracy of interpersonal perception: A general trait? *Journal of Abnormal and Social Psychology, 60,* 20–30.

Craig, K. D., & Lowery, H. J. (1969). Heart-rate components of conditioned vicarious autonomic responses. *Journal of Personality and Social Psychology, 11,* 381–387.

Damasio, A. R. (1994). *Descartes' error*. New York: Putnam.

Darwin, C. (1872). *The expression of emotions in man and animals*. London: Murray.

Davis, M. H., Hull, J. G., Young, R. D., & Warren, G. G. (1987).Emotional reactions to dramatic film stimuli: The influence of cognitive and emotional empathy. *Journal of Personality and Social Psychology, 52,* 126–133.

Duan, C., & Hill, C. E. (1996). The currant state of empathy research. *Journal of Counseling Psychology, 43*(3), 261–274.

Dymond, R. F. (1949). A scale for measurement of empathic ability.*Journal of Consulting Psychology, 14,* 127–133.

Dymond, R. F. (1950). Personality and empathy. *Journal of Consulting Psychology, 14,* 343–350.

Eibl-Eibesfeldt, I. (1970). *Ethology: The biology of behavior*. New York: Holt, Rinehart & Winston.

Eisenberg, N. (2000). Empathy and sympathy. In M. Lewis & J. M. Haviland (Eds.), *Handbook of emotions: Second edition* (pp. 677–691). New York: Guilford.

Englis, B. G., Vaughan, K. B., & Lanzetta, J. T. (1982).Conditioning of counter-empathic emotional responses. *Journal of Experimental Social Psychology, 18,* 375–391.

Feldman, R. S., White, J. B., & Lobato, D. (1982). Social skills and nonverbal behavior in children. In R. S. Feldman (Ed.), *Development of nonverbal behavior in children* (pp. 259–277). New York: Springer-Verlag.

Fenichel, O. (1954). *The psychoanalytic theory of neurosis.* New York: Norton.

Feshbach, N. D. (1978). Studies of empathic behavior in children. In B. A. Maher (Ed.), *Progress in experimental personality research* (Vol. 8, pp. 1–47). New York: Academic Press.

Freud, S. (1921/1950). *Group psychology and the analysis of the ego* (J. Strachey, Trans.). New York: Bantam Books.

Freud, S. (1933/1964). New introductory lectures on psycho-analysis. In J. Strachey (Ed. & Trans.), *The standard edition of the complete psychological works of Sigmund Freud* (Vol. 22, pp. 7–182). London: Hogarth Press.

Grings, W. W., & Dawson, M. E. (1978). *Emotions and bodily responses: A psychophysiological approach.* New York: Academic Press.

Hamilton, M. L. (1972). Imitation of facial expression of emotion.*Journal of Psychology, 80,* 345-350.

Hart, T. (1999). The refinement of empathy. *Journal of Humanistic Psychology, 39*(4), 111–125.

Haskins, J. B. (1984). Morbid curiosity and the mass media: A synergistic relationship. In J. A. Crook, J. B. Haskins, & P. G. Ashdown (Eds.), *Morbid curiosity and the mass media: Proceedings of a symposium* (pp. 1–44). Knoxville, TN: University of Tennessee and the Gannett Foundation.

Henry, J. P. (1986). Neuroendocrine patterns of emotional response. In R. Plutchik & H. Kellerman (Eds.), *Emotion: Theory, research, and experience: Vol. 3. Biological foundations of emotion* (pp. 37–60). Orlando, FL: Academic Press.

Hinde, R. A. (1970). *Animal behaviour: A synthesis of ethology and comparative psychology* (2nd ed.). New York: McGraw-Hill.

Hoffman, M. L. (1973). *Empathy, role-taking, guilt and the development of altruistic motives* (Developmental Psychology Report No. 30). Ann Arbor: University of Michigan.

Hoffman, M. L. (1977). Empathy, its development and prosocial implications. In H. E. Howe, Jr. (Ed.), *Nebraska Symposium on Motivation* (Vol. 25, pp. 169–217). Lincoln: University of Nebraska Press.

Hoffman, M. L. (1978). Toward a theory of empathic arousal anddevelopment. In M. Lewis & L. A. Rosenblum (Eds.), *The development of affect* (pp. 227–256). New York: Plenum Press.

Hoffman, M. L. (1984). Interaction of affect and cognition in empathy. In C. E. Izard, J. Kagan, & R. B. Zajonc (Eds.), *Emotions, cognition, and behavior* (pp. 103–131). Cambridge: Cambridge University Press.

Hoffman, M. L. (1987). The contribution of empathy to justice and moral judgement. In N. Eisenberg, & J. Strayer (Eds.), *Empathy and its development* (pp. 47–80). Cambridge: Cambridge University Press.

Howard, J. L., Reifler, C. B., & Liptzin, M. B. (1971). Effects of exposure to pornography. In *Technical Report of The Commission on Obscenity and Pornography* (Vol. 8, pp. 97–132). Washington, DC: U.S. Government Printing Office.

Humphrey, G. (1922). The conditioned reflex and the elementary social reaction. *Journal of Abnormal and Social Psychology, 17,* 113–119.

Hygge, S. (1976a). *Emotional and electrodermal reactions to the suffering of another: Vicarious instigation and vicarious classical conditioning.* Studia psychologica Upsaliensia 2. Uppsala: Acta Universitatis Upsaliensis.

Hygge, S. (1976b). Information about the model's unconditioned stimulus and response in vicarious classical conditioning. *Journal of Personality and Social Psychology, 33,* 764–771.

Ickes, W. (Ed.). (1997). *Empathic accuracy.* New York: Guilford.

Izard, C. E. (1977). *Human emotions.* New York: Plenum Press.

Katz, R. L. (1963). *Empathy: Its nature and uses.* Glencoe, IL: Free Press.

Keltner, D., & Ekman, P. (2000). Facial expression of emotion. In M. Lewis & J. M. Haviland (Eds.), *Handbook of emotions: Second edition* (pp. 236–249). New York: Guilford.

Kieras, D. (1978). Beyond pictures and words: Alternative information-processing models for imagery effects in verbal memory. *Psychological Bulletin, 85,* 532–554.

Kintsch, W. (1974). *The representation of meaning in memory.* Hillsdale, NJ: Erlbaum.

Kohlberg, L. (1964). Development of moral character and moral ideology. In M. L. Hoffman & L. W. Hoffman (Eds.), *Review of child development research* (Vol. 1, pp. 383–431). New York: Russell Sage Foundation.

Konijn, E. (2000). *Acting emotions: Shaping emotions on stage.* Amsterdam, The Netherlands: Amsterdam University Press.

Laird, J. D. (1974). Self-attribution of emotion: The effects of expressive behavior on the quality of emotional experience. *Journal of Personality and Social Psychology, 29,* 475–486.

Lang, P. J. (1979). A bio-informational theory of emotional imagery. *Psychophysiology, 16,* 495–512.

Lang, P. J. (1984). Cognition in emotion: Concept and action. In C. E. Izard, J. Kagan, & R. B. Zajonc (Eds.), *Emotions, cognition, and behavior* (pp. 192–226). Cambridge: Cambridge University Press.

Lanzetta, J. T., Cartwright-Smith, J., & Kleck, R. E. (1976). Effects of nonverbal dissimulation on emotional experience and autonomic arousal. *Journal of Personality and Social Psychology, 33,* 354–370.

Lanzetta, J. T., & Orr, S. P. (1981). Stimulus properties of facial expressions and their influence on the classical conditioning of fear. *Motivation and Emotion, 5,* 225–234.

LeDoux, J. (1996). *The emotional brain: The mysterious underpinnings of emotional life.* New york: Simon & Schuster.

LeDoux, J. E., & Phelps, E. A. (2000). Emotional networks in the brain. In M. Lewis & J. M. Haviland (Eds.), *Handbook of emotions: Second edition* (pp. 157–172). New York: Guilford.

Leight, K. A., & Ellis, H. C. (1981). Emotional mood states, strategies, and stateg dependency in memory. *Journal of Verbal Learning and Verbal Behavior, 20,* 251–266.

Levenson, R. W., Ekman, P., & Friesen, W. V. (1990). Voluntary facial action generates emotion-specific autonomic nervous system activity. *Psychophysiology,* 27, 363–384.

Levy, J. (1997). A note on empathy. *New Ideas in Psychology, 15*(2), 179–184.

Lipps, T. (1903). *Ästhetik: Psychologie des Schönen und der Kunst: Vol. 1. Grundlegung der Ästhetik* [Aesthetics: Psychology of beauty and art: Vol. 1. Foundation of aesthetics]. Hamburg: Voss.

Lipps, T. (1906). *Ästhetik: Psychologie des Schönen und der Kunst: Vol. 2. Die ästhetische Betrachtung und die bildende Kunst* [Aesthetics: Psychology of beauty and art: Vol. 2. Aesthetic contemplation and educational art]. Hamburg: Voss.

Lipps, T. (1907). Das Wissen von fremden Ichen [Knowledge of other selfs]. *Psychologische Untersuchungen, 1*(4), 694–722.

Lockard, J. S. (Ed.). (1980). *The evolution of human social behavior.* New York: Elsevier.

MacLean, P. D. (1958). The limbic system with respect to self-preservation and the preservation of the species. *Journal of Nervous and Mental Disease, 127,* 1–11.

MacLean, P. D. (1967). The brain in relation to empathy and medical education. *Journal of Nervous and Mental Disease, 144,* 374–382.

MacLean, P. D. (1990). *The triune brain in evolution.* New York: Plenum.

McGaugh, J. L. (1992). Affect, neuromodulatory systems, and memory storage. In S.-Å. Christianson (Ed.), *The handbook of emotion and memory: Research and theory* (pp. 247–268). Hillsdale: NJ: Lawrence Erlbaum Associates.

Mann, J., Berkowitz, L., Sidman, J., Starr, S., & West, S. (1974). Satiation of the transient stimulating effect of erotic films. *Journal of Personality and Social Psychology, 30,* 729–735.

Mathews, K., & Stotland, E. (1973). *Empathy and nursing students' contact with patients.* Mimeo, University of Washington, Spokane.

Mattes, J., & Cantor, J. (1982). Enhancing responses to television advertisements via the transfer of residual arousal from prior programming. *Journal of Broadcasting, 26,* 553–556.

McDougall, W. (1908). *An introduction to social psychology.* London: Methuen.

McDougall, W. (1922). A new theory of laughter. *Psyche, 2,* 292–303.

Mead, G. H. (1934). *Mind, self and society.* Chicago: University of Chicago Press.

Mehrabian, A., & Epstein, N. (1972). A measure of emotional empathy. *Journal of Personality, 40,* 525–543.

Murphy, L. B. (1937). *Social behavior and child personality.* New York: Columbia University Press.

Murphy, L. B. (1983). Issues in the development of emotion in infancy. In R. Plutchik & H. Kellerman (Eds.), *Emotion: Theory, research, and experience: Vol. 2. Emotions in early development* (pp. 1–55). New York: Academic Press.

Öhman, A., & Dimberg, U. (1978). Facial expressions as conditioned stimuli for electrodermal responses: A case of "preparedness"? *Journal of Personality and Social Psychology, 36,* 1251–1258.

Orr, S. P., & Lanzetta, J. T. (1980). Facial expressions of emotion as conditioned stimuli for human autonomic responses. *Journal of Personality and Social Psychology, 38,* 278–282.

Piaget, J. (1948). *The moral judgment of the child.* Glencoe, IL: Free Press.

Plutchik, R. (1980). *Emotion: A psychoevolutionary synthesis.* New York: Harper and Row.

Plutchik, R. (1987). Evolutionary bases of empathy. In N. Eisenberg & J. Strayer (Eds.), *Empathy and its development* (pp. 38–46). Cambridge: Cambridge University Press.

Pylyshyn, Z. W. (1973). What the mind's eye tells the mind's brain: A critique of mental imagery. *Psychological Bulletin, 80,* 1–24.

Rogers, C. R. (1967). *Person to person.* Lafayette, CA: Real People Press.

Saarni, C. (1982). Social and affective functions of nonverbal behavior: Developmental concerns. In R. S. Feldman (Ed.), *Development of nonverbal behavior in children* (pp. 123–147). New York: Springer Verlag.

Saarni, C. (1999). *The development of emotional competence.* New York: Guilford Press.

Saarni, C., & von Salisch, M. (1993). The socialization of emotional dissemblance. In M. Lewis & C. Saarni (Eds.), *Lying and deception in everyday life* (pp. 106–125). New York: Guilford Press.

Sagi, A., & Hoffman, M. L. (1976). Empathic distress in newborns. *Developmental Psychology, 12,* 175–176.

Sapolsky, B. S., & Zillmann, D. (1978). Experience and empathy: Affective reactions to witnessing childbirth. *Journal of Social Psychology, 105,* 131–144.

Scheler, M. (1913). *Zur Phänomenologie und Theorie der Sympathiegefühle und von Liebe und Hass*: Mit einem Anhang über den Grund zur Annahme der Existenz des fremden ich [Contribution to the phenomenology and theory of feelings of sympathy and of love and hate: With a postscript about the reason for the assumption of the existence of the external me]. Halle a. S., Germany: Niemeyer.

Schwartz, G. E. (1977). Biofeedback and patterning of autonomic and central processes: CNS-cardiovascular interactions. In G. E. Schwartz & J. Beatty (Eds.), *Biofeedback: Theory and research* (pp. 183–219). New York: Academic Press.

Scott, J. P. (1959). *Animal behavior.* Chicago: University of Chicago Press.

Scott, J. P. (1969). The social psychology of infrahuman animals. In G. Lindzey & E. Aronson (Eds.), *The handbook of social psychology: Second edition* (Vol. 4, pp. 611–642). Reading, MA: Addison-Wesley.

Seligman, M. E. P. (1970). On the generality of the laws of learning. *Psychological Review, 77,* 406–418.

Simner, M. L. (1971). Newborn's response to the cry of another infant. *Developmental Psychology, 5,* 136–150.

Singer, D. G., & Singer, J. L. (1990). *The house of make-believe: Children's play and the development of imagination.* Cambridge, MA: Harvard University Press.

Singer, J. L. (1980). The power and limitations of television: A cognitive-affective analysis. In P. H. Tannenbaum (Ed.), *The entertainment functions of television* (pp. 31–65). Hillsdale, NJ: Lawrence Erlbaum Assoc.

Smith, A. (1759/1971). *The theory of moral sentiments.* New York: Garland.

Sparks, G. G., & Sparks, C. W. (2000). Violence, mayhem, and horror. In D. Zillmann & P. Vorderer (Eds.), *Media entertainment: The psychology of its appeal* (pp. 73–91). Mahwah, NJ: Lawrence Erlbaum Assoc.

Spitz, R. A., & Wolf, K. M. (1946). The smiling response: A contribution to the ontogenesis of social relations. *Genetic Psychology Monographs, 34,* 57–125.

Stein, E. (1970). *On the problem of empathy* (2nd ed.). The Hague: Nijhoff.

Stotland, E. (1969). Exploratory investigations of empathy. In L. Berkowitz (Ed.), *Advances in experimental social psychology* (Vol. 4, pp. 271–314). New York: Academic Press.

Stotland, E., Mathews, K. E., Jr., Sherman, S. E., Hansson, R. O., &Richardson, B. Z. (1978). *Empathy, fantasy, and helping.* Beverly Hills, CA: Sage Publications.

Sullivan, H. S. (1940). *Conceptions of modern psychiatry.* London: Tavistock Press.

Tagiuri, R. (1969). Person perception. In G. Lindzey & E. Aronson(Eds.), *The handbook of social psychology: Vol. 3. The individual in a social context* (2nd ed., pp. 395–449). Reading, MA: Addison-Wesley.

Tamborini, R. (2003). Enjoyment and social functions of horror. In J. Bryant, D. Roskos-Ewoldsen, & J. Cantor (Eds.), *Communication and emotion: Essays in honor of Dolf Zillmann* (pp. 417–443). Mahwah, NJ: Lawrence Erlbaum Associate.

Tannenbaum, P. H. (Ed.). (1980). *The entertainment functions of television.* Hillsdale, NJ: Lawrence Erlbaum Associates.

Thornton, S., & Thornton, D. (1995). Facets of empathy. *Personality and Individual Differences, 19*(5), 765–767.

Tomes, H. (1964). The adaptation, acquisition, and extinction of empathically mediated emotional responses. *Dissertation Abstracts, 24,* 3442–3443.

Tomkins, S. S. (1962). *Affect, imagery, consciousness: Vol. 1. The positive affects.* New York: Springer-Verlag.

Tomkins, S. S. (1963). *Affect, imagery, consciousness: Vol. 2. The negative affects.* New York: Springer-Verlag.

Tourangeau, R., & Ellsworth, P. C. (1979). The role of facial response in the experience of emotion. *Journal of Personality and Social Psychology, 37,* 1519–1531.

Trevarthen, C. (1984). Emotions in infancy: Regulators of contact and relationships with persons. In K. R. Scherer & P. Ekman (Eds.), *Approaches to emotion* (pp. 129–157). Hillsdale, NJ: Lawrence Erlbaum Associates.

Truax, C. B. (1961). A scale for the measurement of accurate empathy. *Psychiatric Institute Bulletin 1,* No. 12., Wisconsin Psychiatric Institute, University of Wisconsin.

Vaughan, K. B., & Lanzetta, J. T. (1980). Vicarious instigation and conditioning of facial expressive and autonomic responses to a model's expressive display of pain. *Journal of Personality and Social Psychology, 38,* 909–923.

Wagner, H., & Manstead, A. (Eds.). (1989). *Handbook of Social Psychophysiology.* New York: John Wiley.

Washburn, R. W. (1929). A study of the smiling and laughing of infants in the first year of life. *Genetic Psychology Monographs, 6,* 398–537.

Wilson, B. J., Cantor, J., Gordon, L., & Zillmann, D. (1986). Affective response of nonretarded and retarded children to the emotions of a protagonist. *Child Study Journal, 16*(2), 77–93.

Worringer, W. (1908/1959). *Abstraktion und Einfühlung: Ein Beitrag zur Stilpsychologie* [Abstraction and empathy: A contribution to the psychology of style]. München: Piper.

Zajonc, R. B. (1980). Feeling and thinking: Preferences need no inferences. *American Psychologist, 35*(2), 151–175.

Zillmann, D. (1978). Attribution and misattribution of excitatory reactions. In J. H. Harvey, W. J. Ickes, & R. F. Kidd (Eds.), *New directions in attribution research* (Vol. 2, pp. 335–368). Hillsdale, NJ: Erlbaum.

Zillmann, D. (1979). *Hostility and aggression.* Hillsdale, NJ: Lawrence Erlbaum Associates.

Zillmann, D. (1980). Anatomy of suspense. In P. H. Tannenbaum (Ed.), *The entertainment functions of television* (pp. 133–163). Hillsdale, NJ: Lawrence Erlbaum Associates.

Zillmann, D. (1983). Transfer of excitation in emotional behavior. In J. T. Cacioppo & R. E. Petty (Eds.), *Social psychophysiology: A sourcebook* (pp. 215–240). New York: Guilford Press.

Zillmann, D. (1984). *Connections between sex and aggression.* Hillsdale, NJ: Lawrence Erlbaum Associates.

Zillmann, D. (1988). Cognition-excitation interdependencies in aggressive behavior. *Aggressive Behavior, 14,* 51–64.

Zillmann, D. (1989). Effects of prolonged consumption of pornography. In D. Zillmann & J. Bryant (Eds.), *Pornography: Research advances and policy considerations* (pp. 127–157). Hillsdale, NJ: Erlbaum.

Zillmann, D. (1991). Television viewing and physiological arousal. In J. Bryant & D. Zillmann (Eds.), *Responding to the screen: Reception and reaction processes* (pp. 103–133). Hillsdale, NJ: Lawrence Erlbaum Associates.

Zillmann, D. (1996). Sequential dependencies in emotional experience and behavior. In R. D. Kavanaugh, B. Zimmerberg, & S. Fein (Eds.),*Emotion: Interdisciplinary perspectives* (pp. 243–272). Mahwah, NJ: Lawrence Erlbaum Associates.

Zillmann, D. (1998). The psychology of the appeal of portrayals of violence. In J. H. Goldstein (Ed.), *Why we watch: The attractions of violent entertainment* (pp. 179–211). New York: Oxford University Press.

Zillmann, D. (2000). Basal morality in drama appreciation. In I. Bondebjerg (Ed.), *Moving images, culture and the mind* (pp. 53–63). Luton, England: University of Luton Press.

Zillmann, D. (2003). Affective dynamics, emotions and moods. In J. Bryant, D. Roskos-Ewoldsen, & J. Cantor (Eds.), *Communication and emotion: Essays in honor of Dolf Zillmann* (pp. 533–567). Mahwah, NJ: Erlbaum.

Zillmann, D., & Bryant, J. (1975). Viewer's moral sanction of retribution in the appreciation of dramatic presentations. *Journal of Experimental Social Psychology, 11,* 572–582.

Zillmann, D., & Bryant, J. (1986). Shifting preferences in pornography consumption. *Communication Research, 13,* 560–578.

Zillmann, D., & Cantor, J. R. (1976). A disposition theory of humour and mirth. In A. J. Chapman & H. C. Foot (Eds.), *Humour and laughter: Theory, research, and applications* (pp. 93–115). London:Wiley.

Zillmann, D., & Cantor, J. R. (1977). Affective responses to the emotions of a protagonist. *Journal of Experimental Social Psychology, 13,* 155–165.

Zillmann, D., & Gibson, R. (1996). Evolution of the horror genre. In J. B. Weaver & R. Tamborini (Eds.), *Horror films: Current research on audience preferences and reactions* (pp. 15–31). Mahwah, NJ: Lawrence Erlbaum Associates.

Zillmann, D., & Knobloch, S. (2001). Emotional reactions to narratives about the fortunes of personae in the news theater. *Poetics, 29,* 189–206.

Zillmann, D., Mody, B., & Cantor, J. R. (1974). Empathic perception of emotional displays in films as a function of hedonic and excitatory state prior to exposure. *Journal of Research in Personality, 8,* 335–349.

Zillmann, D., & Vorderer, P. (Eds.). (2000). *Media entertainment: The psychology of its appeal.* Mahwah, NJ: Lawrence Erlbaum Associates.

Zillmann, D., & Weaver, J. B. (1996). Gender-socialization theory of reactions to horror. In J. B. Weaver & R. Tamborini (Eds.), *Horror films: Current research on audience preferences and reactions* (pp. 81–101). Mahwah, NJ: Lawrence Erlbaum Associates.

Zuckerman, M., Klorman, R., Larrance, D. T., & Spiegel, N. H. (1981). Facial, autonomic, and subjective components of emotion: The facial feedback hypothesis versus the externalizer-internalizer distinction. *Journal of Personality and Social Psychology, 41,* 929–944.

11

Audience Identification
With Media Characters

Jonathan Cohen
University of Haifa

The essence of entertainment television is that it can delight and enlighten "through the exhibition of the fortunes and misfortunes of others..." (Zillmann & Bryant, 2002, p. 437). In order for the fortunes and misfortunes of others to delight us, they must first arouse our interest. It is necessary, then, that we attend to these others and what happens to them, and this can distract from our own life. The more involved we are with the fictional world and what happens to the people in it, the more likely we are to enjoy being entertained. It is not the mere exposure to entertainment that we enjoy, but the ability of entertainment content to distract us from ourselves and to reveal to us novel and exciting experiences of others. By allowing us to share in the lives of others, entertainment can excite and educate us, can make us imagine, think and feel in ways we may not otherwise have a chance to experience. We expand our emotional and mental lives beyond the scope of our personal experience and participate in community and cultural life. Entertainment is as old as human society, but modern media of communication have multiplied the variety of stories we have access to, and the ways in which these are presented.

The importance of the human capacity to feel for and with others is clear, but people's ability to develop strong feelings for fictional characters is somewhat more perplexing, and part of the larger question about their ability to become engaged in stories. The question of how audiences are able to become emotionally and cognitively involved in stories that they know to be fictional is a long-standing issue in literary theory.[1] Livingstone and Mele (1997) discuss this question as the *paradox of fiction*. Most readers of fiction are sophisticated enough not

---

[1] The term "willing suspension of disbelief" was coined by Coleridge, who saw it as "poetic faith" (Coleridge, 1817/1965, Vol. 2, p. 6). He saw such suspension of disbelief as a result of the poet, writing about the fantastic, providing the reader with a semblance of truth that would allow the reader to ignore the fictionality of the poem. He contrasted this strategy of poetry with that of writing about the unseen beauty and wonder of mundane reality (employed by Wordsworth).

to believe that the events described in fiction are real. Nonetheless, they often come to care deeply about fictional plots and characters. Moreover, readers often possess extra-literary knowledge that should dissipate any suspense about the outcome of a particular plot, but still find that they are anxious and eager to follow the resolution. For example, a child viewing the first of the *Harry Potter* series of films is likely to know that since the series continues, Harry must end up alive and well at the end of the movie. He or she is also likely to have read the book or to have heard about it from friends. And yet, young readers find these films full of suspense, and they continue watching to find out what happens to the young wizard. Serial television and films also provide examples of formulaic stories whose resolution is known to viewers (especially when they are endlessly re-run), and whose fictionality is readily acknowledged, and yet they continue to make up the backbone of network prime-time programming.

Several answers to this paradox have been offered. The idea of the willing suspension of disbelief dates back two centuries to Coleridge (1817/1965), and attempts to explain our capacity to respond emotionally to fiction through our ability to put aside our knowledge that the events we read about are fictional. We put such beliefs aside willingly because we know that otherwise we would not be able to enjoy fiction. It is necessary that we believe that under some conditions (i.e., within the context of the fiction) the events are meaningful in order to care about them and become involved, and hence entertained. Oatley (1994, 1999) proposes that experiencing fiction is similar to performing a mental simulation. We accept a set of assumptions about the fictional world we engage in and try to imagine whether the events and cause–effect sequences described in the plot make sense given these assumptions. Yet another proposition is that our suspension of disbelief is not 'willing' in the sense that it is not active, but rather that our propensity to believe fictional events is natural, and that it requires mental effort and purpose to remind ourselves that what we are viewing or reading is fictional. While the debate about the origin of the paradox of fiction continues, it is clear that audience members care deeply about fictional worlds and the people who inhabit them. This involvement is crucial to the capacity of entertainment to capture our attention, alleviate boredom, and divert us from our daily routines and tedium.

One way in which we develop an interest in fictional events is through identification with characters in a story. Identification provides us with several important keys to fictional involvement: Identifying with a character provides a point of view on the plot; it leads to an understanding of character motivations, an investment in the outcome of events, and a sense of intimacy and emotional connection with a character. It is this concept of identification and its role in entertainment that this chapter explores. Following a conceptual discussion of identification relating it to notions of involvement, this chapter examines studies trying to explain identification with characters, the role of identification in the interpretation and reception of televised texts, and the various ways in which identification is involved in media effects.

DEFINING IDENTIFICATION

Identification with media characters is an imaginative process that is evoked as a response to characters presented in mediated texts. Mediated texts construct worlds in which characters are seen to operate. Viewers often respond to such texts by feeling as if they are part of these fictional worlds, and are experiencing the events occurring in them from within the text. Identifying with a character means feeling an affinity toward the character that is so strong that we become absorbed in the text and come to an empathic understanding for the feelings the character experiences, and for his or her motives and goals. We experience what happens to the characters as if it happens to us while, momentarily at least, forgetting ourselves as audience members, and this intensifies our viewing experience (Cohen, 2001). Thus, identification has both affective (empathy) and cognitive (understanding goals and motives, perspective-taking) components.

In his book *Watching Television*, Wilson (1993) suggests that viewers move in and out of their identification, continually shifting from their role as viewers to their identification with character(s) in the text and the roles they play in the narrative. In a similar way, Galgut (2002) discusses the suspension of disbelief as an active state of mind, but one that is not stable. When one adopts a state of mind that ignores the fictionality of events and becomes immersed in the text, he or she does not lose or discard the ability to discern reality. Similarly, when one identifies with a character, he or she does not lose their identity; they simply suspend it momentarily. Oatley (1994) suggests that the boundary between the reader's entry into the world created by the artist, and the reader's existence as a reader (outside in the real world), is like a semi-permeable membrane. The reader–viewer brings some of, but not all of, him or herself into the experience and is able to mentally move in and out of identification.

Audience members experience texts from multiple positions both within and outside the text. Viewers can simultaneously hold several subject positions, though at any given moment one position is most likely dominant. A viewer may identify with one or more characters, but may know more than any one of the fictional characters know; through the privileged position offered him or her by a director. Viewers can also incorporate knowledge from their extra-textual world into their understanding and experience of the text. Thus, we may feel fear when a character we like is in danger, but immediately remind ourselves that "this is only a movie." While viewing comedy, we may feel bad for Larry and Moe who heatedly engage in an argument because they misunderstand one another (of course, we understand them both), but at the same time we find this conflict funny from our position as viewers. And while watching *Schindler's List*, we may empathize and identify with the horrors the victims endured, but know from our history lessons that those on the list were relatively fortunate.

Through identification with one or more characters, viewers become involved with the text. However, the identification of viewers is influenced by their own perspectives, values, and interests as individuals. Identification, then, requires taking on the perspective of another, walking in another's shoes (Livingstone, 1998), but necessarily doing so through the filters of one's own understanding and experience.

Importantly, identification is filtered not only through any social and psychological distance between viewer and character, but also by the author's or director's art. As viewers, we are often privy to what the director allows us to know, which may be more and different than what the characters are allowed to know. This knowledge, or lack of knowledge, about what has happened and what will happen, impacts our ability to identify and assume a character's position.

That we are able to identify with fictional characters and to empathize with them does not mean we will automatically choose to do so. Viewers may choose to remain distant and maintain their position of detached viewer. Even if viewers are willing to accept the fictional position of a television character, the writing, acting, and directing must be of sufficient quality to engage the viewer. One of the premier challenges of actors and directors is to entice viewers into identification. Inviting identification is partly achieved by offering an illusion of reality, a semblance of how people behave and act in real life, and a consistency of character that resonates with audience members. Another factor that increases involvement and enjoyment is the relevance of the issues and events in the text and the degree to which they resonate with viewers' lives (Cohen & Ribak, 2003). Cooper (1999), for example, found that women tended to like *Thelma & Louise* more than did men partly because the issues of female bonding and gender identity were more relevant to them. Ang (1985) argues that identification can occur not only through realistic representations of events, but also through the "sensitive representation of an identifiable structure of feeling" (p. 45) displayed by the character that the viewer can relate to, and which produces within the viewer an appropriate emotional response.

Identification is only one among many ways viewers respond to characters (Cohen, 2001; Cohen & Perse, 2003). As opposed to parasocial interactions, which develop through simulated interaction in which the character plays to the viewer and requires the viewer to remain outside the text, for the viewer to identify with a character the viewer must enter the textual world. Other responses, such as social comparison, imitation or adoration (fandom), are also common responses to media characters, and often accompany identification, but have different psychological characteristics.

In sum, identification is an active psychological state, but neither stable nor exclusive. It is one of many ways in which we respond to characters, and one of many positions from which we experience entertainment. The development and strength of identification depend on multiple factors: the nature of the character, the viewer, and the text (directing, writing, and acting). Finally, identification is part of a larger set of responses to entertainment, ways in which we become engrossed and delighted by the fortunes and misfortunes of others.

INVOLVEMENT, TRANSPORTATION, AND IDENTIFICATION

Identification brings audiences closer to the mediated worlds created by media performances. Fiske (1989) sees the closeness promoted by identification as a detriment to a critical appraisal of entertainment, as a barrier to resistance to the ideology imbedded in commercial media. However, Fiske says, the "reward for identification is pleasure" (p. 170). Fiske understands the pleasure derived from identification as resulting from wish fulfillment and control over the relationship with the character, but, most likely, it is partly due to its role in allowing us to more completely forget about ourselves and become engaged and involved in the world of the narrative.

Another way to describe the experience of being absorbed and involved in a mediated text is as transportation.[2] Using the travel metaphor, transportation theory suggests that we can become so absorbed in the narrative and focused on the events it portrays, that it is as though we have been transported from our location as viewers into the narrative (Gerrig, 1993). Research has shown that the degree of transportation is related to the effect a text has on an audience (Green & Brock, 2000). It is possible to speculate that identification is positively linked to transportation, though such a link has not yet been empirically tested.[3] Audience members who are more involved in the events of the text are more likely to develop a strong emotional connection with one or more characters, and to be able to be empathic toward their point of view, interests, and goals. Even if this speculation proves true, it is unclear what might be the direction of causality between transportation/absorption/involvement with the text and identification with characters.

It is possible that identifying with characters is one way to become transported by the text and be more fully involved in the viewing experience. As we come to understand and care about the characters, we come to care about what happens to them and thus become involved in the plot and transported by the text into a fictional world. It is equally reasonable to see identification as a result of becoming at least initially transported by the text. If we remain

---

[2]It is not clear how the new psychological concept of transportation differs substantively from the concept of involvement, which has a long history in media research (see, for example, Kim & Rubin, 1997). However, because transportation theory has been used in several recent studies and the transportation scale has become popular, I chose to use both concepts in this review.

[3]Gerrig (1993) argues that identification is neither a necessary, nor a sufficient, condition for transportation (p.170). While he makes a valid argument, he clearly sees identification as a more total experience than the present understanding of identification as a non-stable and non-exclusive mental state. Moreover, it is quite possible to expect that a substantial link exists without arguing for a conditional relationship.

distant from the text and refuse or fail to be transported by it, it is hard to imagine any significant identification developing with characters.

Gerrig (1993) posits that our reaction to stories (speaking specifically of suspense) is "... a natural consequence of the structure of cognitive processing" (p.170). He also points out that when we feel afraid for characters facing dangers of which they themselves are unaware, we are experiencing transportation without identification. However, when we identify with a character, we are necessarily transported, suggesting perhaps that transportation is a condition of identification. Kim and Rubin (1997) also position involvement as preceding a relationship with characters. While these arguments offer some insight into how identification fits within the overall experience of entertainment, these notions still lack clear empirical proof. It seems clear, however, that identification plays a central role in the ability of stories to make us forget about our daily routines and engage in imaginary participation in fictional worlds—in other words, to entertain us.

Antecedents to Identification

Little is known about why people become so involved with characters with which they identify. Studies have explored, however, why audiences choose to identify with a particular character, and what makes them identify more—or less—strongly with a particular character. Many studies have examined preferences for characters (liking), the intensity of parasocial relationships, or other types of relationships with characters, rather than specifically examining identification. However, if we assume that identification is related to other types of character-centered responses, these prior studies can provide suggestive data.

Similarity and Homophily

It seems plausible that viewers would find it easier to relate to characters who are similar to them, and who face similar problems to those they face in their daily lives. Perhaps because similarity and homophily are concepts that have been broadly defined, the data gathered so far on this issue are not definitive. It seems that some forms of similarity promote identification, but others do not.

In discussing why children would choose to identify with a specific character and what factors may impact the intensity of identification, Maccoby and Wilson (1957) suggested that similarity is an important element in explaining identification. They pointed to similarity in age, sex, and social class, as well as consonance between viewer motives and character actions. Their findings suggest that similarity is less important to identification than role modeling; that is, children identify with those whom they wish to be like, rather than with those whom they are like. As Noble (1975) points out, however, Maccoby and Wilson (1957) used a very primitive measure of preference as a substitute for a measure of identification.

Turner (1993) found that similarity in "looks" was not an important factor in character choice, but that attitude homophily was the best predictor of character choice for parasocial relations. Cohen and Perse (2003) found a significant positive correlation between identification and general homophily in attitudes, feelings, and background (r = .45). Feilitzen and Linne (1975) report findings from Scandinavian research on children demonstrating that children most often identified with child characters, and more generally with characters similar to themselves. An exception to this pattern was that girls often chose to identify with male characters (for similar results, see also Eyal & Rubin, 2003, and Hoffner, 1996). They also report that children identified with animals, and that somewhat older children (older than eight) engage in wishful identification with characters that are somewhat older and reflect what they would like to be, more than what they are. Cohen's (1999) study of teens found that though demographic consistencies in favorite character choices were observed, teens often chose opposite-sex characters as favorites based on romantic and sexual attraction. Teens also favored young adult rather than teen characters.

Similarly, Press (1990) found that working class women identified with upper class characters on *Dynasty* more than did middle class women. Gleich (1997) found that about a fifth of German adult men chose a female favorite TV person, as compared with about a third of females who chose a male favorite TV person. Eyal and Rubin (2003) found that aggressive children reported higher homophily and identification with aggressive characters than did non-aggressive children. They also report that homophily and identification were highly correlated (r = .68). This suggests that even if demographic similarity may not be a necessary condition of identification, perceived similarity seems to be an important component of identification. Thus, it seems that whereas similarity in attitudes predicts character choice, simple demographic similarity is not a good predictor. People often identify with characters that represent what they wish to be or to whom they are attracted, rather than what they are. It also seems that psychological similarity is more important than demographic similarity in shaping identification.

Character Traits

Hoffner (1996) found that character gender predicts the traits young viewers used to explain wishful identification (wanting to be like the character). Male characters were liked by both boys and girls for their intelligence, and female viewers also cited sense of humor as an important factor in wanting to be like male characters. In contrast, female characters were judged by both boys and girls solely based upon their looks. These findings suggest that choices of media characters follow social stereotypes, which in turn are reflected in the cast of characters viewers can choose from.

Sanders (2004) found that viewers identified more strongly with heroes than with villains. Hoffner and Cantor (1991) suggest that character traits such as strength, humor and physical attractiveness explain which characters are most liked. Their conclusions suggest that our criteria for choosing television and film characters are identical to the criteria we use in being attracted to people in real life. Though several studies support this notion (e.g., Cohen, 2003; Reeves & Naas, 1996; Tsao, 1996), there is also evidence to support the intuitive assertion that relating to TV characters is not exactly the same as relating to real people, and that some elements of attraction to television characters depend on their fictional nature and the structure of the stories (Livingstone, 1987). For example, Ang (1985) found that many *Dallas* fans reported that J.R. Ewing, the series' ultimate villain, was their favorite character, though they explicitly acknowledged that he was a negative and evil person, and claimed they would not like to be his friend in real life. In sum, we do not yet fully understand to what degree the appeal of television characters is unique, and to what degree it resembles the ways we are attracted to people in our real lives.

Authorial Devices

Besides the role and characteristics of the character, texts can shape the attitudes of audiences toward characters through technical means. In literature, allowing a character to narrate the story can increase identification with that character (Nodelman, 1991). Auter (1992) shows that TV segments in which characters address the viewers directly, as opposed to segments in which characters talk only to each other, produce stronger parasocial interaction. Though he did not conduct a direct test of audience–character relationships, Lombard (1995) found that television screen size had an impact on favorable impressions of people in television. Viewers watching on a medium or large screen had more favorable judgments of characters than those watching on a smaller screen. Interestingly, these effects occurred even though there were no differences in judgments of the viewing environment itself, suggesting that the response was not to the viewing but to the people appearing on TV.

Viewer Characteristics

One issue that has been examined extensively is that of gender differences in the capacity to form relationships with TV characters. Turner (1993) found that women had stronger parasocial interactions with soap opera characters but not with other characters. Cohen (1997, 2004) found in three separate samples that women had stronger parasocial relationships with their favorite TV characters than did men. Gleich (1997) did not find gender differences in parasocial relationships among German adults, but notes that "women put themselves into the place of the TV person to a greater extent than men did" (p. 42), suggesting gender differences in identification. Cohen and Perse (2003) did not find gender differences in identification in two separate samples. McCutcheon, Ashe, Houran and Maltby (2003) found no differences between men and women in attitudes toward celebrities.

In terms of age, it is generally assumed that children identify more strongly and are more influenced than adults by television role models (e.g., Bandura, 1965). However, little direct evidence exists for this claim. More generally, with respect to age, Gleich (1997) found that older adults (over 56) tend to report stronger parasocial relationships than younger adults. Cohen (2003) found that teens report stronger parasocial relationships than adults. In a re-analysis of data collected in both the U.S. and Israel by Cohen and Perse (for details on the data see Cohen & Perse, 2003), no correlation between age and identification with favorite characters was found in either sample.

Psychological traits have also been examined as possible predictors of attitudes and reactions toward characters and celebrities. Most commonly, the hypothesis that identification and parasocial interaction compensate for lack of social interaction has been tested. This hypothesis was advanced by Horton and Wohl (1956) in their seminal article based on their textual analysis of host programs and their intended audience. This hypothesis has often been forwarded by theorists suggesting that TV and TV friendships were an escape and a poor substitute for real life and social interaction. Putnam (2000) goes even further, suggesting that the popularity of TV has eroded social life and replaced community and social activities.

However, numerous subsequent empirical tests have failed to provide empirical support for this claim (e.g., Perse & Rubin, 1990; Tsao, 1996). Rubin, Perse and Powell (1985) found that loneliness did not predict the intensity of parasocial interaction. Turner (1993) found only a weak negative relationship between parasocial relationships with soap opera characters and self-esteem in the realm of communicative abilities. Kanazawa (2002) found that watching more dramas for women and more PBS shows for men was related to slightly more (rather than less) satisfaction with friendships. Tsao (1996) conducted a test to determine whether parasocial interaction was a compensatory mechanism or whether it was part of a more global communicative network that includes interpersonal relationships, which he called the global use hypothesis. He found that the intensity of parasocial relationships correlated with traits – such as empathy and extraversion—that indicate good interpersonal relationships, rather than with those—such as neuroticism—that indicate problems in interpersonal relationships.

Finally, Ashe and McCutcheon (2001) concluded another study by saying:

Our study, in conjunction with those of Perse and Rubin (1990) and Rubin, Perse, and Powell (1985), suggest that the link between two measures of social anxiety and the strength of a parasocial interaction to a celebrity is either very weak or non-existent (p. 130).

Thus, despite multiple studies attempting to demonstrate that relationships with TV characters are what social misfits resort to, it seems that such relationships are in fact quite normal and healthy, and that in many ways they are similar to other human relationships. Some scholars even go further to suggest that our reactions to mediated environments and characters are hard-wired into humans (Reeves & Naas, 1996).

In examining relationships with mediated characters, research has examined how such relationships relate to a variety of personality factors. Studies of attachment styles (i.e., self-report measures of stable attitudes and beliefs that guide close relationships) have shown that people who desire close relationships and develop them easily, but have problems maintaining secure, strong and stable relationships, tend to develop strong parasocial relationships (Cohen, 1997; Cole & Leets, 1999). Thus, it seems that relationships with characters interact with and are integrated into social relationships, but not through simple compensation or deficiencies. In support of Tsao's global use hypothesis, these studies suggest that though the dynamics of social and parasocial relationships are different, similar personality traits guide our responses in both contexts. Interestingly, it is not the secure respondents who were found to have the strongest parasocial feelings, but rather those anxiously attached individuals who desire strong relationships but have trouble developing secure and stable relationships. It is possible that while the desire for parasocial and social relationships stems from similar psychological motives, the different context and dynamics of the two types of relationships lead to differences in how intimacy is developed and maintained. It is probable that people who are good at relating to others – but have trouble trusting others in long-term interpersonal relationships and thus tend to become possessive and drive people away—find TV characters to be more loyal. However, the people who intuitively have the most to gain from relationships with media characters—those who tend to avoid close interpersonal contact and lack the psychological traits required for good interpersonal communication—have been found to have the weakest relationships with TV characters.

Other personality traits have also been examined. Turner (1993) tested the relationships between various aspects of self-esteem and parasocial relationships with television characters from various genres. He reports only two significant correlations: Shyness was negatively correlated with relationships with favorite soap characters ($r = -.32$), and positive self-esteem was positively correlated with relationships with favorite comedian ($r = .37$). McCutcheon et al. (2003) found that celebrity worship was negatively associated with cognitive abilities, though these relationships were small. Finally, Giles and Maltby (2004) found that among school children aged 11–16, normal attachment to celebrities was associated positively and significantly with autonomy, independence, and relationships with peers, and negatively with parental relationships. In sum, it seems that the most significant result of the studies attempting to profile which viewers connect most strongly with TV characters has been to dispel some preconceptions about the uses of such relationships and to point in new directions for the future.

IDENTIFICATION AND RECEPTION

An interesting aspect of identification is the way it affects how we experience texts and construct meaning from them. Identification provides an alternative explanation for data that point to the active interpretation in which viewers are engaged as they view texts. Some of the earliest studies pointing to the importance of interpretation were the studies concerning the impact of the situation comedy *All in the Family* (Brigham & Giesbrecht, 1976; Vidmar & Rokeach, 1974). A surprising finding that emerged in these studies was the failure of Norman Lear's (the creator of the series) efforts to influence and shift audiences' political and social views. Though the series' protagonist, Archie Bunker, and his conservative and racist views were consistently ridiculed, prejudiced viewers liked and admired Archie and saw him as successful.

These findings were originally understood in terms of viewers' power to be selective in their attention and processing. Later, such research was seen as suggesting that viewers' interpretative power should not be underestimated and that viewers sometimes "read against the grain" and resist ideological messages encoded in the text (Cohen, 2002; Papa et al., 2000). Hall (1980) suggests that viewers resist the intended meaning because they bring to their viewing personal cognitive schemas couched in cultural and personal identities, and that the meeting of text and audience is a moment of active interpretation rather than passive reception.

But an additional or alternative explanation for the ideological failure of *All in the Family* involves the importance of viewers' identification with characters and the centrality of this identification to their understanding of the series. If identification involves the taking of a character's perspective, then the conflicts and resolutions in each episode are seen from that character's perspective, and the cognitive mechanisms and biases that are used to interpret reality and protect one's self from negative events are used to interpret the text. In *All in the Family*, Lear imbedded his liberal messages in plot lines in which Archie was revealed as narrow-minded and wrongheaded, but like all good writers, he wrote Archie as a lovable protagonist. What Lear seems to have overlooked is that this love for Archie drove viewers' understanding of the show and ultimately worked against its intended consequences. In line with this explanation, Brigham and Giesbrecht (1976) claim that "liking for and identification with the main characters was strongly related to racial attitudes," though only in part of their sample. Ironically, it was probably the success of the main character that made the series such a popular and financial success and that undermined its intended social effects. If this alternative explanation is correct, prejudiced viewers did not avoid the parts of the message that challenged their views, but rather understood them in a manner that was diametrically opposed to the way they were intended.

Supporting this notion, Cohen (2002) found that identification with *Ally McBeal* was the most important predictor of how viewers understood the show's message. In contrast to Archie's role in *All in the Family*, Ally McBeal was a protagonist that served as the voice for the views of the show's producer, rather than a vehicle to satirize opposing views. And, unlike in the study by Brigham and Giesbrecht (1976), where prejudiced viewers believed Archie was right despite the contrary intention of the producers, viewers of *Ally McBeal* who reported high levels of identification with the main character were most likely to see the show as it was meant by the producers—as a show about the struggles of a strong woman.

Fiske (1989) explains the connection between identification and interpretation by contrasting what Liebes and Katz (1990) call referential and critical readings. Identification is part of a referential approach to a text, which makes the text as an artifact (i.e., its authors, actors, producers, design, etc.) invisible, and in which the viewer is engrossed in the world the text creates. In a referential reading, viewers accept the basic assumptions of the producers and imagine the events described in the text as if they were, or could be, real. One important vehicle for a referential reading is to forget one's role as a viewer and to adopt the perspective of the character, to identify with a character. As a result of such identification, viewers are likely to see and understand the text as they were meant to by the producers and better enjoy the viewing experience (Cohen, 2003). While the link between identification and pleasure is well supported, pleasure can come through negotiation with the text as well as from adopting a dominant interpretation (Cohen and Ribak, 2003). Critical readers, on the other hand, resist the temptation to become involved with the text, and their emotional distance provides them with the ability to critique the show and resist its ideological message, however, they also forgo much of the show's entertainment value.

Perhaps the role of identification in the process of interpretation is easiest to understand through the notion of positioning. Looking at a text as an act of communication between author and audience, it is intended to offer viewers a specific perspective on the events or topics the text addresses. This perspective is often represented in the text by one or more characters, though this representation may not be a direct one. Directors of TV shows use various means in an attempt to position viewers so that they see what the director wants them to see, and in the way the director wants them to see it. Hence, directors attempt to create a structured experience of the text in the hope of achieving a desired impact (to entertain, persuade, etc.). Directors use plot lines and character development, as well as camera angles, set design, and other means, to position audiences.

Positioning allows directors to create a shared discourse with their audiences in which they use certain cultural assumptions, resonant symbols, references, and icons. In news, the desired position may be that of concerned citizens, and the discourse is that of public concern. Sports programs position viewers as fans; talk shows, as outraged or sympathetic bystanders. In contrast, viewers of drama and comedy are asked to enter the plot through identification with characters, rather than experience it only as viewers. They are asked to react emotionally and care about the story, to intimately understand the characters and relate to what they are going through. By identifying with characters, viewers take on the perspective and goals of these characters and position themselves as participants in the evolving plot. If viewers identify with the main character(s), they decode and make meaning from the text as they were meant to and are likely to produce the expected responses and enjoy the experience.

IDENTIFICATION AND EFFECTS

In comparison with the dearth of research into the motives and processes of identification, many investigations have looked into identification with media characters as an effect of media texts, and more often as a mediator of the effects of media texts on audiences. In discussing identification, Morley (1992) argued that: "One can hardly imagine any television text having any effect whatever without that identification" (p. 209). There is evidence to support Morley's assertion in the behavioral, attitudinal, and emotional realms of effects research.

In health communication research, attitudes toward celebrities were found to increase the effect of health messages delivered by these celebrities (Basil, 1996; Brown, Basil, & Bocarnea, 2003). Brown et al. (2003) found that parasocial interaction with the baseball player Mark McGuire led to identification with him and knowledge about and intentions to comply with issues he was known to be involved with (child abuse prevention and using a muscle-building supplement). Papa et al. (2000) also argued that relationships with television characters in an entertainment–education context can be an important agent of change, because they serve as role models and create a social learning environment. They argued that such social learning is prompted by identification with characters, but involves a two-step process that requires subsequent interpersonal conversation.

There are many theoretical arguments explaining why identification amplifies media effects. First, identification creates a unity of perspectives, or what Burke (1950) calls co-substantiality. Burke claims that effective rhetoricians succeed in making their audiences adopt their point of view on a topic. Similarly, Kelman (1961) argues that persuasion through identification leads to an internalization of attitudes that is likely to be stronger and longer lasting than persuasion through other means. More recently, Green and Brock (2000) found that transportation is positively related to the extent of persuasiveness in narrative texts.

At the emotional level, Wied, Zillmann, and Ordman (1994) found that empathy and empathic responses to tragic films increased both the distress viewers experienced during the film and their enjoyment of the film. Tamborini, Stiff, and Heidel (1990) also found that empathy increased emotional reactions to horror films, but in this study these reactions caused avoidance rather than pleasure. Thus, identification seems to engage viewers more deeply in the text, but the outcome of such engagement seems to differ across genres.

A second reason identification is related to media effects is that it can increase exposure and attention to events surrounding the character, or messages delivered by the target of identification. Maccoby and Wilson (1957) found that children who identified with a character remembered more of the events related to that character and messages delivered by, or said to, the character. Rubin and Step (2000) found that parasocial interaction with a talk radio host was positively related to several outcome measures: exposure, perceived importance of information received from the host, perceived influence of the host, and intention to adopt the host's views. Skumanich and Kintsfather (1998) found that the intensity of parasocial relations with television shopping hosts was related to viewing of television shopping programs. Alperstein (1991) conducted in-depth interviews with viewers about their viewing of commercials that featured celebrities. He reports that viewers use such commercials to learn about their favorite celebrities and to form their construction of these celebrities. It seems that viewers pay attention to commercials involving their favorite celebrities and elaborate on the meaning of these commercials.

Another reason that identification with media characters can promote change is that our continuing relationships with media personalities engage our self-identity. In her ethnography of three urban teenagers, Fisherkeller (1997) found that teens do not adopt values portrayed by their television heroes. Rather, she contends, teens tend to adopt the values from their families or close environment, whereas TV role models allow them to explore possible ways of achieving these values and roles. Thus, TV is involved more in teaching the how, rather than the what. In this way, Fisherkeller argues, TV role models play an important role in the shaping of these youngsters' sense of self and provide them with opportunities to better imagine their possible selves. Identification is very important for this type of learning because it involves imagination, and allows viewers to vicariously "try on" various roles, behaviors, and attitudes and imagine what they look and feel like and what their consequences may be.

It is worth noting that not all studies report that identification is related to media effects. Wiegman, Kuttschreuter and Baarda (1992) found that although identification with violent characters was related to viewing more violent content, no relationships existed between identification and violent behavior. Sheehan (1983) reports inconsistent results: In his study, identification with TV characters was related to peer-rated aggression in grades three and four, but not in grade five. In sum, most—but not all—studies point to identification as playing an important role in media effects and suggest several reasons why identification intensifies the effects of media.

CONCLUSION

Identification with media characters is a process that impacts our involvement with, and interpretation of, media texts. It is both a function of and an aid to the human capacity to imagine and to mentally process events at various levels simultaneously. Identification is an important channel through which mediated messages affect our lives and the society in which we live.

This review has attempted to describe, summarize, analyze, and critique what is currently known about identification in the context of entertainment television. Unfortunately, not enough systematic knowledge can be gleaned from existing research. The complexity of the phenomenon, coupled with the conceptual fuzziness in this area of research, leads to conflicted findings that are difficult to disentangle into useful generalizations. It has been argued that in order to understand identification, conceptual clarity is the first order of business (e.g., Cohen, 2001; Cohen and Perse, 2003). Although exploring the differences between identification and related concepts such as parasocial interaction, imitation, transportation, presence, and more, is important for the advancement of theory, it has become clear that conceptual clarity is part and parcel of a more developed theory. What seems more urgent, then, is to develop a more elaborate and precise theory of how, when, and why identification works and the role it plays in reception and media effects. This theory, in turn, must be part of a larger theoretical framework that explains audience involvement. It is within such a theory that the various forms of involvement with characters and other facets of texts will be more fully understood.

While a comprehensive theory is not presently available, this review does suggest some initial conclusions. Identification is an imaginative process in which we adopt a character's point of view and develop an empathic understanding of his or her plight and motivations. It is related to other facets of involvement such as realism and transportation, and it is coupled with increased emotional involvement with texts. There is evidence that identification is related to enjoyment and heavier viewing. There seems to be some tendency to identify with characters we see as both positive and similar to us in some ways, but there are many cases in which this tendency does not hold. There is evidence that our relationships and attitudes toward media characters resemble our relationships with people we meet interpersonally, but we clearly need to know more about how social and mediated relationships differ.

Beyond what is known, a theory of identification should (1) provide a framework to understand the relative importance of viewer characteristics, character traits and roles, and technology in shaping viewer identification with characters; (2) explain how this process differs across genres and how it develops over time; and (3) specify under what conditions it increases effects and leads to dominant interpretations and under which conditions it allows viewers to resist and create alternative interpretations.

Much of the research presented herein was conducted to explore other forms of viewer–character relationships. Another direction for future theoretical development is exploration of the differences in how these various types of relationships operate. It has been convincingly argued that identification is a different psychological process than parasocial relationships, but the nature of this difference has yet to be demonstrated convincingly (Cohen & Perse, 2003). It remains to be seen how various relationships with media characters relate to different media (e.g., film vs. TV), genres (e.g., host shows vs. comedy vs. drama), types of characters (e.g., heroes vs. villains), and different viewers (e.g., gender and age differences, personality types). More studies are necessary before we gain an understanding of this complex array of inter-relating factors.

The research on identification is no longer in its infancy, but it has yielded only limited understanding of this phenomenon. Within the present review, many avenues for future research are implicitly outlined. However, if there are lessons to be learned from the history of research on identification, they are about the importance of theory-driven research and the need to be aware of previous research and to take it seriously, even when it goes against our intuition and preconceived beliefs about the world. The theoretical infancy relative to the vast number of studies that ostensibly deal with identification is a result of many studies that present empirical findings about identification, but do not seriously explore or engage its existing theory. Too much effort has been expended on studies whose results cannot be compared and contrasted with previous research because they used definitions and measures that are not compatible. Other studies have insisted on making assumptions that had previously been proven false by several studies (e.g., the studies looking for compensatory relationships). Nonetheless, the growing interest in relationships with fictional characters, and in entertainment media more generally, promises to advance our understanding of how we become involved with and enjoy entertainment, and how it enhances and detracts from our lives as individuals and more generally affects modern society.

ACKNOWLEDGMENTS

I would like to thank Yariv Tsfati for his helpful comments, Rebekah Tukachinsky for helping to compile the sources for this chapter, and Oren Livio for his invaluable editing and proofreading.

REFERENCES

Alperstein, N. M. (1991). Imaginary social relationships with celebrities appearing in television commercials. *Journal of Broadcasting & Electronic Media, 35*, 43–58.

Ang, I. (1982/1985). *Watching Dallas: Soap opera and the melodramatic imagination* (D. Couling, Trans.). London: Methuen.

Ashe, D. D., & McCutcheon, L. E (2001). Shyness, loneliness, and attitude towards celebrities. *Current Research in Social Psychology, 6.* Retrieved January 12, 2004, from http://www.uiowa.edu/~grpproc/crisp/crisp.6.9.htm

Auter, P. J. (1992). TV that talks back: An experimental validation of a parasocial interaction scale. *Journal of Broadcasting & Electronic Media, 36*, 173–181.

Bandura, A. (1965). Influence of models' reinforcement contingencies on the acquisition of imitative responses. *Journal of Personality and Social Psychology, 1*, 589–595.

Basil, M. D. (1996). Identification as a mediator of celebrity effects. *Journal of Broadcasting & Electronic Media, 40*, 478–495.

Brigham, J. C., & Giesbrecht, L. W. (1976). "All in the Family": Racial attitudes. *Journal of Communication, 26*, 69–74.

Brown, W. J., Basil, M. D., & Bocarnea, M. C. (2003). The influence of famous athletes on health beliefs and practices: Mark McGwire, child abuse prevention, and Androstenedione. *Journal of Health Communication, 8*, 41–57.

Burke, K. (1950). *A rhetoric of motives.* Berkeley: University of California Press.

Cohen, J. (1997). Parasocial relations and romantic attraction: Gender and dating status differences. *Journal of Broadcasting & Electronic Media, 41*, 516–529.

Cohen, J. (1999). Favorite characters of teenage viewers of Israeli serials. *Journal of Broadcasting & Electronic Media, 43*, 327–345.

Cohen, J. (2001). Defining identification: A theoretical look at the identification of audiences with media characters. *Mass Communication & Society, 4*, 245–264.

Cohen, J. (2002). Deconstructing Ally: Explaining viewers' interpretations of popular television. *Media Psychology, 4*, 253–277.

Cohen, J. (2003). Parasocial breakups: Measuring individual differences in responses to the dissolution of parasocial relationships. *Mass Communication & Society, 6*, 191–202.

Cohen, J. (2004). Parasocial break-up from favorite television characters: The role of attachment styles and relationship intensity. *Journal of Social and Personal Relationships, 21*, 187–202.

Cohen, J., & Perse, E. (2003). *Different strokes for different folks: An empirical search for different modes of viewer-character relationships.* Paper presented to the Mass Communication Division at the 53rd annual convention of the International Communication Association (ICA), San Diego, CA, May 24, 2003.

Cohen, J., & Ribak, R. (2003). Gender differences in pleasure from television texts: The case of Ally McBeal. *Women's Studies in Communication, 26*, 118–134.

Cole, T., & Leets, L. (1999). Attachment styles and intimate television viewing: Insecurely forming relationships in a parasocial way. *Journal of Social and Personal Relationships, 16*, 495–511.

Coleridge, S.T. (1817/1965). *Biographia literaria* (Vol. 2). London: Oxford University Press.

Cooper, B. (1999). The relevancy of gender identity in spectators' interpretations of *Thelma & Louise. Critical studies in Mass communication, 16*, 20–41.

Eyal, K., & Rubin, A. M. (2003). Viewer aggression and homophily, identification, and parasocial relationships with television characters. *Journal of Broadcasting & Electronic Media, 47*, 77–98.

Feilitzen, C., & Linne, O. (1975). The effect of television on children and adolescents: Identifying with television characters. *Journal of Communication, 25*, 51–55.

Fisherkeller, J. (1997). Everyday learning about identities among young adolescents in television culture. *Anthropology and Education Quarterly, 28*, 467–492.

Fiske, J. (1989). *Television culture.* London: Routledge.

Galgut, E. (2002). Poetic faith and prosaic concerns. A defense of 'suspension of disbelief.' *South African Journal of Philosophy, 21*, 190–200.

Gerrig, R. J. (1993). *Experiencing narrative worlds.* New Haven, CT: Yale University Press.

Giles, D. C., & Maltby, J. (2004). The role of media figures in adolescent development: Relations between autonomy, attachment, and interest in celebrities. *Personality and Individual Differences, 36*, 813–822.

Gleich, U. (1997). Parasocial interaction with people on the screen. In R. Winterhoff-Spurk & T. H. A. van der Voort (Eds.), *New horizons in media psychology: Research cooperation and projects in Europe* (pp. 35–55). Olpaden, Germany: Westdeutscher Verlag.

Green, M. C., & Brock, T. C. (2000). The role of transportation in the persuasiveness of public narratives. *Journal of Personality and Social Psychology, 79*, 701–721.

Hall, S. (1980). Encoding/Decoding. In S. Hall, D. Hobson, A. Lowe & P. Willis (Eds.), *Culture, media, language* (pp. 128–138). London: Hutchinson.

Hoffner, C. (1996) Children's wishful identification and parasocial interaction with favorite television characters. *Journal of Broadcasting & Electronic Media, 40*, 389–402.

Hoffner, C., & Cantor, J. (1991). Perceiving and responding to mass media characters. In J. Bryant & D. Zillmann (Eds.), *Responding to the screen: Reception and reaction processes* (pp. 63–103). Hillsdale, NJ: Lawrence Erlbaum Associates.

Horton, D., & Wohl, R. R. (1956). Mass Communication and parasocial interaction: Observations on intimacy at a distance. *Psychiatry, 19*, 215–229.

Kanazawa, S. (2002). Bowling with our imaginary friends. *Evolution and Human Behavior, 23*, 167–171.

Kelman, H. C. (1961). Processes of opinion change. *Public Opinion Quarterly, 25*, 57–78.

Kim, J. K., & Rubin, A. M. (1997). The variable influence of audience activity on media effects. *Communication Research, 24*, 107–135.

Liebes, T., & Katz, E. (1990). *The export of meaning: Cross-cultural readings of Dallas.* New York: Oxford University Press.

Livingstone, S. M. (1987). Implicit representation of characters in Dallas: A multidimensional approach. *Human Communication Research, 13*, 399–420.

Livingstone, S. M. (1998). Relationships between media and audiences: Prospects for audience reception research. In T. Liebes & J. Curran (Eds.), *Media, ritual and identity* (pp. 237–255). London: Routledge.

Livingstone, S. M., & Mele, A. R. (1997). Evaluating emotional responses to fiction. In M. Hjort & S. Laver (Eds.), *Emotion and the arts* (pp. 157–176). NY: Oxford University Press.

Lombard, M. (1995). Direct responses to people on the screen: Television and personal space. *Communication Research, 22*, 288–324.

Maccoby, E. E., & Wilson, W. C. (1957). Identification and observational learning from films. *Journal of Abnormal Social Psychology, 55*, 76–87.

McCutcheon, L. E., Ashe, D. D., Houran, J., & Maltby, J. (2003). A cognitive profile of individuals who tend to worship celebrities. *Journal of Psychology, 137*, 309–323.

Morley, D. (1992). *Television, audiences, and cultural studies.* London: Routledge.

Noble, G. (1975). *Children in front of the small screen.* Beverly Hills, CA: Sage.

Nodelman, P. (1991). The eye and the I: Identification and first-person narratives in picture books. *Children's Literature, 19*, 1–30.

Oatley, K. (1994). A taxonomy of the emotions of literary response and a theory of identification in fictional narrative. *Poetics, 23*, 53–74.

Oatley, K. (1999). Meeting of minds: Dialogue, sympathy, and identification, in reading fiction. *Poetics, 26*, 439–454.

Papa, M. J., Singhal, A., Law, S., Pant, S., Sood, S., Rogers, E. M., & Shefner-Rogers, C. L. (2000). Entertainment-education and social change: An analysis of parasocial interaction, social learning, collective efficacy, and paradoxical communication. *Journal of Communication, 50*, 31–55.

Perse, E. M., & Rubin, A. M. (1990). Chronic loneliness and television use. *Journal of Broadcasting & Electronic Media, 34*, 37–53.

Putnam, R. D. (2000). *Bowling alone: The collapse and revival of American community.* New York: Simon & Schuster.

Press, A. L. (1990). Class, gender and the female viewer: Women's responses to *Dynasty*. In M. E. Brown (Ed.), *Television and women's culture: The politics of the popular* (pp. 144–157). London: Sage.

Reeves, B., & Naas, C. (1996). *The media equation: How people treat computers and new media like real people and places.* Cambridge: Cambridge University Press.

Rubin, A. M., Perse, E. M., & Powell, R. A. (1985). Loneliness, parasocial interaction, and local television news viewing. *Human Communication Research, 12*, 155–180.

Rubin, A. M., & Step, M. M. (2000). Impact of motivation, attraction and parasocial interaction on talk radio listening. *Journal of Broadcasting & Electronic Media, 44*, 635–654.

Sanders, M. S. (2004). *Is it a male or female thing?: Identification and enjoyment of media characters.* Paper presented to the Mass Communication Division at the 54th annual convention of the International Communication Association (ICA), New Orleans, LA, May 30, 2004.

Sheehan, P. W. (1983). Age trends and the correlates of children's television viewing. *Australian Journal of Psychology, 35*, 417–431.

Skumanich, S. A., & Kintsfather, D. P. (1998). Individual media dependency relations within television shopping programming: A causal model revisited and revised. *Communication Research, 25*, 200–219.

Tamborini, R., Stiff, J., & Heidel, C. (1990). Reacting to graphic horror: A model of empathy and emotional behavior. *Communication Research, 17*, 616–640.

Tsao, J. (1996). Compensatory media use: An exploration of two paradigms. *Communication Studies, 47*, 89–109.

Turner, J. R. (1993). Interpersonal and psychological predictors of parasocial interaction with different television performers. *Communication Quarterly, 41*, 443–453.

Vidmar, N., & Rokeach, M. (1974). Archie Bunker's bigotry: A study in selective perception and exposure. *Journal of Communication, 24*, 36–47.

Wiegman, O., Kuttschreuter, M., & Baarda, B. (1992). A longitudinal study of the effects of television viewing on aggressive and prosocial behaviours. *The British Journal of Social Psychology, 31*, 147–164.

Wied, M., Zillmann, D., & Ordman, V. (1994). The role of empathic distress in the enjoyment of cinematic tragedy. *Poetics, 23*, 61–106.

Wilson, T. (1993). *Watching television: Hermeneutics, reception, and popular culture*. Cambridge, UK: Polity Press.

Zillmann, D., & Bryant, J. (2002). Entertainment as media effect. In J. Bryant & D. Zillmann (Eds.), *Media effects: Advances in theory and research* (pp. 437–461). Hillsdale, NJ: Lawrence Erlbaum Associates.

12

Involvement

Werner Wirth
University of Zurich

While watching a good movie—and sometimes even hours later—thoughts, feelings, and discussions circle around the plot of the film. Obviously, we are involved, taken away, or even "caught" by good entertainment. Involvement and entertainment therefore appear to be related concepts. It is surprising, however, that involvement usually is not seen in terms of its relation to entertainment. Involvement is probably one of the most successful concepts of communication research and media psychology, but it originated in the context of a different research domain, the social judgment theory (Sherif & Cantril, 1947). Soon, variants of it came up in other areas of research, which led to lively research activities. Today, involvement is considered to play an important role in media usage and its effect. But there is a drawback to its success: Scholars have to cope with a confusing heterogeneity of definitions and operationalizations of involvement.

There is a considerable number of reviews which try to deal systematically with the measurement and the effects of involvement (cf. Antil, 1984; Cohen, 1983; Costley, 1988; Greenwald & Leavitt, 1984; Johnson & Eagly, 1989, 1990; Mitchell, 1981; Muehling, Laczniak, & Andrews, 1993; Petty & Cacioppo, 1990; Rothschild, 1984; Salmon, 1986; Zaichkowsky, 1986). As far as the theoretical background and the discipline are concerned, each of these reviews has a perspective of it own. The goal of this particular review is to describe the most important lines of discussion from a media-psychological perspective and to synthesize them in a comprehensible way. Furthermore, we will consider the relations between entertainment and involvement. In this context, we will try to answer the questions whether involvement can be part of the feeling of entertainment and to what extent users are involved with media entertainment.

MAJOR RESEARCH TRADITIONS OF INVOLVEMENT

In Sherif's work and in the work of his colleagues concerning *social judgment theory*, ego-involvement plays a central role (Sherif & Cantrill, 1947). If a topic activates central values of the self-concept, a person becomes personally involved with the situation. Conversely, ego-involvement deals with the question of how important the issue is for self-identity or a person's self-picture (Salmon, 1986). The importance of ego-involvement for the social judgment theory results from its central hypothesis. According to this hypothesis, a belief change becomes less probable the more a person is involved.

In their later work, Sherif and Hovland (1961) distinguished task-involvement from ego-involvement. In contrast to ego-involvement, task-involvement results from the experimental manipulation performed by the researcher. Participants of experiments may be task-involved, for example, because they are confronted with an issue in which they are personally engaged, but which does not necessarily touch on central aspects of their self-definition. This type of involvement is assumed to be rather volatile and weak (Salmon, 1986). The extent to which central personal values are activated during reception is of critical importance for ego-involvement. During task-involvement we are only dealing with personally-relevant issues (e.g., the planned introduction of tuition fees). In short, involvement can be described in this context as activated relevance for an issue (Salmon, 1986).

For more recent social psychological *theories on persuasion*, involvement is of similar central importance as it is for social judgment theory (Chaiken & Trope, 1999; Johnson & Eagly, 1989, 1990; Petty & Cacioppo, 1979, 1981, 1990; Petty, Priester, & Briñol, 2002). According to the elaboration likelihood model, highly involved recipients are motivated to process the arguments of the message (central route), whereas low-involved recipients select rather peripheral stimuli for a superficial judgment (peripheral route). Similarly, the heuristic systematic model distinguishes between systematic (high involvement) and heuristic processing (low involvement) (Chaiken, 1980; Chen & Chaiken, 1999).

Krugman introduced the concept of involvement in *consumer research* half a century ago. In his definition of involvement, Krugman (1965) did not refer to the social judgment theory with respect to ego-involvement; he implicitly preferred a cognitive approach. For Krugman, involvement is the number of conscious bridging experiences, connections, or personal references that a viewer makes per minute between his or her own life and a stimulus (Krugman, 1965, p. 356). This conceptualization is well matched to the information processing approaches developed after the cognitive turn (cf. e.g. Bransford, 1979; Craik & Tulving, 1975; Graesser & Clark, 1985). Krugman's research stimulated numerous studies and theoretical developments. For example, the rather static hierarchy of effects models were differentiated (Batra & Ray, 1985; Chaffee & Roser, 1986; Ray, 1973). The hierarchy of effects is now different under low involvement conditions than under high involvement conditions. Recently, consumer research dealt with the reciprocal relationships between the effects of TV commercials and the program-induced audience involvement as a context variable (e.g. Bryant & Comisky, 1978; Norris & Colman, 1993, 1994; Park & McClung, 1986; Soldow & Principe, 1981).

Stimulated by consumer research, *audience research* began to study involvement from the perspective of mass communication as well (Donnerstag, 1996). Specifically, media usage under low-involvement conditions seemed to be of particular interest. The usage of electronic media (radio, television) often occurs only incidentally, habitualized and without much attention (e.g., Barwise & Ehrenberg, 1988). Alternatively, within the uses-and-gratifications approach, involvement is explicitly integrated as a part of the concept of audience activity. Levy and Windahl (1984, 1985) suggested a three-phases model. *Before* exposure, involvement is expressed as intent of usage. *During* media exposure, involvement is understood as the perceived connection between an individual and the mass media content on the one hand, and the degree to which the individual interacts psychologically with a medium or its message, on the other. After the person has been exposed to the medium, involvement can be conceptualized as a long-term identification, or parasocial relationship (Levy & Windahl, 1985; Roser, 1990). But involvement fits well with other research traditions, too. Donnerstag (1996) emphasizes the parallels between involvement and agenda setting research. Two important moderator variables, personal concern and need for orientation, are compatible with involvement (McCombs & Reynolds, 2002).

This short look into four of the most important research areas of involvement already illustrates its diversity. Next, we will systematize the various conceptions from different perspectives: the targets of involvement, its locus, its persistence, its components, and its valence (similarly Andrews, Durvasula, & Akhter, 1990; Roser 1990). A synthesis follows at the end of the discussion.

THE TARGETS OF INVOLVEMENT

One cannot imagine involvement without a reference, without a direction! When comparing the results of empirical studies, one always has to consider the objects towards which the involvement measured is directed (Muehling, Laczniak, & Andrews, 1993). It makes a difference whether involvement is directed to the message of the media, a protagonist, a certain program, a series, a medium, the reception situation, the topic of the media message, or to a social problem. The references in consumer research are even more diverse. There, involvement is described as directed toward the message of a commercial, the communicator, the advertised product, the advertised brand, or to the context of the program (Costley, 1988). Principally, one should always mention the target of involvement. From the perspective of entertainment research, in most cases, we are concerned with involvement with entertaining media content. It should be pointed out, however, that we are not dealing with an independent type of involvement (such as entertainment involvement), but, specifically, with involvement with entertainment.

THE LOCUS OF INVOLVEMENT

From a media-psychological perspective, there is no doubt that involvement is a psychological construct located within the individual. Particularly in the early publications on involvement, one can find curtailments, however, which seem to define involvement in terms of object attributes. In his meta-analysis, Salmon (1986) distinguished four different perspectives. Usually, involvement is interpreted as an internal state or process. In some studies, however, involvement is interpreted as a personality trait (see also Roser, 1990; Zaichkowsky, 1986), as salience or relevance of a topic, or as quasi-characteristic of a stimulus, a medium, or of a situation (Rothschild & Ray, 1974). Krugman (1965) and Salomon (1984), for instance, consider television to be a low-involvement medium, whereas magazines and other print media are defined as high-involvement media. Particularly in consumer research, low-involvement products (e.g., tissue paper) are distinguished from high involvement products (e.g., cars).

Of course, such attributions are generalizations: One can postulate that in most instances, people watch TV with rather low involvement and read books with rather high involvement or that the general audience listens to economic topics with greater involvement than with abstract topics such as freedom of the press (Andrews et al., 1990; Shoemaker, Schooler, & Danielson, 1989). Whether these postulates hold true can only be decided on the basis of empirical data. These results have the potential for surprise. In an election study, for example, it was shown that contrary to expectation, TV was not always the low-involvement medium (Shoemaker et al.,

1989). How high the involvement actually is depends on the reception situation and on the audience, as well as on the framing of the message. Therefore, one should not, generalize the locus of involvement.

THE PERSISTENCE OF INVOLVEMENT

Conceptualizations of involvement can be sorted along their temporal persistence and ordered accordingly. To put it simply, one can distinguishe between an enduring, persistent, rather stable form of involvement and a more situational, short-term, and volatile type (Andrews et al., 1990; Houston & Rothschild, 1978).

Enduring involvement is often understood as the (pre-existing) personal relevance a topic or an object has for an individual (Apsler & Sears, 1968; Celsi & Olson, 1988; Chaffee & Roser, 1986; Petty & Cacioppo, 1979, 1981; Richins & Bloch, 1986; Zaichkowsky, 1986). People attribute personal relevance to an object if they feel an ongoing attachment to it. The reason for this may be that it touches central values or goals of the self or of a closely connected person or that, from a subjective view, it has obvious consequences for the self or for a closely connected person (Apsler & Sears, 1968; Havitz & Howard, 1995; Richins & Bloch, 1986). This is largely compatible with the definition of ego-involvement in the social judgment tradition (Sherif & Hovland, 1961). Celsi and Oslon (1988) distinguish intrinsic and situational sources of personal relevance (ISPR and SSPR, respectively). ISPR are relatively stable, enduring structures of personally relevant knowledge, derived from past experience and stored in the memory. In contrast, SSPR are a wide variety of specific stimuli, cues, and contingencies in an individual's immediate environment that activate or are closely associated with self-relevant consequences, goals, and values. Situational personal relevance, therefore, is a situational activation of long-term goals or values that gradually guide action during reception, and thus influence information processing. Houston and Rothschild (1978) see process-oriented response involvement as a result of the interaction between enduring and situational involvement (see also Andrews et al., 1990; Burnkrant & Sawyer, 1983; Patwardhan, 2004; Richins & Bloch, 1986; Richins, Bloch, & McQuarrie, 1992).

Many authors regard involvement as an attribute of information processing itself (e.g., Cameron, 1993; Greenwald & Leavitt, 1985; Putrevu & Lord, 1994; Rubin & Perse, 1988; Step, 1998; see later section on intensity). The process-oriented definition of involvement directly refers to the phase of media usage and encompasses the intensity of an individual's cognitive, emotional, or conative engagement with the media message.

Finally, other authors conceptualize involvement as a mode of reception, as opposed to a distanced or analytical mode of media usage (Liebes & Katz, 1986; McQuail, 1985; Vorderer, 1993). These authors agree that viewers—at least temporarily—emotionally and cognitively "live within" the world presented by the media, a phenomenon which has been recently referred to as non-mediation, or presence (Lombard & Ditton, 1997; Vorderer, Klimmt, & Ritterfeld, 2004).

By all means, involvement as a reception mode is compatible with the conceptualization of involvement as a process. Each reception mode should be connected with corresponding thoughts and elaboration. They express the individual's involvement either in a story or in the message as an aesthetic artifact and can be negative as well as positive (see the section about the valence of involvement below). Referring to Liebes and Katz (1986), Sood (2002) distinguishes between referential reflection (thoughts about the media content and about related personal experiences) and critical reflections (thoughts about the aesthetic construction of the media program).

In both Tan (1996) and Oatley (1994), one can find a corresponding distinction on the affective level. The emotions experienced in the reception mode 'involvement' are termed "fiction emotions" (Tan, 1996) or "within-emotions" (Oatley, 1994) and are contrasted with those feelings which the recipient has concerning the nature of a media message as an artifact ("artifact emotion" in Tan; "without-emotions" in Oatley).

THE COMPONENTS OF INVOLVEMENT

Many authors partition involvement into cognitive, affective (or emotional), and conative components. This distinction can be traced back to Rothschild and Ray (1974) who compared various definitions.

The *cognitive component* can be found most frequently in the literature on involvement (e.g. Bryant & Comisky, 1978; Flora & Maibach, 1990; Lo, 1994; Perse, 1990a, 1990b, 1990c, 1990d; Rubin & Perse, 1988). Krugman (1965, 1966) already defined involvement as a "bridging experience" or as "personal connections" between the stimulus and the experiences and views of the person. Salomon (1981), too, conceptualized his "AIME" ("amount of invested mental effort") as cognitive involvement. The basis for cognitive involvement, however, is the cognitive response approach (Petty, Ostrom, & Brock, 1981). Its rationale is as follows: How enduring an attitude change will be is determined by the extent to which the arguments of a media message are considered and the quality of these considerations. From this, Petty and Cacioppo developed their Elaboration likelihood model of persuasion (Petty & Cacioppo, 1981, 1986). A person who is highly motivated and capable is going to be highly involved during media usage, that is, s/he will exert extensive and careful thoughts and considerations about the media message and its meaning for him/herself. During such elaborations, incoming information is connected with existing knowledge and images. In this process, connotative and associative meanings are attached as well (Perse, 1990c). Because attention is a presupposition for this elaborative information processing, some authors include attention as part of the cognitive component of involvement as well (Perse, 1998, see also Greenwald & Leavitt, 1984). Cameron (1993) locates cognitive involvement within spreading activation theory (e.g., Anderson, 1983; Collins & Loftus, 1975; for critical comments see Ratcliff & McKoon, 1981). For Cameron (1993), cognitive involvement is increased cognitive activation. Because increased knowledge can result from intense cognitive engagement with a media message, some authors interpret knowledge and recall as cognitive involvement (e.g., Chaffee & Roser, 1986; Greenwald & Leavitt; 1984; Perse, 1990b). Ray (1973) subsumes attention, awareness, comprehension, and learning under cognitive involvement. Such a broad conceptualization of cognitive involvement, however, is a dead end. It seems to make more sense to restrict cognitive involvement to higher-order thinking processes (elaborations, thoughts). Awareness and attention should be seen as prerequisites for cognitive involvement, whereas knowledge, learning, and recall should be seen as *consequences* of cognitive involvement (e.g., Shoemaker et al., 1989).

Affective (emotional) involvement is studied slightly less often than cognitive involvement. If affective involvement is the issue, then usually a theoretical and/or empirical distinction is made between cognitive and affective involvement (e.g., Hoffman & Batra, 1991; Park & McClung, 1986; Zaichkowsky, 1987). Park and McClung (1986) consider emotional involvement to mean the subjective experience of media usage (e.g., of a commercial spot), whereas cognitive involvement means the processing of issue-oriented information. Some researchers subsume parasocial interaction under affective involvement (Levy & Windahl, 1985). Recently, however, scholars in this field argue that parasocial interaction itself encompasses affective, cognitive, *and* conative components (Hartmann, Schramm, & Klimmt, 2004; Sood, 2002; see also Chapter 17 by Klimmt, Hartmann, & Schramm, in this volume).

Chaudhuri's and Buck's approach (e.g. Buck, Chaudhuri, Georgson, & Kowta, 1995; Chaudhuri & Buck, 1995a, 1995b) locates affective involvement in the frame of the cognitive response approach (similarly MacInnis & Jaworski, 1989). According to this approach, affective responses are post-communicative self-statements about the felt intensity of selected emotions. Perse (1990a, 1990b, 1990c, 1990d) and Step (1998) conceptualize emotional involvement as intensity of felt emotions. In reference to social judgment research, Shoemaker et al. (1989) use attitude extremity as a criterion for affective involvement; Chaffee and Roser (1986) additionally use perceived risk (see also Roser, 1990).

The *conative component* of involvement is seldom integrated into research. According to Ray (1973), conative involvement encompasses intention, behavior, and action. The consumer researchers Richins et al. (1992) mention the following categories of conative involvement: active search for information, giving advice to others, and communication about the experiences one has with a product. In practical research, media usage on specific topics is often used as an indicator for behavioral involvement (e.g., Chaffee & Roser, 1986; Rubin & Perse, 1988; Shoemaker et al., 1989).

Conceptualizations which stress the *motivational character* of involvement put an even stronger emphasis on the intentional aspect (Slater, 2002). According to Mitchell (1981), involvement is an interaction between the goals of an individual and a stimulus. This interaction determines the attention and the intensity with which a media message is processed and evaluated, or with which inferences are drawn and elaborations are constructed. Buck and coworkers (Buck, 1988; Buck et al., 1995) emphasize that the motivational component is relevant to affective involvement as well, because emotions prepare us for appropriate actions and suggest, for example, either approach or avoidance behavior. Emotion can be seen as a process through which motivational potential is realized or read out when activated by a stimulus (Chaudhuri & Buck, 1995b).

In summary, cognitive involvement is conceptualized in the literature as active and intensive information processing which extends from perception and attention to intensive thinking, elaboration and recall. Basically, in the context of cognitive involvement, we are talking about elaborative activities. Affective or emotional involvement is the sensation of intense feelings or affective statements. In this context, conative involvement specifically means the search for information in media and the interpersonal exchange of information about a topic. Motivational components emphasize the action intention and the informational character of involvement for the individual.

THE INTENSITY OF INVOLVEMENT

In some approaches, the intensity or the level of involvement are not only measured empirically, but are also theoretically differentiated. Krugman (1965, 1966) simply distinguished between low and high involvement and derived different processes of advertising effectiveness for low- and high-involvement products or media. In dual-process theories (see aforementioned), the authors also distinguish between high and low involvement: Researchers who rely on Craik's and Lockhart's (1972) level-of-processing framework suggest a broader differentiation (Burnkrant & Sawyer, 1983). According to Greenwald and Leavitt (1984, 1985), there are four levels of (cognitive) involvement, which differ in their abstractness of symbolic activity that occurs in the analysis of an incoming message: preattention, focal attention, comprehension, and elaboration. The higher the level of involvement, the more cognitive capacity is utilized and the more attentional effort is exerted. At the same time, increasingly complex representational systems are required. Moreover, higher levels of involvement lead to stronger recall effects (see also Perloff, 1984). The authors point out that we are dealing with qualitatively different levels and not with a continuum. Therefore, the conceptual knowledge of the user is exclusively applied to the highest level. Thus, it becomes evident that many involvement processes of interest during media usage only occur on the highest level. That is why the approach does not seem to be sufficiently differentiated. MacInnis and Jaworski (1989) even propose *six* levels that differentiate involvement more broadly, especially on the higher involvement levels: feature analysis, basic categorization, meaning analysis, information integration, role-taking, and constructive processes. The authors describe cognitive as well as emotional aspects for each of these levels. An approach that separates involvement from attention and perception, and which differentiates more clearly, specifically on the higher cognitive levels (elaborations, thoughts), seems to make more sense. Cameron (1993), for instance, connects the intensity of involvement with the spreading activation theory. According to this theory, involvement is a continuum that emerges in the interaction between cognitive knowledge structures (prior knowledge) and arousal. Because Cameron (1993) defines involvement as a cognitive process, he is able to treat arousal, prior knowledge, and interest as separate concepts.

THE VALENCE OF INVOLVEMENT

In general, involvement is seen as a positive action, that is, as an action which is directed towards an object or a person. According to the cognitive response theory (Petty et al., 1981) and to the elaboration likelihood model (Petty & Cacioppo, 1981, 1986), the valence of involvement is not determined. Given a high motivation and an ability to process, negative, non-favorable thoughts concerning the message may emerge, for instance, if a person strongly distrusts (Kramer, 1998). Kamins, Assael, and Graham (1990) found different functions of supporting arguments and counterarguments, as well as positive and negative source responses, during high versus low involvement. During affective involvement, feelings or emotions may emerge against protagonists or against the media message (Perse, 1990c, 1998). Hartmann et al. (2004) explicitly consider negative parasocial interaction with the media actors, too. Conative involvement can also be defensive. Take, for example, the facial expression of negative emotions such as anger, hatred, abhorrence, or disgust. Obviously, even negative motivational involvement is possible. Negative motivation is directed toward escaping from, or preventing, some undesirable outcome. To be negative motivationally involved means to try hard to prevent something. This may be caused by fear (e.g., Frijda, 1986; Izard, 1993). The intensity of a "negative motivation" increases as a feared object or event approaches. Of course, negative motivation may lead to avoidance behavior and, thus, tends to decrease if it is satisfied.

CONCEPTUALIZING INVOLVEMENT: A SYNTHESIS

So, what exactly is involvement? It is obvious that the diversity of the approaches to the study of involvement make it difficult to find a common definition that takes all aspects into account. If you are willing to accept some kind of fuzziness, one can distinguish between a broader and a more confined concept of involvement. Involvement, as a metaconcept, encompasses a family of related, though distinct, concepts that inform us how users are occupied with the media and its content, and how they engage with them in a cognitive, affective, conative, and motivational way (Salmon, 1986, p. 244). However, one should not expect that different aspects of this involvement family lead to uniform effects on recall, persuasion, or entertainment. In this view, involvement is rather a framework or a research perspective, its multiple interactions with other concepts of media usage and effects can be studied and related to each other systematically. Of particular importance is the distinction between rather enduring, persistent attitudes or traits on the one hand, and processes which are more directed to the processing and evaluation of objects (e.g., media content) during the communicative phase on the other hand (e.g. Levy & Windahl, 1985; Richins & Bloch, 1986). This second conceptualization can be understood as involvement in a more confined sense. It is not necessarily an affirmative, positive effort; the critical media usage, too, can be considered as one form of involvement in this sense.

How can situation-specific and situation-unspecific forms of involvement be connected? Involvement as a situation-unspecific attitude or trait is transformed into a process variable during media usage if corresponding internal (thoughts, emotions) or external cues (media, task) lead to an activation of the attitude (see also Celsi & Olson, 1988; Gotlieb & Sarel, 1991; Richins et al., 1992). Thus, involvement means motivated strategic processing, which is not only influenced by processing goals, but also by situational and stimulus factors and respectively salient features of these factors as well (Muehling, Laczniak, & Andrews, 1993; see also Higgins, 1996). It is important to consider qualitative differences of involvement with different target objects next: According to Slater (2002, see also Sood, 2002), depending on the processing goal, different processing styles may emerge: value-protecting, value-affirmative, outcome-based, dedicated processing, information-scanning, hedonic processing.

Thus, on the basis of qualitatively different involvement processes, we reach a better understanding of the moment-to-moment engagement with the media message. Process involvement implies, on the one hand, elaborative and energetic features in the spirit of Cameron (1993). On the other hand, involvement varies between approaching and distancing in the spirit of Vorderer (1993), too. As it has been argued, both subprocesses can be regarded as independent of each other. Finally, after exposure, the reception process is transformed into the compact form of an experience. Depending on how much the enduring and processing goals are satisfied or new goals are stimulated, a modification of the enduring involvement may occur. In sum, involvement can be conceptualized as a temporally structured, multi-component concept (see also Richins & Bloch, 1986).

INVOLVEMENT AS A FEELING OF ENTERTAINMENT

Up to now, we have concentrated on a comprehensive review of the literature on involvement. Now, we will focus on the connections between involvement and entertainment. First of all, one may ask: Isn't involvement an aspect of entertainment? For Vorderer (1998), the affective processes occurring during reception, in particular, are expressions of the entertainment experience. He distinguishes between ego-emotional and socio-emotional engagement. Whereas ego-emotional engagement means that a topic touches central values, goals, or own life experiences, socio-emotional engagement, or empathy, refers to voluntary and transient sympathizing with the protagonist's experiences (Zillmann, 1991, 1994). Both processes easily can be connected with the affective component of involvement (see aforementioned). In this view, ego-involvement (central values) and affective involvement can be interpreted as an aspect of entertainment.

Various phenomena from media psychology as well can be conceptualized in this sense. In Zillmann's (1994) affective disposition theory, conditions of socio-emotional engagement in the protagonist's fate are mentioned. Hereby, spectators interact parasocially with the actors. During repetitive encounters with the actors (e.g., in daily soaps) the spectators engage in parasocial relations with them (Perse & Rubin, 1989; Vorderer, 1996).

Affective involvement has been shown in the usage of a wide variety of informational content, such as health campaigns (Chaffee & Roser, 1986), news (Perloff, 1989; Perse, 1990d), or the search for information in the internet (Patwardhan, 2004). Chaudhuri and Buck (1995b) conducted comparative studies on emotional involvement in TV and print commercials.

Perse (1990c) found that emotional and cognitive involvement correlate. Feeling angry and sad were linked to elaboration on crime and government-related news. Moreover, persons who use news for entertainment indicated that they felt happy and satisfied. Therefore, positive emotional involvement can be seen as closely connected to entertainment motives.

What effects does affective involvement have on phenomena such as learning? It sounds fascinating that learning and knowledge acquisition should benefit from a person's being entertained (Vorderer et al., 2004). It is well known that cognitive involvement has positive effects on knowledge acquisition (e.g., Cameron, 1993). Could this be true for affective involvement as well? The results are not very encouraging. Shoemaker et al. (1989) did not find any correlations between affective involvement and recall or recognition for individuals who rely on newspaper or television. Chaffee and Roser (1986) did not find any correlations between perceived risk of heart disease as an indicator for affective involvement and the consistency between attitude, knowledge, and behavior in the case of a long-term health intervention campaign. Lo (1994) did not find a correlation between attitude extremity as an indicator for affective involvement and knowledge about the (first) Gulf War. In the latter two studies, the authors assumed that the indicators were not good enough to measure affective involvement. One can have a moderate attitude toward the Gulf War and yet be highly affectively touched (Lo, 1994, p. 51).Conversely, one can subjectively perceive a high risk of heart disease without *feeling* affectively very deeply touched (Chaffer & Roser, 1989, p. 391). Shoemaker et al. (1989) came to similar results. They studied recall and recognition of election information and again used attitude extremity as an indicator of affective involvement. As an indicator of emotional involvement, Perse (1990d) used 15 emotional reaction items from which three emotional factors (happiness, anger, sadness) could be extracted. But again, not one of the factors correlated with knowledge about news.

The heterogeneity of the indicators for affective involvement suggests the connection between affective involvement and entertainment should be examined more closely. There are some relevant results in the context of consumer research. In explorative studies, Norris and colleagues (Norris & Colman, 1994; Norris, Colman, & Aleixo, 2001) have shown that involvement and entertainment/enjoyment only have a small portion of common variance and, moreover, lead to different advertising effects (see also Furnham, Gunter, & Walsh, 1998).

Recently, some theoretical foundations concerning enjoyment as the core of entertainment have been presented. In these contributions, enjoyment is conceptualized as an attitude (Nabi & Krcmar, 2004), as coalescence of social norms, viewing situations, and program content (Denham, 2004), or as dependent on multiple prerequisites (Vorderer et al., 2004). Affective involvement such as empathy or non-mediation, therefore, is only one of various possible prerequisites (Vorderer et al., 2004). In sum, affective involvement seems to be related to enjoyment but the two concepts are not identical (Nabi & Krcmar, 2004).

INVOLVEMENT WITH ENTERTAINMENT CONTENT

To which extent can involvement be observed during the usage of entertainment content? There are a number of pertinent studies within the context of the uses-and-gratifications approach. According to this approach, affective, cognitive, and behavioral involvement are important aspects during the usage of TV programs and of soap operas on the radio (e.g. Kim & Rubin, 1997; Rubin & Perse, 1988; Rubin & Step, 2000).

Additionally though, surprisingly few studies deal with involvement encompassing exposure to entertainment programs. Bente and Feist (2000; Feist, 2000) studied the motives for viewing German talk shows. As one of four motive factors, parasocial interaction explained most of the variance of talk show exposure. Further, they found that involvement as immersion, critical thinking, and thinking about the talk show (elaborations) were seen to correlate highly with each other. Thinking about other things which had nothing to do with either the talk show or the self proved to be the antipode to involvement. Such "anti-involvement" thoughts were found particularly during non-arousing, boring contributions.

Two interesting studies report about the moment-to-moment variation in involvement. Vorderer (1993) measured involved vs. analytical reception (self-report: "living within the movie" versus "reflecting upon the movie") in one-minute intervals during two action-oriented detective movies. Both modes of reception were highly correlated with one another. Therefore, they do not seem to be independent of each other (Vorderer, 1993, see also Bente & Feist, 2000; Feist, 2000). Additionally, Vorderer (1993) compared the two measured reception modes with official television ratings. It is interesting to note that involved reception, but not analytic reception, decreased several minutes before the movies were stopped or interrupted.

In an earlier study, after every minute of action-adventure program segments, Bryant and Comisky (1978) measured reaction times (signal detection paradigm) in order to operationalize attention and asked their subjects to give ratings on three involvement indicators (subjective absorption, interest, and cognitive involvement). The results show that attention and all three involvement indicators yielded their highest values immediately before and immediately following the climax, as well as after the resolution of the story.

Andringa (1996) tried to answer the question to which extent narrative structures have an effect on the different forms of involvement.Specifically, she investigated the effect of an open narrative commentary on involvement. Open narrative commentaries are background information given by the narrator. According to Andringa (1996), this should result in a narrative distance. Like Tan (1996), the author measured A-emotions and several kinds of F-emotions; he found that A-emotions were positively correlated with one kind of F-emotion. The narrative distance, however, only had a small influence on F-emotions as an indicator of emotional involvement.

In a qualitative, cross-cultural study on the reception of a Dallas sequel, Liebes and Katz (1986) found differentiated patterns of involvement. On the basis of group discussions, they collected data concerning: (1) referential (similar to F-emotions) versus critical statements (similar to A-emotions); (2) realistic (indicative in form) versus playful (subjunctive in form) thoughts; (3) normative versus value-free statements about the actors; and (4) statements with distancing "they"-referents versus statements with involved "we"-referents. Most of the statements were based on "we"-referents and were formulated in the indicative; on the basis of theoretical considerations, the authors interpreted this as typical for the involved reception mode. The many critical statements pointed toward a distanced reception. The authors have not tested, though, whether the participants actually felt involved or distanced.

Vorderer, Cupchik, and Oatley (1997) showed that experientially and emotionally loaded texts engage readers in different ways than action-oriented and descriptively oriented texts do. Readers rated texts as more involving and richer in meaning when the protagonist's experiences were the center of interest, as compared to texts in which the action was central.

Results on involvement in online-activities were reported, too. In a cross-cultural study, Patwardhan (2004) found that entertainment websites are visited with moderate involvement. In contrast to information search and communication applications, cognitive and affective involvement with internet entertainment were rather low. The results were valid for Americans as well as for Asian Indians. In addition, cognitive involvement levels were generally slightly higher as compared to emotional involvement.

Entertainment education is an important new research domain. Here too, postcommunicative effects of involvement with entertainment programs are intensively discussed and investigated. Entertainment programs can have positive effects by influencing beliefs, attitudes, and behavior. Sood (2002) found a referential-affective and a critical-cognitive dimension of involvement in the statements about entertainment-education radio soap operas in India. Both dimensions of involvement lead to an increased self-efficacy and collective efficacy. Additionally, referential-affective involvement stimulated discussions about the talk show. Usually, Bandura's (1977) social-cognitive theory (e.g., Sood, 2002) is the theoretical basis of entertainment-education. In contrast, Slater and Rouner (2002) refer to persuasion theories. They propose a modified elaboration likelihood model in which involvement is not conceptualized as personal relevance or as issue involvement, but instead as engagement with the narration and its characters, or as immersion involvement. This form of involvement is similar to the above mentioned non-mediation phenomena; here, we are talking about the immersion into the narrative world (Lombard & Ditton, 1997; Vorderer, 1993; Vorderer et al., 2004). Green and colleagues (Green & Brock, 2000; Green, Brock, & Kaufmann, 2004) proposed another non-mediation phenomenon: transportation, which can be interpreted as being based on involvement. They found that the extent of transportation correlated negatively with the number of critical thoughts, thus facilitating persuasive effects. In a study by Stephenson and Palmgreen (2001), it was shown that narrative processing (also cognitive involvement with the narration) of short public service announcements was most closely connected to the rejection of drugs.

A branch of newer research prominent in consumer research is dedicated to the influence of involvement with entertainment programs on the effects of TV commercials placed within. The subjects' ratings of the programs as involving, entertaining, and enjoyable were correlated positively with subsequent ratings of the advertised brands and purchase intentions (Norris & Colman, 1993, 1994). The authors explained the effect as a carry-over effect from the program, or a mood-congruency effect (Bower, 1981). However, they found a negative relation between program-induced viewer involvement and recall of the advertisements (Norris & Colman, 1993). A similar relation between the perceived entertainment value and the advertisement effectiveness could not be shown (Norris & Colman, 1994). The authors explained this by pointing out the differences between feeling entertained and being involved. Involving TV programs absorb attention and mental capacity in a way that interferes with the recall of the TV commercial. Enjoying television programs does not absorb the viewer's attention in the same way as involvement does. Recently, these findings elaborated and differentiated in a further study (Norris, Colman, & Aleixo, 2001, 2003).

THE FUTURE OF INVOLVEMENT AND ENTERTAINMENT

If media are entertaining, they are involving too! Even if this proposition sounds extremely plausible, it should be considered as premature. First of all, it is not yet sufficiently clear what the specific differences between entertainment and involvement are. The reason for this is that entertainment experience is not sufficiently explained theoretically and, conversely, involvement is part of too many diverse theories. More entertainment theories that integrate forms of involvement are necessary. (e.g., affective disposition theory, Zillmann, 1994). Following theoretical integration, one has to test the concepts empirically. We do not have enough empirical evidence in order to verify the above-mentioned hypothesis. For this reason, it would be advantageous to restrict the discussion of the concept of involvement to the confined sense. Involvement, then, could be defined as the intense engagement with an object, comprised of cognitive, affective, and conative components, as well as of a temporal structure. In the communicative phase, involvement could be conceptualized as an intensive process of either approaching (positive involvement) or distancing (negative involvement) an object or event. A concept that includes cognitive responses, felt emotions, attention, recall, information seeking, *and* discussions about the topic is not very useful. If involvement were all-encompassing, we could easily abandon the concept.

REFERENCES

Anderson, J. R. (1983). A spreading activation theory of memory. *Journal of Verbal Learning and Verbal Behavior, 22*, 261–295.

Andrews, J., Durvasula, S., & Akhter, S. H. (1990). A framework for conceptualizing and measuring the involvement construct in advertising research. *Journal of Advertising, 19*, 18–26.

Andringa, E. (1996). Effects of 'narrative distance' on readers' emotional involvement and response. *Poetics. Journal of Empirical Research on Literature, the Media and the Arts, 23*, 431–452.

Antil, J. (1984). Conceptualization and operationalization of involvement. *Advances in Consumer Research, 11*, 203–209.

Apsler, R., & Sears, D. O. (1968). Warning, personal involvement, and attitude change. *Journal of Personality and Social Psychology, 9*, 162–168.

Bandura, A. (1977). *Social learning theory.* Englewood Cliffs: NJ: Prentice Hall.

Barwise, P., & Ehrenberg, A. (1988). *Television and its audience.* London, Great Britain: Sage.

Batra, R., & Ray, M. L. (1983). Operationalizing involvement as depth and quality of cognitive response. *Advances in Consumer Research, 10*, 309–313.

Batra, R., & Ray, M. L. (1985). How advertising works at contact. In L. F. Alwitt, & A. A. Mitchell (Eds.), *Psychological processes and advertising effects* (pp. 13–43). Hillsdale, NJ: Lawrence Erlbaum Associates.

Bente, G., & Feist, A. (2000). Affect talk and its kin. In D. Zillmann, & P. Vorderer (Eds.), *Media entertainment. The psychology of its appeal* (pp. 113–134). Mahwah, NJ: Lawrence Erlbaum Associates.

Bower, G. H. (1981). Mood and memory. *American Psychologist, 11*, 11–13.

Bransford, J. D. (1979). *Human cognition. learning, understanding, and remembering.* Belmont, CA: Wadsworth.

Bryant, J., & Comisky, P. W. (1978). The effect of positioning a message within differentially cognitive involving portions of a television segment on recall of a message. *Human Communication Research, 5*, 63–75.

Buck, R. (1988). *Human motivation and emotion.* New York: John Wiley.

Buck, R., Chaudhuri, A., Georgson, M., & Kowta, S. (1995). Conceptualizing and operationalizing affect, reason, and involvement in persuasion: The ARI model and the CASC scale. *Advances in Consumer Research, 22*, 440–447.

Burnkrant, R. E., & Sawyer, A. G. (1983). Effects of involvement and message content on information-processing intensity. In R. J. Harris (Ed.), *Information processing research in advertising* (pp. 43–64). Hillsdale, NJ: Lawrence Erlbaum Associates.

Cameron, G. L. (1993). Spreading activation and involvement: An experimental test of a cognitive model of involvement. *Journalism Quarterly, 70*, 854–867.

Celsi, R. L., & Olson, J. C. (1988). The role of involvement in attention and comprehension processes. *Journal of Consumer Research, 15*, 210–224.

Chaffee, S. H., & Roser, C. (1986). Involvement and the consistency of knowledge, attitudes, and behaviors. *Communication Research, 13*, 373–400.

Chaiken, S. (1980). Heuristic verus systematic information processing and the use of source versus message cues in persuasion. *Journal of Personality and Social Psychology, 39*, 752–766.

Chaiken, S., & Trope, Y. (Eds.). (1999). *Dual-process theories in social psychology.* New York, London: Guilford.

Chaudhuri, A., & Buck, R. (1995a). Affect, reason, and persuasion: Advertising strategies that predict affective and analytic-cognitive responses. *Human Communication Research, 21*, 422–441.

Chaudhuri, A., & Buck, R. (1995b). Media differences in rational and emotional responses to advertising. *Journal of Broadcasting and Electronic Media, 39*, 109–125.

Chen, S., & Chaiken, S. (1999). The heuristic-systematic model in its broader context. In S. Chaiken, & Y. Trope (Eds.), *Dual-process theories in social psychology* (pp. 73–97). New York, London: Guilford.

Cohen, J. B. (1983). Involvement and you: 1000 great ideas. *Advances in Consumer Research, 10*, 325–328.

Collins, A., & Loftus, E. (1975). A spreading-activation theory of semantic processing. *Psychological Review, 82*, 407–428.

Costley, C. (1988). Meta analysis of involvement research. *Advances in Consumer Research, 15*, 554–562.

Craik, F. I. M., & Lockhart, R. S. (1972). Levels of processing: A framework for memory research. *Journal of Verbal Learning and Verbal Behavior, 11*, 671–684.

Craik, F. I. M., & Tulving, E. (1975). Depth of processing and the retention of words in episodic memory. *Journal of Experimental Psychology: General, 104*, 268–294.

Denham, B. E. (2004). Toward an explication of media enjoyment: The synergy of social norms, viewing situations, and program content. *Communication Theory, 14*, 370–387.

Donnerstag, J. (1996). *Der engagierte Mediennutzer. Das Involvementkonzept in der Massenkommunikationsforschung* [The engaged media user. The concept of involved in mass communication research]. Munich, Germany: Fischer.

Feist, A. (2000). *Emotionale Wirkungen von Fernsehtalkshows*. Aachen: Shaker Verlag. Trans: [Emotional effects of tv talk shows]

Flora, J., & Maibach, E. (1990). Cognitive responses to AIDS information: The effects of issue involvement and message appeal. *Communication Research, 17*, 759–774.

Frijda, N. H. (1986). *The emotions*. New York: Cambridge University Press.

Furnham, A., Gunter, B., & Walsh, D. (1998). Effects of programme context on memory of humorous television commercials. *Applied Cognitive Psychology, 12*, 555–567.

Gotlieb, J. B., & Sarel, D. (1991). Comparative advertising effectiveness: The role of involvement and source credibility. *Journal of Advertising, 20*, 38–45.

Graesser, A. C., & Clark, L. F. (1985). *Structures and procedures of implicit knowledge*. Norwood, NJ: Ablex.

Green, M. C., & Brock, T. C. (2000). The role of transportation in the persuasiveness of public narratives. *Journal of Personality and Social Psychology, 79*, 701–721.

Green, M. C., Brock, T. C., & Kaufmann, G. F. (2004). Understanding media enjoyment: The role of transportation into narrative worlds. *Communication Theory, 14*, 311–327.

Greenwald, A. G., & Leavitt, C. (1984). Audience involvement in advertising: Four levels. *Journal of Consumer Research, 11*, 581–592.

Greenwald, A. G., & Leavitt, C. (1985). Cognitive theory and audience involvement. In L. F. Alwitt, & A. A. Mitchell (Eds.), *Psychological processes and advertising effects. Theory, research, and applications* (pp. 221–240). Hillsdale, NJ: Lawrence Erlbaum Associates.

Hartmann, T., Schramm, H., & Klimmt, C. (2004). Personenorientierte Medienrezeption: Ein Zwei-Ebenen Modell parasozialer Interaktionen [Person-oriented media usage: A two-level model of parasocial interactions]. *Publizistik, 49*(1), 25–47.

Havitz, M. E., & Howard, D. R. (1995). How enduring is enduring involvement? A seasonal examination of three recreational activities. *Journal of Consumer Psychology, 4*, 255–276.

Higgins, E. T. (1996). Knowledge activation: Accessibility, applicability, and salience. In E. T. Higgins, & A. W. Kruglanski (Eds.), *Social psychology: Handbook of basic principles* (pp. 133–168). New York: The Guilford Press.

Hoffman, D. L., & Batra, R. (1991). Viewer response to programs: Dimensionality and concurrent behavior. *Journal of Advertising Research, 31*(4), 46–56.

Houston, M. J., & Rothschild, M. L. (1978). Conceptual and methodological perspectives on involvement. In S. C. Jain (Ed.), *Research frontiers in marketing: dialogues and directions* (pp.184–187). Chicago, IL: American Marketing Association.

Izard, C. E. (1993). Organizational and motivational functions of discrete emotions. In M. Lewis & J. M. Haviland (Eds.), *Handbook of emotions* (pp. 631–641). New York: Guilford Press.

Johnson, B. T., & Eagly, A. H. (1989). Effects of involvement on persuasion: A meta-analysis. *Psychological Bulletin, 106*, 290–314.

Johnson, B. T., & Eagly, A. H. (1990). Involvement and persuasion: Types, traditions, and the evidence. *Psychological Bulletin, 107*, 375–384.

Kamins, M. A., Assael, H., & Graham, J. L (1990). Cognitive response involvement model of the process of product evaluation through advertising exposure and trial. *Journal of Business Research, 20*, 191–215.

Kim, J., & Rubin, A. (1997). The variable influence of audience activity on media effects. *Communication Research, 24*, 107–135.

Kramer, R. M. (1998). Paranoid cognition in social systems: Thinking and acting in the shadow of doubt. *Personality and Social Psychology Review, 2*, 251–275.

Krugman, H. E. (1965). The impact of television advertising: Learning without involvement. *Public Opinion Quarterly, 29*, 349–356.

Krugman, H. E. (1966). The measurement of advertising involvement. *Public Opinion Quarterly, 30*, 584–585.

Levy, M. R., & Windahl, S. (1984). Audience activity and gratifications: A conceptual clarification and exploration. *Communication Research, 11*, 51–78.

Levy, M. R., & Windahl, S. (1985). The concept of audience activity. In K. E. Rosengren, L. E. Wenner, & P. Palmgreen (Eds.), *Media gratifications research–current perspectives* (pp. 109–122). Beverly Hills, CA: Sage.

Liebes, T., & Katz, E. (1986). Patterns of involvement in television fiction: A comparative analysis. *European Journal of Communication, 1*, 151–171.

Lo, V. (1994). Media use, involvement, and knowledge of the gulf war. *Journalism Quarterly, 71*, 43–54.

Lombard, M., & Ditton, T. (1997). At the heart of it all: The concept of presence. *Journal of Computer Mediated Communication, 3*(2) [Online]. Available: http://209.130.1.169/jcmc/vol3/issue2/lombard.html.

MacInnis, D. J., & Jaworski, B. J. (1989). Information processing from advertisements: Toward an integrative framework. *Journal of Marketing, 53*, 1–23.

McCombs, M., & Reynolds, A. (2002). News influence on our pictures of the world. In Bryant, J., & D. Zillmann (Eds.), *Media effects. Advances in theory and research* (pp. 1–19). Mahwah, NJ: Lawrence Erlbaum Associates.

McQuail, D. (1985). With the benefit of hindsight. Reflections on uses and gratification research. In M. Gurevitch, & M. R. Levy (Eds.), *Mass communication review yearbook* (Vol. 5, pp. 125–141). Newbury Park, CA: Sage.

Mitchell, A. A. (1981). The dimensions of advertising involvement. *Advances in Consumer Research, 8,* 25–29.

Muehling, D. D., Laczniak, R. N., & Stoltman, J. J. (1991). The moderating effects of ad message involvement: A reassessment. *Journal of Advertising, 20,* 29–38.

Muehling, D. D., Laczniak, R. N., & Andrews, J.C. (1993). Defining, operationalizing, and using involvement in advertising research: A review. *Journal of Current Issues and Research in Advertising, 15,* 21–57.

Nabi, R. L., & Krcmar, M. (2004). Conceptualizing media enjoyment as attitude: implications for mass media effects research. *Communication Theory, 14,* 288–310.

Norris, C. E., & Colman, A. M. (1993). Context effects on memory for television advertisements. *Social Behavior and Personality, 21,* 279–296.

Norris, C. E., & Colman, A. M. (1994). Effects of entertainment and enjoyment of television programs on perception and memory of advertisments. *Social Behavior and Personality, 22,* 365–376.

Norris, C. E., Colman, A. M., & Aleixo, P. A. (2001). Context effects of cognitively involving, entertaining and enjoyable television programmes on two types of advertisments. *Social Psychological Review, 3,* 3–24.

Norris, C. E., Colman, A. M., & Aleixo, P. A. (2003). Selective exposure to television programmes and advertising effectiveness. *Applied Cognitive Psychology, 17,* 593–606.

Oatley, K. (1994). A taxonomy of the emotions of literary response and a theory of identification in fictional narrative. *Poetics 23,* 53–74.

Park, C. W., & McClung, G. W. (1986). The effect of TV program involvement on involvement with commercials. *Advances in Consumer Research, 13,* 544–548.

Patwardhan, P. (2004). Exposure, involvement and satisfaction witrh online activities. *Gazette: The International Journal for Communication Studies, 66,* 411–436.

Perloff, R. M. (1984). Political involvement: A critique and a process-oriented reformulation. *Critical Studies in Mass Communication, 1,* 146–160.

Perloff, R. M. (1989). Ego-involvement and the third-person effect of televised news coverage. *Communication Research, 16,* 236–262.

Perse, E. M. (1990a). Audience selectivity and involvement in the newer media environment. *Communication Research, 17,* 675–697.

Perse, E. M. (1990b). Cultivation and involvement with local television news. In N. Signorielli, & M. Morgan (Eds.), *Advances in cultivation analysis* (pp. 51–69). Newbury Park, CA: Sage.

Perse, E. M. (1990c). Involvement with local television news: Cognitive and emotional dimensions. *Human Communication Research, 16,* 556–581.

Perse, E. M. (1990d). Media involvement and local television news effects. *Journal of Broadcasting & Electronic Media, 34,* 17–36.

Perse, E. M. (1998). Implications of cognitive and affective involvement for channel changing. *Journal of Communication, 48,* 49–68.

Perse, E. M., & Rubin, R. B. (1989). Attribution and para-social relationships. *Communication Research, 16,* 59–77.

Petty, R. E., & Cacioppo, J. T. (1979). Issue involvement can increase or decrease persuasion by enhancing message-relevant cognitive responses. *Journal of Personality and Social Psychology, 37,* 1915–1926.

Petty, R. E., & Cacioppo, J. T. (1981). Issue involvement as a moderator of the effects on attitude of advertising content and context. *Advances in Consumer Research, 8,* 20–24.

Petty, R. E., & Cacioppo, J. T. (1986). *The Elaboration Likelihood Model of persuasion.* New York: Academic Press.

Petty, R. E., & Cacioppo, J. T. (1990). Involvement and persuasion: Tradition versus integration. *Psychological Bulletin, 107,* 367–374.

Petty, R. E., Ostrom, T. M., & Brock, T. C. (Eds.). (1981). *Cognitive Responses in Persuasion.* Hillsdale, NJ: Sage.

Petty, R. E., Priester, J. R., & Briñol, P. (2002). Mass media attitude change: Implications of the elaboration likelihood model of persuasion. In J. Bryant, & D. Zillmann (Eds..), *Media effects. Advances in theory and research* (pp. 155–198). Mahwah, NJ: Lawrence Erlbaum Associates.

Putrevu, S., & Lord, K. R. (1994). Comparative and noncomparative advertising: Attitudinal effects under cognitive and affective involvement conditions. *Journal of Advertising, 23,* 77–90.

Ratcliff, R., & McKoon, G. (1981). Does activation really spread? *Psychological Review, 88,* 454–462.

Ray, M. L. (1973). Marketing communicaiton and the hierarchy-of-effects. In P. Clarke (Ed.), *New Models for Communication Research* (pp. 147–176). Beverly Hills, CA: Sage.

Richins, M. L., & Bloch, P. H. (1986). After the new wears off: The temporal context of product involvement. *The Journal of Consumer Research, 13,* 280–285.

Richins, M. L., Bloch, P. H., & McQuarrie, E. F. (1992). How enduring and situational involvement combine to create involvement responses. *Journal of Consumer Psychology, 1,* 143–153.

Roser, C. (1990). Involvement, attention, and perceptions of message relevance in the response to persuasive appeals. *Communication Research, 17*, 571–600.

Rothschild, M. L. (1984). Perspectives on involvement: Current problems and future directions. *Advances in Consumer Research, 11*, 216–217.

Rothschild, M. L., & Ray, M. L. (1974). Involvement and political advertising effect. An exploratory experiment. *Communication Research, 1*, 264–285.

Rubin, A. M., & Perse, E. M. (1988). Audience activity and soap opera involvement: A uses and effects investigation. *Human Communication Research, 14*, 246–268.

Rubin, A. M., & Step, M. M. (2000). Impact of motivation, attraction, and parasocial interaction in talk radio listening. *Journal of Broadcasting & Electronic Media, 44*, 635–654.

Salmon, C. T. (1986). Perspectives on involvement in consumer and communication research. In B. Dervin, & M. J. Voigt (Eds.), *Progress in communication sciences* (pp. 243–268). Beverly Hills, CA: Sage.

Salomon, G. (1981). Introducing AIME: The Assessment of Children's Mental Involvement with Television. In H. Kelly, & H. Gardner (Eds.), *Viewing children through televison* (pp. 181–198). San Francisco, CA: Jossey-Bass.

Salomon, G. (1984). Television is 'easy' and print is 'tough': The differential investment of mental effort in learning as a function of perceptions and attributions. *Journal of Educational Psychology, 4*, 647–658.

Sherif, M., & Cantril, H. (1947). *The psychology of ego-involvement: Social attitudes and identifications*. New York: Wiley.

Sherif, M., & Hovland, C. I. (1961). *Social judgment: Assimilation and contrast effects in communication and attitude change*. New Haven, CT: Yale University Press.

Shoemaker, P. J., Schooler, C., & Danielson, W. A. (1989). Involvement with the media. Recall versus recognition of election information. *Communication Research, 16*, 78–103.

Sood, S. (2002). Audience involvement and entertainment-education. *Communication Theory, 12*, 153–172.

Slater, M. D. (2002). Involvement as goal-directed strategic processing. Extending the elaboration likelihood model. In J. P. Dillard, & M. Pfau (Eds.), *The persuasion handbook. Developments in theory and practice* (pp. 175–194). Thousand Oaks, CA: Sage.

Slater, M. D., & D. Rouner (2002). Entertainment-education and elaboration likelihood: Understanding the processing of narrative persuasion. *Communication Theory, 12*, 173–191.

Soldow, G. F., & Principe, V. (1981). Response to commercials as a function of program context. *Journal of Advertising Research, 21*(2), 59–65.

Step, M. M. (1998). *An emotional appraisal model of media involvement, uses, and effects*. Unpublished doctoral dissertation, Kent State University.

Stephenson, M. T., & Palmgreen, P. (2001). Sensation seeking, message sensation value, personal involvement, and processing of anti-drug PSAs. *Communication Monographs 68*, 49–71.

Tan, E. S. (1996). *Emotion and the structure of narrative film. Film as an emotion machine*. Mahwah, NJ: Lawrence Erlbaum Associates.

Vorderer, P. (1993). Audience involvement and program loyalty. *Poetics. Journal of Empirical Research on Literature, the Media and the Arts, 22*, 89–98.

Vorderer, P. (1996). *Fernsehen als "Beziehungskiste". Parasoziale Beziehungen und Interaktionen mit TV-Personen* [Television as, 'relationship box.' Parasocial interactions and relationships with TV actors]. Wiesbaden, Germany: Opladen.

Vorderer, P. (1998). Unterhaltung durch Fernsehen: Welche Rolle spielen parasoziale Beziehungen zwischen Zuschauern und Fernsehakteuren? In W. Klingler, G. Roters, & O. Zoellner (Eds.), *Fernsehforschung in Deutschland. Themen—Akteure–Methoden* [Television research in Germany. Issues—actors—methods] (pp. 689–707). Baden-Baden, Germany: Nomos.

Vorderer, P., Cupchik, G. C., & Oatley, K. (1997). Encountering the literary landscapes of experience and action from self-oriented and spectator perspectives. In S. Totosy de Zepetnek (Ed.), *The systemic and empirical approach to literature and culture as theory and application, Vol. 7* (pp. 559–571). Siegen, Germany: LUMIS-Publications.

Vorderer, P., & Klimmt, C., & Ritterfeld, U. (2004). Enjoyment: At the heart of media entertainment. *Communication Theory, 14*, 388–408.

Zaichkowsky, J. L. (1986). Conceptualizing involvement. *Journal of Advertising, 15*, 34–14, 34.

Zaichkowsky, J. L. (1987). Emotional aspects of product involvement. *Advances in Consumer Research, 14*, 32–35.

Zillmann, D. (1991). Empathy: Affect from bearing witness to the emotions of others. In J. Bryant, & D. Zillmann (Eds.), *Responding to the screen. Reception and reaction processes* (pp. 135–168). Hillsdale, NJ: Lawrence Erlbaum Associates.

Zillmann, D. (1994). Mechanism of emotional involvement with drama. *Poetics. Journal of Empirical Research on Literature, the Media and the Arts, 23*, 33–51.

13

Dramaturgy for Emotions
From Fictional Narration

Dolf Zillmann
University of Alabama

After clarifying several issues that have unnecessarily complicated and confused the analysis of emotion arising from fiction, this chapter integrates research-supported psychological and physiological paradigms of emotion to explain the diversity of affective reactions to dramatic fictional formats. Cinematic presentation is of focal interest, but alternative forms of presentation are given attention also. The three-factor theory of emotion and the excitation-transfer paradigm are employed to account for the elicitation of emotional reactions and for the intensity of these reactions. In an analysis of excitatory functions, the escalation of affect intensity by dramaturgic means is given special consideration. A theory of the formation of affective dispositions and their consequences for empathic responding is based on the analysis of cognitive functions. In this connection, a model of the dispositional override of empathy is featured to shed light on seemingly inappropriate, malicious, if not sadistic, joyous reactions to others' demise. Cognitive functions are further explored in the emotional effect of moral sanction.

PRELIMINARY CONSIDERATIONS

We all know how compellingly a good tale can engross and emotionally stir an audience. We all have experienced being touched and roused by dramatic stories that, upon reflection, we had little cause to construe as veridical accounts of actual happenings. And we all have witnessed others succumb to the same experiences. Oddly, however, many of those who ventured to subject the circumstances of such reactivity to rational analysis responded with bewilderment and came to consider the evocation of emotion by fictional narratives wanting in plausibility.

Responding to Fictional Events
as Though They Were Real

The analyses in question tend to start with exemplars of the unreality of fictional portrayals. Holland (2003), for instance, in the process of pondering why we respond to fictional events as though they were real, relates how he and others cried their hearts out when, during exposure to the motion picture *Love Story*, the college girl Jenny Cavalieri dies of leukemia. He asks himself why he and the others cared, as much as they apparently did, about the plight of a make-believe creature, invented by a writer, played by a healthy actress, recorded by an intrusive camera, and projected onto a screen as a fleeting flicker. The circumstances are undoubtedly artificial in the highest degree, and one should expect that rational beings cannot help becoming cognizant of this artificiality, have a hard time ignoring it, and ultimately be unable to respond as if Jenny were a real person of their acquaintance.

Some time ago, the British writer Coleridge (1817/1960) thought to resolve the irritating recognition of 'fictional unreality' by stipulating a faith requirement for reaching what he called the poetic truth of literary works. Specifically, he asked from his readers a "willing suspension of disbelief for the moment" (p. 169). Given the romantic nature of Coleridge's poetry, the requested suspension of disbelief may well have aided the making of somewhat realistic meaning of his often supernatural settings. In the face of modern cinema, with its high-fidelity representation of emotional circumstances and expressed experiences, however, it rings hollow to expect viewers to intently discard their concerns about 'the unreality of plays of light on the screen,' along with other artificialities, in order to allow emotional reactions to materialize. Fictional films apparently have the capacity to evoke emotions without the stipulated self-deceptive cognitive effort toward the dismissal of artificiality within the presentation. In the absence of evidence that would favor the volitional suspension of disbelief as a necessary condition for emotional reactivity, the persisting broad acceptance of Coleridge's suggestion in many literary circles can only be considered astounding.

Holland (2003) recently attempted to provide firmer grounds for Coleridge's formula. Based on neurological speculations about aesthetic sensation and the conception of beauty (Tooby & Cosmides, 2001), he essentially argues that, as the respondent to fiction is typically confined to a restful sitting position, the brain's highly developed prefrontal cortices are busy inhibiting overt actions and their planning—such planning being what these structures usually do—and that they therefore fail to generate cognizance of the artificiality of the emotion-inducing circumstances. Although it is left unclear why the inhibition of action planning would place greater demands on information processing in the prefrontal structures than would the actual planning and execution of specific plans, it is thus suggested that the necessity of action inhibition renders the prefrontal regions incapable of cautioning us about the fictional nature of presentations, and that we consequently fall prey to the lure of fiction, mistaking it for reality, with the result that our emotions are left to the unchecked powers of archaic brain structures. Holland then extended his reasoning by insinuating a fictional gullibility for children. As their prefrontal cortices are not yet fully developed and the perception of reality is immature, they are seen as yielding readily to the impressions provided by the archaic structures. Moreover, he applied his neural-immaturity model to adulthood, arguing, in accordance with earlier developed psychoanalytical interpretations of his (Holland, 1968), that the consumption of fiction triggers a mental regression to childhood, and that the reality illusion is part and parcel of this regression.

Alternative efforts to elucidate the perplexing illusion of reality in fiction may seem less contrived, but also fail to provide a convincing account of the illusory process (cf. Tan, 1996; Turvey, 1997). The more coherent explanations simply proclaim the existence of a mental faculty for converting the physical stimuli of fiction to mental representations that are unbound by reality concerns and that function as the pivotal causal agents for emotions. Carroll (1988, 1990), for instance, argues that the assessment of veridicality is immaterial for the elicitation of emotions. Mental representations or thought constructions are deemed capable of triggering emotions, irrespective of the truth-value of these representations and constructions. Smith (1995a, 1995b, 1997), referring to such mental representations as imaginations, similarly contends that emotion elicitation does not depend upon a commitment to the actuality of events. Carroll and Smith, along with others (e.g., Allen, 1993; Peters, 1989), thus hold that fiction need not be mistaken for reality in order to arouse emotions, and that imaginary processes, once instigated by fiction, are supremely capable of evoking genuine emotions. The epistemic illusion is therefore considered expandable baggage in the explanation of emotion from fiction. Allen (1997) considers this illusion "thoroughly debunked" (p. 79).

Such debunking did not prevent, however, that a new illusion emerged and took center stage in recent fiction theories. The new illusion focuses on fictional absorption as a reality that mediates emotion. Allen (1993, 1997), for example, envisions a 'projective illusion;' that is, an illusion in which respondents succumb to sensory deception by a presentation although, at the onset of exposure, they may have been cognizant of inherent unrealities. The process has come to be known as *experiencing fiction from within*. In this scheme of things, it is 'imagining fiction from the inside' that is expected to produce the variety of emotions that fiction is capable of producing (Smith, 1997; Walton, 1990). Such perception from the inside is also construed as 'central imagining,' in contrast to decentralized perception from the standpoint of no one in particular (Wollheim, 1987). Analogously, perception of this kind has been characterized as 'personal,' in contrast to impersonal (Currie, 1995). Observers unfamiliar with the nomenclature and conceptualizations of this literary discourse may wonder whether 'experiencing fiction from within' is not much the same as what Coleridge must have had in mind when he called for the abandonment of all disbeliefs upon entrance into a fictional world.

Both Currie (1995) and Wollheim (1987) also speak of *simulation* as a process of adopting and sharing the beliefs, dispositions, and experiences of fictional others, this to the point of disconnecting from habitual reactions under less artificial circumstances. Walton (1990) similarly proposes that 'imagination from the inside' involves imagining oneself to be a person other than oneself. These and similar constructions (Gaut, 1999) usually focus on *mimesis* as a process of permeating and taking-on others' mental and bodily traits (Oatley, 1995). They entail essential elements of psychoanalytic reasoning and thus may be viewed as a variant of the Freudian concept of *identification* (Freud, 1921/1964a, 1923/1964b), if not as a renewed embrace of the original concept. In the case of the earlier reported intense emotional reactions to seeing the heroine of *Love Story* suffer and die, emotionality would be interpreted as the result of viewers' deliberate or possibly involuntary imaginary penetration and usurpation of the heroine's experiential state. The fact that this process of 'taking the place of another person' is thought to be contingent upon, as well as concurrent with, the temporary abandonment of self-consciousness, is again reminiscent of Coleridge's implicitly stipulated denial of the prevailing actual, situational reality. One can only wonder how, under these conditions of 'feeling with others from within,' respondents to fiction ever manage to experience emotions as their own.

The problem with these interpretations is that they are chiefly intuitive, at best supported by selective personal experiences and informal observations of others' behavior. Pertinent contributions of contemporary psychology are largely ignored and certainly not meaningfully integrated in the indicated theorizing (Konijn, 1999). As a result, the construction of new rationales often amounts to a rephrasing of earlier expressed ideas in philosophy and fiction scholarship; such as, for instance, Allen's (1993) projective illusion that can be construed as a reformulation of Burch's (1979) notion of being present in, and part of, the fictional environment, the so-called *diegetic effect*.

The proposal that emotions from fiction are mediated by mental processes of sorts is not so much amiss as it is incomplete and vague. The issue is to discern and implicate specific mediating processes and to assemble them into a coherent mechanism of emotion evocation via exposure to fiction. It would seem that, rather than relying on philosophy and pre-empirical psychology, this can be best accomplished by constructing theories in accordance with current evidence from psychological and neuroendocrinological explorations into emotional reactivity. The construction of theories in these terms is consequently attempted in the discourse to follow.

Discounting Apparent Reality: A Reversal of Coleridge's Formula

Fictional narration manifests itself in two distinct semiotic modalities. It is either *iconic*, in which case the stimuli that represent mimic the physical features of the stimuli that are represented; or it is *symbolic*, in which case the relationship between representing and represented stimuli is morphologically arbitrary and representation must be arranged by consent. Symbolic representation typifies conventional, natural languages. Iconic representation, evolutionarily speaking the older one of the two formats, is manifest in copies of the represented, these copies having sufficient resemblance with the represented to identify it without necessitating additional explanation. Cinematic presentations epitomize iconic representation within the visual and auditory perceptual domains. However, iconic representation applies to all forms of sensory means of information conveyance, including olfaction and tactility. Needless to say, fictional presentations liberally combine iconic and symbolic modalities of representation; that is, verbal presentations may feature images, and spoken language typically permeates predominantly iconic presentations.

The fact that representations that are characterized by extreme degrees of iconicity are essentially indistinguishable from the physical stimulus conditions that they represent has momentous implications for emotional reactivity. The reason for this is that, if the *physical reality* is capable of triggering emotional reactions, so should their iconically mediated *apparent reality*. If, for example, an actual encounter with a poisonous snake in striking position strikes fear, so should its encounter in a perfect iconic representation. The argument that respondents to iconic representations would be cognizant of the mediational artificiality, and thus could not respond to it emotionally, at least not as strongly as they would to the represented physical reality, is not necessarily compelling. This, because recent neurophysiological research revealed that substructures of the limbic system, mostly the amygdala, continually monitor the environment for indications of threats and dangers, and that upon their encounter, emotional reactions are triggered before the information is passed on to the neocortex; that is, prior to awareness of the specific emotion-inducing conditions (LeDoux, 1996; LeDoux & Phelps, 2000). It has been demonstrated, moreover, that the amygdala not only signals detected threats, but also estimates their severity and thereby determines the intensity of emotional reactivity. Most important here, any analysis and scrutiny of the presentational or representational status of emotion-inducing stimuli can commence only after autonomic and incipient behavioral reactions have been initiated. In other words, emotional reactions elusive of volitional control, including those associated with sympathetic excitation as the pivotal determinant of emotional intensity, have been set in motion before 'reality' could be discerned as actual or fictional.

The time priority of amygdaloid response mediation over neocortical stimulus evaluation challenges Coleridge's (1817/1960) conception of creating reality from fiction. It would appear that a 'willing suspension of disbelief' is utterly unnecessary for the evocation of emotion via fiction, at least via high-fidelity iconic representations of fictional events. Not only is it unnecessary to get rid of doubts and intently embrace an illusion to experience emotions, but the reverse applies in that *genuine emotions are to be discounted by rising awareness of the artificiality of their induction.* Iconically represented reality functions as actual reality that may or may not be immediately faulted and degraded as pseudo-reality. The iconic representation of images and sounds of an onrushing wall of water, for instance, thus should stir our emotions as apparent reality, much as their physical reality would, and only upon reflection should we appreciate, perhaps inevitably so, the artificiality of the induction of our emotions. The sequence of events, therefore, is not that cognizance of the pseudo-reality of presentations has to be suppressed before emotions can occur, but that emotions are first induced by apparent reality, which then may be discounted as artificial.

The evocation of emotion via symbolic representations of fictional events is obviously less direct in that apparent realities need to be ideationally constructed. Put simply, symbolic representations, usually conveyed by spoken and written language, have to be translated, through immense associative activity in the neural networks of the neocortex, to mental representations of any kind of presented reality (Damasio, 1994; Lang, 1979, 1984). This activity calls upon experiences that pertain to the symbolic input and thereby personalizes the rendering of its meaning. Via direct connections between amygdala and hippocampus, emotional experiences are afforded long-term storage in the latter structure. The associative pursuit of experiences related to the symbolic input also calls upon this store and activates salient emotional memories. Their activation tends to reinstate, at least in part, the autonomic and somatic manifestations of the focal emotional experiences. Such revival of emotional memories further personalizes the rendering of the input's meaning. However, notwithstanding such seemingly elaborate conversion of symbolic representations to 'apparent reality,' ideational representations, once constructed, should mediate emotions much as the apparent reality of iconic representations. This is to say that symbolic representations of fictional events also induce genuine emotions that, despite their likely higher degree of subjectification than those induced by iconic representations, again may be discounted by rising awareness of the artificiality of their induction.

Irrespective of the semiotic mode of fictional presentations, the autonomic reactivity associated with the emotions that these presentations evoke is largely independent of volition. Behavioral reactivity, in contrast, is subject to volitional control. This control manifests itself in the inhibition of most, if not all, goal-directed responses that would be meaningful if the presented events were to happen in the respondent's actual environment. But inhibition does not extend to all emotion-linked movements that are given meaning by a presentation. In fact, incipient movements, evident in jerks of the body and limbs in correspondence with presented events, may be construed as indicators of genuine emotional action preparedness. In appreciation of the inappropriateness of any prepared actions, however, their execution is quickly suppressed.

This analysis suggests that respondents to fiction are both (a) lost in the apparent reality of presentations, indeed responding to it as if it were real; and (b) cognizant of its artificiality, by inhibiting actions that would be meaningful only within the represented reality. The two states are non-concurrent, however. It is suggested that respondents liberally enter and exit the one or the other state. Fictional presentations abound with indications of artificiality (cf. Tan, 1996), and these indications function as cues that discount apparent reality and demand the inhibition of overt action. It would seem likely, then, that respondents to fiction can be held emotionally captive by apparent reality, but not for any length of time. The opportunity for opting out seems ever present.

On Getting Carried Away in Fictional Environments

The apparent exception to the inhibition rule is the rather common performance of communicative acts, the address of fictional characters in particular. Respondents to fiction are known to talk to these characters as if they were present in the flesh. When sufficiently engaged emotionally, they routinely warn protagonists of imminent danger and suggest protective action. It has been observed, for instance, that members of the audience of the horror film *Friday the 13th*, when seeing the heroine stalked and cornered by the infamous hockey-masked killer Jason, yelled out to her: "Watch out! Behind you!" and "Take the ax. Hit him! Hit him!" (Zillmann, Weaver, Mundorf, & Aust, 1986). Approval of effective coping was analogously expressed in exclamations like "That's the way!" or "That will show him!" Anybody who ever watched children respond to puppet shows will appreciate that young audiences are particularly expressive in these terms. Scenes of a likable protagonist who is apparently ignorant of the fact that an ill-willed crocodile, lion, or dragon is sneaking up on him, have been used around the world and through the ages to tease children into frantically screaming warnings to their hero, in efforts of saving him. Reactions of this kind have been observed in children as young as four years of age (Zillmann & Bryant, 1975). Children seem to be truly lost, if only for the moment, in the apparent reality of fiction. However, a sensitivity to discounting cues, enabling them to elude the spell of such reality, is bound to develop and mature soon enough.

The sketched communicative actions by respondents to fiction (actually, to nonfiction as well) have been used to challenge the broadly used Freudian concept of *identification* (Freud, 1921/1964a, 1923/1964b). A detailed account of this challenge may be found elsewhere (Zillmann, 1995). Suffice it here to focus on identification in the context of fiction and its implications for emotional reactivity.

Freud (1905–1906/1987) addressed this limited issue most directly in a treatise on stage play. The play is thought to create a pseudo-world that allows the spectator, characterized as "a poor soul to whom nothing of importance seems to happen, who some time ago had to moderate or abandon his ambition to take center stage in matters of significance, and who longs to feel and to act and to arrange things according to his desires" (pp. 656–657), to attain the fulfillment of his thwarted wishes. Following Freud, the spectator "wants to be a hero, if only for a limited time, and playwrights and actors make it possible for him through *identification* with a hero" (p. 657).

The noted film director Martin Scorsese, in televised commercials promoting motion-picture entertainment, expressed the broad generalization of this conception most succinctly in pointing to the wealth of intriguing fictional characters and proclaiming to the audience that, as you watch the films, "You are them!" Identification, then, is taken to mean that respondents to fictional drama experience other beings 'from the inside,' thinking and feeling as though they were them.

The usefulness of this conceptualization has been severely challenged by both Tan (1995, 1996) and Zillmann (1995). This challenge is in large measure based on the already indicated communicative behavior that respondents to fiction direct at its characters. Such behavior does not suggest that respondents, if only for the moment, believe to perform actions in a fictional character's stead or to exist as a particular character of a play. Instead, this communicative behavior compellingly reveals that emotionally engaged respondents are *witnesses* to the events before them. As these events define 'apparent reality,' the respondents may be accused of having succumbed to the illusion of being party to depicted happenings in the sense of being present as a witness. This conception may be considered a form of the earlier mentioned *diegetic effect* (Burch, 1979; Tan, 1996) that manifests itself in the illusion of being present in, and therefore part of, the fictional environment.

Taking a witness perspective to fiction removes the mysticism connected with the idea of 'becoming one with another being,' in one form or another, from the discourse about induced emotions. It does not remove concepts such as emulation, however. In fact, the desire to want to be like some others, to assume their features and habits, is now understood as a consequence of the emotional reactivity of witnesses, especially of their sympathetic feelings toward particular characters.

It should be noticed that, in dealing with fiction and emotional reactivity, the witness perspective is entirely compatible with the paradigm of *parasocial interaction* (Horton & Strauss, 1957; Horton & Wohl, 1956). This paradigm addresses the formation of dispositions of sympathy or antipathy toward repeatedly observed fictional characters in television programs, such as soap operas or situation comedies. Ample evidence has been aggregated that characters are eventually treated, including being talked to, as if they were actual friends or enemies (e.g., Fabian, 1993; Gleich & Burst, 1996; Isotalus, 1995; Perse & Rubin, 1989; Rubin & McHugh, 1987). Clearly, this paradigm manifests the witness perspective and entails a denial of identification (Giles, 2002). Equally clear is that parasocial interaction is not limited to the repetitive and prolonged encounter of characters in fiction, but also applies to the limited encounter of characters during the course of a single play.

The following analysis of emotional reactivity to fiction is cast entirely within the witness perspective and should give evidence of its epistemic efficacy.

EXCITATORY PROCESSES IN EMOTIONAL EXPERIENCE

The dynamics of emotion that govern responses to actual situations, to their iconic or symbolic representation, and to the presentation of fictional events may be much the same. There is research evidence, in fact, that demonstrates considerable commonality in the mediation of affect by these different formats (Zillmann, 2000a; Zillmann & Knobloch, 2001). However, one principal condition exists that sets cinematic storytelling apart from alternative means of relating chains of events, and this condition proves to be pivotal in considering the creation and modification of emotional reactions. The condition in question is simply that cinematic narrative invariably compresses the time course of the happenings that make up a story and then, in delivering the story, imposes continuance of reception in real time (Bordwell, 1985; Branigan, 1992; Carroll, 1990; Tan, 1996).

Consequences of Continuance of Information Uptake

Emotions evoked in actuality by personal success or failure are usually allowed to run their course. A person, after achieving an important goal, may be ecstatic for minutes and jubilant for hours. Alternatively, a grievous experience may foster despair or sadness that similarly persists for comparatively long periods of time. Mostly for physiological reasons, but also as a result of reflection, emotions are not momentary experiences. But cinematic narrative treats them as if they were. As a rule rather than the exception, featured events that instigate emotions are followed by the presentation of other events long before all relevant aspects of the instigated emotions have subsided. Such compression of emotional and nonemotional events has, as we will see, intriguing implications for emotional experience.

It should be noted at this point that the compression of events in cinematic narrative does not necessarily extend to fiction generally. Written prose allows readers to pause when emotionally stirred and to continue reading only after recovery. All presentational formats that permit the pacing of information uptake afford recipients a degree of control over their affective responding. All formats that dictate the pace of uptake, whether concerning fiction or nonfiction, do not; or they do so in a most limited way. It is conceivable, for instance, that cinematic presentations are occasionally halted when viewers are unduly distressed by featured events, and that exposure is continued later as composure is regained. Readers of prose may similarly violate presumed pacing norms by continuing to read despite being emotionally agitated from exposure to immediately preceding text. In case such uninterrupted, continuous reading occurs, the contiguity of emotion instigation approximates that of formats imposing continuance of reception. It is by no means identical, however, as prose cannot present episodic occurrences in real time. Such discrepancies in continuous reception notwithstanding, the cinematically imposed continuous information uptake in real time, as well as the occasionally self-imposed undisrupted information uptake by avid readers of prose, entails unique means of evoking and escalating emotional experience. The paradigm that addresses these means focuses on the transfer of excitation from an initial emotional reaction to subsequent ones, primarily to the immediately following reaction.

Excitation Transfer in the Experience of Emotion

Cognitive activity does not sufficiently define emotional experience. It is generally thought that emotions entail a stirring, rousing, and driving component. This component of the emotions has been labeled arousal or excitation, and it has been conceived of in bodily terms. Two-factor theories have suggested an interaction between cognition and arousal, with cognition determining emotions in kind and arousal their experiential intensity and behavioral urgency (Hebb, 1955; Schachter, 1964). This conception has been elaborated in a three-factor theory that more fully accounts for the mostly involuntary evocation of excitatory activities as well as for their waxing and waning over time (Zillmann, 1978, 1983).

Three-factor theory distinguishes between dispositional, excitatory, and experiential components of emotion. Both the dispositional and excitatory components integrate reflexive response tendencies with reactions acquired through learning, whereas the experiential component involves cognition in the service of behavioral guidance and response correction. Basic emotions, such as specific fears and aggressive impulsions, often defy rationality and are not instigated by reflection. Excitation associated with these emotions is obviously not controlled by contemplation either. Rather archaic mechanisms mediate these reactions, whether they are made in response to actual situations or to their iconic representations. However, cognitive elaboration can function as a corrective, and diminish and shortcut emotions that are recognized as inappropriate and groundless. It also can exacerbate and even initiate emotions by fostering comprehension and evaluation of relevant circumstances.

This brings us to emotional misreactions that are not recognized as such, misreactions that are regularly and often deliberately created in cinematic presentations. On the well-founded premise that cognitive adaptation to stimulus change is rapid and quasi-instantaneous, whereas excitatory adaptation is sluggish and time-consuming, it can be expected that persons will quickly switch cognitively from situation to situation, while excitation instigated by a first situation will persist through a second one and possibly through yet others (Zillmann, 1996a). It is established beyond doubt that excitation, once triggered, decays rather slowly. For all practical purposes, it takes at least three minutes, often ten or more minutes, on occasion hours for excitation to return to normal levels. This is for reasons of humoral mediation. Specifically, excitatory reactions are instigated by the release of adrenal hormones (the catecholamines epinephrine, norepinephrine, and dopamine, in particular) and, to a lesser degree, of gonadal steroids (mostly testosterone) into systemic circulation that persists until these agents are metabolized (Zillmann & Zillmann, 1996). Excitation in response to particular stimuli, then, is bound to enter into subsequent experiences. In case of contiguously placed discrete emotions, residual excitation from the first thus will intensify the immediately subsequent emotion, regardless of taxonomical differences in these emotions. Moreover, depending on the strength of the initial excitatory reaction and the time separation of emotions elicited at later times, residual excitation may intensify experiences further down the line. This is the principle of excitation transfer.

Before considering the effects of the cinematic chunking of emotion-inducing episodes, let us take a look at common experiences of emotional overreaction in situations of rapid cognitive but sluggish excitatory adjustment to changing conditions. Let us imagine, for a moment, a lady who steps on a snake in the grass of her backyard. Deep-rooted survival mechanisms, organized in the brain's limbic system (LeDoux, 1996; LeDoux & Phelps, 2000), will be activated and make her jump back and possibly scream. Following this initial reaction, she might find the time to construe her emotional behavior as fear and panic. She might also notice herself shaking and thus realize that she is greatly excited. Let us imagine further that, upon looking once more at the object of her terror, she recognizes that the snake is a rubber dummy, in all probability planted by her mischievous son who rushes onto the scene, laughing his head off. This recognition, a result of instant cognitive adjustment to changing circumstances, proves her initial emotion of fear groundless and invites a new interpretation of her experiential state. If she is annoyed with her son for giving her such a scare, she is likely to become infuriated. But after fully comprehending the prank, she might consider being angry inappropriate and cognitively adjust once more, this time joining in his laughter and appraising her experience as amusement. Throughout this cognitive switching from experiential state to state, the excitatory reaction to the detected danger in the grass persisted to varying degrees. It initially determined the intensity of the fear reaction. The residual excitation from this reaction then intensified the emotion of anger and the experience of amusement. Had the lady acted out her anger, she would have overreacted in punishing her son. But transfer-intensified reacting might also have expressed itself in fits of laughter bordering on the hysterical.

Emotional overreactions of this sort are commonplace. At one time or another, everybody seems to have experienced the extraordinary intensity of frustration after rousing efforts, of joy upon the sudden resolution of nagging annoyances, of gaiety after unfounded apprehensions, or for that matter, of sexual pleasures in making up after acute conflict (Zillmann, 1998a). Irrespective of personal experiences, however, ample research evidence exists that shows the transfer intensification of all so-called active emotions (i.e., emotions associated with increased levels of excitation), of their experiential states as well as of their behavioral manifestations (Zillmann, 1983, 1996a). It has been demonstrated experimentally, for instance, that residual excitation from sexual excitement can intensify anger and aggressive behavior, but also altruistic feelings and supportive actions. Moreover, residual excitation from either sexual excitement or disgust has been found to facilitate such diverse emotional experiences as the enjoyment of music, the appreciation of humor, and feelings of sadness. Residues from feelings of sadness and fear, in turn, have been found to intensify joyous reactions to fortuitous happenings. Frustration has been observed to intensify euphoric as well as angry subsequent feelings. Even nonemotionally induced, hedonically neutral excitation was found to transfer into subsequent states. Specifically, it has been demonstrated that residual excitation from strenuous physical exercise can enhance feelings of anger and aggressive behavior, intensify sexual arousal, promote help-giving, elicit feelings of grandiosity and elation, foster favorable reactions to advertisements, and facilitate sexual attraction.

In summary, then, residual excitation from essentially any excited emotional reaction is capable of intensifying any other excited emotional reaction. The degree of intensification depends, of course, on the magnitude of residues prevailing at the time. Figure 13.1 presents this paradigm of the intensification of contiguous emotions in graphic form.

Transfer in the Experience of Emotions From Cinematic Narration

In the empirical exploration of the enjoyment of cinematic presentations, the excitation-transfer paradigm has been employed, primarily, to explain the suspense paradox. Why is it that

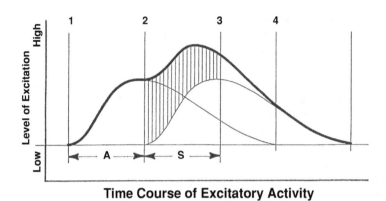

Time Course of Excitatory Activity

FIG. 13.1.   A model of excitation transfer in which residual excitation from a preceding excitatory reaction combines additively with the excitatory reaction to current stimulation. An antecedent stimulus condition (A), persisting from time 1 to time 2, is assumed to produce excitatory activity that has entirely decayed only at time 4. Similarly, a subsequent stimulus condition (S), persisting from time 2 to time 3, is assumed to produce excitatory activity that has entirely decayed only at time 5. Residual excitation from condition A and excitation specific to condition S combine from time 2 to time 4. The extent to which the transfer of residues from condition A increases the excitatory activity associated with condition S is shown in the shaded area.

emotional distress from witnessing protagonists in peril can be converted to joy upon suspense resolution? Moreover, how can it be that greater initial distress fosters more joy in the end (Carroll, 1990; Vorderer & Knobloch, 2000; Zillmann, 1980)? Research with both children and adults has firmly established this long presumed relationship (cf. Zillmann, 1996b).Excitation transfer is its explanation. Specifically, the distressing experience of suspense is arousing, and residues of this arousal linger through resolution and intensify the experience of relief and euphoria. Again, cognitive adjustment to the changed circumstances featured in the resolution is rapid, whereas excitatory adjustment is drawn out. The more intense the suspense-induced distress, the greater the excitatory residues that come to energize joyous reactions to the satisfying outcomes of the resolution.

The transfer logic is perhaps best illustrated by narratives that present seemingly doomed protagonists who struggle against hostile environments and who merely manage to survive. There may be little heroism, if any, in this getting away with dear life. The resolution thus offers little to celebrate and to be jubilant about. Such minimal-heroism resolution can be intensely enjoyed, however, when appropriately preceded by empathic torment. This overreaction to minimal heroism is simply more obtrusive than that triggered by resolutions featuring great heroic accomplishments. In neither case, however, can the euphoric reaction be considered the result of experienced relief, because the reaction is not contingent upon a sharp decline in excitedness but, if anything, hinges on a boost of excitation.

Another narrative domain that has received some attention in these terms is tragedy (de Wied, Zillmann, & Ordman, 1995; Zillmann, 1998b). It has been observed that highly empathic persons are more distressed by exposure to dramatized tragic happenings than are their less empathic counterparts, and that the former group literally sheds more tears about these happenings than does the latter group. The resolution again offers little that might be celebrated. But those who are particularly distressed by the tragic events take whatever redeeming value there may be in the resolution as a cue for contentment, experience such feelings more intently because of excitatory residues, and report greater enjoyment of tragic drama overall.

Other research has shown the transfer intensification of humor (Zillmann, 2000b). The concept of comic relief obviously focuses on relief. Its cinematic form may well serve this purpose in preventing excessive distress on occasion (King Jablonski & Zillmann, 1995; Zillmann, Gibson, Ordman, & Aust, 1994). But the concept can also be construed as one that maximizes mirth in response to comic situations. Comic material may be mediocre, but is bound to produce strong reactions after tense, arousing scenes. These arousal-enhanced reactions of amusement, especially when overtly expressed in laughter, may have a cumulative effect and result in assessments of greater enjoyment of drama that provides frequent opportunities for comic relief.

Toward a Dramaturgy for the Transfer Facilitation of Emotions

Much research on the transfer-facilitation of emotions has been conducted independent of cinematic considerations (cf. Zillmann, 1979, 1996a, 1998a). Nonetheless, most demonstrations, such as that residual arousal from distress can facilitate subsequent sexual excitement or that excitatory residues from fear can intensify feelings of sympathy and support, are directly applicable to cinematic dramaturgy. Scenes can be aggregated in ways that maximize emotional reactivity to some and minimize it to others. For instance, arousing violence preceding the display of sexual behavior will intensify reactions to the sexual scenes, and arousing distressing torture will energize jubilation and applause to the punitive brutalization of the torturer. Transfer theory projects such facilitation for all scene-evoked affective reactions, provided that the afore-placed scenes produce arousal and that residues of the arousal outlast these scenes.

In developing a dramaturgy of excitation transfer more formally, the following propositions can be expressed;

(a) Arousing scenes from which excitatory residues are to be transferred into subsequent scenes must terminate before appreciable dissipation of excitation can manifest itself. Ideally, arousing scenes conclude with arousal at a maximum;

(b) The intensification of affect in response to subsequent scenes is a function of the magnitude of excitation elicited by these scenes plus that of residual excitation from preceding scenes;

(c) Affect facilitation is stronger, the more immediate the placement of subsequent scenes;

(d) Affect facilitation is stronger, the less drawn out the subsequent scenes;

(e) In case both antecedent and subsequent scenes are strongly arousing, affect facilitation can escalate. The escalation is limited, however, by experiential maxima for excitation. A law of initial values specifies that excitatory contributions from arousing scenes are inversely proportional to the height of prevailing levels of arousal (Wilder, 1957). In other words, as experienced arousal levels increase, successively less excitation from arousing scenes can be added. The law thus renders the aggregation of highly arousing scenes comparatively inefficient for transfer;

(f) The facilitation of affect in response to scenes that are separated from preceding arousing scenes by unarousing scenes is stronger the shorter the time of the separating scenes. Facilitation terminates, of course, with the complete dissipation of residual excitation; and

(g) The facilitation of affect in response to subsequent scenes is prevented by delaying their placement until excitatory residues from preceding scenes have completely dissipated.

Figure 13.2 presents principal transfer situations in simplified graphic form.

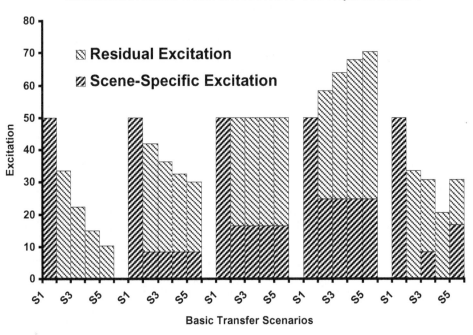

FIG. 13.2.   Principal forms of the intensification of emotional reactions to a sequence of scenes as a function of excitatory residues from an immediately preceding arousing scene. Detailed explanation may be found in the text.

The first schema, at the far left, shows an extremely arousing scene (S1) followed by four unarousing scenes (S2–S5). Residual excitation is presumed to dissipate by one third from scene to scene. As can be seen, reactions to S2, whatever its contents, will be highly emotional despite the fact that this scene does not contribute arousal. Transfer intensification is successively weaker for the subsequent scenes.

The second schema indicates transfer under the same conditions, except that S2–S5 are now presumed to contribute minor degrees of arousal (one sixth of S1). These contributions, as can be seen, retain excitation at comparatively high levels for S2–S5. The logic is that of compounded interest. S2 combines excitation from its stimuli with residues from S1. S3 also combines excitation from its stimuli with residues from its antecedent, S2; the S2 residues, however, combine those of S1 and S2. All later scenes analogously benefit from the combined residues of their antecedents.

In the third schema, S2–S5 are presumed to supplement the amount of excitation, one third, that is lost to decay. As shown, high levels of excitation can be maintained by such supplementation.

The fourth schema displays excitatory escalation for subsequent scenes by making them contribute half the excitation of S1, the extremely arousing initial scene. As can be seen, the escalation is negatively accelerated, leveling out eventually (as the result of the law of initial values).

The last schema, at the far right, illustrates excitation transfer in an arbitrarily varied situation. S2 and S4 are presumed to be unarousing, S3 to be mildly arousing (one sixth of S1), and S5 to be moderately arousing (one third of S1). The schema indicates that intermixed nonarousing scenes can be made to appear considerably arousing and that the occasional usage of mildly and moderately arousing scenes can maintain excitation and therefore affect intensity at comparatively high levels.

Strategies for the creation of scene compositions with optimal emotional effects can be constructed by applying propositions (a) through (g). Such strategies are also apparent from inspection of the schemata presented in Figure 13.2.

Our analysis of arousing events in cinematic presentations focused on narration. There is no question, however, that numerous non-textual cinematic techniques exist that are capable of generating excitation. A discussion of these techniques and their possible effects on emotion has been provided by Tan (1996). It should suffice here to point out that, to the extent that the indicated techniques produce arousal independent of connected narrational scenes, their contribution to the intensity of evoked emotions would be in accord with expectations based on excitation-transfer theory.

COGNITIVE PROCESSES IN EMOTIONAL EXPERIENCE

The primary function of cognition is, of course, to provide guidance for our interaction with the physical and social environment. Largely mediated by activity in the prefrontal cortices, cognition mediates the appraisal of conditions that impact our own welfare, and it projects courses of action to get us what we need and to bypass what would harm us. In so doing, it assesses the effectiveness of our efforts and, if need be, corrects their direction.

All this must seem irrelevant for the consideration of emotional reactivity to fiction. Respondents, partaking in an apparent environment as witnesses, need not respond in the indicated manner. But, although their cognitive apparatus does not serve the preparation of overt action in an environment that can be altered by such action, this apparatus is actively preparing the respondents for meaningful interaction, much as if such interaction was imminent. Some of these preparations may be reflexive and not require cognitive elaboration. Moving the head out of the way of an apparently rapidly approaching object would be a case in point. Less obvious are the necessary elaborations in the formation of emotional dispositions. Although the respondents to fiction shall never meet the cast of characters, cognition nonetheless prepares them for interacting with them. As suggested by the already discussed parasocial-interaction model, respondents come to like or dislike fictional characters and then treat them as if they were real friends or enemies. Characters in a play analogously become friend- or enemy-like to varying degrees. This formation of affective dispositions, along with its consequences for emotional reactivity, will now be considered in greater detail. Specifically, we will explore its dependence on the continual monitoring of the characters' behavior, along with the moral assessment of this behavior. However, we shall first give some attention to the more direct ways of instigating emotions by exposure to apparent environments.

Emotions From Monitoring Apparent Environments

Evolutionary psychology has emphasized the fight–flight reaction (Cannon, 1932; MacLean, 1990). Individuals are thought to continually monitor their environment for danger and to respond with attack or escape when detecting it. A burst of energy is needed to respond in such fashion, and the immediate instigation of sympathetic excitation serves this purpose. The emotions of anger and fear, then, are energized in preparation for action. Such action is obviously not called for when responding to cinematic representations of danger. However, because these reactions are organized in archaic brain structures, the amygdala in particular, cinematic scenes of danger, in defiance of rationality, still trigger excitatory reactions (cf. LeDoux, 1996; Zillmann, 1998c, 2003).

The fight-flight dichotomy was eventually expanded to a response trichotomy that includes sexual preparedness. Sexuality, serving the preservation of the species rather than self-preservation, is similarly deep-rooted evolutionarily. Sexual activity, organized in the septum, also requires energy for bouts of exertion, and this energy is likewise provided by sympathetic excitation. As in the case of danger, the cinematic presentation of others' sexual opportunities and actions still elicits sexual excitedness, notwithstanding the fact that sexual targets for consummatory behavior are not immediately available (Zillmann, 1986).

Iconic representations of danger and sexual opportunities, then, may be considered basic stimulus conditions that reliably arouse and that, because of it, will foster responses that are construed as affective experiences or emotions. It would seem to be a grave error, however, to consider displays of perilous happenings and of erotic enticements the only, or even the primary, conditions for the creation of excitation and emotions. Cinematic narratives invariably involve, and are built around, people and other animated entities. Floods, quakes, and fires, but also poisonous snakes, snarling leopards, and murderous villains, threaten others; that is, they threaten the narratives' main cast of characters. On occasion, these threats are visually presented as if they were about to victimize the viewer personally (cf. Smith, 1997). But even when presented in this manner, they still only supplement the display of others in peril. For instance, an avalanche presented as rushing at the viewers, or a snarling dog presented as snapping at them, may prove arousing because they more closely than alternative presentations replicate the stimulus conditions of being personally threatened. Such displays thus may be used to create arousal. But they also may serve to provide viewers with a better appreciation of the dangers facing those who are seen coping with them.

But cinematic narratives undoubtedly evoke emotions primarily by featuring others' confrontation with threatening conditions and fortuitous circumstances, as well as by displaying these others' reactions, including emotional ones, to their demise or to their enrichment as such outcomes materialize. Unless the narrative is interactive and makes respondents active participants in its flow (Grodal, 2000; Vorderer, 2000), they remain mere witnesses to the fate of others (Tan, 1995; Zillmann, 1995). Given that, spectators to fictional narratives respond nonetheless with emotions, at times with emotions of extreme intensity, to the fortunes and misfortunes they see others enjoy or suffer. In order to explain such strong emotional involvement with others and their fate, the concept of empathy has been invoked and employed to good effect.

Empathic Evocation of Emotion

Empathy can be construed as an archaic mechanism that, through the millennia, served emotional contagion and the coordination of action (Buck & Ginsburg, 1997; Hoffman, 1978, 1987; Plutchik, 1987; see also Zillmann, this volume, Chapter 10). It ultimately served the preservation of individuals and their species. In a group's confrontation with danger, for instance, it undoubtedly was adaptive to get jointly excited and thus prepared for vigorous action. The contagious effect of one individual's expression of fear could instantly permeate the group, readying all for flight, or, the expression of anger and assertive behavior could instantly foster preparedness for concerted resistance and attack.

The conditions of life in contemporary times have, of course, deprived empathy of much of such utility. However, as a mechanism of excitatory contagion, empathy has been retained in the paleomammalian structures of the brain (MacLean, 1967). If this were not the case, it would be difficult to explain, for instance, why observers experience distress when seeing a construction worker fall off the scaffolding and hit the ground, cringing in pain; or for that matter, when watching a movie that shows the protagonist cling with his fingertips to a cliff, apparently about to fall to his death.

Common observation and research evidence (e.g., Eisenberg & Strayer, 1987; Stotland, 1969) leave no doubt about the fact that people, in responding to the emotions displayed by others in actual situations or in fictional presentations, tend to experience emotions that are hedonically similar to those witnessed and that often have considerable depth. Some time ago, Adam Smith (1759/1971), in connection with his theory of moral sentiments, recognized the lack of ulterior benefits from such emotional investment. In his words:

> How selfish soever man may be supposed, there are evidently some principles in his nature, which interest him in the fortune of others, and render their happiness necessary to him, though he derives nothing from it except the pleasure of seeing it (p. 1).

Empathy with others' experience and expression of emotion is by no means a necessary response, however. There obviously exist circumstances under which empathic sensitivities diminish or are entirely abandoned and overpowered by alternative response mechanisms. Under these circumstances, those who witness others' misfortunes are free to take pleasure in these others' demise. The circumstances in question have been well understood since antiquity.

Regarding dramatic narratives, Aristotle articulated them succinctly, although in negative form (Aristotle, ca 330 BCE/1966). Specifically, he found fault with two principal narrative transitions, deeming them utterly unenjoyable. In his *Poetica* he stipulated that:

1. a good man must not be seen passing from happiness to misery; or
2. a bad man from misery to happiness.

By implication, he recommended as joy-producing plots those that feature (1) a good person passing from misery to happiness, or (2) a bad person from happiness to misery. Whether presented in negative or positive form, however, the propositions concerning negatively judged persons indicate the absence of empathic reactions to the projected outcome. Apparently, only good characters warrant empathic concerns. Bad characters do not. Bad characters' joy from coming to glory cannot be affectively shared. Their joy may prove distressing, instead. Analogously, their pains from coming to harm are not to be shared. Those who witness the demise of bad characters can freely applaud it, instead.

Aristotle thought it self-evident that the narrative transitions on which he had focused could not foster joy. He simply stated that these transitions would be odious. In discussing tragic plots, however, he articulated his reasons for projecting reactions of displeasure and vexation. Aristotle specifically implicated moral judgment with the mediation of reactions of joy versus revulsion to the resolution of various forms of dramatic narrative. He essentially argued that persons pursuing good causes (i.e., consensually approved causes) are considered good people, and that good people are judged deserving of good fortunes. Analogously, persons pursuing bad causes (i.e., consensually condemned causes) are bad people, and bad people are judged deserving of bad fortunes; or, at the very least, undeserving of good fortunes. Outcomes in accord with moral considerations thus can be enjoyed. In contrast, outcomes that violate moral considerations are those thought to squelch enjoyment and to foster irritation and contempt, instead.

One is inclined, therefore, to expand on Smith's reflections about empathy and complete his thought by considering the abandonment of empathy, transitory as this abandonment may be.

> There are evidently some principles in human nature that make individuals take an interest in the fortunes of others and that, in case good fortunes are judged unwarranted and bad fortunes are deemed just and called for, render these others' demise necessary, although onlookers derive nothing from it except the pleasure of seeing it.

Considerations of morality have assumed a central position in drama theory ever since (e.g., Bordwell, 1985; Carroll, 1990; Tan, 1996), and in the form of moral sanctions they have entered into the contemporary psychology of drama appreciation as well (Jose & Brewer, 1984; Zillmann & Bryant, 1975). In particular, moral assessments have become an integral, pivotal part of the disposition theory of emotion that has been employed to explain the enjoyment of drama in subordinate plots as well as in major, overarching plots (Zillmann, 2000a).

Dispositional Mediation of Emotion

The indicated intertwined operation of moral judgment and emotional disposition is outlined in Figure 13.3. Witnessed behavior, as can be seen, is assessed in moral terms (i.e., good vs. bad, to varying degrees), and such assessment is expected to determine emotional dispositions. The approval of actions and their apparent purpose is thought to prompt dispositions of liking and caring. Their disapproval, in contrast, is thought to prompt dispositions of disliking and resenting. Liking defines protagonists, disliking antagonists. Character development thus is considered a function of moral evaluation. Without such evaluation, dispositions of indifference would prevail, and witnesses to social happenings would show little emotional involvement, if any. Witnesses to socially relevant events in cinematic narratives may be thought of as untiring moral monitors. Their continually rendered verdicts are bound to yield the approval of the conduct and the resultant adoration of some characters, and the disapproval of the conduct and the resultant detestation of others. The interdependence between moral assessment and emotional disposition is further apparent in loop *c* of the figure, which indicates the possibility of feedback from disposition to judgment. It has been observed that liking invites overly favorable, forgiving assessments, whereas disliking biases in the opposite direction.

Emotional dispositions, once firmly established, are thought to foster anticipatory emotions. These anticipatory emotions are either positive or negative, their hedonic valence reversing as a function of morally determined dispositions. As the figure shows, positive dispositions foster hopes for positive, rewarding happenings along with fears about negative, punitive ones. Negative dispositions foster the opposite hopes and fears. If and when the hoped for or feared events materialize, the evoked emotions will be in accord with anticipations. Specifically, hoped for and morally sanctioned outcomes (i.e., rewarding events for protagonists and punitive events for antagonists) will foster euphoric, joyous reactions, whereas feared and morally unwarranted outcomes (i.e., rewarding events for antagonists and punitive events for protagonists) will prompt reactions of dysphoria, discontent, disappointment, and contempt.

Positive emotional dispositions are known to foster hedonically compatible reactions to events that evoke emotions in witnessed persons. Negative emotional dispositions, in contrast, are those that relax and overwhelm empathic inclinations, and that enable witnesses to rejoice in response to esented others' misfortune and agony. Negative dispositions also get in the way of empathizing with gratified others who are deemed undeserving of such fortune. In fact, the perception of undeserved gratification can stir intense emotions of righteous indignation. Anti- or counter-empathic emotional reactions of this kind are obviously the result of moral considerations. Villains are to get their just deserts, and concerns about their welfare would amount to misinvested efforts at emotion control. Villains, moreover, are simply not entitled to good fortunes. Oddly, then, it is morality that liberates observers, allowing them to take pleasure from the punitive torment of others who are judged deserving of such fate. But it is morality also that fosters infuriation and indignation upon witnessing the benefaction of those who are deemed utterly undeserving of being rewarded.

These considerations lead to the following predictions of euphoric and dysphoric emotions in response to the resolution of dramatic conflict in cinematic narratives. The classification

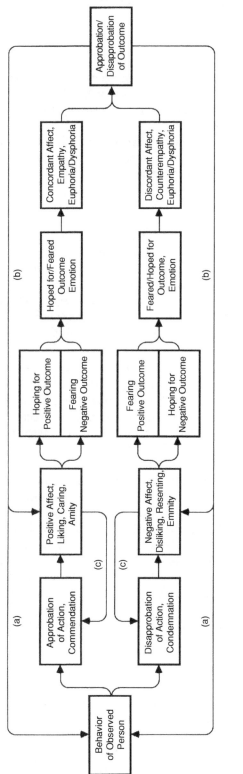

FIG. 13.3. A model of the dispositional mediation of emotion from witnessing the actions and contingent emotional experiences of others. Stages 2 and 7 indicate the involvement of moral considerations in the formation of emotional dispositions, and stages 3 and 4 the resulting emotional dispositions and their influence on anticipatory emotions. Stages 5 and 6 specify emotional reactions to pertinent outcomes, such as gratification or aversion, and to their expressive consequences, such as elation or distress. Feedback loop *c* indicates the influence of formed dispositions on moral judgment, such as amity fostering tolerance and enmity fostering strictness. Loop *b* suggests a similar influence of witnessed outcomes through their impact on dispositions. Loop *a* indicates that the process described in stages 1 through 7 is recursive and can be chained to arbitrary length (i.e., short dramatic plots can be chained within overarching dramatic plots).

in moral terms is to highlight the significance of moral assessments in the mediation of the emotions of witnesses.

Conditions of Justice

1. Witnessing the victimization of a disliked antagonist at the hands of a liked protagonist fosters delight, the experiential intensity of which increases with (a) the liking of the protagonist; (b) the disliking of the antagonist; and (c) the extent to which the antagonist is deemed deserving of a particular victimization.
2. Witnessing the benefaction of a liked protagonist fosters delight, the experiential intensity of which increases with (a) the liking of the protagonist; and (b) the extent to which the protagonist is deemed deserving of a particular benefaction.

Conditions of Injustice

3   Witnessing the victimization of a liked protagonist at the hands of a disliked antagonist fosters repugnance, the experiential intensity of which increases with (a) the liking of the protagonist; (b) the disliking of the antagonist; and (c) the extent to which the protagonist is deemed undeserving of a particular victimization.
4   Witnessing the benefaction of a disliked antagonist fosters repugnance, the experiential intensity of which increases with (a) the disliking of the antagonist; and (b) the extent to which the antagonist is deemed undeserving of a particular benefaction.

Support for these predictions comes from research on the enjoyment of a variety of dramatic formats (Zillmann, 1996b; Zillmann & Knobloch, 2001; Zillmann & Paulus, 1993). It comes, obviously, from the exploration of drama proper, but also from the exploration of specific genres (such as suspenseful narrative or comedy) and genre-like nonfictional exposition (such as sports and the news). Suffice it here to exemplify the outlined moral-dispositional mechanisms with two selected investigations.

The most direct demonstrations of the power of moral judgment in the mediation of emotion in response to others' emotion come from empathy research (Wilson, Cantor, Gordon, & Zillmann, 1986; Zillmann & Cantor, 1977). School children were exposed to specially produced films in which either a loved or a hated character was developed, and in which this character was either victimized or benefited during resolution. His victimization showed him in excruciating pain, his benefaction in extreme joy. The children's facial reactions to these final scenes were unobtrusively recorded and then scrutinized. The findings were entirely in line with the specifications of stages 5 and 6 of Fig. 3. Respondents empathically cringed when the beloved character was in pain, and they exhibited joy when he was euphoric. They responded counter-empathically, however, to the behavioral displays of the resented character. They cringed when he jumped for joy, and they expressed pleasure when he was hurt. In the latter condition, he apparently got what he deserved; in the former, the outcome was unjust and hence annoying and detestable.

A parallel investigation with mentally challenged children demonstrated that, when the capacity for moral judgment at the level of equitable retribution is not developed, empathy becomes mechanical. In particular, counter-empathic reactivity does not materialize. Such mentally challenged children invariably expressed joy in response to witnessed joy, and they invariably expressed distress in response to witnessed distress. Whether the witnessed emotions were exhibited by a beloved or by a resented character was immaterial.

This latter investigation shows compellingly that empathy functions as a basic default mechanism that, if not opposed and overpowered by affective dispositions that derive from assessments of deservingness, governs emotional reactivity to the observed fate of others. The condemnation of others' conduct and the resulting disliking, then, are indeed prerequisite to joy over their demise as well as to distress over their good fortunes.

Moral Sanction of Resolutions to Dramatic Conflict

In dramatic narratives, plots are known to dwell on hostile confrontation and conflict. Conflict is almost always resolved, however, usually promptly so. Both in minor plots (i.e., minor in terms of duration) and in major plots (i.e., those that span large portions of narratives, if not their entirety), the parties in conflict are disengaged in ways that are more fortuitous to one party than to others. Resolution may simply consist of the cessation of hostility or endangerment. More likely, it entails glorious victory for one party and humiliating defeat for another. In emotional terms, resolutions provide at the very least relief from empathic distress. More characteristically, however, fully embellished resolutions, especially those overarching a narrative, evoke emotions of considerable intensity. Depending on dispositions toward the victorious and defeated parties, respondents will experience happiness or sadness; or at least, emotions with affinity to these experiences. But dispositions are not the only factor that influences emotions in response to resolutions. Resolutions must be sanctioned morally to have their intended effect on emotion. Feelings of joy in response to a protagonist's triumph can be spoiled by his or her actions that are deemed inappropriate, if not deplorable. Analogously, feelings of sadness will suffer impairment if the protagonist's imperfections, her or his 'tragic flaw,' prove intolerable.

The emotions evoked by the resolution of conflict in drama are undoubtedly pivotal to the enjoyment of cinematic narratives. Given that, along with the fact that these emotions hinge on moral considerations and are readily compromised, closer examination of the concept of moral sanction would seem to be warranted.

The assessment of what is morally correct under given circumstances may be a deliberate, reflective process yielding specific verdicts. It may stipulate a particular punishment for a particular transgression or indicate a particular reward for a particular accomplishment. Moral sanction is not thought to have such a high degree of specificity. It is not considered to prescribe and demand particular outcomes. Rather, moral sanction is conceived of as a readiness to accept, in moral terms, observed outcomes. It may well happen that, on occasion, specific harm, such as torture and death, is deliberately wished upon a brutal villain. But, as a rule, expectations of punishment and reward are not specific to particular treatments and outcomes. Moral sanction is characterized, instead, by considerable latitude in accepting punitive and rewarding actions and events. Respondents to drama that features rape, for instance, may in a round-about fashion wish harm upon the rapist, but be satisfied when seeing him either caught and convicted, contract a debilitating disease, or crippled by a falling tree. The latitude of retribution is not unlimited, however. The respondents would probably be distressed if the only punitive consequence was that one of the rapist's victims managed to bloody his nose. The respondents might be similarly distressed when seeing him subjected to castration or having his arms chopped off. Transgression during conflict and punishment during resolution must be roughly commensurate for the punishment to be morally sanctioned and deemed emotionally satisfying. Punishments that fall outside the latitude of sanction leaves the respondents' sense of justice disturbed, which ultimately diminishes the enjoyment of resolutions. The same applies to accomplishments for which the rewards fall outside the latitude of moral sanction.

Also, the exercise of moral sanction is not presumed to involve the use of formal systems of moral judgment, such as Kant's categorical imperative (Kant, 1785/1922) or Bentham's utilitarian formula (Bentham, 1789/1948), and to necessitate the violation of derived precepts in order for punitive happenings to become sanctionable. Reminiscent of Aristotle's afore discussed suggestions, the morality thought to be involved is truly basic in prescribing good fortunes to good people (i.e., people who are good because they do good deeds) and bad fortunes to bad people (i.e., people who are bad because they do bad things). If moral judgment is thus conceived of as a not entirely systematic evaluation of situational behavior, that is, as verdicts of good versus bad or right versus wrong in idiosyncratic terms, we must expect profound individual differences in moral assessments. For instance, some will consider the death penalty fair retribution for taking the life of a fellow human; others will consider this penalty a crime against humanity. Some will consider sexual preference a moral entitlement; others will deem specific ones morally indefensible. Some will think it good and right to save the big redwoods in California and Oregon; others will think it good and right to sacrifice a bit of nature to ensure continued income and, perhaps, a better life for the loggers. Some will see fit to honor and defend the national flag because it is thought to signify the political doctrine of equal justice for all; others will be ready to burn the flag because they deem this political doctrine wanting in its administration of social justice. Some will embrace the morality manifest in prevailing social conventions; others will consider these traditions decadent and declare them morally bankrupt, thereby elevating themselves into a moral elite that is called upon to challenge the morality of those deemed morally inferior.

Moral judgment is simply not monolithic, as some ethicists would have us believe. It would seem futile, in fact, to treat people's moral sanction of drama as uniform and normative. Recipients bring their idiosyncratic morality to the screen, sanction or condemn witnessed actions and agents in accord with it, and then experience emotions as a result of their assessment. As moral assessments vary, so will the respondents' emotions. In constructing theories of drama appreciation that involve moral sanction as an essential mechanism, it is imperative to recognize, and to make allowances for, the diversity of basic morality in strata of the population at large. In order to predict more accurately which retributive events foster delight and which repugnance in whom, it will be necessary to stake out existing morality subcultures and to determine the judgmental properties that characterize and distinguish them.

In the face of the indicated profound diversity in moral assessments, common ground should not be overlooked, however. Considering coercive and socially-supportive actions, in particular, the members of different subcultures are likely to render similar judgments. Additionally, apprehensions about others being granted access to gratifications that are denied us, may be widespread and nearly universal. Such apprehensions might also explain why we can take pleasure from witnessing the punishment and torment of those we think have taken unfair advantage of situations and, hence, have done wrong. Perhaps the overarching theme of enjoyable fictional exposition is conveyed in the projection of social justice in the sense that gratifications have to be earned by all our fellow humans just as we, by our own efforts, have to earn them; and that none of our follow humans be exempt from the punitive contingencies that govern our own lives. Violations of this conception of justice will strike us as repugnant, whereas exposition within these principles will delight us.

EPILOGUE ON THE DRAMATURGY
OF GOOD AND EVIL

Our discussion of the evocation of emotion by fictional narratives appears to render these narratives' morality plays of one kind or another. Moral monitoring is thought to foster approval or disapproval of the actions of the characters of plays and thereby yield feelings of sympathy toward the well-behaved protagonists and antipathy toward the ill-behaved antagonists. Within this good-versus-evil dichotomy, the strength of these affective dispositions is expected to determine the depth of empathy or counter-empathy, of the anticipatory emotions of hope or fear, and of joyous emotions as hoped for outcomes materialize versus distressing emotions as feared outcomes do. Throughout the display of relevant actions, the depth of the recipients' emotional reactions is clearly a function of the magnitude of dispositional involvement. Poorly developed characters (that is, characters whose actions and apparent intentions prompt neither applause nor condemnation) will not be engaging. In contrast, the recipients' emotions are bound to be engaged by characters who do and intend to do things that, for whatever particular moral reason, are deemed supportive, courageous, brave, and simply wonderful, on the one hand, or arrogant, malicious, brutal, and plainly evil, on the other. The more we can love or hate the characters that the narrative develops, the more we shall enjoy outcomes that show those we love triumph over those we hate. In fact, if our emotions are sufficiently engaged, we shall applaud the cruelest destruction of evil characters without having moral misgivings about it. We could, after all, morally sanction the brutality involved.

It would seem, then, that those cinematic narratives that develop the most admirable protagonists and the most terrifying antagonists, all within the limits of dramatic credibility, are likely to evoke the strongest emotions. The greatest dispositional separation between protagonists and antagonists promises the most intense emotions in response to the resolution of conflict. Joy will be at a maximum as the best of good triumphs over the worst of evil. And should evil get the better of good, as it does in tragic resolutions, the deepest reactions of disappointment, dejection, and sadness can be expected.

These observations seem to question the wisdom of developing and featuring complex characters; that is, characters who exhibit an admixture of laudatory and evil traits without a clear dominance of the one over the other. Such character complexity, as it violates the purity of good or evil, must be considered a detriment to drama that focuses on the evocation of strong emotions. What should be recognized, however, is that the evocation of emotion is not the only objective of drama, not even necessarily the most desirable one. Drama may captivate and intrigue us in cognitive terms (Zillmann, 1991). It can be thought-provoking and inspiring. Rather than stir our emotions to the fullest, it may gently touch us. Drama that combines the indicated elements, that both touches our hearts and challenges our minds, may arguably define a genre of superior entertainment value.

REFERENCES

Allen, R. C. (1993). Representation, illusion, and the cinema. *Cinema Journal, 32*(2), 21–49.

Allen, R. C. (1997). Looking at motion pictures. In R. Allen & M. Smith (Eds.), *Film theory and philosophy* (pp. 76–94). Oxford: Clarendon Press.

Aristotle. (330 BCE/1966). De Poetica (I. Bywater, Trans.). In *The works of Aristotle* (Vol. 11). Oxford: Clarendon.

Bentham, J. (1789/1948). *An introduction to the principles of morals and legislation.* New York: Hafner.

Bordwell, D. (1985). *Narration in the fiction film.* Madison: University of Wisconsin Press.

Branigan, E. (1992). *Narrative comprehension and film.* London: Routledge.

Buck, R., & Ginsburg, B. (1997). Communicative genes and the evolution of empathy. In W. Ickes (Ed.), *Empathic accuracy* (pp. 17–43). New York: Guilford.

Burch, N. (1979). *To the distant observer.* Berkeley: University of California Press.

Cannon, W. B. (1932). *The wisdom of the body.* New York: Norton.

Carroll, N. (1988). *Mystifying movies: Fads and fallacies in contemporary film theory.* New York: Columbia University Press.

Carroll, N. (1990). *The philosophy of horror or the paradoxes of the heart.* New York: Routledge.

Coleridge, S. T. (1817/1960). *Biographia Literaria or biographical sketches of my literary life and opinions.* London: J. M. Dent & Sons.

Currie, G. (1995). *Image and mind: Film, philosophy, and cognitive science.* New York: Cambridge University Press.

Damasio, A. R. (1994). *Descartes' error.* New York: Putnam.

de Wied, M., Zillmann, D., & Ordman, V. (1995). The role of empathic distress in the enjoyment of cinematic tragedy. *Poetics, 23,* 91–106.

Eisenberg, N., & Strayer, J. (Eds.). (1987). *Empathy and its development.* Cambridge: Cambridge University Press.

Fabian, T. (1993). Fernsehen und Einsamkeit im Alter: Eine empirische Untersuchung zu parasozialer Interaktion [Television and loneliness in the elderly: An empirical investigation of parasocial interaction]. Münster, Germany: LIT.

Freud, S. (1921/1964a). Group psychology and the analysis of the ego. In J. Strachey (Ed. and Trans.), *The standard edition of the complete psychological works of Sigmund Freud* (Vol. 18, pp. 69–143). London: Hogarth Press.

Freud, S. (1923/1964b). The ego and the id. In J. Strachey (Ed. and Trans.), *The standard edition of the complete psychological works of Sigmund Freud* (Vol. 19, pp. 13–66). London: Hogarth Press.

Freud, S. (1905/1906/1987). Psychopathische Personen auf der Bühne [Psychopathic persons on stage]. In A. Richards (Ed.), *Sigmund Freud: Gesammelte Werke. Nachtragsband: Texte aus den Jahren 1885 bis 1938* (pp. 655–661). Frankfurt am Main: Fischer Verlag.

Gaut, B. (1999). Identification and emotion in narrative film. In C. Plantinga & G. M. Smith (Eds.). *Passionate views: Film, cognition, and emotion* (pp. 200–216). Baltimore: Johns Hopkins University Press.

Giles, D. C. (2002). Parasocial interaction: A review of the literature and a model for future research. *Media Psychology, 4*, 279–305.

Gleich, U., & Burst, M. (1996). Parasoziale Beziehungen von Fernsehzuschauern mit Personen auf dem Bildschirm [Parasocial relationships of television viewers with persons on the screen]. *Medienpsychologie, 8*(3), 182–200.

Grodal, T. (2000). Video games and the pleasures of control. In D. Zillmann & P. Vorderer (Eds.), *Media entertainment: The psychology of its appeal* (pp. 197–213). Mahwah, NJ: Lawrence Erlbaum Associates.

Hebb, D. O. (1955). Drives and the C.N.S. (conceptual nervous system). *Psychological Review, 62*, 243–254.

Hoffman, M. L. (1978). Toward a theory of empathetic arousal and development. In M. Lewis & L. A. Rosenblum (Eds.), *The development of affect* (pp. 227–256). New York: Plenum Press.

Hoffman, M. L. (1987). The contribution of empathy to justice and moral judgement. In N. Eisenberg, & J. Strayer (Eds.), *Empathy and its development* (pp. 47–80). Cambridge: Cambridge University Press.

Holland, N. N. (1968). *The dynamics of literary response.* New York: Oxford University Press.

Holland, N. N. (2003, Jan 22). The willing suspension of disbelief: A neuro-psychoanalytic view. *PsyArt: Online Journal for the Psychological Study of the Arts, 7.* Retrieved February 8, 2004, from http://www.clas.ufl.edu/ipsa/journal/2003/hollan06.htm

Horton, D., & Strauss, A. (1957). Interaction in audience-participation shows. *American Journal of Sociology, 62*, 579–587.

Horton, D., & Wohl, R. R. (1956). Mass communication and para-social interaction: Observations on intimacy at a distance. *Psychiatry, 19*, 215–229.

Isotalus, P. (1995). Friendship through screen. *The Nordicom Review of Nordic Research on Media and Communication, 1*, 59–64.

Jose, P. E., & Brewer, W. F. (1984). Development of story liking: Character identification, suspense, and outcome resolution. *Developmental Psychology, 20*(5), 911–924.

Kant, I. (1785/1922). Grundlegung zur Metaphysik der Sitten [Metaphysical foundation of morality]. In *Immanuel Kant's sämtliche Werke* (Vol. 5). Leipzig: Inselverlag.

King Jablonski, C., & Zillmann, D. (1995). Humor's role in the trivialization of violence. *Medienpsychologie: Zeitschrift für Individual- und Massenkommunikation, 7*(2), 122–133, 162.

Konijn, E. A. (1999). Spotlight on spectators: Emotions in the theater. *Discourse Processes, 28*, 169–194.

Lang, P. J. (1979). A bio-informational theory of emotional imagery. *Psychophysiology, 16*, 495–512.

Lang, P. J. (1984). Cognition in emotion: Concept and action. In C. E. Izard, J. Kagan, & R. B. Zajonc (Eds.), *Emotions, cognition, and behavior* (pp. 192–226). Cambridge: Cambridge University Press.

LeDoux, J. E. (1996). *The emotional brain: The mysterious underpinnings of emotional life.* New York: Simon & Schuster.

LeDoux, J. E., & Phelps, E. A. (2000). Emotional networks in the brain. In M. Lewis & J. M. Haviland (Eds.), *Handbook of emotions: Second edition* (pp. 157–172). New York: Guilford.

MacLean, P. D. (1967). The brain in relation to empathy and medical education. *Journal of Nervous and Mental Disease, 144*, 374–382.

MacLean, P. D. (1990). *The triune brain in evolution.* New York: Plenum.

Oatley, K. (1995). A taxonomy of the emotions of literary response and a theory of identification in fictional narrative. *Poetics, 23*(1/2), 53–74.

Perse, E. M., & Rubin, A. M. (1989). Attribution in social and parasocial relationships. *Communication Research, 16*, 59–77.

Peters, J. M. (1989). Het filmische denken [Cinematic thought]. Leuven, Belgium: Acco.

Plutchik, R. (1987). Evolutionary bases of empathy. In N. Eisenberg & J. Strayer (Eds.), *Empathy and its development* (pp. 38–46). Cambridge: Cambridge University Press.

Rubin, R. B., & McHugh, M. P. (1987). Development of parasocial interaction relationships. *Journal of Broadcasting and Electronic Media, 31*, 279–292.

Schachter, S. (1964). The interaction of cognitive and physiological determinants of emotional state. In L. Berkowitz (Ed.), *Advances in experimental social psychology* (Vol. 1, pp. 49–80). New York: Academic Press.

Smith, A. (1759/1971). *The theory of moral sentiments.* New York: Garland.

Smith, M. (1995a). *Engaging characters: Fiction, emotion, and the cinema.* Oxford: Clarendon Press.

Smith, M. (1995b). Film spectatorship and the institution of fiction. *Journal of Aesthetics and Art Criticism, 53*(2), 113–127.

Smith, M. (1997). Imagining from the inside. In R. Allen & M. Smith (Eds.), *Film theory and philosophy* (pp. 412–430). Oxford: Clarendon Press.

Stotland, E. (1969). Exploratory investigations of empathy. In L. Berkowitz (Ed.), *Advances in experimental social psychology* (Vol. 4, pp. 271–314). New York: Academic Press.

Tan, E. S. (1995). Film-induced affect as a witness emotion. *Poetics, 23,* 7–32.

Tan, E. S. (1996). *Emotion and the structure of narrative film: Film as an emotion machine.* Mahwah, NJ: Erlbaum.

Tooby, J., & Cosmides, L. (2001). Does beauty build adapted minds? Toward an evolutionary theory of aesthetics, fiction, and the arts. *SubStance, 94/95,* 6–27.

Turvey, M. (1997). Seeing theory: On perception and emotional response in current film theory. In R. Allen & M. Smith (Eds.), *Film theory and philosophy* (pp. 431–457). Oxford: Clarendon Press.

Vorderer, P. (2000). Interactive entertainment and beyond. In D. Zillmann & P. Vorderer (Eds.), *Media entertainment: The psychology of its appeal* (pp. 21–36). Mahwah, NJ: Erlbaum.

Vorderer, P., & Knobloch, S. (2000). Conflict and suspense in drama. In D. Zillmann & P. Vorderer (Eds.), *Media entertainment: The psychology of its appeal* (pp. 59–72). Mahwah, NJ: Erlbaum.

Walton, K. L. (1990). *Mimesis as make-believe: On the foundations of the representational arts.* Cambridge: Harvard University Press.

Wilder, J. (1957). The law of initial values in neurology and psychiatry: Facts and problems. *Journal of Nervous and Mental Disease, 125,* 73–86.

Wilson, B. J., Cantor, J., Gordon, L., & Zillmann, D. (1986). Affective response of nonretarded and retarded children to the emotions of a protagonist. *Child Study Journal, 16*(2), 77–93.

Wollheim, R. (1987). *Painting as an art.* London: Thames & Hudson.

Zillmann, D. (1978). Attribution and misattribution of excitatory reactions. In J. H. Harvey, W. J. Ickes, & R. F. Kidd (Eds.), *New directions in attribution research* (Vol. 2, pp. 335–368). Hillsdale, NJ: Lawrence Erlbaum Associates.

Zillmann, D. (1979). *Hostility and aggression.* Hillsdale, NJ: Lawrence Erlbaum Associates.

Zillmann, D. (1980). Anatomy of suspense. In P. H. Tannenbaum (Ed.), *The entertainment functions of television* (pp. 133–163). Hillsdale, NJ: Lawrence Erlbaum Associates.

Zillmann, D. (1983). Transfer of excitation in emotional behavior. In J. T. Cacioppo & R. E. Petty (Eds.), *Social psychophysiology: A sourcebook* (pp. 215–240). New York: Guilford Press.

Zillmann, D. (1986). Coition as emotion. In D. Byrne & K. Kelley (Eds.), *Alternative approaches to the study of sexual behavior* (pp. 173–199). Hillsdale, NJ: Lawrence Erlbaum Associates.

Zillmann, D. (1991). The logic of suspense and mystery. In J. Bryant & D. Zillmann (Eds.), *Responding to the screen: Reception and reaction processes* (pp. 281–303). Hillsdale, NJ: Lawrence Erlbaum Associates.

Zillmann, D. (1995). Mechanisms of emotional involvement with drama. *Poetics, 23,* 33–51.

Zillmann, D. (1996a). Sequential dependencies in emotional experience and behavior. In R. D. Kavanaugh, B. Zimmerberg, & S. Fein (Eds.), *Emotion: Interdisciplinary perspectives* (pp. 243–272). Mahwah, NJ: Lawrence Erlbaum Associates.

Zillmann, D. (1996b). The psychology of suspense in dramatic exposition. In P. Vorderer, H. J. Wulff, & M. Friedrichsen (Eds.), *Suspense: Conceptualizations, theoretical analyses, and empirical explorations* (pp. 199–231). Mahwah, NJ: Lawrence Erlbaum Associates.

Zillmann, D. (1998a). *Connections between sexuality and aggression: Second edition.* Mahwah, NJ: Lawrence Erlbaum Associates.

Zillmann, D. (1998b). Does tragic drama have redeeming value? *Siegener Periodicum zur Internationalen Empirischen Literaturwissenschaft, 17*(1), 4–14.

Zillmann, D. (1998c). The psychology of the appeal of portrayals of violence. In J. H. Goldstein (Ed.), *Why we watch: The attractions of violent entertainment* (pp. 179–211). New York: Oxford University Press.

Zillmann, D. (2000a). Basal morality in drama appreciation. In I. Bondebjerg (Ed.), *Moving images, culture and the mind* (pp. 53–63). Luton, England: University of Luton Press.

Zillmann, D. (2000b). Humor and comedy. In D. Zillmann & P. Vorderer (Eds.), *Media entertainment: The psychology of its appeal* (pp. 37–57). Mahwah, NJ: Lawrence Erlbaum Associates.

Zillmann, D. (2003). Affective dynamics, emotions and moods. In J. Bryant, D. Roskos-Ewoldsen, & J. Cantor (Eds.), *Communication and emotion: Essays in honor of Dolf Zillmann* (pp. 533–567). Mahwah, NJ: Lawrence Erlbaum Associates.

Zillmann, D., & Bryant, J. (1975). Viewer's moral sanction of retribution in the appreciation of dramatic presentations. *Journal of Experimental Social Psychology, 11*, 572–582.

Zillmann, D., & Cantor, J. R. (1977). Affective responses to the emotions of a protagonist. *Journal of Experimental Social Psychology, 13*, 155–165.

Zillmann, D., Gibson, R., Ordman, V. L., & Aust, C. F. (1994). Effects of upbeat stories in broadcast news. *Journal of Broadcasting and Electronic Media, 38*(1), 65–78.

Zillmann, D., & Knobloch, S. (2001). Emotional reactions to narratives about the fortunes of personae in the news theater. *Poetics, 29*, 189–206.

Zillmann, D., & Paulus, P. B. (1993). Spectators: Reactions to sports events and effects on athletic performance. In R. N. Singer, M. Murphey, & L. K. Tennant (Eds.), *Handbook of research on sport psychology* (pp. 600–619). New York: Macmillan.

Zillmann, D., Weaver, J. B., Mundorf, N., & Aust, C. F. (1986). Effects of an opposite-gender companion's affect to horror on distress, delight, and attraction. *Journal of Personality and Social Psychology, 51*, 586–594.

Zillmann, D., & Zillmann, M. (1996). Psychoneuroendocrinology of social behavior. In E. T. Higgins & A. W. Kruglanski (Eds.), *Social psychology: Handbook of basic principles* (pp. 39–71). New York: Guilford Press.

14

Mood Management Theory, Evidence, and Advancements

Silvia Knobloch-Westerwick
The Ohio State University

This chapter lays out the basics of mood management theory and describes related empirical research. It also aims to pinpoint some gaps and inconsistencies in the existing evidence to inspire future empirical work. For instance, many studies that have been linked to the mood management framework did not precisely look at moods or content choices as postulated by the theory, or they tested hypotheses that were not included in the original theoretical claim. On the other hand, some of the original suggestions have only rarely been investigated in rigorous terms. Furthermore, some recent advancement from studies on mood-based media choices and related motivations are discussed, especially inasmuch as they help to address challenges from empirical findings to the original theory.

ASSUMPTIONS OF CLASSIC MOOD MANAGEMENT THEORY

Mood Management Goals of Selective Media Exposure

The idea of selecting media content—more specifically, in the interest of enhancing one's feeling—state has been proposed by Zillmann and Bryant (1985) and Zillmann (1988). Initially, these assumptions were referred to as theory of affect-dependent stimulus arrangement (Zillmann & Bryant, 1985), but subsequently gained more prominence under the label of mood management theory (Zillmann, 1988). The idea of media use being guided by motivations to reduce tensions can be traced back to Festinger (1957; see also Zillmann, 1988), who postulated that individuals avoid messages that do not converge with existing attitudes. Although this theory of cognitive dissonance was primarily laid out in cognitive terms, hedonic and emotional aspects can be linked to it. Apparently, the avoidance of dissonant information is not always functional. Oftentimes, new information may be important for adaptation, and even survival, although it can create dissonance. Then efforts to avoid such information should be rooted in hedonic motivations.

The core prediction of mood management theory claims that individuals seek out media content that they expect to improve their mood. Mood optimization in this sense relates to levels of arousal—plausibly, individuals are likely to avoid unpleasant degrees of arousal, namely boredom and stress. By selecting media content, media users can regulate their own mood with regard to arousal levels. For example, after a stressful workday, media consumers will enjoy unwinding by watching a televised travel magazine. Furthermore, the valence of desirable moods is evidently positive. Accordingly, up-cheering media messages are preferable in light of mood optimization interests, thus media consumers should always appreciate fare such as a happy-ending movie or a joyful comedy. In addition, whatever might remind the individual of sources of negative current moods is likely to impair the feeling state and should, according to the theory, be avoided. For example, a student who has experienced an academic set-back might circumvent a campus-centered movie.

The specific hypotheses of mood management theory have been summarized as follows by Zillmann (2000a, p. 104):

> The indicated hedonistic objective is best served by selective exposure to material that (a) is excitationally opposite to prevailing states associated with noxiously experienced hypo- or hyperarousal, (b) has positive hedonic value above that of prevailing states, and (c) in hedonically negative states, has little or no semantic affinity with the prevailing states.

Mood-Impacting Characteristics of Media Stimuli

Mood management theory has been developed to explain and predict media choices, although its principles apply to the broader realm of mood optimization (see, for instance, Thayer, 1996). In order to connect mood states and media content choices, it is important to specify characteristics of the media stimuli that may be sought out or avoided. Although some empirical research that aimed to investigate mood management assumptions merely looked at amounts of media exposure (see following), the theory clearly links mood states to selective exposure to specific kinds of media content. Thus, criteria to describe media stimuli are indispensable. Zillmann (1988) suggested four dimensions on which media content can be differentiated in light of mood management theory.

The first mood-impacting attribute of media messages is the *excitatory potential* (Zillmann, 1988), which is relevant with regard to influences on arousal level. For example, fast-paced music is likely to increase arousal, whereas slow songs usually have a calming effect. Similarly, fast cuts in audio-visual media generally heighten arousal and can thus be considered to have a high excitatory potential, in contrast to video stimuli with fewer cuts and smooth transitions between takes. Besides format, content also relates to the excitatory potential because violent or erotic depictions tend to be more exciting than a classical music concert.

The second dimension to distinguish media messages in the context of mood management theory is the *absorption potential* (Zillmann, 1988). A given feeling state is sustained by ruminating negative incidents or recurring positive events. Media consumption, though, can interfere with a preexisting mood. Depending on the absorption potential of media content, the media exposure will alter the mood—the higher the absorption potential, the more effectively the mood alteration will be. The most absorbing messages will interrupt the rehearsal of thoughts related to the original mood, allowing for a mood change. Accordingly, when in a negative feeling state, individuals should prefer content that effectively disrupts this state.

For instance, when one's spirits are low, this feeling state might be bettered by watching an intriguing mystery because the involvement in the plot dissipates the preexisting mood. In contrast, a positive feeling state should result in seeking non-absorbing messages, or in avoidance of media consumption altogether.

Possibly intervening with this, a third consideration comes into play with *semantic affinity* (Zillmann, 1988) between preexisting mood and media message. An example is the connection between a mood induced by being abandoned by a romantic partner, on the one hand, and a romantic movie on the other hand. If such affinity is high, then media consumption cannot induce a mood change and is more likely to contribute to rehearsal of mood-related thoughts. Such high affinity might be desirable when the current mood is positively valenced, but would noxiously reinforce a negative mood. Accordingly, romantically deprived persons might circumvent any romantic media content because they do not want to be reminded of romantic failures. Although the avoidance of media content associated with sources of negative mood might bring the broad concept of escape via media consumption (Katz & Foulkes, 1962; Pearlin, 1959) to mind, escapism has referred to general states of alienation from individual life circumstances instead of situational mood states.

Lastly, the *hedonic valence* (Zillmann, 1988) of a media message plausibly relates to mood impacts resulting from exposure. Pleasant stimulation via media depictions raises an individual's spirits, whereas sad and disagreeable stimuli evidently should have a down-bringing effect. Complications arise, though, from idiosyncratic interpretations of what presents a positively or negatively valenced message. Oftentimes, a message that is 'good news' to some means 'bad news' to others (Zillmann & Knobloch, 2001). For example, if a political candidate wins an election, this will produce euphoria among his/her partisans but distress for his/her adversaries. Similarly, movies depicting non-traditional female heroines may yield discomfort among viewers with traditional gender roles while being greatly enjoyed by those with more progressive attitudes.

To conclude these considerations, it can simply be noted that mood states and the desired alteration thereof, together with mood-impacting characteristics of media messages, will produce selective exposure patterns of media consumer choices, according to mood management theory. As explained, four dimensions of mood-relevant characteristics of media messages have been suggested, yet it is to be conceded that they are not entirely distinct; for example, an exciting message is likely to be absorbing, too.

Awareness of Mood Management Processes

On the part of the media user, awareness of mood optimization needs does not have to be assumed. It has been argued that, during media consumption, mood management processes may go by-and-large unnoticed by those who act on them—at least very little cognitive elaboration usually takes place (Zillmann, 1985; 1988). Instead of deliberate action, learning processes similar to operant conditioning have been said to create mood management processes (Zillmann & Bryant, 1985). After having experienced mood enhancements through media use, which may have initially been accidental or adopted from observed others, traces of such an experience are stored in an individual's memory. Later encounters of similar media options in similar mood states should then trigger choices that have been reinforced by bettered mood earlier. However, even if higher levels of awareness are assumed, the same mood management processes can be predicted (Knobloch, 2003a), although their origins then require different conceptualizations. Regardless of considerations on awareness levels, empirical evidence has been accumulated for mood management predictions. Yet assumptions on levels of awareness with regard to mood management have great relevance for the methodological design of empirical research. If it is argued that media users are not fully aware of their own mood-management interests, then self-reports appear to be only the last resort for researchers to study related theoretical notions empirically. With such potential limitations in mind, pertinent investigations are reviewed in the following.

EMPIRICAL EVIDENCE

Experiments

A large part of mood management research has been conducted experimentally. In line with Zilllmann's assumption (1985) that such selective exposure patterns oftentimes do not acquire much cognitive attention, methodological examinations frequently relied on unobtrusive observation of behavior instead of introspection of media users. A study by Bryant and Zillmann (1984) provides an excellent example.

Overcoming Boredom and Stress

This investigation examined *arousal states* and induced boredom or stress by having participants either perform tedious manual tasks or complete intellectual exam tasks under time pressure. After this first part of the procedure, participants ostensibly went through a waiting period in which they had the opportunity to watch television. The provided TV set offered six programs from which participants sampled via a technical device. Unbeknownst to participants, this device recorded their viewing selection across time. The featured TV segments had been pre-categorized into three soothing and three stimulating programs. The observed viewing patterns revealed that the stressed participants dedicated about the same portion of the scheduled ostensible waiting period to both program types, whereas bored participants watched next to no relaxing TV content and allotted almost the entire quarter hour to exciting, stimulating TV content. Bryant and Zillmann (1984) interpreted these findings as support for the mood management hypothesis on arousal regulation via media consumption.

Improving Negative Moods

Empirical examination of the mood regulation hypothesis referring to *hedonic valence* of the material is less easy to identify than research on media-aided arousal regulation. Valence of media content often depends on individual interpretations, for example, the same news content can be enjoyed by some but disliked by others (Zillmann & Knobloch, 2001). Also, a sad love story can be depressing to those happy in love but uplifting for the romantically deprived because they see that they are not the only ones facing romantic disenchantment (Knobloch & Zillmann, 2003). Moreover, valence can be confounded with other mood-impacting content characteristics, for instance, fast-paced music might tend to signal positive spirits but also high levels of energy and arousal. These complexities have been reflected in results on the valence dimension of mood management.

For an example, in an experiment by Knobloch and Zillmann (2002), participants were free to choose from pop music after they received feedback about an alleged test on social skills, which was actually a mood induction to create bad, mediocre, or good feeling states. The songs, offered via a computerized jukebox, had been pre-tested for levels of joyfulness and energy of the musical expression—yet evaluations of these properties were correlated in the pretest of the top-charts musical selections. Participants who had been placed in a bad mood spent more time on energetic-joyful music than those in a mediocre mood, who, in turn, also dedicated more time to uplifting music than the respondents in the good-mood condition. Due to the association between "energy," which could resemble the absorption potential of music, and "joyfulness" (valence) of the musical stimuli, it is not entirely clear whether participants aimed to disrupt or to improve negative moods—or do both simultaneously—by favoring energetic-joyful songs.

Arousal Management Versus Mood Enhancement

Two studies (Biswas, Riffe, & Zillmann, 1994; Knobloch, 2002) actually pre-tested the presented media stimuli to ensure that they differed with regard to valence while being similarly "interesting" as an item that can be connected to absorption potential. Both used the same mood induction procedure (adopted from Zillmann, Hezel, & Medoff, 1980), although the more recent investigation employed a computerized version of it. In the study by Biswas et al. (1994), only female participants' behavior complied with mood management predictions, whereas males in negative moods tended to favor negative content, if anything. In Knobloch's (2002) investigation, participants in a negative mood in fact spent more time on positively valenced web pages than participants in a mediocre mood did. However, counter to expectations, respondents in good moods were apparently less concerned with the valence aspect of their media choices— the time they spent on positive web content fell between the amounts of the experimental conditions for mediocre and negative moods and showed a more erratic pattern across time. In brief, it appears that hedonic valence of media stimuli relates to other mood-impacting characteristics and may not be the most important feature, at least not when a media user is in a good mood anyway. This might be the case when respondents in a "good mood" condition lack options for "selective exposure to material that . . . has positive hedonic value *above* that of prevailing states" (Zillmann, 2000a, p. 104, emphasis added).

Disrupting Negative Moods—Anger and Fear

Yet more complexities emerged in mood management studies that pertained to the third hypothesis of the theory on the role of *semantic affinity*. Here, further differentiation of stimuli and kinds of mood—beyond positive and negative valence—is crucial. An experiment by Zillmann, Hezel, and Medoff (1980) employed a mood induction procedure (later adopted by Biswas et al., 1994; Knobloch & Zillmann, 2002, and Knobloch, 2002) and then offered participants to selectively view sitcom, action drama, or game show programs. Although the initial mood also impacted entertainment choices significantly, selective exposure patterns were not fully in line with the mood regulation hypothesis referring to valence of the material. For instance, those in a negative mood favored comedies the least. Zillmann et al. (1980) interpreted this as a result from the excitatory potential and the absorption potential that the programs probably featured besides the different valences of these genres. Moreover, they reasoned that semantic affinity to the origins of a negative mood might have played a role. As much of the humor in comedy stems from disparagement and hostile actions (Zillmann, 2000b), people in a bad mood may interpret this to be connected to a negative experience they just had during the mood induction phase. Thus the confoundings of mood-impacting characteristics of media content form a considerable complication when testing assumptions of mood management theory.

To shed more light on the processes underlying Zillmann et al.'s (1980) results, Medoff (1979, 1982) distinguished between provocation and frustration as sources of negative moods, while including a neutral control group also, and encountered great gender differences in subsequent selections of hostile and non-hostile comedy. Women's behavior mostly corroborated mood management hypotheses, as they generally preferred positively valenced comedy without hostility. In contrast, only men in the neutral control group favored non-hostile comedy, but frustrated men spent the most time with hostile comedy and provoked men abstained from comedy consumption altogether. These contradictions in males' selection behavior have been considered bewildering (Zillmann, 1988) and constitute one of the frequent observations that the genders may employ different mood regulation approaches.

In addition to aggressive inclinations, fear and its influence on media choices have also been investigated. Wakshlag, Vial, and Tamborini (1983) made fears salient to half of their research participants by showing a crime documentary, whereas the control group saw an innocuous film. This fear manipulation proved to be successful, although women were more fearful in general. After the treatment, participants chose seven out of 14 film descriptions, and they were told they would have a part in the choice for subsequent viewing. The film descriptions were manipulated to signal movie content with different levels of justice restoration and of victimization. Raters had evaluated the indication of these dimensions in the film descriptions in order to assign victimization and justice restoration scores accordingly to them. The results showed that participants with increased fear levels preferred films with low victimization scores and with high justice scores. According to Wakshlag et al. (1983), these findings show that media users aim to minimize stimuli that are associated with the source of a negative mood. This could be interpreted in two ways. First, media users favored comparatively positive content that featured less violence and more justice, and was thus better in terms of hedonic valence, which is in line with the second mood management hypothesis. The second approach is somewhat more problematic. One could argue that the participants avoided content with semantic affinity to the source of their fear, in line with the third mood management hypothesis. Yet all media choices offered had some affinity with that source, although with varying degrees. Accordingly, the design used by Wakshlag et al. (1983) does not lend itself to testing this hypothesis, which was also not their key interest. Yet their findings have been interpreted within the framework of mood management (Zillmann & Bryant, 1985).

Quasi-Experiments

Relief for the Lonely and the Lovelorn

The importance of media content's semantic affinity with adversary moods has been touched upon in a few quasi-experimental studies—respondents reported personal strains and could subsequently choose from media offerings. For example, young adults were inquired about their romantic happiness as part of an ostensible campus socialization questionnaire (Knobloch & Zillmann, 2003; Knobloch, Weisbach, & Zillmann, 2004). Then, they sampled from eight love songs, presented via computer, of which four featured sad love lyrics and four songs contained happy romantic lyrics. Young adults who were romantically deprived exposed themselves significantly longer to sad love music, apparently to avoid hearing about more successful peers. A study by Mares and Cantor (1992), which investigated loneliness of elderly viewers and their viewing preferences, yielded similar results. Lonely elderly viewers preferred to see TV programs on lonely and unhappy elderly protagonists, whereas the opposite applied to socially integrated participants.

Yet, at closer inspection, these studies offer only limited insight into Zillmann's suggestions that "the indicated hedonistic objective is best served by selective exposure to material that . . . [,] in hedonically negative states, has little or no semantic affinity with the prevailing states" (Zillmann, 2000a, p. 104). Participants were only offered media content that actually had semantic affinity with one's own situation, though depicting this life domain in positive or negative light. Thus personal circumstances instead of moods were examined, and furthermore participants had no way to fully circumvent related content. With these procedures, the postulation of avoidance of media content that has to do with origins of a current negative mood cannot be tested rigorously.

When Utility Overrides Hedonism

Two other studies did include content choices without semantic affinity to troubling life aspects. Both ascertained individual satisfaction for a variety of life domains and then offered research participants to select from media content associated with these domains. Trepte, Zapfe, and Sudhoff (2001) conducted a survey among adolescents who reported perceived problems due to parents, peers, romantic relationships, and their own looks. When choosing from TV talk show vignettes that featured titles related to these sources of pressure, the adolescents favored talk shows on their current problems according to simple correlations reported by Trepte et al. (2001). In a quasi-experiment by Hastall, Rossmann, and Knobloch (2004), students completed a life satisfaction questionnaire before they browsed online news sections, which were related to the life domains covered in the questionnaire. Participants who reported low satisfaction for specific domains (studies, romantic relationship) spent more time on content about these areas. These findings contradict Zillmann's notion of avoidance of one's one problems as depicted in the media. It seems more as if media users aim to find guidance (Atkin, 1973) from the media in order to solve their problem. The described studies presented information content as stimuli, which should also qualify as fare suitable for mood management purposes (e.g., Biswas et al., 1994). Yet Zillmann (1988, 2000a) referred to negative mood states instead of more enduring problems. Thus it may present a leap if troubling life situations are connected to current moods.

Field Studies

Relief From Stress Through Entertainment Use

In contrast to the studies described above, field studies on mood management examined selective exposure to both entertainment and information fare, with participants reporting moods and TV consumption in diaries at specific times of the day for an extended period. For instance, Anderson, Collins, Schmitt, and Jacobvitz (1996) found that stress resulted in longer TV consumption. Stressed participants spent more time on TV entertainment, such as comedy, and neglected news and documentaries under such circumstances. Men and women differed in their choices when under stress—stressed women watched more game shows and variety programs, and stressed men preferred action programs. Brosius, Roßmann, and Elnain (1999) used a similar design for a German sample and also came to the conclusion that stressed viewers favor entertainment while viewers without such strains spend more time on information.

Total Exposure Versus Content Matters

Interpretations of field study findings occasionally overlooked that mood management theory does not relate to *total* amounts of media consumption but, instead, links mood states with choices of media content *categories*. In other words, based on the original theory, no predictions can be made on how much time is spent with television or another medium while not differentiating the type of favored content. Yet Schmitz, Alsdorf, Sang, and Tasche (1993) called mood management into question in light of results from their diaries-based field study. However, this examination did not distinguish kinds of TV content, and only looked at total TV consumption. Similarly, Donsbach and Tasche (1999) placed their investigation in the context of mood management theory, but connected reported mood states only to total extent of TV exposure, which does not allow rigorous testing of mood management assumptions. The problem of lack of TV content differentiation also applies to studies that employed the experience sampling method (Kubey, 1986; Kubey & Csikszentmihalyi, 1990)—although moods and TV exposure were ascertained at random times during the day, the data cannot shed light on mood management processes due to a conceptual mismatch.

Surveys

Investigating mood management phenomena via surveys is problematic because lay rational-izations and social desirability of some responses are likely to bias self-reports (Zillmann, 1985). However, the extent of this problem certainly depends on how transparent the proce-dure might be to respondents. Occasionally, respondents have been placed in very contrived situations, for instance, they were asked to imagine a setting and to describe the kind of music they would like to hear in this setting with semantic differentials (e.g., Gembris, 1990). Other examples include studies in the uses-and-gratifications context that regularly ask survey re-spondents whether they consume various media in order to relax, overcome boredom, and so fourth (e.g., Rubin, 1984). It can be queried that such data are valuable for mood management research because they possibly reflect lay theorizing on mood-related media choices more than actual mood management behavior, however, other survey procedures did not reveal the research interest this openly. Interestingly, biological factors such as hormone levels and per-sonality traits (which are considered as rooted in biological structures by Eysenck, 1990, and Zuckerman, 1991), have emerged as linked to selective media exposure. These patterns are not surprising when adopting the notion that moods result from biological origins (Thayer, 1996).

Overcoming Exasperation From Hormonal Phases

For instance, Meadowcroft and Zillmann (1983) were interested in whether the menstrual cycle influenced females' preferences for TV genres. In this survey, female college students reported their interest in seeing various TV programs, presented in a list, on the same evening. At the end of the questionnaire, they were asked to provide information regarding their menstrual cycle. Premenstrual and menstrual women indicated stronger interest for comedy than other females who were in the middle of their cycle. The researchers inferred that women aim to overcome noxious mood states, resulting from hormonal phases, by watching comedy. This survey corroborates mood management assumptions based on unobtrusive measures. Helregel and Weaver (1989) presented parallel results for mood states during pregnancy.

Mood Tendencies Reflected in Personality Traits

Another unobtrusive approach to study mood management is the connection of selective media use with personality traits, because it should not be transparent to respondents that mood management processes are under investigation. Yet this approach, again, suffers from the limitation that specific situations and related moods cannot be examined, although this is what mood management theory is about. Inferences on mood management based on personality data depend on the assumption that certain personality types are more likely to experience certain moods, and thus experience of these moods is also more likely when choosing from media fare.

An investigation by Dillman Carpentier, Knobloch, and Zillmann (2003) ascertained partici-pants' rebelliousness, and associated these personality measures with actual selective exposure to rebellious and non-rebellious music. Individuals who scored high in proactive rebelliousness spent more time on rebellious music. Even though such music generally stresses conflict and tension, specific personality types preferred listening to negative lyrics.

More frequently, though, personality traits have been linked to reported genre preferences—either music types (e.g., Litle & Zuckerman, 1986; Dollinger, 1993) or television formats (e.g., Burst, 1999; Brosius & Weaver, 1994). For instance, Burst (1999) connected the 'Big Five' traits (Costa & McCrae, 1985) and sensation seeking (Zuckerman, 1984) with TV genre preferences. Only openness to experience was (positively) associated with use of information genres. Sensation seeking was positively associated with use of erotica on TV, as well as with a preference for thriller/action/sci-fi genres. On the contrary, neuroticism, agreeableness, and, to a smaller degree, conscientiousness showed a negative correlation with this genre bundle. Agreeableness was also positively linked with relationship genres (e.g., soaps and romance). These findings could be interpreted as follows: Individuals with a high sensation-seeking motive are more frequently bored, and thus turn to arousing TV genres more often than other people do. In contrast, individuals who score high on traits such as neuroticism or agreeableness favor lower levels of arousal and tension than sensation seekers do, thus their avoidance of action genres stands to reason from a mood management perspective. In short, Burst's (1999) findings are in line with the mood management theorizing, although Burst employed trait measures instead of state measures, while the theory focuses on situations.

RECENT DEVELOPMENTS

Mood Management Through News Consumption

Field studies were not the first approach to mood management by selective exposure to media content including news and information. Zillmann (1988) had noted that the theory as such encompasses communication genres and thereby includes selections of news, documentaries, and sports. Whereas Mares and Cantor (1992) had employed documentaries, the first investigation that studied the proposal that selective exposure to news depends on feeling states was presented by Biswas, Riffe, and Zillmann (1994). Their research procedure was mentioned earlier—after a mood induction, participants were asked to select six out of twelve news-magazine article pages for subsequent reading. For these articles, a clear valence had been established, either positive or negative, in a pretest that also ensured equal levels of 'interesting' ratings for all selections. As mentioned, women in negative moods favored positive messages more than females in a positive feeling state. Yet males' preferences did not fall in line with mood management predictions because men in negative states showed an unexpected interest in negative messages, although this tendency did not reach significance. While these findings corroborate mood management assumptions, the gender differences are not accounted for by the theory.

Mood Management Through Internet Use

With the advent of the World Wide Web, media content is now readily available around-the-clock for convenient selection (Wirth & Schweiger, 1999). Only two studies have examined mood management during WWW use thus far.

An investigation by Mastro, Eastin, and Tamborini (2002) tested the assumptions that stressed individuals would have more WWW page hits and would favor more relaxing content than bored participants. After a mood induction, participants were free to surf the Internet during an ostensibly unrelated research session. Visited sites and exposure times were logged, coders rated the sites as more or less stimulating/relaxing. Mastro et al. (2002) found no support for assumptions that can be inferred directly from mood management theory—bored surfers did not prefer content that coders considered to be more stimulating. Yet the other hypothesis gained evidence, because bored participants visited significantly more sites than their stressed counterparts. Although frequency of content sampling has not been addressed in original mood management theorizing (Zillmann, 1988), another study also found it to be related to feeling states—Knobloch and Zillmann (2002) revealed that, when in a bad mood, music listeners made fewer selections from the same set of choices during the same time interval than others in better moods. Apparently, differences in the valence dimension yield different selective exposure patterns than arousal variations.

The second investigation on mood management while surfing online was a field study presented by Knobloch (2002). After a computerized version of Zillmann et al.'s (1980) mood induction procedure, respondents were asked to probe a new Internet portal in a purportedly unrelated study. This portal featured eight sites in a navigation bar, each illustrated with a thumbnail-size screenshot image. The sites had been categorized in a pretest and included two positively valenced entertainment sites, two negatively valenced entertainment sites (e.g., horror), two positively valenced information sites, and two negatively valenced information sites. As mentioned earlier, negative moods resulted in longer exposure on positively valenced web-pages than mediocre moods, but in contrast to mood management theory, exposure to positive content of those in a good mood fell between the amounts of the conditions for mediocre and negative moods. A hypothesis that goes somewhat beyond the original theory, though, was corroborated—respondents in negative mood states favored entertainment sites more than the mediocre mood condition, which, in turn, spent more time on entertainment than the experimental group for good mood.

Thus the two studies on mood management through WWW use did not produce unequivocal support for original assumptions of mood management theory. The reasons could be seen in the coding of how stimulating web pages are—the validity of such coding can be called into question—and in the relative importance of the various mood-impacting stimuli characteristics (as discussed earlier). Yet, two derived considerations were confirmed, namely the selection frequency being dependent on arousal levels and the genre choice—information versus entertainment—as function of mood valence.

CHALLENGES AND ADVANCEMENTS

The theoretical proposition of mood management theory has been faced with challenges. Especially the use of negative media content and the oftentimes emerging gender differences in mood-related media choices call the general applicability of mood management theory into question (see also Zillmann, 2000a; Oliver, 2003).

Exposure to Down-Bringing Content

Negative media genres include tragedies that evoke sadness, horror movies that instigate fear and disgust, as well as news on disasters and enduring problems. Although other genres also produce emotions of distress, such as suspense from thrillers and uncertainty from mystery, they do not pose a great paradoxon for mood management explanations. It can be argued that media consumers know about the enjoyment of relief and resolution that are to come about eventually for these genres (Knobloch, 2003b), which should even be magnified by the prior distress (Zillmann, 1996). But exposure to the other formats mentioned above—tragedy, horror, and deplorable news— are not readily explained with mood management considerations because they typically do not provide positive endings (King & Hourani, 2000; Metzger, 2000; Oliver, 1993). Any negative media content that does not provide a satisfying outcome violates hedonistic principles. Some possible explanations for selective exposure to such media stimuli is closely connected to mood management theory and related considerations outlined above.

As already discussed, the mood-impacting characteristic of media depictions will often be confounded in real-life stimuli. This also applies to negative messages because they tend to create higher arousal and are probably more absorbing in comparison to positive messages (see Frijda, 1988). Accordingly, media users who indulge in negative media depictions might actually do so because they seek to heighten their arousal or forget sources of on-going negative feeling states.

Exposure to content that is negative throughout, including the ending, might also just happen unintentionally because the media consumer chooses a message while not anticipating a tragic or horrifying outcome. Such an occurrence would not contradict mood management theory because its assumptions pertain to media use motivation and do not include actually attained effects. These effects may not always be what the individual had anticipated and desired.

Yet a variety of additional explanations with linkages to mood management have been brought up in the literature. Catharsis might be the most prominent one, as the notion has been around since the days of Aristotle (350 BC/1961). According to this concept, either sadness or aggressions are purged by vicarious experience of others tragic experiences or violent behavior. Such a mood-bettering effect, possibly attained via media use, would be of great interest for an individual that aims to terminate an unpleasant mood state. In spite of the enduring appeal to lay theorizing, no convincing empirical support has been presented for it (for a review, see Zillmann, 1998b). The catharsis idea also contradicts the notion of priming (Roskos-Ewoldsen, Roskos-Ewoldsen, & Dillmann Carpentier, 2002) according to which a stimulus evokes and intensifies related thoughts instead of diminishing them.

Other suggested approaches can be grouped into two broad domains—(a) feeling positively about something that is generally deemed negative, and (b) accepting or even seeking unpleasant feeling states for the sake of their functionality in a given situation (mood adjustment).

Seeking out the Negative to Feel Good

One explanation can be derived from Zillmann's affective disposition notion (e.g., Zillmann, 1994). Media users enjoy to witness the fortunes of liked protagonists and also the misfortunes of disliked antagonists. Thus any media depiction—comedy, thriller, news, and so on—should produce mood enhancement as long as likable characters are shown to overcome challenges. Although some characters may suffer terribly in a usually deplorable fashion, this would lift the viewers' spirits nonetheless, as long as these characters are disliked. In this sense, even negative events as shown in the media may be interpreted as positive. In fact, comedies employ this principle abundantly—why else may we find ourselves laughing about fairly brutal slapstick shows (Zillmann, 2000b)? No empirical data on the effects of these phenomena on actual media selections are available, although it seems plausible that the intuitive knowledge thereof intervenes with mood management processes.

Another explanation for selective exposure to negative media content was first explored by Wills (1981), which actually instigated a whole new line of research in social psychology. Wills (1981) suggested that the ample amount of negative news is enjoyable because it offers an opportunity for what he called downward comparison. Extending Festinger's (1954) concept of social comparison, this idea stresses the effect of mood enhancement that could result from comparing oneself with others who are in a worse situation. The news certainly depicts ample individuals in the most deplorable circumstances, so anyone in a normal living situation might derive relief from news consumption—after all, one's own problems should appear trivial in comparison. Although several scholars agreed with the interpretation of downward comparison and extended it to entertainment genres (Vorderer, 1996; Oliver, 1993), only little empirical evidence on actual content selection is at hand. The findings by Mares and Cantor (1992) and Knobloch and collaborators (Knobloch & Zillmann, 2003; Knobloch, Weisbach, & Zillmann, 2004; see section "Relief for the Lonely and the Lovelorn") render support to the downward comparison principle in media content selections. Much social–psychological research has linked self-esteem to downward comparison choices (e.g., Wills, 1991), thus a study by Knobloch-Westerwick and Hastall (submitted) gives this explanation for selective exposure to negative depiction even stronger support. This investigation found that respondents' self-esteem influenced selective exposure to news stories about individuals in unambiguously positive versus negative circumstances. These findings fit assumptions on social comparison, where seeing others in worse conditions than the self can better one's own feeling states. On the other hand, it can be argued that seeing others in positive situations gives hope and instigates empathic positive feelings. More research is needed on this phenomenon.

Some scholars have suggested that depictions of negative situations in the media can produce negative emotions, and then, subsequently, positive appraisals thereof. Although almost no data on actual media content-choices are at hand, the various suggestions and implications for mood management will be addressed in the following. Oliver (1993) explained enjoyment of sad movies with a concept labeled meta-emotions. According to her, an experience of sadness may be perceived as valuable and gratifying by some media users, who, in turn, should then seek out tragedies more often than others. A leaning to find sadness rewarding on a meta-level, as cognitive appraisal, appears likely for those who have been rewarded for showing sadness and empathic distress—typically women (Oliver, 2003). Mills (1993) suggested that feeling sad due to watching a tragedy might also serve to confirm that the self features empathy as favorable trait. In this case, media users would indeed employ media in order to 'work' on their self and their identity, as Vorderer (1996) indicated. Zillmann (2000a) also pondered the idea that some individuals may assess the indulgence in light-hearted media entertainment to be despicable, and thus favor downcast content to gain a positive concept of their selves.

Mood Adjustment

Circumstances where more-or-less disagreeable moods appear functional have been discussed by Knobloch (2003a). Essentially, individuals may perceive a task or situation as such that associated requirements are more easily met in unpleasant moods. Examples include taking an exam, where quiet concentration promises a better outcome than pleasant distraction, or an encounter with a disliked supervisor to address a conflict of interests, during which a sarcastic, humorous mood might be very enjoyable but also very dysfunctional. Oftentimes, not mood optimization, but mood adjustment, to current or upcoming situational requirements could guide mood regulation motives during media consumption. Parameters in the related media use behavior have been addressed by Knobloch (2003a)—perceptions and anticipations of situational requirements regarding mood, suitability of media stimuli to induce a desired mood, as well as use of time spans to adjust moods, should all influence the mood adjustment process.

Evidence from several studies corroborate these conceptualizations (see Knobloch, 2003a). An illustration can be drawn from Knobloch-Westerwick and Alter (in press). Research participants were placed in negative mood states by ostensible test feedback from a confederate, and either led to believe they would get a chance to retaliate against the provoker, or did not receive such an indication. After these experimental inductions, participants were free to sample form online news in a purportedly unrelated study. The presented news stories had been categorized into bad news and good news on empirical grounds. The results showed that women who anticipated a retaliation opportunity preferred good news, whereas men favored more bad news before the retaliation opportunity. The genders did not differ in the experimental group that had not received an announcement on a retaliation opportunity. These findings can be interpreted as follows: Women prevent aggressive behaviors by soothing their moods, because such behavior is not socially acceptable for females and probably contradicts values women were socialized to adhere to. In contrast, Western societies deem it 'unmanly' not to pay back a provoker. Thus, men aim to maintain levels of anger if they foresee a retaliation opportunity.

Mood management studies have frequently revealed different selective exposure patterns for the genders. As explained above, mood-dependent selections of good and bad news (Biswas et al., 1994), comedy selections (Medoff, 1979), and, moreover, mood adjustment processes (Knobloch-Westerwick & Alter, in press) differed by gender. As a general rule, women comply with the assumptions of mood management theory, whereas men occasionally do not. More examples on gender differences in mood management could be drawn from other experiments, as well as field studies. In light of mood adjustment considerations, it can be speculated that oftentimes anticipated mood requirements interfere with mood optimization motivation and that these mood requirements differ by gender because of emotion-related gender stereotypes (Fischer, 2004). This reasoning could explain the frequent gender differences in mood-dependent media selections.

IDEAS ON FURTHER DEVELOPMENT OF MOOD MANAGEMENT RESEARCH

Mood management theory has inspired a whole body of research. Given its charming parsimony, it is astonishing that, at closer inspection, some of its claims have not been tested rigorously up to this point. Some mixed evidence exists for the hypothesis that media users prefer content with hedonic valence above current moods, yet, for the most part, valence and excitatory potential and absorption potential were not clearly differentiated in pretests of the employed material. This may be due to the complex nature of media messages, where clear-cut categorization is often quite challenging. Furthermore, there is apparently no single study at hand that strictly tested the claim on semantic affinity and produced unequivocal findings. All investigations that can be linked to this claim either turned up bewildering gender differences, looked at life strains instead of moods, or did not offer selections that had absolutely no affinity with the mood or strains. It is obvious that an unclouded investigation of the third mood management hypothesis is lacking badly.

Beyond this necessary groundwork, some other directions can be derived from the presented studies. Extensions of original mood management theory were presented in investigations that demonstrated how selection frequency, as well as preferences regarding entertainment and information, depend on mood states. Additional research might yield further dependent variables of interest.

Some criticism can be raised regarding the setup of mood management experiments, in hopes of inspiring designs of future research on mood management. It is important to emphasize the conceptual difference between an emotion and a mood state. Emotions are object-related and relatively short-lived, whereas causes of the more enduring moods are more obscure (Schwarz & Clore, 1996). In other words, specific events, possibly internal, instigate emotions. Once the emotion has dissipated, it may still leave traces that linger on in the individual's mood. While mood management theory has been devised with regard to moods, it could be argued that some of the related empirical studies were more concerned with actual emotions. One might consider that the so-called mood induction methods were more likely to instigate emotions that related to particular events. The common procedure to declare the mood induction part as an unrelated study might accomplish that research participants perceive a change of situation. Yet, it can be called into question if these procedures in fact impacted moods, instead of prompting an emotional reaction. Whether the produced affective states fall into one or the other category—emotion or mood—would depend on whether participants were still pondering the object that triggered the affective change. As Knobloch and Zillmann (2002) noted, a computerized procedure is preferable because it hardly instigates motivational implications towards the source of mood. On the other hand, in studies like the one presented by Medoff

(1982) or Biswas et al. (1994), for instance, an experimenter induced negative moods by provoking respondents and then conducted the session part where respondents were to select media stimuli. Thus, the object that evoked the negate affective state remained present, and participants were probably in an emotional state linked to that object (the experimenter) instead of a mood state with only diffuse motivational implications. It is possible that participants, especially men, were more in an angry emotion than in a diffuse negative mood. In this case, the employed research procedures are not ideal for testing mood management predictions.

For further development of mood management theory, I consider the differentiation of negative moods as crucial. Negative moods are those that primarily call for mood repair efforts and thus lend themselves to further investigation. In the original postulations on mood management by Zillmann (1988), the valence dimensions simply referred to negative versus positive moods. However, it appears highly plausible that different kinds of negative moods will prompt different mood enhancing strategies. For instance, fear is associated with 'flee' instincts, whereas anger triggers 'fight' instincts. Psychological research has started to look further into the implications of distinct negative emotions (Lerner & Dacher, 2000), and related differentiations are likely to be fruitful for mood management research, too.

While ample evidence has supported mood management notions, various gaps still exist and should be filled in the future. Advancements and extensions of the original theory have been proposed and also deserve further investigation and discussion. Above all, it is important that media scholars take note of the strong relation that mood states have to media consumption— large amounts of communication research have neglected the role of affect.

REFERENCES

Anderson, D.R., Collins, P.A., Schmitt, K.L., & Jacobvitz, R.S. (1996). Stressful life events and television viewing. *Communication Research, 23*(3), 243–260.

Aristotle. (1961). *Poetics* (S. H. Butcher, Trans.). New York: Hill & Wang. (original work published 350 BC)

Atkin, C. (1973). Instrumental utilities and information seeking. In P. Clarke (Ed.), *New models for mass communication research* (pp. 205–242). Beverly Hills, CA: Sage.

Biswas, R., Riffe, D., & Zillmann, D. (1994). Mood influence on the appeal of bad news. *Journalism Quarterley, 71*(3), 689–696.

Brosius, H.-B., & Weaver, J.B. (1994). Der Einfluss der Persönlichkeitsstruktur von Rezipienten auf Film- und Fernsehpräferenzen in Deutschland und den USA [translation: The influence of viewers' personality structure on movie and televison preferences]. In L. Bosshart & W. Hoffmann-Riem (Eds.), *Medienlust und Mediennutz. Unterhaltung als öffentliche Kommunikation* (pp. 284–300). Munich, Germany: Ölschläger.

Brosius, H.-B., Roßmann, R., & Elnain, A. (1999). Alltagsbelastung und Fernsehnutzung [translation: Daily strains and televison use]. In U. Hasebrink & P. Rössler (Eds.), *Publikumsbindungen: Medienrezeption zwischen Individualisierung und Integration* (pp. 167–187). Munich, Germany: Reinhard Fischer.

Bryant, J., & Zillmann, D. (1984). Using television to alleviate boredom and stress: Selective exposure as a function of induced excitational states. *Journal of Broadcasting, 28*(1), 1–20.

Burst, M. (1999). Zuschauerpersönlichkeit als Voraussetzung für Fernsehmotive und Programmpräferenzen [translation: Viewer personality as determinant of viewing motives and program preferences]. *Medienpsychologie, 11*(3), 157–181.

Collins-Standley, T., Gan, S., Jessy Yu, H.-J., & Zillmann, D. (1996). Choice of romantic, violent, and scary fairy-tale books by preschool girls and boys. *Child Study Journal, 26*(4), 279–301.

Costa, P. T., & McCrae, R. R. (1985). *The NEO personality inventory*. Odessa: Psychological Assessment Resources.

Dillman Carpentier, F., Knobloch, S., & Zillmann, D. (2003). The rebellion in rock and rap. A comparison of traits predicting selective exposure to rebellious music. *Personality & Individual Differences, 35*(7),1643–1655.

Dollinger, S. (1993). Research note: Personality and music preference: Extraversion and excitement seeking or openness to experience? *Psychology of Music, 21*, 73–77.

Donsbach, W., & Tasche, K. (1999, August). *When mood management fails. A field study on the relationships between daily events, mood, and television viewing.* Paper presented at the convention of the International Communication Association, San Francisco.

Eysenck, H. J. (1990). Biological dimensions of personality. In L. A. Pervin (Ed.), *Handbook of personality: Theory and research* (pp. 244–276). New York: Guilford

Festinger, L. (1954). A theory of social comparison processes. *Human Relations, 7*, 117–140.

Festinger, L. (1957). *A theory of cognitive dissonance.* Stanford, CA: Stanford University Press.

Fischer, A. (2004). *Gender and emotion.* Cambridge, UK: Cambridge University Press

Frijda, N.H. (1988). The laws of emotion. *American Psychologist, 43*(5), 349–358.

Gembris, H. (1990). Situationsbezogene Präferenzen und erwünschte Wirkungen von Musik. *Jahrbuch Musikpsychologie, 7*, 73–95.

Hansen, C. H., & Hansen, R. D. (2000). Music and music videos. In D. Zillmann & P. Vorderer (Eds.), *Media entertainment: The psychology of its appeal* (pp. 175–196). Mahwah, NJ: Lawrence Erlbaum Associates.

Hastall, M., Rossmann, M., & Knobloch, S. (2004, May). *Approach or avoidance? Selective exposure to information on distressing issues.* Paper presented at the International Communication Association conference, New Orleans.

Helregel, B. K., & Weaver, J. B. (1989). Mood-management during pregnancy through selective exposure to television. *Journal of Broadcasting & Electronic Media, 33*(1), 15–33.

Herzog, H. (1944). What do we really know about daytime serial listeners? In P. F. Lazarsfeld & F. N. Stanson (Eds.), *Radio research, 1942–1943* (pp. 3–33). New York: Duell, Sloan & Pearce.

Katz, E., & Foulkes, D. (1962). On the use of the mass media as escape: Clarification of a concept. *Public Opinion Quarterly, 26*(3), 377–388.

King, C., & Hourani, N. (2000, July/Aug.). *Don't tease me: Effects of ending type on horror film enjoyment.* Paper presented at the Congress of the International Society for the Empirical Study of Literature (IGEL), Toronto.

Knobloch, S. (2002). 'Unterhaltungsslalom' bei der WWW-Nutzung: Ein Feldexperiment [translation: 'Crisscrossing through entertainments' while surfing the WWW: A field experiment]. *Publizistik, 47*(3), 309–318.

Knobloch, S. (2003a). Mood adjustment via mass communication. *Journal of Communication, 53*(2), 233–250.

Knobloch, S. (2003b). Suspense and mystery. In J. Bryant, D. Roskos-Ewoldsen, & J. Cantor (Eds.), *Communication and emotion* (pp. 379–395). Mahwah, NJ: Lawrence Erlbaum Associates.

Knobloch, S., & Zillmann, D. (2002). Mood management via the digital jukebox. *Journal of Communication, 52*(2), 351–366.

Knobloch, S., & Zillmann, D. (2003). Appeal of love themes in popular music. *Psychological Reports, 93*, 653–658.

Knobloch, S., Weisbach, K., & Zillmann, D. (2004). Love lamentation in pop songs: Music for unhappy lovers? *Zeitschrift für Medienpsychologie, 16*(2), 116–124.

Knobloch-Westerwick, S., & Alter, S. (in press). *Mood adjustment to social situations through mass media use: How men ruminate and women dissipate angry moods. Human Communication Research.*

Knobloch-Westerwick, S., & Hastall, M. R. (submitted). *Selective social comparison with media personae: How gender, age, and self-esteem influence selective exposure to good and bad news.*

Kubey, R. W. (1986). Television use in everyday life: Coping with unstructured time. *Journal of Communication, 36*(6), 108–123.

Kubey, R., & Csikszentmihalyi, M. (1990). Viewing as cause, as effect, and as habit. In R. Kubey & M. Csikszentmihalyi (Eds.), *Television and the quality of life: How viewing shapes everyday experience* (pp. 119–148). Hillsdale, NJ: Lawrence Erlbaum.

Lerner, J. S., & Keltner, D., (2000). Beyond valence: Toward a model of emotion-specific influences on judgement and choice. *Cognition & Emotion, 14*(4), 473–493.

Litle, P., & Zuckerman, M. (1986). Sensation seeking and music preferences. *Personality and individual differences, 7*(4), 575–578.

Mares, M.-L., & Cantor, J. (1992). Elderly viewers' responses to televised portrayals of old age. *Communication Research, 19*(4), 459–478.

Mastro, D.E., Eastin, M.S., & Tamborini, R. (2002). Internet search behaviors and mood alterations: A selective exposure approach. *Media Psychology, 4*, 157–172.

Meadowcroft, J.M., & Zillmann, D. (1987). Women's comedy preferences during the menstrual cycle. *Communication Research, 14*(2), 204–218.

Medoff, N.J. (1979). *The avoidance of comedy by persons in a negative affective state: A further study in selective exposure.* Unpublished doctoral dissertation, Indiana University.

Medoff, N. J. (1982). Selective exposure to televised comedy programs. *Journal of Applied Communication Research, 10*(2), 117–132.

Metzger, M.J. (2002). When no news is good news: Inferring closure for news issues. *Journalism & Mass Communication Quarterly, 77*(4), 760–787.

Mills, J. (1993). The appeal of tragedy: An attitude interpretation. *Basic and Applied Social Psychology, 14*(3), 255–271.

Oliver, M.B. (1993). Exploring the paradox of the enjoyment of sad films. *Human Communication Research, 19*(3), 315–342.

Oliver, M. B. (2000). The respondent gender gap. In D. Zillmann & P. Vorderer (Eds.), *Media entertainment: The psychology of its appeal* (pp. 215–234). Mahwah, NJ: Lawrence Erlbaum Associates.

Oliver, M. B. (2003). Mood management and selective exposure. In J. Bryant, D. Roskos-Ewoldsen, & J. Cantor (Eds.), *Communication and emotion* (pp. 85–106). Mahwah, NJ: Lawrence Erlbaum Associates.

Pearlin, L.I. (1959). Social and personal stress and escape television viewing. *Public Opinion Quarterly, 23*(2), 255–259.

Roskos-Ewoldsen, D. R., Roskos-Ewoldsen, B., & Dillman Carpentier, F. (2002). Media priming: A synthesis. In J. Bryant & D. Zillmann (Eds.), *Media effects: Advances in theory and research* (pp. 97–120). Mahwah, NJ: Lawrence Erlbaum Associates.

Rubin, A.M. (1984). Ritualized and instrumental television viewing. *Journal of Communication, 34*(3), 67–77.

Schmitz, B., Alsdorf, C., Sang, F., & Tasche, K. (1993). Der Einfluss psychologischer und familialer Rezipienten-merkmale auf die Fernsehmotivation [translation: The influences of individual and family-related characteristics on television viewing motivations]. *Rundfunk & Fernsehen, 41*(1), 5–19.

Schwarz, N., & Clore, G. L. (1996). Feelings and phenomenal experiences. In E. T. Higgins & A. W. Kruglanski (Eds.), *Social psychology: Handbook of basic principles* (pp. 433–465). New York: The Guilford Press.

Sparks, G. G., & Sparks, C. W. (2000). Violence, mayhem, and horror. In D. Zillmann & P. Vorderer (Eds.), *Media entertainment: The psychology of its appeal* (pp. 73–91). Mahwah, NJ: Lawrence Erlbaum Associates.

Tamborini, R. (2003). Enjoyment and social functions of horror. In J. Bryant, D. Roskos-Ewoldsen, & J. Cantor (Eds.), *Communication and emotion* (pp. 417–443). Mahwah, NJ: Lawrence Erlbaum Associates.

Thayer, R.E. (1996). *The origin of everyday moods.* New York: Oxford University Press.

Trepte, S., Zapfe, S., & Sudhoff, W. (2001). Orientierung und Problembewältigung durch TV-Talkshows: Empirische Ergebnisse und Erklärungsansätze [translation: Orientation and coping through television talk shows]. *Zeitschrift für Medienpsychologie, 13*(2), 73–84.

Vorderer, P. (1996). Rezeptionsmotivation: Warum nutzen Rezipienten mediale Unterhaltungsangebote? [Media use motivation: Why do recipients choose entertainment content?] *Publizistik, 41*(3), 310–326.

Wakshlag, J., Vial, V., & Tamborini, R. (1983). Selecting crime drama and apprehension about crime. *Human Communication Research, 10*(2), 227–242.

Wills, T. A. (1981). Downward comparison principles in social psychology. *Psychological Bulletin, 90,* 245–271.

Wills, T. A. (1991). Similarity and self-esteem in downward comparison. In J. Suls & T. A. Wills (Eds.), *Social comparison. Contemporary theory and research* (pp. 51–78). Hillsdale, NJ: Lawrence Erlbaum Associates.

Wirth, W., & Schweiger, W. (1999). *Selektion im Internet* [translation: Selection on the Internet]. Opladen: Westdeutscher Verlag.

Zillmann, D. (1985). The experimental exploration of gratifications from media entertainment. In K. E. Rosengren, L. A. Wenner, & P. Palmgreen (Eds.), *Media gratifications research: Current perspectives* (pp. 225–305). Beverly Hills/London/New Delhi: Sage.

Zillmann, D. (1988). Mood management through communication choices. *American Behavioral Scientist, 31*(3), 327–340.

Zillmann, D. (1994). Mechanisms of emotional involvement with drama. *Poetics, 23,* 33–51.

Zillmann, D. (1998a). The psychology of the appeal of portrayals of violence. In J. H. Goldstein (Ed.), *Why we watch: The attractions of violent entertainment* (pp. 179–211). New York: Oxford University Press.

Zillmann, D. (1998b). Does tragic drama have redeeming value? *Siegener Periodicum zur Internationalen Empirischen Literaturwissenschaft SPIEL, 17*(1), 4–14.

Zillmann, D. (2000a). Mood management in the context of selective exposure theory. In M. F. Roloff (Ed.), *Communication yearbook 23* (pp. 103–123). Thousand Oaks, CA: Sage.

Zillmann, D. (2000b). Humor and comedy. In D. Zillmann & P. Vorderer (Eds.), *Media entertainment: The psychology of its appeal* (pp. 37–57). Mahwah, NJ: Lawrence Erlbaum Associates.

Zillmann, D., & Bryant, J. (1985). Affect, mood, and emotion as determinants of selective exposure. In D. Zillmann & J. Bryant (Eds.), *Selective exposure to communication* (pp. 157–190). Hillsdale, NJ: Lawrence Erlbaum Associates.

Zillmann, D., & Knobloch, S. (2001). Emotional reactions to narratives about the fortunes of personae in the news theater. *Poetics, 29,* 189–206.

Zillmann, D., Hezel, R.T., & Medoff, N.J., (1980). The effect of affective states on selective exposure to televised entertainment fare. *Journal of Applied Psychology, 10,* 323–339.

Zillmann, D., Weaver, J. B., Mundorf, N., & Aust, C. F. (1986). Effects of an opposite-gender companion's affect to horror on distress, delight, and attraction. *Journal of Personality and Social Psychology, 51,* 586–594.

Zuckerman, M. (1984). Sensation seeking: A comparative approach to a human trait. *The Behavioral and Brain Sciences, 7,* 413–471.

Zuckerman, M. (1991). *Psychobiology of personality.* Cambridge: Cambridge University Press.

15

Social Identity Theory

Sabine Trepte

Universitat Hamburg

INTRODUCTION

Thirty-something women like 'Sex and the City,' but men go for sports. Youngsters watch MTV, whereas the older generation tunes in for the 'Golden Girls' (Cassata & Irwin, 2003). Germans, Mexicans, and South-Koreans have something in common: They all prefer television produced in their home country over international programming (Waisbord, 2004). Although perhaps intuitive, ratings and shares clearly show that media consumers prefer entertainment that refers to the social groups they belong to—be it gender (Oliver, 2000; Oliver, Weaver & Sargent, 2000; Trepte, 2004), age (Haarwood, 1999) or culture (Greenberg & Atkin, 1982; Zillmann et al., 1995). In particular, they seek out entertainment that favors their 'in-group,' sometimes even drawing a sharp line to distinguish them from other 'out-group' people. We can assume that social identity influences the selection of media entertainment, because people are creating their personal media profile to support their own identity. Also, it is likely that processes of social identity come into play during the reception process and that they determine the effects of watching entertaining fare.

Social identity theory (SIT) focuses on "the group in the individual" (Hogg & Abrams, 1988, p. 3) and assumes that one part of the self-concept is defined by our belonging to social groups. According to Tajfel and Turner (1979), people categorize themselves and others as belonging to different social groups and evaluate these categorizations. Membership, alongside the value placed on it, is defined as the social identity. To enhance their self-esteem, people want to develop a positive social identity. To do so, they show all kinds of different behavior that might also be observed in the context of entertainment selection and reception.

Social identity has been shown as a plausible theoretical background for identity related gratifications in the uses-and-gratifications approach to understanding media use (Blumler, 1979). Blumler (1985) stated that "[. . .] little attention has been paid to the social group memberships and affiliations, formal and subjective, that might feed audience concerns to maintain and strengthen their social identities through what they see, read, and hear in the media" (p. 50). However, over the next twenty years, the idea of social identity in terms of SIT has only very superficially 'inspired' research on identity processes. Only then did scholars systematically begin research that was based on assumptions of social identity theory (Harwood, 1999; Trepte, 2004; Zillmann et al., 1995). SIT is, therefore, still a comparatively new social-psychological theory in terms of how often it has been applied to problems and questions in media effects and entertainment research.

In this chapter, social identity theory and social categorization theory will be outlined and its application in entertainment research will be considered. SIT has been used to show that selective exposure can be determined by group memberships and to prove effects of identification in the context of models such as media dependency and third-person effect. Also, SIT has been used to explain inter-group processes in computer mediated communication. Finally, future developments and a research agenda for SIT in entertainment research will be discussed.

SOCIAL IDENTITY THEORY

Social identity theory (SIT) was first proposed by Tajfel (1978, 1979) and later by Tajfel and Turner (1979). It is a social-psychological theory that attempts to explain cognitions and behavior with the help of group-processes. SIT assumes that we show all kinds of "group" behavior, such as solidarity, within our groups and discrimination against out-groups as a part of social identity processes, with the aim to achieve positive self-esteem and self-enhancement (Abrams & Hogg, 1988).

Research on SIT has mainly been stimulated by thoughts about social settings and groups (Tajfel & Turner, 1979). Tajfel (1979) proposed the "minimal group paradigm" and showed that mere categorization to one group or another makes people discriminate against the designated out-group and favor their in-group. Groups in the early experiments had no face-to-face interaction and the categorization was randomly assigned (Tajfel, Flament, Billig, & Bundy, 1971). Even these minimal conditions led the members to in-group favoritism and discrimination against the out-group. They tried to maximize the difference of rewards between in-group and out-group, whereas maximizing their own, in-group profit was less important to them.

Thus, in comparison to most other social psychological theories, SIT does not begin with assumptions considering the individual, but rather with assumptions referring to a social group. A social group consists of a number of people who feel and perceive themselves as belonging to this group and who are said to be in the group by others (Tajfel & Turner, 1979, p. 40). Intermember interaction can take place, but is by no means a presupposition for the perception of its members as belonging to the same group. For instance a football club can be considered a social group, as well as a group consisting of all of the women in an organization. Also, members do not have to share interdependent goals or a similar understanding of a concurrent out-group to make them a social group. Tajfel (1979) structures the definition of a group alongside a cognitive component (knowing about the group membership), an evaluative component (positive or negative evaluation of group membership) and an emotional component (positive or negative emotions associated with the group membership and its evaluation). Based on his understanding of social groups outlined here, he suggested four underlying principles of SIT, which are social categorization, social comparison, social identity and self-esteem. All four will be elaborated in the following paragraphs. But before, very briefly self-categorization theory as a further development of SIT will be outlined.

Further Developments on Social Identity: Self-Categorization Theory

SIT has further been developed in co-operation with numerous scholars, particularly by Tajfel's colleague Turner (1987), who later proposed the self-categorization theory (for an overview see Turner, 1987, 1999; Turner & Onorato, 1999). Both theories share the idea of social identity, but in self-categorization theory (SCT) social identity is seen as the process that changes interpersonal to inter-group behavior. SCT does not define interpersonal and inter-group behavior as the poles of one continuum, but suggests that personal and social identity represent different levels of self-categorization. It is the "relative" salience of different levels of self-categorization which determines the degree to which behavior expresses individual differences or collective similarities (Turner, 1999). In some situations even both, personal and social identity, can become salient (Turner & Onorato, 1999).

In research on media entertainment, however, it is hard to draw a clear line between SIT and SCT, because rather than trying to "prove" one or the other theory, scholars have addressed processes of social categorization and social identity found in both of the theories. Some authors particularly address both theories (Mastro, 2003; Zillmann et al., 1995), but most of them refer to the more general SIT and then reflect developments that have been made in the tradition of SCT. The self-esteem hypothesis (Hogg & Abrams, 1990), ideas on salience (Oakes, 1987), and also the accentuation principle have initially been triggered and suggested by Tajfel (1978, 1979), but have further been developed under the aegis of Turner (1987). To accommodate this progress, the chapter will consider the principles of social identity outlined in SIT and it will be complemented by theoretical developments in the realm of SCT.

Social Categorization

Tajfel (1979) states that due to reduced capacities in processing information we define categories and schemes to encode and decode messages. Similar to other entities in our surrounding, we categorize people into groups to simplify our understanding of the world and to structure social interaction. For instance, we use categories such as 'Punk' or 'Skater' to describe groups with similar and specific clothing style and habits, and we have certain expectations, hopes and fears about people belonging to social categories. Tajfel and Turner (1979) summarize: "Social categorizations are conceived here as cognitive tools that segment, classify, and order the social environment, and thus enable the individual to undertake many forms of social action. [. . .] They create and define the individual's place in society" (p. 40).

Based on group categorization, differences between categories (interclass differences) are accentuated and differences between members within the same category (intraclass differences) are underestimated or restrained. This "accentuation principle" is more pronounced when the categorization is salient, important and of immediate relevance to the individual (for the first theoretical ideas on the accentuation principle see Tajfel, 1959; for the first experiment see Tajfel and Wilkes, 1963; see also Hogg & Abrams, 1988, p. 20).

If social categorizations are shared by all group members, they function as "social stereotypes" and help interpret, explain, and even justify our behavior (Tajfel, 1981). In addition, the goal of SIT scholars in social psychology is to define consequences and behavioral outcomes of the processes underlying social identity. Since Tajfel published his article "Cognitive Aspects of Prejudice" in 1969, one of the major issues on the research agenda has been social stereotyping (Oakes, 1996). It has triggered a tremendous amount of research in the field of applied social psychology (for reviews see Hamilton & Sherman, 1994; Oakes, Haslam & Turner, 1994).

Of course, we also categorize *ourselves* into groups such as sporting clubs, fans of certain TV series, or members of a university. All people belong to a number of different groups, but they are not of the same importance at the same time. In SIT, Tajfel (1979) adds to the idea of minimal groups that the group membership has to be "salient" to initiate behavior in terms of social identity. Although, there is a long tradition of research on salience that goes back to Festinger's idea of religious identity in 1947, there is no theoretical development in line with SIT. Oakes (1987) proposed that salience means that group membership influences perception and behavior. This notion of "psychological salience" is differentiated from "stimulus salience" that might be some situational cue reminding of a group membership and that functions as a causal antecedent of psychological salience (Oakes, 1987, p. 118). Early conceptions of salience proposed that group memberships are more likely to be salient, if the differences between categories are "clear". Conversely, Oakes (1987) suggests that salience occurs if a social categorization is accessible *and* best fits the information available. "Accessibility refers to the relative "readiness" of a given category to become activated; the more accessible the category, the less input is required to invoke the relevant categorization [...]" (Oakes, 1987, p. 127). Accessibility is determined by the relative centrality or importance of a group membership and by its current emotional or value significance to a person. The fit of a categorization is termed as the degree to which observed similarities and differences between people correlate with the expected social categories (Oakes, Turner & Haslam, 1991). Hence, it is a fit between input and category specifications. Given the same accessibility, the category will become salient that guarantees the best fit between an observed stimulus (e.g., a person with Iroquois haircut) and a predefined (and stereotypical) idea of a category (e.g., a Punk).

Social Comparison

The first type of behavior that is triggered by social categorization is social comparison. To define an individual's place in society, social categorizations are evaluated in comparison with other groups. SIT assumes that we not only categorize ourselves and others, but that we evaluate the groups. To get an idea of the superiority or inferiority of our group and of how reasonable and adequate our belonging to it is, we compare it with other groups, their characteristics, members, and benefits. This concept is based on Festinger's (1954) theory of social comparison. Festinger assumes that we have a need to compare our opinions and abilities with others, particularly if there are no objective standards that we can refer to.

The aim of social comparison, in terms of SIT, is to evaluate the social groups to which we and others belong. Social comparison usually takes place with groups that are similar to one's own group, and refers to dimensions that compose the group. Both the other group's similarity and the dimensions, on which inter-group comparisons take place, define the relevance of inter-group comparison. The "closer" the other groups are to ourselves in terms of the dimensions on which we compete, the more relevant the social comparison gets and the more we "need" and want a positive outcome. The outcome of social comparisons largely determines our social identity and self-esteem.

There are three premises for social comparison (Tajfel & Turner, 1979, p. 41). First, individuals must have internalized their group membership as a part of their self-concept. They must be identified with their in-group. Second, the situation must allow social comparison. Third, the out-group must be relevant in terms of similarity and proximity (Hinkle & Brown, 1990).

Social Identity

Tajfel (1978) defines social identity as "that part of an individual's self-concept which derives from his knowledge of his membership of a social group (or groups) together with the value and emotional significance attached to that membership" (p. 63). Thus, social identity is based on more or less favorable comparisons between the in-group and a relevant out-group.

By differentiating between in-group from out-group on dimensions in which the in-group falls at the evaluative positive pole, the in-group acquires a "positive distinctiveness," and thus a relatively positive social identity in comparison to the out-group. The main aim of individuals is to achieve *positive* social identity. Positively discrepant comparisons produce positive social identity, and negatively discrepant comparisons produce negative social identity.

As groups and their performance and status change, social comparison takes place constantly and social identity is negotiable. "Categories come and go (prior to the mid-twentieth century there was not such occupational category as 'computer programmer'), their defining features alter (historical modifications of stereotypes of North American Blacks), their relations with other categories change (inter-group relations between the sexes), and so on." (Hogg & Abrams, 1988, p. 14). For that reason, the motivation to reach positive social identity is always present and in progress.

Tajfel and Turner (1979) conceptualized different belief structures and associated strategies to reach a positive social identity. If inferiority of their own group cannot be denied, members might leave a group and join a higher status group. This strategy is based on the belief structure of "social mobility." It is restricted by the perceived characteristics, group boundaries, strength of objections, and sanctions of the groups. If social mobility is not possible, or the confrontation with the dominant group has to be avoided, group members might adhere to the belief structure of "social change". This implies strategies such as "social competition" or "social creativity". Social comparison with lower status groups is used to emphasize the in-group's superiority on the relevant dimensions of comparison. Social creativity implies redefining the value associated with the low-status criterion, focusing on additional dimensions of comparison, or comparing with a different group. For an overview on the strategies and belief structures see Tajfel & Turner (1979, p. 42 ff.) and Hogg & Abrams (1988, p. 27).

Self-Esteem

SIT suggests a fundamental individual motivation for self-esteem (Tajfel & Turner, 1979; Turner, 1982). In his early work, Tajfel (1969) stated that the motivation underlying positive social identity is to preserve the integrity of the self image and, later, he assumed that the main drive is to reach self-enhancement (Tajfel, 1972). The formal theoretical statement on SIT finally refers to self-esteem as the motivation underlying inter-group behavior. The idea stems from social comparison theory (Festinger, 1954; see prior) and implies that people strive to confirm aspects of their own self-definition. In terms of SIT the need for positive self-esteem is satisfied by a positive evaluation of one's own group (Turner, Brown & Tajfel, 1979). If a group membership is crucial to one's self-concept (e.g., football club membership to a professional player), social comparison should lead to positive social distinctiveness and enhance self-esteem.

However, in SIT the motivation for self-esteem is considered a premise and has not clearly been integrated in the processes of social identity. Also, the empirical status remains unclear and the combination of cognitive (categorization) and motivational (self-esteem) constructs have not fully been discussed (Abrams & Hogg, 1988; Hogg & Abrams, 1990, p. 32). For that reason, Abrams and Hogg (1988) elaborated on this issue and suggest "the self-esteem hypothesis." It states two corollaries: First, that successful inter-group discrimination leads to increased self-esteem (self-esteem as a dependent variable); second, that low or threatened self-esteem motivates increased out-group discrimination (self-esteem as independent variable). Research on both corollaries does not unambiguously support one of each. Several reasons for lacking support have been discussed (Abrams & Hogg, 1988; Hogg & Abrams, 1990). It has been criticized that in most of the studies global self-esteem has been measured, although SIT refers to self-esteem as "the esteem in which specific self images are held" (Hogg & Abrams, 1990, p. 38; see also Turner, 1982). Another reason discussed is that self-esteem might be one, but not always *the* motive. There are supposed to be other, competing needs that are to be fulfilled with inter-group comparison and evolving positive social identity (Hogg & Abrams, 1990); for instance, the general motive of self-knowledge or self-actualization. In this sense, self-esteem is seen as a consequence of having fulfilled the need to know more about oneself. Other motives might be to construct meaning or to reach consistency by self-categorization. Also, exhibiting power or control and reaching self-efficacy have been discussed as motives to enhance self-esteem in the first place. Summing up research on the idea of self-esteem in SIT, Hogg and Abrams (1990) state that "while it clearly does play an important role, self-esteem may be one of a number of motives and effects of different forms of group behavior. Possibly more fundamental is some form of self-evaluative motive" (p. 46).

SOCIAL IDENTITY IN MEDIA EFFECTS AND ENTERTAINMENT RESEARCH

In communication studies, social identity was considered to be of relevance to selective exposure relatively early on (Blumler, 1979, 1985; McQuail, 2000). Despite this early recognition, however, SIT has long been neglected in media psychology and entertainment research. Only within the last decade empirical studies have been done based on its assumptions and in its methodological tradition. Alongside with empirical data on the subject matter, theoretical models that conceptualize processes of social identity before, during, and after the consumption of media entertainment have been developed (Reid, Giles & Abrams, 2004; Trepte, 2004).

In the following sections, ways in which SIT can contribute to the psychology of entertainment will be elaborated. Studies and models in this area of research usually either address how social identity determines media selection and media preferences or how the media affect social identity. The following sections will deal with these two perspectives, respectively. Additionally the last section will address how SIT has been applied to reception processes in computer-mediated communication.

MEDIA PREFERENCES GUIDED BY SOCIAL IDENTITY

To find out why people choose certain entertainment products, a variety of theories have been proposed. Very different kinds of models in terms of their paradigmatic background now serve as explanations for selective exposure (for an overview see: Vorderer, Wulff & Friedrichsen, 1996; Zillmann & Bryant, 1985; Zillmann & Vorderer, 2000). In the late 1990s, social identity and SIT were added to these models. SIT contributes the idea that we choose entertainment in concordance with certain group memberships and connects social settings and individual motivations in entertainment consumption. This process is made easier as entertainment diversifies and attempts to serve the needs of the vast variety of groups. Particularly in today's successful entertainment programs, such as casting shows (e.g., "American Idol"), game shows (e.g., "Who Wants to Be a Millionaire"), and talk shows, all kinds of different social backgrounds, even the rarest types of social groups, are shown and addressed (Trepte, 2005). Also, technical developments such as digital video recorders and digital cable simplify this specialization (Harwood & Roy, 2005).

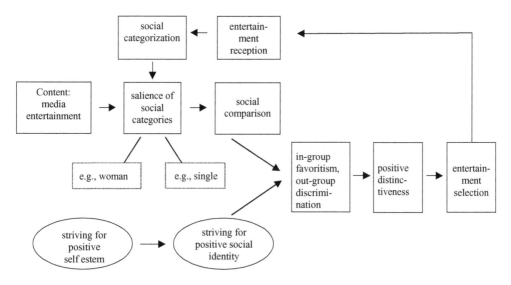

FIG. 15.1. Processes of social identity in entertainment consumption

The idea behind the influence of social identity on entertainment media preferences has been conceptualized by Trepte (2004; see figure one). The process shown can either be situated before or during media reception, given that recipients make choices—either to stay with it or to switch. The model starts with recipients of entertainment content making their choices (see box labeled 'Media Entertainment'). As an example, consider a woman who wants to pick an entertainment program on television. She either browses the channels or looks at the short descriptions in the TV Guide. By doing so, certain social categories she belongs to (such as young, urban, female, and single) will become salient, because some shows (such as "Sex and the City" or "Ally McBeal") fit to these categories. For her, the shows offer the opportunity to "meet" members of her in-group and to compare herself with out-groups. Given that she has the motive to attain positive self-esteem and positive social identity, she might select one of the shows, because her in-group (female single) is depicted in a rather favorable way. Also, the story offers content that might discriminate the out-group (e.g., married women or men). The woman's goal to reach positive distinctiveness might be met and she will pick the show, because of this promising outcome. Consequently, her social identity will be strengthened, because a positive evaluation of the in-group takes place during media reception. A redefinition of the categories' characteristics might take place and succeeding media choices will be influenced by this experience.

The model shows that working on one's social identity and enhancing self-esteem might be a motive for media selection. A number of studies have been done to empirically prove how group memberships determine media selection. As categories or group memberships they address the recipients' age (Harwood, 1997, 1999), their gender (Knobloch et al., 2005; Oliver, 2000; Oliver, Weaver & Sargent, 2000; Trepte, 2004), their culture, national and ethnic background (Zillmann et al., 1995; Mastro, 2003; Trepte, 2004). Also, institutionalized groups such as school classes have been investigated (Tarrant, North & Hargreaves, 2001). The empirical studies considering the categories of age, gender, and culture will be outlined in the following.

Harwood (1997) employed a content analysis to code fictional, prime-time television shows according to the age of all characters with speaking roles. Compiling the results with the Nielsen ratings of these shows revealed that young (0–20 years old), middle-aged (21–60 years old), and older (over 60 years old) viewers prefer shows featuring lead characters of their own age. In a second study, Harwood (1997) manipulated short descriptions he took out of TV-guides considering the protagonists' age, and he asked students to rate how often they might choose to view the show. Although he found the overall pattern that participants preferred shows with same-age characters, participants did not rate all of the shows in the expected direction. Only for six of the twelve shows could significant results be yielded, all of which had a romantic couple as a key element. Other shows, such as police dramas, were not preferred by the youngsters, even if they featured younger characters. It seems that for younger viewers' age identities particularly come into play if relationships and romance are considered. In a corollary study, Harwood (1999) tried to replicate these results. However, the age group identification was not associated with preferences for shows featuring protagonists of their own age, and only weak correlations could be yielded between age group identification and age identity gratifications in particular (such as, "I like watching people of my own age"). Also, a rather weak relationship between self-esteem, age identity, and viewing behavior was found. To summarize, there is a preference for characters of one's own age that can be observed in ratings and shares, but media selection itself is not fully determined by needs evolving from age identifications. To support the results based on the Nielsen ratings, further research employing more than one age group is needed. In addition, experiments in this line of research should entail media selections that is not only based on short descriptions, but that is followed by the actual viewing of an entertainment program and, thus, have a higher external validity.

One group membership that has many cues in entertainment programming is certainly gender. The "battle of the sexes" has been one of the most frequented topics in entertainment. Ratings and shares clearly show that women— and even little girls—prefer entertainment programs that feature female characters, whereas men and boys prefer male characters (Knobloch et al., 2005; Oliver, 2000). Additionally, the preference for certain content differs with gender (Oliver, 2000; Oliver, Weaver & Sargent, 2000; Trepte, 2004). Identification with gender (measured with Bem's Sex Role inventory, 1974), influences how much people watch "sad films" and how they evaluate them (Oliver, 2000; Oliver, Weaver & Sargent, 2000).

In terms of SIT, the distinct lines drawn between men and women on various entertainment programs should elicit processes of social comparison. It can be assumed that because of the accessibility of gender categories and its emotional value, they very easily become salient. Additionally, entertainment programs offer people of both sexes multitudinous material for social comparison and the opportunity to attain positive social identity. Studies on SIT and gender (Trepte, 2004; Trepte & Krämer, in preparation) prove this notion. Trepte (2004) manipulated short descriptions of ten entertaining TV series, according to the gender of the protagonists. She showed that participants from Germany and the United States clearly preferred the shows featuring characters of their own gender. This result could partly be replicated for an experiment conducted in the UK and Germany (Trepte & Krämer, in preparation). In both countries, women evaluated series with female characters more positively than those with male characters; however, this effect was not found for men. An interesting side-effect occurring in both studies was that, although salience was manipulated, women were not affected by the manipulation in their evaluation of an entertaining TV series. They rated the "female series" more positively, whether or not gender was given as a prior cue. This highlights the psychological salience of gender, which easily affects attitudes and behavior. Especially women, who are still under-represented as lead characters in entertaining programs of many countries (Elasmar, Hasegawa & Brain, 1999), are very aware of their gender, although they picking entertainment programs to fit their needs and watch television much more selectively (Reid, Giles & Abrams, 2004).

Culture, nation, and ethnic background are other categories that have stimulated a lot of research in the area of SIT (Hamilton & Sherman, 1994; Oakes, Haslam & Turner, 1994), and that have later been adopted in the research of social identity in media effects. Even though they are crucial variables for social identity, is has also been difficult to empirically link them to entertainment.

Zillmann et al. (1995) exposed African-American and White high-school students to music videos featuring popular rock, nonpolitical rap, or radical political rap. They assumed that African-Americans would enjoy radical political rap, which articulates African-American defiance, more than the other genres, because rap considers the ethnicity of African-American performers and themes, and because it affronts agencies of power that oppress African-Americans. The authors assumed that Whites would not appreciate this music genre (Zillmann et al., 1995). In terms of SIT, listeners prefer music favoring the in-group. As expected, African-American students enjoyed rap more than rock, and Whites enjoyed rock more than rap. Other than these results, the authors' expectations on the influence of listening to "in-group music" on self-esteem, social cohesion, or ethnic support, were not met. For that reason, they suppose that "Rap—radical rap in particular—appears to be a momentary, fleeting delight for African-American audiences" (Zillmann, et al., 1995, p. 21), rather than delivering positive self-esteem. Zillmann, et al. (1995) remark that an experiment reflects only one very brief media experience compared to all the music listening that has been done before and is experienced daily. They suppose that any effect of media consumption on self-esteem might have already materialized before and cannot be demonstrated by later exposure.

Trepte (2004) carried out a study on media choices guided by nationality, focusing on whether entertaining TV-series with a salient nationality (U.S. or German) would lead to media preference. Student participants were asked to rate whether they found short descriptions of series, which they were told to have been produced in either Germany or the U.S., entertaining and if they would like to watch them. The results did not differentiate between countries. The author suggests that there is not enough experience in the United States with internationally produced TV series and that U.S. citizens would have a hard time conceptualizing what a German program would be like. To support this interpretation of the results, the study was replicated in the U.K. and in Germany, because British viewers are more likely to be exposed to German entertainment programming (Trepte & Krämer, in preparation). Results showed that Germans would prefer the series produced in Germany over a British series, but British people would watch a series from either country. Also, if people identify strongly with their own country, their choices are not necessarily influenced in the expected direction. Trepte and Krämer (in preparation) assume that the genre of TV series might not be a relevant dimension to compare on. Unlike televised soccer and other sports programming, entertaining TV series might not trigger social comparisons between people coming from different countries. In terms of SIT: The psychological salience of categories such as nation and culture is not high, if TV series coming from different countries are to be evaluated.

Mastro (2003) addressed the effect of ethnic in-group favoritism on the judgment of media content. She assumes that racial identification among Whites would decrease their propensity to justify a Latino's criminal behavior featured in a movie. Additionally, she assumed that Whites exposed to Latino criminality on television would report higher self-esteem than those exposed to White criminality. To test her hypotheses, Mastro (2003) employed two studies. First, she made White participants read a script for a TV drama. Second, she showed them excerpts from a TV police drama that varied only in the race/ethnicity of the criminal. As dependent variables, she employed the justification of the criminals' behavior (measured by six items such as "considering the circumstances, the crime was justified") and self-esteem (measured with Rosenberg's self-esteem scale, 1991). Limited support was provided for both hypotheses. Whites with high racial identification justified the White characters' criminal behavior, but there was no relationship between their racial identification and the justification of the Latinos' criminal behavior. However, watching the TV drama did have a positive effect on participants' self-esteem. The Latino's criminal behavior did increase the Whites' self-esteem.

This is, however, the only result that significantly proves effects of media entertainment on self-esteem. In previous research, scholars repeatedly tried to find moderating variables that increase the likelihood to select media entertainment based on group membership. In particular, they used measures of self-esteem and identification with the group. However, results almost unequivocally indicate that neither self-esteem and nor identification moderate the relationship. As shown above, young adults did not show increased self-esteem after they had the chance to choose age-related shows (Harwood, 1999). There is no relationship between the selection of shows featuring characters of the own sex and gender-specific self-esteem (Trepte & Krâmer, in preparation). The same results emerge for categories of nation and culture. Self-esteem is not enhanced if home-grown TV series have been selected (Trepte & Krämer, in preparation), nor if people of a certain ethnic group observe that an ethnic out-group is discriminated in an entertainment program (Mastro, 2003). Also, listening to musical genres that are associated with the ethnic in-group, and articulate their defiance, does not enhance self-esteem (Zillmann et al., 1995).

The difficulties in realizing effects of self-esteem might be attributed to the fact that self-esteem can hardly be manipulated by a single, brief exposure to media entertainment in an experimental setting. People are exposed daily to media entertainment that is similar to those offered in the experiments. Consequently, all effects on self-esteem are reflected in the participants' behavior before they enter the experimental setting. It is highly unlikely that the experimental manipulation would have an effect on measures of self-esteem. Additionally, research on self-esteem hypothesis suggests that there are multiple motives for social comparison (Hogg & Abrams, 1990; see prior). Particularly for media entertainment, it has been shown that self-knowledge and self-actualization might be motives that drive the audience's media choices (Trepte, 2005).

Similarly, only limited support emerges from applying the extent of identification with the in-group as moderating the relationship between group memberships and media selection. In different studies, scholars tried to show that high identification with the in-group increases the relationship of group membership and media choices. For instance, it was assumed that high gender or age-group identification would intensify the relationship between the gender or age-group viewers belong to and their appreciation of a show featuring characters of their own gender or age (Harwood, 1999; Trepte & Krämer, in preparation). However, none of the hypotheses on the moderating effects of identification—be it nation, culture, age or gender—achieved support. These results are in line with fundamental research on SIT. Hinkle and Brown (1990) refer to different experiments and show that the motivational impetus for positive social identities does not automatically increase with high in-group identification.

In the experiments on media entertainment, groups with high psychological salience, such as gender, culture, or age, have been applied. Results showed that the dimensions of comparison were relevant to the group, because participants selectively chose media content referring to the in-group. Hence, lacking salience of the group, or an irrelevant dimension of social comparison, cannot be the reason for the results. However, the reason might be found in the nature of experiments in media entertainment. Maybe, high-identifiers, in comparison to low-identifiers, do not show a stronger need to favor the in-group and discriminate the out-group in every social comparison they perform, but they use different ways and settings to compare, and by that show their motivation for higher social identity. Media entertainment might just be one setting amongst others that allows for social comparison. Further research should compare the drive for social comparison in different settings such as work-life, family, and peers.

EFFECTS OF MEDIA ENTERTAINMENT ON SOCIAL IDENTITY

In a world where entertainment products are very diversified and customized to all kinds of different social groups, it seems to be crucial to ask for its effects in stabilizing and loosening identities. Reid, Giles and Abrams (2004) as well as Harwood and Roy (2005) primarily address the consequences of media consumption on strategies of social mobility, social creativity, and social comparison, as consequences of media consumption. They suggest that "the media occupy a causal role in inter-group relations that can function to maintain the status quo (i.e., act as a force for social stasis), or act as a producer of social change" (Reid, Giles & Abrams, 2004, p. 20). In the core of their social identity model of media uses and effects, Reid, Giles and Abrams (2004) assume that media users engage with the media and form their identities and their strength of association with a group, as well as their beliefs of how stable the group's status is, how permeable the group boundaries are, and their perceptions of group vitality from that engagement. Depending on the audience's conception of these variables, it might either exhibit strategies of social mobility, social creativity, or social competition, to create a positive distinctiveness. Strategies of social mobility are likely being chosen if the identification with the group is low, the group has a stable and legitimate status, and boundaries are permeable and in-group vitality is low; whereas strategies of social competition are likely to be chosen if the identification with the in-group is high and it has a rather unstable status, impermeable boundaries, and high group vitality.

The most important idea behind the model is that people not only select the media to fit their identities and to "deliver" identity related gratification, but also—and very importantly—media entertainment functions as a source of information on groups and their legitimate status. However, as media content is influenced and filtered by interests of money, power, and dominant private interests, so also is the perception of groups, their boundaries and impact (Reid, Giles & Abrams, 2004, p. 22).

How the media influence inter-group attitudes and beliefs about out-groups has been shown in research on media effects (Bryant & Zillmann, 2002). For instance, the coverage of sporting events is usually influenced by ethnocentric views and therefore determines the viewers' feelings of nationality (Horak, 2003). Stereotyped portrayals of minorities, such as Latinos or African-Americans, transported by entertainment programming influence, their self-perception and how they are perceived by others (for an overview see Greenberg, Mastro, & Brand, 2002). Armstrong, Neuendorf, and Brentar (1992) showed that exposure to TV entertainment was associated with the belief that African-Americans have a higher socio-economic status according to income, social class, and education; whereas exposure to TV news led to the perception that African-Americans were worse off in contrast to Whites.

Similarly, individual's ideas of group vitality might be influenced because of entertainment. Groups that are underrepresented in television, such as the elderly and women (Elasmar, Hasegawa & Brain, 1999), might perceive group boundaries as impermeable and their status as rather fixed. The opposite is also possible: Minority media may encourage social mobility. Similarly, mainstream entertainment is able to influence the belief structures of social mobility and social change. For instance, telenovelas in Latin American countries provided images of poor people ascending to higher status groups and having access to symbols of capitalism (Straubhaar, 1991). And studies demonstrate that shows such as "The Cosby Show" or "Sesame Street" were able to improve perceptions of interracial relationships (Harwood & Roy, 2005). In fact, media effects research offers numerous studies that prove effects of stereotyped portrayals in entertainment on group identification and group vitality. However, these studies do not explicitly consider the processes underlying social identity or try to apply SIT.

Apart from that, there is previous research based on SIT to theoretically explain media effects. Duck, Hogg, and Terry (1999) investigated how social identity affects the perception of media persuasion and they based their studies on SIT, SCT, and the third person effect. The third person effect implies that people typically think that negative media content has more impact on others rather than on themselves (Davison, 1983). It was expected that the salience of self-categorization accentuates the perceived similarity between self and in-group others and thus the third person effect will have a stronger impact. In the study, 58 Australian students watched 11 commercials that showed risks and effects of HIV. The results revealed that students who identified with the student community expected the AIDS advertisements to have the same impact on themselves and on others. In contrast, low-identifiers perceived the ads to be influential for the student community and other people in general, but not for themselves. Duck, Hogg, and Terry (1999) assume that media impact is only acknowledged if it is socially accepted. This implies identification with the social group in question and the acceptance of the media message within that group.

Morton and Duck (2000) investigated the relationship of gay media use, identification with the gay community and media dependency to gay media messages. Media dependency means that media effects vary according to how dependent audiences are on them as sources of information and whether messages are linked to the satisfaction of their goals (Ball-Rokeach, 1985). Morton and Duck (2000) suppose that in terms of SIT, people not only categorize their world according to group memberships, but also learn stereotypic characteristics and norms of the in-group, and exhibit theses norms in subsequent behavior. The authors expected that members of the gay community would perceive gay media content as relevant to their social identity, and for that reason display the strongest dependencies on gay media for sexual health information. Also, they assumed that dependency on gay media would produce effects on personal safe sex attitudes for high-identifiers. Their study of 76 gay men showed that identification with the gay community has no impact on dependency on gay media. But high-identifiers, who displayed more intense dependence on gay media, had rather positive personal attitudes toward safe sex. Thus, if identification and media dependency are related, media messages being communicated in in-group media have stronger impact on group members. And contrarily, gay men who are not as concerned about or don't identify as much with the gay community are not directly affected by its media messages, even if they rate high in media dependency.

SOCIAL IDENTITY IN COMPUTER MEDIATED COMMUNICATION

Theories of social identity and social categorization have been applied to processes of communication in computer-mediated communication (CMC). In fact, they have fundamentally changed the belief that CMC must lead to a leveling of hierarchies and power as well as to anti-normative behavior. Spears and Lea (1994; see also Reicher, Spears, & Postmes, 1995) suggested the social identity model of de-individuation (SIDE). They grounded their framework in social identity theory and "argue that this framework is able to account for both the liberating and repressive potential within CMC systems and predict when and why each will occur" (Spears & Lea, 1994, p. 427).

The main idea of the SIDE model is that during anonymous CMC a user's personal or social identity can be more or less salient. When a social identity becomes salient, and the person identifies with the group, conformity to an internalized group norm will be strong. The normative and even stereotyping effect is thought to be more accentuated than in face-to-face interaction because individual characteristics of other users cannot be identified. In contrary, when personal identity is salient, de-individuation in CMC increases and activates individual norms. In a 2x2 experiment Spears, Lea, and Lee (1990) manipulated the salience of personal or social identity, and fixed settings that differed according to anonymity and de-individuation. In the de-individuated condition, participants located in different rooms discussed topics such as nuclear power stations, or governmental subsidies for theatres via CMC. In the individuation condition, participants, who were located in the same room and instructed not to talk to each other, discussed the same topics online. After the discussion, participants were asked for their attitudes on the issues. The results show an interaction effect: de-individuated discussants, who were immersed in a group, produced greater polarization in the direction of a pre-established group norm than participants who were addressed as individuals. Also, there was a greater change toward the norm in the case of the de-individuated condition as opposed to the individuated-group condition. In the individuated condition there was no difference between subjects with salient social and personal identity. Hence, we can summarize that the more people are immersed in groups under conditions of anonymity and de-individuation, the more likely they are to follow group norms. In contrast, people with a salient personal identity, who find themselves in an anonymous setting of CMC, are least likely to obey group norms.

The SIDE model suggests a variety of applications for entertaining CMC, such as chatrooms, online computer games, and fan forums. However, only little has been done to prove the SIDE model's implications for entertainment applications. Utz (1999) has investigated multi-user-dungeons (MUDs) in which people meet virtually to play (text-based) fantasy games. She assumes that people participating in MUDs are interested in fantasy games and role plays in general and hope to meet companions. For that reason, their social identity should be salient right when they enter the community. In a survey with 206 German MUDers, Utz (1999) accordingly showed that newcomers, who don't know MUD-members in physical life, identify stronger with the group. Also, a newbie overestimates the homogeneity of the group.

It is likely that virtual groups can be invented on the Net and create the same processes of social identity as they would do offline. Also, the SIDE model may explain social influence exhibited in CMC. This field of research is particularly promising, because previously the anonymity and violence of CMC has been emphasized as a danger (Kiesler, Siegel & McGuire, 1984). The SIDE model offers an approach on which the meaning of the anonymous social community to online gamers can be understood.

SIT AND ENTERTAINMENT: A RESEARCH AGENDA

The application of SIT in entertainment research as shown in this chapter is a developing field to explain selective exposure based on group membership, to understand the influence of entertainment on identity-related issues and to follow up on communication processes taking place in computer mediated communication. Although SIT has only recently been applied in the area of media and communication studies, it has already shown to explain some aspects of media usage and effects. There are shortcomings, however, that can be observed in the way SIT has been applied to entertainment research. In addition, not all possibilities of using the theory have been employed. In these terms, previous research and theorizing on SIT and entertainment show three major issues that should be on the agenda of future research.

First, it seems crucial to find out more about the motivational variables in SIT. What drives people to choose entertainment due to their social categories—is it self-esteem, self-actualization, or self-knowledge? And what role does social identification play? Particularly in exposure to entertainment, the motivation variable is under-researched in terms of theoretical and empirical implications. In former studies shown in this chapter, the first step has been taken. It has been shown that group membership is associated with preferences for certain kinds of entertainment, for example young people like shows featuring people of their own age groups (Harwood, 1997) and women like series with female lead characters (Trepte, 2004). The second step will be to elaborate on psychological processes that motivate behavior like that.

Very closely related to this problem is the fact that social comparison has not fully been integrated into empirical research on SIT and entertainment. There are no experiments in which participants have been observed to actually compare themselves with their out-group. All studies are based on the idea of evaluating the in-group versus the out-group. More research is needed that asks for details on social comparison, either by observing it within the situation of processing media input or by designing experiments that go beyond mere evaluations and imply relevant operationalizations of in-group favoritism. As long as this second step of broadening our understanding about the psychological and motivational variables in SIT has not been taken, we can only know for sure that: "Outcomes from an identity based television selection may not reflect SIT's traditional social comparison/self-esteem link, but rather a more basic solidarity/affiliation effect" (Harwood 1999, p. 130).

Second, in previous research on SIT and entertainment, social categories with very high psychological salience have been employed such as gender, cultural, or ethnic background. However, it seems that the dimensions of comparisons were not always relevant. Do people from different countries compare themselves on dimensions such as the 'quality of entertaining TV series'? There might be audiences that are very much involved with television entertainment (such as the authors of this book) that might consider this a relevant dimension, but the regular recipient might not share this notion. Hence, to create reasonable effects, dimensions of comparison should be chosen that actually mean something to the audiences. The international song contest "World Idol" brought together the winners of shows such as "American Idol" (U.S.), "Deutschland sucht den Superstar" (Germany), "Pop-Idol" (U.K.), and nine other countries. "World Idol" was very successful in almost all European countries. However, comparing unknown, teenage, amateur singers does not define the success of this show. Rather, it can be attributed to the popularity the shows got in the different countries before. It can be assumed that the international audiences of this entertainment event knew the singer of their own country by the time "World Idol" was broadcast. Hence, the dimension of comparison has already been learned by the viewers due to prior media experiences.

Another relevant dimension of comparison in entertainment is sports programming, which is deeply rooted in the media experiences of many viewers. There are multitudinous applications for looking at sports in entertainment research. It is likely that watching sports is contributing to positive or negative social identity. Nationality, but also other kinds of groups such as region, club, age or gender are addressed in sporting events. Hence, in further experiments, dimensions of comparison should accommodate entertainment media and the audience in question.

Third, a very promising field that has widely been neglected in research on SIT is the effects of media entertainment on minorities. Reid, Giles and Abrams (2004) address this issue in their "social identity model of media uses and effects." Nowadays, there are more chances than ever for minority groups to find their platform of entertainment, because all kinds of media—especially television and the Internet—are diversifying (Greenberg, Mastro & Brand, 2002). But, at the same time, certain groups are underrepresented in the mainstream media. These are either sizable societal groups, such as the older generation, or women (Elasmar, Hasegawa & Brain, 1999), but also minority groups, such as homosexuals or African-Americans (Greenberg

& Atkins, 1982). The fact that some groups are underrepresented in mainstream entertainment could have a number of effects. Media literate audiences might find their personal entertainment in niche media and experience positive social identity because they are able to find entertainment that reflects their understanding of which the group to they belong, and does not discriminate against them. In contrast, there are audiences that do not have access to minority media and widely rely on mainstream offers. The "digital divide" may have strong effects on groups and their identities. Further research is needed to find out how media entertainment affects identity-related behavior and what entertainment producers, and also governmental institutions, could actually do to enhance positive identity for minorities and for groups that are underrepresented in the media.

REFERENCES

Abrams, D., & Hogg, M. A. (1988). Comments on the motivational status of self-esteem in social identity and inter-group discrimination. *European Journal of Social Psychology, 18*, 317–334.

Armstrong, B., Neuendorf, K. A., & Brentar, J. E. (2001). TV entertainment, news, and racial perceptions of college students. *Journal of Communication, 42*(3), 152–176.

Ball-Rokeach, S. J. (1985). The origins of individual media-system dependency: A social framework. *Communication Research, 12*(4), 485–510.

Bem, S. L. (1974). The measurement of psychological androgyny. *Journal of Consulting Clinical Psychology, 42*(2), 155–162.

Blumler, J. G. (1979). The role of theory in uses and gratifications studies. *Communication Research, 6*, 9–36.

Blumler, J. G. (1985): The Social Character of Media Gratifications. In K. Rosengren, Wenner, L. & Palmgren, P. (Eds.), *Media Gratifications Research. Current Perspectives* (pp. 41–60). Beverly Hills.

Bryant, J., & Zillmann, D. (Eds.). (2002). *Media Effects. Advances in Theory and Research.* Mahwah, NJ: Lawrence Erlbaum Associates.

Cassata, M. & Irwin, B. (2003). *Going for the Gold: The Golden Girls are a hit!* http://www.medialit.org/reading_room/article411.html [25.08.2004]

Davison, W. P. (1983). The third-person effect in communication. *Public Opinion Quarterly, 47*, 1–15.

Duck, J. M., Hogg, M. A., & Terry, D. J. (1999). Social Identity and perceptions of media persuasion: Are we always less influenced than others? *Journal of Applied Social Psychology, 29*(9), 1879–1899.

Elasmar, M., Hasegawa, K., & Brain, M. (1999). The portrayal of women in US prime time television. *Journal of Broadcasting & Electronic Media, 43*, 20–34.

Festinger, L. (1954). A theory of social comparison process. *Human Relations, 7*, 117–140.

Greenberg, B. S., & Atkin, C. (1982). Learning about minorities from television: A research agenda. In G. Berry & C. Mitchell-Kernan (Eds.), *Television and the socialization of the minority child* (pp. 251–243). New York: Academic Press.

Greenberg, B. S., Mastro, D., & Brand, J. E. (2002). Minorities and the mass media: television into the 21st century. In J. Bryant & D. Zillmann (Eds.), *Media effects: Advances in theory and research* (pp. 333–352). Mahwah, NJ: Lawrence Erlbaum Associates.

Harwood, J. (1997). Viewing age: Lifespan Identity and television viewing choices. *Journal of Braodcasting and Electronic Media, 41*, 203–213.

Harwood, J. (1999). Age identification, social identity gratifications, and television viewing. *Journal of Broadcasting & Electronic Media, 43*(1), 123–136.

Harwood, J., & Roy, A. (2005). Social identity theory and mass communication research. In J. Harwood, & H. Giles (Eds.), *Inter group communication: Multiple perspectives* (pp. 189–212). New York: Peter Lang.

Hamilton, D. L., & Sherman, J. W. (1994). Stereotypes. In R. S. Wyer & T. K. Srull (Eds.), *Handbook of social cognition* (2nd ed., pp. 1–68). Hillsdale, NJ: Erlbaum Associates.

Hinkle, S., & Brown, R. (1990). Inter-group comparisons and social identity: Some links and lacune. In D. Abrams & M. A. Hogg (Eds.), *Social Identity Theory: Constructive and critical advances* (pp. 48–70). New York: Harvester Wheatsheaf.

Hogg, M. A., & Abrams, D. (1988). *Social identifications. A social psychology of inter-group relations and group processes.* London: Routledge.

Hogg, M. A., & Abrams, D. (1990). Social motivation, self-esteem and social identity. In D. Abrams & M. A. Hogg (Eds.), *Social Identity Theory: Constructive and critical advances* (pp. 28–47). New York: Harvester Wheatsheaf.

Horak, R. (2003). Sport space and national identity. *American Behavioral Scientist, 46*, 1506–1518.

Keillor, B. D. & Hult, G. T. M. (1999). A five country study of national identity. Implications for international marketing research and practice. *International Marketing Review, 16*(1), 65–82.

Kiesler, S., Siegel, J., & McGuire, T. W. (1984). Social psychological aspects of computer-mediated communication. *American Psychologist, 39*, 1123–1134.

Knobloch, S., Callison, C., Chen, L., Fritzsche, A., & Zillmann, D. (2005). Children's sex-stereotyped self-socialization through selective exposure to entertainment fare: cross cultural experiments in Germany, China, and the United States. *Journal of Communication, 55*(1), 122–138.

Mares, M. L., & Cantor, J. (1992). Elderly Viewers Responses to televised portrayals of old age. Empathy and Mood Management versus Social Comparison. *Communication Research, 19*(4), 459–478.

Mastro, D. E. (2003). A social identity approach to understanding the impact of television messages. *Communication Monographs, 70*(2), 98–113.

McQuail, D. (2000). *McQuail's mass communication theory* (4th ed.). London: Sage.

Morton, T. A., & Duck, J. M. (2000). Social Identity and media dependency in the gay community. The prediction of safe sex attitudes. *Communication Research, 27*(4), 438–460.

Oakes, P. (1987). The salience of social categories. In J. C. Turner (Ed.), *Rediscovering the social group: A self-categorization theory* (pp. 117–141). Oxford: Basil Blackwell.

Oakes, P. J. (1996). The categorization process: Cognition and the group in the social psychology of stereotyping. In W. P. Robinson (Ed.), *Social groups and identities: Developing the legacy of Henri Tajfel* (pp. 95–120). Oxford: Butterworth-Heinemann.

Oakes, P. J., Haslam, A., & Turner, J. C. (1994). *Stereotyping and social reality*. Oxford: Blackwell.

Oakes, P. J., Turner, J. C., & Haslam, A. (1991). Perceiving people as group members: The role of fit in the salience of social categorizations. *British Journal of Social Psychology, 30*, 127–144.

Oliver, M. B. (2000). The respondent gender gap. In D. Zillmann & P. Vorderer (Eds.), *Media entertainment. The psychology of its appeal* (pp. 215–234). Mahwah, NJ: Lawrence Erlbaum Associates.

Oliver, M. B., Weaver, J. B. & Sargent, L. (2000). An examination of factors related to sex differences in enjoyment of sad films. *Journal of Broadcasting & Electronic Media, 44*(2), 282–300.

Reicher, S. D., Spears, R., & Postmes, T. (1995). A social identity model of deindividuation phenomena. In W. Stroebe & M. Hewstone (Eds.), *European reviews of social psychology* (vol. 6, pp. 161–197). Chichester: Wiley.

Reid, S. A., Giles, H., & Abrams, J. R. (2004). A social identity model of media usage and effects. *Zeitschrift für Medienpsychologie, 16*(1), 17–25.

Rosenberg, M. (1991). The self-esteem scale. In J. Robinson, P. Shaver, & L. Wrightsman (Eds.), *Measures of personality and social psychological attitudes* (pp. 121–123). San Diego, CA: Academic Press.

Spears, R., & Lea, M. (1994). Panacea or panopticum? The hidden power in computer-mediated communication. *Communication Research, 21*(4), 427–459.

Spears, R., Lea, M., & Lee, S. (1990). De-individuation and group polarization in computer-mediated communication. *British Journal of Social Psychology, 29*, 121–134.

Straubhaar, J. D. (1991). Beyond media imperialism: Assymetrical interdependence and cultural proximity. *Critical Studies in Mass Communication, 8*, 39–59.

Tajfel, H. (1959). The anchoring effects of value in a scale of judgements. *British Journal of Psychology, 50*, 294–304.

Tajfel, H. (1969). Cognitive aspects of prejudice. *Journal of Social Issues, 25*, 79–97.

Tajfel, H. (1972). Social categorisation. In S. Moscovici (Ed.), *Introduction à la Psychologie Sociale* (Vol. 1). Paris: Larousse.

Tajfel, H. (1978). *Differentiation between social groups*. London: Academic Press.

Tajfel, H. (1979). Individuals and groups in social psychology. *British Journal of Social and Clinical Psychology, 18*, 183–190.

Tajfel, H. (1981). *Human groups and social categories*. Cambridge: University Press.

Tajfel, H., & Turner, J. (1979). An integrative theory of inter-group conflict. In J. A. Williams & S. Worchel (Eds.), *The social psychology of inter-group relations* (pp. 33–47). Belmont, CA: Wadsworth.

Tajfel, H., Flament, C., Billig, M. G., & Bundy, R. F. (1971). Social categorization and inter-group behavior. *European Journal of Social Psychology, 1*, 149–178.

Tajfel, H., & Wilkes, A. L. (1963). Classification and quantitative judgement. *British Journal of Psychology, 54*, 101–114.

Tarrant, M., North, A. C., & Hargreaves, D. J. (2001). Social categorization, self-esteem, and the estimated musical preferences of male adolescents. *The Journal of Social Psychology, 141*(5), 565–581.

Trepte, S. (2004). Soziale Identität und Medienwahl. Eine binationale Studie zum Einfluss von Gender-Identität und nationaler Identität auf die Auswahl unterhaltender Medieninhalte [Social identity and media choices. The influence of gender identity and national identity on selective exposure]. *Medien & Kommunikationswissenschaft, 52*(2), 230–249.

Trepte, S. (2005). Daily Talk as Self-Realization. An empirical study on lay participation in daily talk shows. *Media Psychology, 7*(2), 165–189.

Trepte, S., & Krämer, N. (in preparation). Expanding social identity theory to research in media effects.

Turner, J. C. (1982). Toward a cognitive redefinition of the social group. In H. Tajfel (Ed.), *Social identity and inter-group relations* (pp. 15–40). Cambridge: Cambridge University Press.

Turner, J. C. (Ed.). (1987). *Rediscovering the social group: A self-categorization theory.* Oxford: Basil Blackwell.

Turner, J. C. (1999). Some current issues in research on social identity and self-categoization theories. In N. Ellemers, R. Spears, & B. Dossje (Eds.), *Social identity: Context, commitment, content* (pp. 6–34). Oxford: Blackwell Publishers.

Turner, J. C., Brown, D., & Tajfel, H. (1979). Social comparison and group interest in in-group favouritism. *European Journal of Social Psychology, 9,* 187–204.

Turner, J. C., & Onorato, R. S. (1999). Social identity, personality, and the self-concept: A self-categorization perspective. In T. R. Tyler, R. M. Kramer, & O. P. John (Eds.), *The psychology of the social self* (pp. 11–46). Mahwah, NJ: Lawrence Erlbaum Associates.

Utz, S. (1999). *Soziale Identifikation mit virtuellen Gemeinschaften—Bedingungen und Konsequenzen.* [Social identification with virtual communities—antecedents and consequences] Lengerich: Pabst.

Utz, S. (2002). Interaktion und Identität in virtuellen Gemeinschaften.[Interaction and identity in virtual communities] In B. Bente, N. C. Krämer, & A. Petersen (Eds.), *Virtuelle Realitäten* (pp. 159–180). [Virtual realities]. Gättingen: Hogrefe.

Vorderer, P., Wulff, H. J., & Friedrichsen, M. (Eds.). (1996). *Suspense. Conceptualizations, theoretical analyses, and empirical explorations.* Mahwah, NJ: Erlbaum.

Waisbord, S. (2004). McTV: Understanding the global popularity of television formats. *Television and New Media, 5*(4), 359–383.

Zillmann, D. (1988). Mood management: Using entertainment to full advantage. In L. Donohew, H. E. Sypher, & E. T. Higgins (Eds.), *Communication, social cognition, and affect* (pp. 147–171). Hillsdale, NJ: Lawrence Erlbaum.

Zillmann, D., Aust, C. F., Hoffman, K. D., Love, C. L., Ordman, V. L., Pope, J. T., Seigler, P. D., & Gibson, R. (1995). Radical Rap: Does it further ethnic division? *Basic and Applied Social Psychology, 16*(1&2), 1–25.

Zillmann, D., & Bryant, J. (Eds.). (1985). *Selective exposure to communication.* Hillsdale, NJ: Lawrence Erlbaum Associates.

Zillmann, D., & Vorderer, P. (Eds.). (2000). *Media entertainment. The psychology of its appeal.* Mahwah, NJ: Lawrence Erlbaum Associates.

16

Equity and Justice

Manfred Schmitt
University of Koblenz-Landau

Jürgen Maes
University of Trier

Surprise, suspense, immediate emotional involvement, and pleasure are the most important ingredients of entertaining activities such as reading a novel, watching a movie, or playing a game (Bryant, Roskos-Ewoldsen, & Cantor, 2003; Zillmann & Vorderer, 2000). Activities are pleasurable if they satisfy the actor's values and needs. The focus of this chapter is on a value and need that has been linked with entertainment in Aristotle's drama theory (*Poetica),* disposition theory (Zillmann, 2000; Zillmann & Bryant, 1975), moral sanction theory (Raney & Bryant, 2002), and structural affect theory (Brewer & Lichtenstein, 1982): justice.

THE NEED FOR JUSTICE

Lerner (1980a) proposed and demonstrated in a series of experimental studies (Lerner & Miller, 1978; Lerner, Miller & Holmes, 1976) that people have a strong desire to live in a just world, a world where people get what they deserve, and deserve what they get. This desire motivates people to believe in a just world. Believing in justice bestows the person with a sense of security, control, and trust. A world in which individuals, groups, and institutions interact on the basis of justice is a safe and predictable world. Knowledge of principles of fairness enables people to predict and control the consequences of their own behavior and the behavior of others. A person who knows what deeds will be rewarded and what acts will be punished can maximize positive and minimize negative outcomes by obeying the rules. Furthermore, people can feel safe if they trust that others will also follow the rules.

The belief in a just world has positive consequences for well-being because its implications—security, control, and trust—are core components of mental health (Bandura, 1977; Rotter, 1980; Seligman, 1975). Several recent studies have shown that the belief in justice serves as a personal resource. It lowers vulnerability to critical life events and contributes to well-being and satisfaction over and above other factors such as extraversion, neuroticism, and optimism (Dalbert, 2001). Research on job satisfaction and absenteeism is also consistent with Lerner's need for justice construct. Many studies in organizational psychology have revealed what has been called the fair-process effect. Fair treatment at the workplace has a positive impact on job satisfaction and decreases the number of sick days employees take (e.g., Schmitt & Dörfel, 1999).

Besides serving as a personal resource, the belief in justice is a fundamental precondition for the successful functioning of groups and societies. If members of a society did not believe in the moral integrity of their fellow citizens, they would not trust in the reliability of social contracts. Furthermore, if citizens lost their belief in the uprightness of institutions, authorities, and political leaders, their national identification would diminish (Tyler & Smith, 1998). In fact, anomia is the ultimate consequence of disbelief in justice (Arts, Hermkens, & van Wijck, 1995). Last but not least, by committing group members to social norms and contracts, the need for justice lowers the necessity of social control and frees resources the group can use for advancing its productivity. Direct support for this conjecture comes again from research in organizational psychology (Tyler, 1986). Several studies have found that the violation of justice principles by superordinates or fellow employees decreases organizational citizenship behavior and increases anti-citizenship behavior such as stealing, absenteeism, malicious gossip, and sabotage (e.g., Skarlicki & Folger, 1997).

Lerner's (1977) explanation for this effect follows up on Piaget's (1932) theory of moral development. Lerner argues that moral autonomy and self control are based on the capacity to delay gratification and on believing that desirable outcomes are contingent upon appropriate behavior. As soon as these two attributes have emerged in the process of development and socialization, children begin to make *personal contracts*. They offer moral behavior in return for desirable outcomes such as approval of peers or academic success. The belief in a just world appears as a result of two sorts of generalizations. Children generalize from concrete contracts to abstract principles and from their own reasoning and behavior to the logic and conduct of other people.

COPING WITH THREATS TO THE BELIEF IN JUSTICE

We all know, of course, that our world is not just, or it is at least less just than we wish it were. In ordinary folk language: Shit happens. A glance at the front page of any newspaper or a glimpse at any news channel is sufficient to prove it. Each of us can easily recall incidents where we felt treated unfairly (Mikula, Petri, & Tanzer, 1990; Scherer, Wallbott & Summerfield, 1986). We all can name a friend, a relative, or a colleague who recently complained about mistreatment or betrayal at work or in a private relationship. From a *rational* point of view, we must admit that injustice is all over the place. This rational knowledge implies an ongoing threat to our desire for justice. How do we cope with this threat?

Lerner has described several strategies people employ to defend or restore their belief in justice. Helping the innocent victim and punishing the victimizer are straightforward and attractive strategies because they not only affirm the belief in justice but also give the bystander a sense of power and efficacy. Unfortunately, direct interventions are often impossible or costly. That is why people employ less noble strategies. They reinterpret the situation to make it appear just, after all, or at least less unjust than it seemed on first sight. They blame victims for their misfortune to make it appear deserved (Maes, 1994; Ryan, 1971). They derogate the victim by making inferences from the archaic principle that good people deserve good fates and bad people bad fates (Heider, 1958). Ascribing responsibility or bad character to the suffering victim frees bystanders from their obligation to help (Lerner & Simmons, 1966; Montada &

Schneider, 1989). A fourth strategy employs a shift in time perspective. A bystander might admit that injustice indeed occurred, yet assume that eventually the victim will be compensated and the perpetrator punished (Maes, 1998). These, and additional mechanisms, explain why people are able to hold on to their justice belief even in the face of considerable counterevidence.

In his recent writing, Lerner (2003) offers an additional explanation for the persistence of the belief in justice and the power of the justice motive. In line with other two-process theories (Devine, 1989; Strack & Deutsch, 2004), Lerner contends that individuals process justice-related information via two distinct systems that operate simultaneously, but often independently. In the controlled information processing mode, people elaborate the incoming information *explicitly* and analyze it rationally. When confronted with an obvious case of injustice, they acknowledge that injustice occurred, express concern, and reflect upon ways to restore justice. If they cannot come up with a reasonable way of restoring justice, justifications for the situation and other defensive logic will occur at the highest level of intellectual sophistication at which the person is capable. In the automatic or *implicit* mode, by contrast, the person processes justice-related information schematically and without conscious awareness. The range of intuitive impulses includes helping the victim, aggressing against the perpetrator, and activating defense schemata such as derogating the victim.

The distinction between implicit and explicit justice motives and beliefs has important consequences for the entertaining potential of justice themes. This is true because the implicit and the explicit belief in justice are often *dissociated*. Consistent with the skewed distribution of responses to items of just world scales (Schmitt, 1998), Lerner (1980) argues that most adults consider belief in justice naïve and false. Despite their rational disbelief in justice, however, they behave *as if* the world was just. They continue to make personal contracts (such as trading virtue for happiness), they trust in social contracts, they invest in long term goals, and so forth. According to Lerner (1980), this happens (1) because people implicitly deny injustice even though they *explicitly* admit it; and (2) because the experience of injustice does not annihilate the justice motive. Instead, the experience of injustice augments the desire for justice. People continue *wanting* to believe in a just world and because they implicitly do so, the rational admittance of injustice does not stop them from acting as if the world was just.

JUSTICE AS A PLEASURE FACTOR IN THE MEDIA: GENERAL HYPOTHESIS AND UNSYSTEMATIC EVIDENCE FROM CONTENT ANALYSES OF CASE STUDIES

Because justice is a core need, justice themes grab attention and have great potential to generate emotion and suspense (Mikula, Scherer, & Athenstaedt, 1998; Montada, 1993; Zillmann, 2003). The desire for justice makes people prefer information that affirms their belief in justice over information that casts doubt on it (Lerner, 1980). Accordingly, episodes conveying a contingency between just behavior and reward versus unfair behavior and punishment are intrinsically satisfying (Brewer, 1985; Raney & Bryant, 2002; Zillmann, 2000). We can expect on the basis of justice motive theory and concordant empirical observations (1) that justice is a frequent issue in many literature and movie genres; (2) that the arousal and subsequent satisfaction of the justice need is a powerful entertainment tool that will be employed regularly by a large variety of media and; (3) that this mechanism is effective even if the audience is not consciously aware of justice being an issue.

Indeed, a content analysis of fairy tales, novels, and various movie genres is consistent with these assumptions (Chatman, 1978; Friedman, 1975). Rubin and Peplau (1973) were the first to link fairy tales with the development of the justice motive. They observed, as others had before, that most fairy tales come down to a simple message: Goodness wins, badness loses. Rubin and Peplau assume that this message plays an important role in the development of the child's understanding of, desire for, and belief in, justice. In order to make the message most powerful, fairy tales often disturb the sense of justice by letting the antagonist take pleasure in the protagonist's suffering. The suffering of the protagonist and the Schadenfreude of the antagonist are inconsistent with the child's intuitive sense of right and wrong (Zillmann & Cantor, 1977). Furthermore, protagonists are portrayed in ways that make it easy for children to identify with them (Perrine, 1959). Creating a unit-relationship (Lerner, 1980) between children and the protagonist augments children's empathy (Zillmann, 1991) which in turn amplifies their justice sensitivity (Miller, 1998). Children wish for an end to the protagonist's suffering because they suffer vicariously. In addition to wanting relief from pain, their moral intuition tells them that a good character should not suffer and a bad character does not deserve pleasure. In other words, children want equity. There is no need to say that most fairy tales satisfy children's need for justice—in the end, goodness wins and badness loses. The brave knight who overwhelmed the dragon will marry the beautiful princess and be given half of her father's kingdom on top.

Even though presented more subtly and often veiled behind a multifaceted narrative, "the macrostory of all drama seems to be that all injustice necessarily results in some restoration of justice." (Raney & Bryant, 2002, p. 404). A good example is Tolkien's *Lord of the Rings*, a trilogy of unparalleled complexity. Its overarching theme, however, is simple: good fights bad. The good guys, represented by Frodo, are desperately dedicated to save their just world, which the bad guys, controlled by Sauron, want to destroy. Sauron is defeated, but the triumph of virtue over evil has a tremendous price—proving that justice is precious and worth the lives of thousands of brave soldiers. They died for a just cause. Just causes are good causes, perhaps the best causes for an audience with a fragile belief in a just world.

The urge for justice increases with the innocence of the victims. Children are the most innocent of victims. Peter Hoeg's *Smilla's Sense of Snow* may serve to illustrate the point. Jesaja, witness of his father's death, a man who had been abused by a syndicate of scrupulous scientists, adventurers, and would-be emperors, is killed in a faked accident to wipe out his knowledge. Smilla, an Inuit, snow researcher, and loser like Jesaja, can tell from Jesaja's footprints in the snow that his death was not an accident. Realizing that the police are uninterested in finding the truth and doing justice, she takes the case into her own hands. Although the book draws its thrilling potential from a variety of fictional ingredients and narrative techniques, justice is the overarching theme and driving force behind the story. The need for justice keeps the reader restless and wishing for a moral completion of the story. Like most writers, Hoeg satisfies his readers' need for justice: At the risk of her own life, Smilla hunts down the villains.

Given the entertaining potential of justice, it comes as no surprise that justice is a theme in a wide range of motion picture genres, sometimes latent and subtle, sometimes obvious and blatant (Raney, 2002, 2003a). Justice plays a role in virtually every Walt Disney movie, for instance *King of the Lions*, in the classic Western such as Fred Zinnemann's *High Noon*, and in criminal trial dramas like Billy Wilder's *Witness for the Prosecution*. Last but not least, justice is *the* fundamental issue in revenge movies.

In his chapter, *Clint Eastwoold and Equity*, Miller (1998) devotes a brilliant treatise on revenge to this genre. Taking classic movies such as Michael Winner's *Death Wish* or Jerry Hogrewe's *Dirty Harry* as examples, Miller's analysis comes down to a straightforward conclusion: "The modern revenge film is about justice, doing justice." (p. 170). Miller identifies several subtypes of revenge narratives. Of those, two can be distinguished by the hero's role. The first kind of avenger pays back, like Paul Kersey in *Death Wish*, the wrong that was brought on him. The second type of hero takes revenge on behalf of someone else. *Smilla's Sense of Snow* represents this category and serves as a good example for another pattern that Miller identified. Revenge stories often imply a criticism of the state-delivered justice, for instance by portraying the police as corrupt or incompetent, the courts as inefficient or overly mild with the transgressor. From the perspective of justice-motive theory, such narrative techniques are effective for two reasons. They amplify the need for justice by imposing an additional threat to the justice belief over and above the wrong that was done by the villain. In addition, they provide justification for the avenging heroes for violating the state's monopoly on retributive violence. Most interestingly perhaps, from a social justice perspective, and in line with observations made by other authors (Zillmann, 2000), Miller proposes that avengers possess moral authority only as long as their actions meet the equity criterion. Avengers who retaliate insufficiently leave the sense of justice disturbed and risk losing the respect of their audience. Avengers who overact and ignore normative constraints fail to satisfy the audience's need for justice as well. In fact, little seems more disappointing for an audience than a knightly hero who turns into a desperado and trades honor for bloody lust.

Our next sketchy analysis is devoted to sports, a very important and one of the oldest sources of entertainment. Again, a few examples may suffice to demonstrate that justice is an important pleasure factor in this domain as well (Raney, 2003b). Members and fans of the losing team will be outraged if victory was due to unfair play or an obviously false judgment of a referee. At the same time, the burden of unfairness will limit the joy, pride, and satisfaction of the winning team and those who identify with it. Doping is a second example. Ben Johnson, Johann Mühlegg, and some other of the world's best athletes never made it back to the top again after they had been convicted of doping. Their failure to come back was probably less due to physical insufficiency but to the stigma of needing dope to win. What could be more demoralizing and *mentally* handicapping for someone who runs for honor than knowing that every future success will raise doubtful questions? Finally, let us consider winners whose victories grant the deepest satisfaction in spectators. These are winners who succeed even though the odds are definitely against them. Wilma Rudoph, winner of three gold medals at the 1960 Rome Olympics (100 m, 200 m, 4 × 100 m), is perhaps the best example. At the age of 5, Wilma contracted polio and lost the use of her left leg. Although her family was poor, they never gave up. Brothers and sisters took turns massaging the paralyzed leg, her mother drove Wilma 90 miles roundtrip to a hospital for the weekly therapy. Years of determined treatment and training helped Wilma not only to become a normal child but a basketball star and eventually even the world's best sprinter. According to justice motive theory, the deep satisfaction her success gives the observer results from its deservedness. If the most disadvantaged can win, the world cannot be unjust.

Finally, the justice-finality principle also shapes the hedonic value of nonfiction drama displayed by news media (Zillmann, Taylor, & Lewis, 1998). A true case may serve to illustrate the principle. Less than a year ago, a young woman and her daughter drove in their medium-size car on the Autobahn near Stuttgart, Germany. Driving at approximately 120 km/h, they were approached from behind by a sports car speeding at 220 km/h. Flashing his headlights, the speeder tried to chase the woman out of his lane. The frightened mother pulled right, her car turned over, and hit a tree. Mother and daughter were killed. The accident was witnessed by other drivers who later testified in the trial. Unfortunately, the accident had occurred in darkness. None of the witnesses was able to read the license plate of the speeding car. The event was in the newspapers, on the radio, and on TV for weeks. Coverage of the case did not end after a test driver of Daimler-Crysler was suspected and arrested. In a long trial, he was eventually convicted on the basis of indirect evidence the court considered beyond a reasonable doubt. During the trial and even after the sentencing, the case provoked a fervent discussion both in the public and the media. We propose that a justice dilemma fueled the debate and its long media coverage. Right after the accident, every voice expressed the compassionate wish that the speeder be identified and punished. Accordingly, public relief was tremendous when the suspect was arrested. However, the inquiry of the witnesses did not reveal unanimous evidence of his guilt—leaving the public need for justice unsatisfied. Moreover, cautious voices, a clear minority in the debate, reminded people of the presumption of innocence principle. Not surprisingly, this reminder added to the frustration of the majority who wished nothing more than certain justice. Despite its great tragedy, the case would have attracted much less attention if the speeder had been identified immediately and without doubt—or if the suspect had admitted his guilt. The need for justice would have been satisfied and the case would have disappeared quickly from the media.

WHAT IS JUST? ANSWERS FROM SOCIAL JUSTICE RESEARCH

So far, we have only argued that justice and deserving play a significant role in entertainment. We have not yet addressed what it is exactly that novel and newspaper readers, movie watchers, theatre audiences, and spectators of sports events consider as just. Maintaining that good people deserve good outcomes and that evildoers deserve punishment is only a crude answer to this question. It may be sufficient as a rule of thumb for our basal sense of morality (Zillmann, 2000). From a scientific point of view, however, more detailed knowledge about *what* exactly contributes to our sense of justice and its entertaining effects seems desirable. Such knowledge would allow for a sophisticated analysis of justice episodes and their narrative architecture. Detailed knowledge is necessary for untying the structural and substantive elements of narratives and for determining their general and conditional effects on drama appreciation, story-liking, and the entertaining value of episodes (Raney & Bryant, 2002). Fortunately, such knowledge is available.

Over the last 40 years, social scientists have exerted considerable research effort to understand when and why people feel treated fairly and how they react to injustice. This work was mostly done in social psychology and sociology, some of it also in political science and economy. In contrast to the *normative* solutions to justice problems sought in philosophy, theology, and law, social sciences are devoted to investigate empirically how subjective standards of justice affect judgment, emotion, and behavior. Although normative and empirical approaches overlap and feed into each other (Sabbagh, 2001), the latter are more relevant for the present purpose because what matters in entertainment are subjective feelings.

Types of Justice

The social justice literature suggests that a conceptual distinction among four types of justice is necessary for capturing the full range of justice issues that come up in social interaction: distributive justice, procedural justice, interactive justice, and retributive justice. Parts of this distinction were proposed already by Aristotle in his *Nicomachean Ethic*.

Distributive justice is concerned with the fair allocation of goods among entitled recipients. Systematic distributive justice research was initiated 40 years ago by Adams' (1965) proposal of equity theory. Equity theory submits that people feel rewarded fairly whenever the outcome-input ratio is equal across recipients. Any deviation from the proportionality principle causes distress. Allocators fear being accused of favoritism, the over-rewarded feel guilty, whereas the under-rewarded react with anger and indignation. All three emotions motivate their beholders to correct the distribution. If that is not possible, allocators and recipients will reassess inputs and outcomes so as to make them subjectively equitable. This can be achieved by the under-rewarded, for instance, by downgrading their own inputs, downgrading outcomes of the other recipient, upgrading their own outcomes, or upgrading the other recipient's inputs. If this strategy fails, for instance because inequity is too marked, recipients will eventually quit the exchange relationship. Its appealing simplicity made equity theory popular and influential (Berkowitz & Walster, 1976). However, experimental and survey studies soon revealed that people consider other distribution principles, such as the principle of equality and the need principle, often as more just than the equity principle. Furthermore, several studies supported Deutsch's (1975, 1985) conjecture that preferences for distribution principles depend on the social context and the resource to be distributed (Schwinger, 1980; Törnblom, 1992). Equity is considered most fair in achievement contexts such as sports and labor, parity is preferred in close relationships, and need is considered just when resources are devoted to the welfare of recipients who cannot care for themselves.

Procedural justice deals with the decision-making process, rather than with its outcome. Researchers' attention was attracted to procedural justice issues by research findings showing that advantageous outcomes are sometimes less satisfying than unpleasant outcomes (Lind & Tyler, 1988). This happens when positive outcomes result from unfair procedures and unpleasant outcomes from fair decisions. Tyler's group value model offers an explanation for this pattern (Tyler, Degoey, & Smith, 1996). Superordinates who make decisions according to fair procedures communicate respect and esteem to superordinate group members. What procedures are fair then? Thibaut and Walker (1975), who conducted their studies in the legal context, contended that procedural fairness consists of two broad components, process control and outcome control. Based on theoretical reasoning and empirical observations, Leventhal (1980) proposed to break down these components into more narrow ones: consistency (in the application of decision criteria), nonpartiality (in the treatment of individuals), accuracy (in the collection and integration of information), correctability (of the decision or the criteria), representativeness (of the information that was used), and conformity (of the decision rules) with ethical standards. Some studies in organizational psychology were indeed able to trace the fair process effect (see above) to the fulfillment versus violation of these criteria (e.g., Schmitt & Dörfel, 1999).

Interactional justice qualifies the way decision makers interact with those who are affected by their decisions. In a series of theoretical and empirical papers, Bies (1987; Bies & Moag, 1986; Bies & Shapiro, 1987) proposed that the *manner* in which procedures are executed and decisions published needs to be differentiated from the procedures itself and from the *criteria* on which these were based. Bies predicts that decisions are experienced as more fair when the decision maker communicates them in a friendly and respectful way, shows empathic concern for the effects the decision might have, explains why it was made and justifies unfavorable consequences. Bies demonstrated in his own research that interactional justice measures contribute uniquely to the fair process effect. From the perspective of Tyler's group value model, procedural and interactional justice are two sides of the same coin because both the procedures and how they are communicated and carried out share justice values as a common factor. This common factor is what matters to the targets who use whatever information is available to assess fairness.

Retributive justice comprises the principles that regulate when, how, and for what moral reasons a wrongdoer is sanctioned (Miller, 2001; Vidmar, 2001). Sanctions include retaliation and vengeance executed by the victim or his kin, punishment by the group to which the perpetrator belongs, and formal sentencing according to a legal code by a moral authority. Research on retributive justice has dealt with a large range of issues such as: the motives (justice vs. control) and aims (getting even versus reintegrating the fallen group member into the group) of punishment; the influence of shared vs. unshared group membership; characteristics of the victim, the victimizer, and their relationship; the perpetrator's behavior after the wrongful act (Oswald, Hupfeld, Klug, & Gabriel, 2002; Wenzel, 2002). Of the rich body of findings from this research, a few will be mentioned for illustrative purposes. It was found, for instance, that group members are often punished more harshly than non-members (Marques & Páez, 1994). From the perspective of social identity theory (Tajfel & Turner, 1986), this black sheep effect indicates a rebuttal against the threat a group member's wrongdoing implies for the social identity and self-esteem of the remaining group members. A second set of studies has shown that the victim's willingness to forgive and abstain from retaliation depends on the perpetrator's behavior toward the victim after the transgression. Most victims are willing to forgive perpetrators who apologize, regret their behavior, show remorse, offer compensation, and promise honestly to comply with the violated norm in the future (Schmitt, Gollwitzer, Förster &Montada, 2004). A third research example deals with a major reason for the conceptual distinction between retributive and distributive justice as first proposed by Aristotle. In modern terms, he argued that retributive punishment and the allocation of goods are qualitatively distinct matters that cannot be projected on a bipolar continuum. Therefore, they require different standards of justice. Recent research on the positive–negative asymmetry shows that lay judges behave as Aristotle requested (e.g., Sabbagh & Schmitt, 1998). Most people are more concerned with erring on the side of the withdrawal of goods, such as freedom, money, or honor, than in the award of such goods. This asymmetry resembles principles in legal justice such as the presumption of innocence and the necessity of evidence beyond a reasonable doubt. The idea of killing an innocent person who was accused of first-degree murder seems much more terrifying for most people, jurors, and judges than letting a true murderer free for lack of evidence.

Individual Differences in Justice Judgment and Behavior

At the time of their proposal, all major social justice theories claimed far reaching generalizability. Equity theory, for instance, predicted that all people will react with anger to under-reward, and with guilt to over-reward. In the same vein, Lerner assumed that all people have a need and belief in a just world. In contrast to such parsimonious assumptions, large proportions of variance remained unexplained in experimental justice research. This drew scholars' attention to cultural (e.g., Gergen, Morse, & Gergen, 1980), gender (e.g., Major & Deaux, 1982), and individual differences (e.g., Rubin & Peplau, 1973). Research interest in individual differences stimulated the construction of justice measures that helped us to understand better than was possible with experimental studies alone when and why specific individuals feel treated unfairly in specific situations and react with a specific response pattern. As a result of this line of research, we know that individuals differ systematically in

- the extent of their just world belief (e.g., Lipkus, 1991; Rubin & Peplau, 1973),
- their preferences for principles of distributive justices such as equity, need, parity (e.g., Davey, Bobocel, Hing, & Zanna, 1999; Sabbagh, Dar, & Resh, 1994),
- their attitudes towards principles of procedural justice (e.g., Schmitt & Dörfel, 1999),
- their punitiveness towards transgressors (e.g., Schmitt, Neumann, & Montada, 1995),
- the value they place on justice (e.g., Dalbert, Montada, & Schmitt, 1987),
- the strength of their emotional reaction to injustice (e.g., van den Bos, Maas, Waldring, & Semin, 2003),
- their sensitivity to the violation of equity (e.g., Huseman, Hatfield & Miles, 1985),
- their sensitivity as victims of deprivation and injustice (e.g., Dar & Resh, 2001, Schmitt, 1996),
- their sensitivity as beneficiaries of injustice (e.g., Montada & Schneider, 1989; Schmitt, Behner, Montada, Müller, & Müller-Fohrbrodt, 2000),
- their sensitivity as observers of injustice (e.g., Fetchenhauer & Huang, 2004; Schmitt, Gollwitzer, Maes, & Arbach, 2005).

Person x Situation Interactions in Justice Judgment and Behavior

Research from other psychological domains suggests that justice judgment, emotion, and behavior are jointly shaped by attributes of the situational context and the personality of the involved individuals. Although this proposal sounds like a truism, social justice research has devoted little attention to person x situation interactions. The few studies in which the joint impact of personality and situation factors was considered, however, identified an effect pattern that Schmitt, Eid, and Maes (2003) termed synergistic interaction of functionally equivalent person and situation factors. Factors are functionally equivalent if they affect the same outcome variable for the same psychological reason. A synergistic interaction occurs if two factors amplify each other such that the impact of one factor correlates positively with the value of the other factor.

Synergistic interactions of this kind were identified both in research on the justice motive and in distributive justice studies. Regarding the former, the synergistic model predicts that general effects of the justice motive, such as helping innocent victims or derogating them, should be stronger for observers with a high belief in justice (justice motive) than for observers with a low belief in justice (justice motive). This pattern of effect was indeed found in a few— albeit not in all—studies on the joint impact of situational and individual justice motive factors (Schmitt, 1998).

Regarding distributive justice, the synergistic model implies that the subjective deservedness of outcomes does not depend only on recipients' inputs (equity principle), or needs (need principle), but also on their attitudes toward the distribution principle at issue. More specifically, the model predicts that achievements will make a larger difference for individuals with a favorable attitude toward equity than it will for individuals with an unfavorable attitude. Similarly, recipients with a positive attitude toward the need principle should be more responsive to need differences than will be recipients with a negative attitude. Exactly this pattern was obtained in a study by Herrmann and Winterhoff (1980). Furthermore, attitudes toward equality should have the opposite effect: Achievement or need differences should be less relevant for recipients with a positive attitude toward equality than for individuals with a negative attitude. Given the bipolar nature of attitudes, the preceding effect can be reframed into a synergistic principle: The impact of any situational information that justifies unequal distribution will increase with increasing *negativity* of a person's attitude toward equality. A vignette study by Schmitt et al. (2003) and two experiments by Schmitt and Sabbagh (2004) obtained results that were in perfect agreement with this prediction.

JUSTICE AS AN ENTERTAINMENT FACTOR: SPECIFIC HYPOTHESES

Obviously, the rich pattern of results social justice research has produced over the last 40 years goes far beyond the simple principle that good behavior deserves reward and bad behavior punishment (Raney & Bryant, 2002). Rather, *innumerous specific hypotheses* can be derived from combining the justice facets we have addressed, namely:

- the type of justice (distributive, procedural, interactive, retributive),
- the situational context (achievement, friendship, nurturance; group membership; relationship between victim and perpetrator; behavior of perpetrator toward victim after transgression),
- the personality of the individuals who are involved in a justice episode as actors or observers (belief in a just world; attitude toward various distribution principles; attitude

toward various procedural principles; centrality of justice; equity sensitivity; victim sensitivity; perpetrator/beneficiary sensitivity; observer sensitivity),
- the personality x situation combination (equal or unequal values on functionally equivalent justice factors).

A single example may suffice for illustrating the specificity of hypotheses that can be derived from combining these facets. Let's use a fairy tale that contains a distribution conflict among siblings. An elderly couple owns a small farm and wants to hand it over to their sons, Peter and Paul. Let's assume they decide to divide their property equally. Whether or not this decision is considered just by the observer depends on many factors and their interactions. If Peter and Paul are portrayed as rivals, if achievement issues are made salient, and if both sons differ substantially in how much they have helped to run the farm, the distribution will appear unjust to most readers. If Peter and Paul are portrayed as best friends and if harmony issues are made salient, the distribution will be approved as fair by most observers. If need issues are made salient by portraying Peter as a fragile person who needs a stable environment and depends on getting the farm to make a decent living, whereas Paul is an energetic, creative, and independent man who could have any successful career, most witnesses would be bothered by the decision—unless Paul gave Peter his half of the farm. Each of these effects will depend on the observer's attitudes toward equality, equity, and need as possible principles of distributive justice. Furthermore, the impact of mismatches between the observers' attitudes and the decision will be a function of their justice sensitivity and their belief in a just world. Regarding the latter, their sense of justice will depend on which way they believe injustice predominates. Their sense of injustice will be disturbed more if the belief in immanent justice prevails. The opposite effect can be expected for readers who prefer to believe in ultimate justice. We could go on with splitting these hypotheses into even more specific ones. We stop because the principle should be clear by now. To put it in Zillmann's (2000) words: "Moral judgment, then, is highly diverse and in no way uniform and monolithic" (p. 60).

JUSTICE AS A PLEASURE FACTOR IN THE MEDIA: SELECTED SYSTEMATIC RESEARCH EVIDENCE

In relation to the vast knowledge social justice research has gathered over the last 40 years, and the great number of specific hypotheses that can be derived from this knowledge, scientific evidence on the role of justice in drama appreciation and media entertainment is scarce in quantity and limited in scope. This is true because research has been focused almost exclusively on retributive justice and the contingency among character morality and outcome valence. Nevertheless, these studies are valuable because they test the fundamental presumption of justice-motive theory and the conclusion from it, namely that justice is a factor of drama appreciation and shapes the entertaining value of mediated narratives. Representative examples of the most relevant studies will be reviewed in chronological order.

Zillmann and Bryant (1975) presented a videotaped fairy tale to children of two age groups (4-year olds vs. 7-/8-year olds). The tale told a story of two sibling princes, of whom one was portrayed as honest and the other as deceitful. The latter hated his brother, tried to steal his royal rights and intended to send him to exile. His hateful plan was stopped by parts of the royal guard who were loyal to his brother, overwhelmed the bad prince's forces, and put the good prince in the position to retaliate. The degree of retaliation was varied experimentally (under-retaliation, fair retaliation, over-retaliation). Facial expressions and children's answers to questions asked at the end of the viewing session served as indicators of story-liking and liking of the princes. The main result of the study was that younger children enjoyed over-retaliation more than fair retaliation, and fair retaliation more than under-retaliation, whereas older children preferred fair retaliation the most. This pattern is in line with theories of cognitive and moral development (e.g., Montada, 1980) and suggests that 4-year-old children are not yet able to perform the mental arithmetic necessary for comparing the empirical input–outcome ratio with the normative input–outcome ratio (equity). The pattern of results obtained for the 7- and 8- year-old children clearly supports justice motive theory.

Fein (1976) showed the participants of her study (kindergarten/1st graders and 3rd/4th graders) a videotaped sequence of a girl engaging in moral versus immoral behavior, followed by a fortunate versus unfortunate outcome. The episode consisted of four segments. Segment one portrayed a girl helping a friend. Segment two depicted a girl stealing candy from a peer. Segment three presented a fortunate outcome: the girl finds $10. Segment four showed an unfortunate outcome: a shelf of books falls on the girl. Eight different combinations of these segments were shown to eight experimental groups. Groups one through four saw one of the four segments only. Groups five through eight saw the four possible combinations of character value (good vs. bad girl) and outcome value (good vs. bad fate). Children from the last four groups and those who saw only the outcome segment were asked to rate the girl and the outcome. Fein expected that incongruent combinations of character value and outcome value would threaten the belief in justice and motivate the children to engage in defensive judgment either by upgrading the fortunate girl, by downgrading the unfortunate girl, or by judging the outcome as more-or-less fortunate depending on the girl's behavior. Results were only in partial agreement with these predictions. Whereas the fortunate bad girl was upgraded in comparison to the bad girl whose behavior was unconnected to fate, an unfortunate fate did not lead to a similar downgrade of the good girl. A similar asymmetry appeared when the girl had to be rated on the basis of outcome information alone. The effects on outcome evaluation were also in partial agreement with just-world theory. As predicted, outcomes following bad behavior were judged lower (less positive in the reward condition, more negative in the punishment condition) than outcomes following good behavior. Quite surprisingly, however, this effect occurred only in younger children. Although Fein did not obtain indicators of character and story-liking, studies on the effect of character value on character and story-liking suggest that good behavior and good outcomes had a positive impact on enjoyment of the videos.

Hormuth and Stephan (1981) predicted that viewers of the television series "Holocaust" would tend to blame the victims if the viewers identified with the Nazis. No defensive reactions were expected for viewers who identified with the Jewish people and for non-viewers. The authors argued that "Holocaust" passes on an implicit accusation only to those who identify with the Nazis. Consequently, this group of viewers must fear for the positive outcomes they expect on the basis of their personal contracts. Making themselves believe that the victims share some responsibility for their fate is a means of coping with their fear. Because non-viewers or viewers who identify with the Jewish people are not exposed to or do not experience a similar accusation, they feel no need for defensive responsibility attribution. The pattern of results was highly similar for German and American participants and, more importantly, perfectly consistent with Hormuth and Stephans's prediction. Although these authors did not link their findings with drama appreciation, it seems likely that the small number of viewers who identified with the Nazis were able to watch the series only because they successfully appeased their guilty conscious by blaming the victims.

Jose and Brewer (1984) read a set of stories to their participants who belonged to three age groups (2nd graders, 4th graders, 6th graders). Four experimental factors were varied: gender of character; age of character (child vs. adult); character valence (good vs. bad) and; outcome valence (fortunate vs. unfortunate). Gender and age were varied in order to increase or decrease children's identification with the character, which in turn was assumed to affect their empathy and concern for justice. Based on previous findings of developmental change (see prior text),

Jose and Brewer expected that the effect of character–outcome–consistency would increase as a function of age. Ten dependent variables were measured: perceived similarity (of child with character); liking of the character; identification with the character; suspense; liking of outcome; liking of story; caring about the character; excitement; surprise; and sadness. Consistent with justice motive theory and the authors' developmental hypothesis, character–outcome–consistency had a significant effect on outcome liking and this effect varied as a function of age. It was weakest in the group of 2nd graders and strongest in the group of 6th graders. Further support for the authors' structural affect theory was obtained from age group specific path analyses among the dependent variables. Three results are most relevant in the present context: (1) Character–outcome–consistency had an indirect effect on story-liking. (2) This effect was mediated by outcome liking. (3) The direct effect of character–outcome–consistency differed in pattern and strength between the age groups as follows: (a) In the youngest group, character value had a main effect on liking of character which in turn predicted story-liking directly and indirectly via outcome liking. Outcome value had an independent effect on outcome liking and via this mediator an indirect effect on story-liking. However, character–outcome–consistency, that is, the interaction of character value and outcome value, had no incremental effect on outcome liking over and above the main effect of outcome value; (b) In the middle age group, this incremental effect of character—outcome–consistency appeared, but it was both weak and accompanied by a still strong main effect of outcome value on outcome liking; (c) In the oldest group of children, finally, the main effect of outcome value on outcome liking disappeared and gave way to a strong interaction between character value × outcome value.

Using a less complex design, a replication of the Jose and Brewer (1984) study was submitted by Brewer (1996). The two stories his participants had to read and rate capture Lerner's justice-motive notion more directly than perhaps any other narrative. In one of the stories, a young widow (innocent victim!) wants to be the first person to swim across the Gibraltar Strait. She had trained for five years (large input!), hoping to make a living for herself and her children (good mother!) from the book and movie rights. In the justly ending version, the woman's willpower (virtue!) makes her overcome head currents and painful stings by jellyfish (unfair counterforce!). She makes it across the Strait of Gibraltar. In the unfairly ending version, the woman is overwhelmed by pain and exhaustion and has to give up. A second story was of similar dramatic make up. Ratings for overall liking, story-liking, outcome liking, story completeness, and arrangement were obtained as dependent variables. In line with just-world theory and the author's story completion model, significantly higher ratings of outcome liking and completeness were obtained for the justly ending versions than for the unfairly ending versions of the stories. The author concludes from his research that "The results support the hypothesis that narratives with bad endings will be perceived as lacking completeness, presumably due to the fact that the just-world beliefs of the readers lead them to expect that the story should continue until the wrong has been righted" (p. 270).

The work by Raney and Bryant (2002) and Raney (2002) goes beyond previously described experiments because it takes individual differences into account. Raney and Bryant (2002) proposed an integrated model of enjoyment for crime drama. The model assumes that the enjoyment of mediated justice sequences depends on the "degree of correspondence between the viewer's sense of justice and the statement about justice made in the drama" (p. 407). Empathy with the victim is assumed to amplify the impact of this correspondence on enjoyment. More importantly in the present context, the authors assume that the viewer's sense of justice is shaped by two personality variables, viewers' punitiveness and their vigilantism. Both can be linked theoretically to the level and kind of retribution a viewer considers appropriate for righting the wrong that was committed by the perpetrator. The Raney and Bryant (2002) study tests parts of the model. Following exposure to a video clip containing a typical crime drama, participants were administered a questionnaire containing items for the measurement of three variables: (1) enjoyment; (2) sympathy for the victim and; (3) fairness of the punishment the perpetrator had received. Consistent with the authors' expectations, both predictors contributed uniquely to the explanation of enjoyment variance. In addition to the three variables measured by Raney and Bryant (2002), Raney's (2002) study included measures for empathy, punitiveness, and vigilantism. Furthermore, using clips from Michael Caton-Jones' *Rob Roy* revenge movie, deservedness of punishment was manipulated experimentally. This was achieved by keeping the retribution (killing of the villain) constant across two crime severity levels. Raney (2002) expected that the impact of the personality variables would vary across deservedness, being weaker in cases of appropriate retaliation and stronger in cases of over-retaliation and under-retaliation. Given that only two levels of deservedness were realized in the experiment, a complete test of this hypothesis was not possible. Partial support for the conjecture was obtained, however. Compared to the appropriate retaliation condition, the personality variables had stronger total effect on enjoyment in the over-retaliation condition. In clear disagreement with justice-motive theory, however, deservedness had a negative effect on enjoyment.

Despite this unexpected and irritating finding, the Raney and Bryant (2002) and Raney (2002) studies are steps in the right direction according to our point of view. Given the large set of individual difference in justice attitudes and beliefs (see prior text) and their likely interaction with message content and narrative structure factors, it seems time to go beyond the overly simple 2 × 2 designs of experimental drama appreciation research. We hope that our brief review of the social justice literature offers a guideline for the specification of the many research questions that remain to be posed and answered.

BEYOND PLEASURE AND ENTERTAINMENT: MEDIA AS MORAL EDUCATORS

Up to this point, we have considered only the entertaining effects of justice in the media. Entertaining effects are not the only effects that are worth psychological analysis, however. Mediated communication of justice episodes is likely to contribute to the development of justice beliefs and justice norms. Classic drama theories viewed drama as a tool of moral education. Tragedies, for instance, were supposed by Aristotle to induce pity and fear, to purify affect via catharsis, and to induce of sense of belonging to the community via shared emotion. Similarly, Gotthold Ephraim Lessing interpreted pity and fear as products of identification with the characters of a tragedy. He viewed these and other emotions as cathartic agents who enabled spectators to broaden their capacity for compassion, contributed to their moral purification, and eventually increased the moral autonomy of mankind. Even more emphatically, Friedrich Schiller demanded that theatre serve as a moral institution making ethical issues visible and promoting ideal solutions to moral dilemmas.

Translating these normative ideas of classic drama theories into Melvin Lerner's justice-motive theory suggests that the mediated communication of justice episodes contributes to shaping, stabilizing, and validating the personal contracts of the audience. Personal contracts are shaped by justice episodes because the protagonists of these episodes are usually portrayed in a way that makes it easy to identify with them. Identification, in turn, eases social learning. In other words, protagonists of justice episodes serve as models whose personal contracts are adopted by the audience. Personal contracts are stabilized by justice episodes because the protagonists of most justice episodes eventually enjoy some sort of gratification. This gratification serves as a vicarious reinforcement of the readers' or spectators' own personal contract. This is true at least as long as the moral heros' personal contract is similar to the personal contract of the receiver. This similarity is likely given that personal contracts are socialized by moral authorities who belong to the same moral community as the protagonist. Finally, personal contracts are validated because the protagonists of mediated justice episodes tend to fulfill their own contracts, and these are similar to those of the audience for the reasons outlined earlier. Taken together, justice episodes in the media contribute to the establishment and to the maintenance of a moral community.

Note that shifting the focus of our analysis from the entertaining effects of justice in the media to its socializing effects has broadened the scope of questions that need to be answered by empirical research. Whereas justice was a facet of the independent variable in studies on story liking and enjoyment, we now need to include justice as a facet of both the independent variable and the dependent variable. Moreover, our thoughts on the socializing effects of media require a move from cross-sectional research designs to longitudinal studies that include indicators of justice and enjoyment at different points in time. These studies would help to clarify how individual differences in justice beliefs and norms influence the selection of and preference for certain media products and how these products in turn shape, change, and stabilize moral orientations. To the best of our knowledge, this type of research has not yet been done.

AUTHOR NOTE

We thank Jane Thompson for helpful comments on an earlier version of the paper.

REFERENCES

Adams, J. S. (1965). Inequity in social exchange. In L. Berkowitz (Ed.), *Advances in experimental social psychology* (Vol. 2, pp. 267–299). New York: Academic Press.

Arts, W., Hermkens, P., & van Wijck, P. (1995). Anomie, distributive justice and dissatisfaction with material well-being in Eastern Europe. *International Journal of Comparative Sociology, 34*, 1–16.

Bandura, A. (1977). Self-efficacy: Toward a unifying theory of behavioral change. *Psychological Review, 84*, 191–215.

Berkowitz, L., & Walster, E. (Eds.) (1976). *Equity Theory: Toward a General Theory of Social Interaction* (Advances in Experimental Social Psychology, Vol. 9). New York: Academic Press.

Bies, R. J. (1987). Beyond "voice": The influence of decision maker justification and sincerity on procedural fairness judgments. *Representative Research in Social Psychology, 17*, 3–14.

Bies, R. J., & Moag, J. S. (1986). Interactional justice: Communication criteria of fairness. *Research on Negotiation in Organizations, 1*, 43–55.

Bies, R. J., & Shapiro, D. L. (1987). Interactional fairness judgments: The influence of causal accounts. *Social Justice Research, 1*, 199–218.

Brewer, W. F. (1996). Good and bad story endings and story completeness. In R. J. Kreuz & M. S. MacNealy (Eds.), *Empirical approaches to literature and aesthetics* (pp. 261–274). Westport, CT: Ablex Publishing.

Brewer, W. F., & Lichtenstein, E. H. (1982). Stories are to entertain: A structural-affect theory of stories. *Journal of Pragmatics, 6*, 473–486.

Bryant, J., Roskos-Ewoldsen, D., & Cantor, J. R. (Eds.) (2003). *Communication and emotion: Essays in honor of Dolf Zillmann.* Mahwah, NJ: Lawrence Erlbaum Associates.

Chatman, S. (1978). *Story and discourse: Narrative structure in fiction and film.* Ithaca, NY: Cornell University Press.

Dalbert, C. (2001). *The justice motive as a personal resource.* New York: Kluwer Academic/Plenum Publishers.

Dalbert, C., Montada, L., & Schmitt, M. (1987). Glaube an eine gerechte Welt als Motiv: Validierungskorrelate zweier Skalen [Belief in a just world as a motive: Validation and correlates of two scales]. *Psychologische Beiträge, 29*, 596–615.

Dar, Y., & Resh, N. (2001). Exploring the multifaceted structure of sense of deprivation. *European Journal of Social Psychology, 31*, 63–81.

Davey, L. M., Bobocel, D. R., Hing, L. S. S., & Zanna, M. P. (1999). Preference for the Merit Principle Scale: An individual difference measure of distributive justice preferences. *Social Justice Research, 12*, 223–240.

Deutsch, M. (1975). Equity, equality, and need: What determines which value will be used as the basis of distributive justice? *Journal of Social Issues, 31*, 137–149.

Deutsch, M. (1985). *Distributive Justice: A Social Psychological Perspective.* New Haven, CT: Yale University Press.

Devine, P. G. (1989). Stereotypes and prejudice: Their automatic and controlled components. *Journal of Personality and Social Psychology, 56*, 5–18.

Fein, D. (1976). Just world responding in 6- and 9-year-old children. *Developmental Psychology, 12*, 79–80.

Fetchenhauer, D., & Huang, X. (2004). Justice sensitivity and behavior in experimental games. *Personality and Individual Differences, 36*, 1015–1031.

Friedman, N. (1975). *Form and meaning in fiction.* Athens, GA: University of Georgia Press.

Gergen, K. J., Morse, S. J., & Gergen, M. (1980). Behavior exchange in cross-cultural perspective. In H. Triandis & R.W. Brislin (Eds.), *Handbook of cross-cultural psychology* (Vol. 5, pp. 121–153). Boston, MA: Allyn & Bacon.

Heider, F. (1958). *The Psychology of Interpersonal Relations.* New York: Wiley.

Herrmann, T., & Winterhoff, P. (1980). Leistungsbezogenes Aufteilen als situationsspezifische Korrektur von Gerechtigkeitskonzepten—Zum Einfluß von Personenmerkmalen auf die Gewinnaufteilung [Achievement-related allocation of goods as situation-specific correction of general justice notions—On the influence of personality on allocation behavior]. *Zeitschrift für Sozialpsychologie, 11*, 259–273.

Hormuth, S. E., & Stephan, W. G. (1981). Effects of viewing "Holocaust" on Germans and Americans: A just-world analysis. *Journal of Applied Social Psychology, 11*, 240–251.

Huseman, R. C., Hatfield, J. D., & Miles, E. W. (1985). Test for individual perceptions of job equity: Some preliminary findings. *Perceptual and Motor Skills, 61*, 1055–1064.

Jose, P. E., & Brewer, W. F. (1984). Development of story liking: Character identification, suspense, and outcome resolution. *Developmental Psychology, 20*, 911–924.

Lerner, M. J. (1977). The justice motive in social behavior. Some hypotheses as to its origins and forms. *Journal of Personality, 45*, 1–52.

Lerner, M. J. (1980). *The belief in a just world. A fundamental delusion.* New York: Plenum Press.

Lerner, M. J. (2003). The justice motive: Where social psychologists found it, how they lost it, and why they may not find it again. *Personality and Social Psychology Review, 7*, 388–399.

Lerner, M. J., Miller, D. T., & Holmes, J. G. (1976). Deserving and the emergence of forms of justice. In L. Berkowitz (Ed.), *Advances in Experimental Social Psychology* (Vol. 9, pp. 133–162). New York: Academic Press.

Lerner, M. J., & Miller, D. T. (1978). Just world research and the attribution process: Looking back and ahead. *Psychological Bulletin, 85*, 1030–1050.

Lerner, M. J., & Simmons, C. H. (1966). The observer's reaction to the "innocent victim." Compassion or rejection? *Journal of Personality and Social Psychology, 4*, 203–210.

Leventhal, G. S. (1980). What should be done with equity theory? New approaches to the study of fairness in social relationships. In K. J. Gergen, M. S. Greenberg, & R. H. Willis (Eds.), *Social exchange* (pp. 27–55). New York: Plenum.

Lind, A. E., & Tyler, T. R. (1988). *The social psychology of procedural justice.* New York: Plenum.

Lipkus, I. M. (1991). The construction and preliminary validation of a global belief in a just world scale and the exploratory analysis of the multidimensional belief in a just world scale. *Personality and Individual Differences, 12*, 1171–1178.

Maes, J. (1994). Blaming the victim—belief in control or belief in justice? *Social Justice Research, 7*, 69–90.

Maes, J. (1998). Immanent justice and ultimate justice: Two ways of believing in justice. In L. Montada, & M. J. Lerner (Eds.), *Responses to victimization and belief in a just world* (pp. 9–40). New York: Plenum Press.

Major, B., & Deaux, K. (1982). Individual differences in justice behavior. In J. Greenberg, & R.L. Cohen (Eds.) (1982). *Equity and justice in social behavior* (pp. 43–76). New York: Academic Press.

Marques, J. M., & Paéz, D. (1994). The 'Black Sheep effect': Social categorization, rejection of ingroup deviates, and perception of group variability. In W. Stroebe, & M. Hewstone (Eds.), *European Review of Social Psychology* (Vol. 5, pp. 37–68). Chichester, UK: Wiley.

Mikula, G., Petri, B., & Tanzer, N. (1990). What people regard as unjust: Types and structures of everyday experiences of injustice. *European Journal of Social Psychology, 20*, 133–149.

Mikula, G., Scherer, K. R., & Athenstaedt, U. (1998). The role of injustice in the elicitation of differential emotional reactions. *Personality and Social Psychology Bulletin, 24*, 769–783.

Miller, D. T. (2001). Disrespect and the experience of injustice. *Annual Review of Psychology, 52*, 527–553.

Miller, W. I. (1998). Clint Eastwood and equity: Popular culture's theory of revenge. In A. Sarat, & T. R. Kearns (Eds.), *Law in the domains of culture* (pp. 161–202). Ann Arbor: The University of Michigan Press.

Montada, L. (1980). Developmental changes in concepts of justice. In G. Mikula (Ed.), *Justice and social interaction* (pp. 257–284). New York: Springer.

Montada, L. (1993). Understanding oughts by assessing moral reasoning or moral emotions. In G. Noam, & T. Wren (Eds.), *The moral self* (pp. 292–309). Boston: MIT-Press.

Montada, L., & Schneider, A. (1989). Justice and emotional reactions to the disadvantaged. *Social Justice Research, 3*, 313–344.

Oswald, M. E, Hupfeld, J., Klug, S. C, & Gabriel, U. (2002). Lay-perspectives on criminal deviance, goals of punishment, and punitivity. *Social Justice Research, 15*, 85–98.

Perrine, L. (1959). *Story and structure.* New York: Harcourt & Brace.

Piaget, J. (1932). *Je jugement moral chez l'enfant* [The moral judgment of the child]. Paris: Alcan.

Raney, A. A. (2002). Moral judgment as a predictor of enjoyment of crime drama. *Media Psychology, 4*, 305–322.

Raney, A. A. (2003a). Disposition-based theories of enjoyment. In J. Bryant, D. Roskos-Ewoldsen, & J. R. Cantor (Eds.), *Communication and emotion: Essays in honor of Dolf Zillmann* (pp. 61–84). Mahwah, NJ: Lawrence Erlbaum Associates.

Raney, A. A. (2003b). Enjoyment of sport spectatorship. In J. Bryant, D. Roskos-Ewoldsen, & J. R. Cantor (Eds.), *Communication and emotion: Essays in honor of Dolf Zillmann* (pp. 397–416). Mahwah, NJ: Lawrence Erlbaum Associates.

Raney, A. A., & Bryant, J. (2002). Moral judgment and crime drama: An integrated theory of enjoyment. *Journal of Communication, 52*, 402–415.

Rotter, J. B. (1980). Interpersonal trust, trustworthiness, and gullibility. *American Psychologist, 35*, 1–7.

Rubin, Z., & Peplau, L. A. (1973). Belief in a just world and reactions to another's lot: A study of participants in the National Draft Lottery. *Journal of Social Issues, 29*(4), 73–93.

Ryan, W. (1971). *Blaming the victim*. New York: Pantheon Books.

Sabbagh, C. (2001). A taxonomy of normative and empirically oriented theories of distributive justice. *Social Justice Research, 14*, 237–263.

Sabbagh, C., Dar, Y., & Resh, N. (1994). The structure of social justice judgments: A facet approach. *Social Psychology Quarterly, 57*, 244–261.

Sabbagh, C., & Schmitt, M. (1998). Exploring the structure of positive and negative justice judgments. *Social Justice Research, 12*, 381–396.

Scherer, K. R., Wallbott, H. G., & Summerfield, A. B. (Eds.) (1986). *Experiencing emotion: A cross-cultural study*. Cambridge: Cambridge University Press.

Schmitt, M. (1996). Individual differences in sensitivity to befallen injustice. *Personality and Individual Differences, 21*, 3–20.

Schmitt, M. (1998). Methodological strategies in research to validate measures of belief in a just world. In L. Montada & M. J. Lerner (Eds.), *Responses to victimization and belief in a just world* (pp. 187–215). New York: Plenum Press.

Schmitt, M., Behner, R., Montada, L., Müller, L., & Müller-Fohrbrodt, G. (2000). Gender, ethnicity, and education as privileges: Exploring the generalizability of the existential guilt reaction. *Social Justice Research, 13*, 313–337.

Schmitt, M., & Dörfel, M. (1999). Procedural injustice at work, justice sensitivity, job satisfaction and psychosomatic well-being. *European Journal of Social Psychology, 29*, 443–453.

Schmitt, M., Eid, M., & Maes, J. (2003). Synergistic person x situation interaction in distributive justice behavior. *Personality and Social Psychology Bulletin, 29*, 141–147.

Schmitt, M., Gollwitzer, M., Förster, N., & Montada, L. (2004). Effects of objective and subjective account components on forgiving. *The Journal of Social Psychology, 144*, 465–485.

Schmitt, M., Gollwitzer, M., Maes, J., & Arbach, D. (2005). Justice Sensitivity: Assessment and Location in the Personality Space. *European Journal of Psychological Assessment, 21*, 202–211.

Schmitt, M., Neumann, R., & Montada, L. (1995). Dispositional sensitivity to befallen injustice. *Social Justice Research, 8*, 385–407.

Schmitt, M., & Sabbagh, C. (2004). Synergistic person × situation interaction in distributive justice judgment and allocation behavior. *Personality and individual Differences, 37*, 359–371.

Schwinger, T. (1980). Just allocations of goods: Decisions among three principles. In G. Mikula (Ed.), *Justice and social interaction* (pp. 95–125). Bern: Huber.

Seligman, M. E. P. (1975). *Helplessness: On depression, development, and death*. San Francisco: Freeman.

Skarlicki, D. P., & Folger, R. (1997). Retaliation in the workplace: The roles of distributive, procedural, and interactional justice. *Journal of Applied Psychology, 82*, 434–443.

Strack, F., & Deutsch, R. (2004). Reflective and impulsive determinants of social behavior. *Personality and Social Psychology Review, 8*, 220–247.

Tajfel, H., & Turner, J. C. (1986). The social identity theory of intergroup behaviour. In S. Worchel, & W. G. Austin (Eds.), *Psychology of intergroup relations* (2nd ed.; pp. 7–24). Chicago, IL: Nelson-Hall.

Thibaut, J. W., & Walker, L. (1975). *Procedural justice: A psychological analysis*. Hillsdale, NJ: Lawrence Erlbaum Associates.

Törnblom, K. Y. (1992). The social psychology of distributive justice. In K. Scherer (Ed.), *Justice: Interdisciplinary perspectives* (pp. 175–236). Cambridge: Cambridge University Press.

Tyler, T. R. (1986). Procedural justice in organizations. In R. Lewicki, M. Bazerman, & B.H. Sheppard (Eds.), *Research on negotiation in organizations* (Vol. 1, pp. 7–73). Greenwich, CT: JAI Press.

Tyler, T. R., Degoey, P., & Smith, H. (1996). Understanding why the justice of group procedures matters: A test of the psychological dynamics of the group-value model. *Journal of Personality and Social Psychology, 70*, 913–930.

Tyler, T. R., & Smith, H. (1998). Social justice and social movements. In D.T. Gilbert, S.T. Fiske, & G. Lindzey (Eds.), *The handbook of social psychology* (Vol. II, pp. 595–629). Oxford: Oxford University Press.

van den Bos, K., Maas, M., Waldring, I., & Semin, G. P. (2003). Toward understanding the psychology of reactions to perceived fairness: The role of affect intensity. *Social Justice Research, 16*, 151–168.

Vidmar, N. (2001). Retribution and revenge. In J. Sanders & V. L. Hamilton (Eds.), *Handbook of justice research in law* (pp. 31–63). New York: Kluwer Academic/Plenum Publishers.

Wenzel, M. (2002). The impact of outcome orientation and justice concerns on tax compliance: The role of taxpayers' identity. *Journal of Applied Psychology, 87*, 629–645.

Zillmann, D. (1991). Empathy: Affect from bearing witness to the emotions of others. In J. Bryant, & D. Zillmann (Eds.), *Responding to the screen: Reception of reaction and processes* (pp. 135–167). Hillsdale, NJ: Lawrence Erlbaum Associates.

Zillmann, D. (2000). Basal morality in drama appreciation. In I. Bondebjerg (Ed.), *Moving images, culture, and the mind* (pp. 53–63). Luton, England: University of Luton Press.

Zillmann, D. (2003). Theory of affective dynamics: Emotions and moods. In J. Bryant, D. Roskos-Ewoldsen, & J. R. Cantor (Eds.), *Communication and emotion: Essays in honor of Dolf Zillmann* (pp. 533–567). Mahwah, NJ: Lawrence Erlbaum Associates.

Zillmann, D., & Bryant, J. (1975). Viewer's moral sanction of retribution in the appreciation of dramatic presentations. *Journal of Experimental Social Psychology, 11*, 572–582.

Zillmann, D., & Cantor, J. R. (1977). Affective responses to the emotions of a protagonist. *Journal of Experimental Social Psychology, 13*, 155–165.

Zillmann, D., Taylor, K., & Lewis, K. (1998). News as nonfiction theatre: How dispositions toward the public cast of characters affect reactions. *Journal of Broadcasting and Electronic Media, 42*, 153–169.

Zillmann, D., & Vorderer, P. (Eds.) (2000). *Media entertainment: The psychology of its appeal.* Mahwah, NJ: Lawrence Erlbaum Associates.

17

Parasocial Interactions and Relationships

Christoph Klimmt
Hanover University of Music and Drama

Tilo Hartmann
Hanover University of Music and Drama

Holger Schramm
University of Zurich

The vast majority of entertainment media is about people. Movies portray the faith of characters, talk shows host communicative guests, sports broadcasts feature competing athletes and sympathetic commentators. These people obviously add to the entertainment value of the media offerings in which they appear and in many cases, the success of the product totally depends on one single person. Without Jay Leno, watching the "Tonight Show" would be a boring experience. Without Lara Croft, the "Tomb Raider" computer games would be much less fun to play. Apparently, using media entertainment is most often about observing how people in the media look, what they are saying, and what they are doing (e.g., Flora, 2004). Given the dominating position of people in most entertainment media, it is more than probable that such observations (can) breed enjoyment.

Not surprisingly, several approaches to entertainment theory underline the importance of responding to media characters. Disposition-based theories, for example (Zillmann, 1996; Raney, 2003), model media users' emotions as reactions toward media characters and formulate a mechanism of how such affective responses contribute to enjoyment (Raney, this volume; Zillmann, this volume). Cohen (this volume) argues that the concept of identification with media characters also holds explanatory value for entertainment experiences.

Given the enormous diversity of people who appear in media entertainment, and the large differences in their looks, functions, actions, and statements, communication researchers have searched for broad accounts to user responses to media characters that can cover a variety of processes and mechanisms through which entertainment audiences experience enjoyment. One promising concept has been introduced by Horton and Wohl (1956). They coined the term "parasocial interactions" as a label for TV viewers' responses to people on the screen. Horton and Wohl (1956) were among the first who systematically identified similarities between the social situations that media characters attempt to establish for their audience, and real-life encounters. They argued that many media "personae" address viewers more-or-less directly, and that viewers presumably respond to such behaviors just the way they would if the characters were standing in front of them. The peculiarities of the medium limit the range of possible interactions between persona and viewer, however, so that viewers' responses were called "parasocial." Nevertheless, Horton and Wohl (1956) assumed that viewers establish some sort of relationship with (some) media people over time, which increases the relative importance of media characters for viewers' social life.

The assumption that users of media offerings respond to media people similarly to how they feel, think and behave in real-life encounters has inspired many communication researchers. They have created a remarkable history of inquiry about parasocial interactions (PSI) and parasocial relationships (PSR). Many studies have implicitly or explicity connected PSI/PSR to media entertainment (e.g., Perse & Rubin, 1988; Vorderer, 1998), although no systematic elaboration of those conceptual links has been proposed. Such a theoretical endeavour promises to be fruitful for entertainment research, because PSI and PSR cover a larger portion of the conceivable user responses to media characters than most other existing theories. Based on a brief review of the literature on PSI/PSR research, we offer a process-oriented reconceptualization of PSI as a dynamic set of various affective, cognitive, and behavioural responses to media personae and discuss the relationships between PSI processes, PSR and enjoyment. Finally, we outline directions for future research on the entertainment experience that arise from (parasocially) interacting with media personae.

PAST THEORY AND RESEARCH ON PSI AND PSR

Reasons for Conceptual Diversity

Many researchers have picked up Horton's and Wohl's (1956) considerations, redefined, modified, and interpreted them, and employed them in empirical investigations. However, the development of the PSI and PSR concepts traces back to a very heterogeneous history.

First, culturally diverse scientific communities, such as Anglo-American, German, and Scandinavian researchers, have worked on PSI and PSR. It is probable that the conceptual development of both PSI and PSR took at least slightly different ways within these different communities which, in turn, might have lead to different understandings of both constructs.

Second, distinct scientific disciplines, such as communication, social psychology, media psychology, film studies and the arts are involved with PSI and PSR research. Naturally, they have developed different definitions. Uses-and-gratifications researchers, for example, regard PSI and PSR primarily as motivations of selective exposure (e.g. Palmgreen, Wenner, & Rayburn, 1980) or as a special type of "interpersonal involvement" (Rubin, Perse, & Powell, 1985, p. 157) that combines different phenomena such as interaction, identification, and long-term identification" (Rubin et al., 1985, p. 156) with media personae. Such definitions do not separate PSI and PSR analytically, but, rather, treat the terms interchangeably as a continuous interpersonal involvement with media figures that includes different phenomena and may take place in situations of media exposure as well as in pre- or post-exposure situations.

In addition to uses-and-gratifications research, PSI and PSR have also been explicated in other domains, mostly by German researchers, for example, in the tradition of symbolic interactionism (Teichert, 1973; Ellis, Streeter, & Engelbrecht, 1983; Krotz, 1996), semiotics (Wulff, 1992, 1996), and media psychology (Gleich, 1997; Giles, 2002; Hartmann, Schramm, & Klimmt, 2004a, 2004b). In contrast to the notion of PSI and PSR held by uses-and-gratifications research, most of these approaches argue for a clear distinction between PSI and PSR (cf. Krotz, 1996; Schramm, Hartmann & Klimmt, 2002; Vorderer, 1998). PSI, in this sense, specifically means the one-sided process of media person perception during media exposure. In contrast, PSR stands for a cross-situational relationship that a viewer or user holds to a media person, which includes specific cognitive and affective components (cf. Krotz, 1996, Hartmann et al., 2004b). In sum, the terms, definitions, and models explicating PSI and PSR differ across scientific backgrounds and traditions, which impedes the application of the concepts to entertainment research.

Third, research on PSI and PSR differs extensively to the degree it relies on theoretical assumptions versus empirical-inductive inferences. A broad range of German PSI/PSR-related publications are solely of theoretical nature. In contrast, Anglo-American research of PSI and PSR focuses much more on empirical assessments, which, in turn, implies a rather data-driven evolution of the conceptualization of both constructs (e.g., Houlberg, 1984; Levy, 1979). In our view, those rich theoretical ruminations and empirical findings on PSI/PSR are still in need of conceptual integration.

Fourth, conceptualizations of PSI and PSR differ because they are often bound to specific media contexts. Beyond TV and radio, PSI and PSR have been applied to the internet (Hoerner, 1999), computer-games (Hartmann, Klimmt, & Vorderer, 2001), or instructional computer software (Bente, Krämer, & Petersen, 2002). Moreover, the type of media persona, as well as the analyzed format, varies across investigations. In the past, PSI and PSR have been investigated in relation to diverse media figures including politicians (Gleich, 1999), TV shopping hosts (Grant, Guthrie, & Ball-Rokeach, 1991), soap characters (Rubin & Perse, 1987; Visscher & Vorderer, 1998), TV talkshow guests (Thallmair & Rössler, 2001), TV talk how hosts (Trepte, Zapfe, & Sudhoff, 2001), comedians (Auter, 1992), actresses (Rubin & McHugh, 1987), radio hosts (Rubin & Step, 2000), virtual avatars in computer-games or on the internet (Schramm, Hartmann, & Klimmt, 2004), comic figures (Hoffner 1996), or characters of audio stories (Ritterfeld, Klimmt, Vorderer, & Steinhilper, 2005). As both PSI and PSR are phenomena of interaction, their structure relies on user characteristics as well as on the type of encountered media personae. Thus, it remains doubtful if what is addressed as PSI/PSR is exactly similar across investigations conducted in different contexts with distinct types of personae.

The great diversity in theoretical approaches and traditions that have been adopted for research on PSI/PSR holds the potential for both fruitful integration and failure to achieve conceptual progress. In addition to the mentioned problems of past PSI/PSR theory, the lack of systematic connections to theories from social psychology must be mentioned. Recent work by Giles (2002) and Cohen (2003, 2004) has elaborated some approximations between PSI and general theories of social psychology, especially with respect to the relationship between PSI and real-world interaction (Giles, 2002) and between parasocial relationships and attachment styles (Cohen, 2004). But, in general, the substantial knowledge from social psychology on human interaction and relationships has not been incorporated to a sufficient extent, which appears to be – aside of the integration of the different lines of theory listed above – the most important challenge for PSI/PSR theory (Schramm et al., 2002).

Empirical Findings

In spite of the heterogeneity in the evolution of PSI and PSR research and the focus on positively evaluated persona-involvement, the existing empirical literature provides some fundamental insight on the phenomenon. Some results are particularly relevant for entertainment research. Our review primarily builds on studies that used the terms PSI and PSR interchangeably. Therefore, we refer to the assessed construct as "PSI/PSR".

What fosters a positive (and enjoyable) PSI/PSR? Similar to real-life interactions, both attractiveness (physical, social, and task attractiveness) and perceived similarity of the media character seem to be important antecedents (Cohen, 2001; Hartmann et al., 2001; Rubin & McHugh, 1987; Rubin & Rubin, 2001; Turner, 1993; Visscher & Vorderer, 1998). Personae often exhibit character traits that receivers admire and would like to hold as well (Caughey,

1984, 1986). Gender also plays a role in this respect. Vorderer (1996) found that female viewers tend to admire the attractiveness of personae more than male viewers and thus tend to hold stronger PSR in the sense of worships. PSI/PSR to female TV-characters are generally more intense than PSI/PSR to their male counterparts (Vorderer, 1996; Vorderer & Knobloch, 1996). However, females display greater admiration for male characters and vice versa (Vorderer & Knobloch, 1996; Hartmann et al., 2001), which again points out to the importance of attractiveness for the formation of PSI/PSR. These findings should be generalized with care, because they may only be valid for specific characters and maybe only for a specific tonality of PSI/PSR (e.g., a romantic worship).

Findings regarding the effects of age and education are mixed. Vorderer (1996, 1998) found that PSI/PSR to stars of TV series are intense among older, less educated individuals, who frequently turn to television, but are also intense among adolescent fans of TV-series, in contrast to the middle-age groups. Gleich (1997) and Levy (1979) found that stronger PSI/PSR were associated with the age of the onlookers. In turn, Giles and Maltby (2004) point out the important role of PSI/PSR among adolescents (cf. also McCutcheon, Ashe, Houran, & Maltby, 2003). Other researchers found no significant effects of age or educational level (Grant et al., 1991). Nordlund (1978) reported that the intensity of PSI/PSR increased with less variety and a lesser number of spare time alternatives, which is a typical process in people who are aging. The intensity of media use, which might also be reflected in media dependency, is important, as it was positively associated with strong PSI/PSR in various investigations (Gleich, 1997; Grant et al., 1991; Levy, 1979; Nordlund, 1978; Rubin et al., 1985; Vorderer, 1996). Accordingly, Rubin and Perse (1987, see also Perse and Rubin, 1988) reported that PSI/PSR increased with a stronger affinity to soaps and intensity of exposure to soaps. Therefore, the frequency of selective exposure might be crucial, whereas the effect that older people tend to report stronger PSI/PSR might be diminished if the amount of television usage is controlled.

It is not perceived loneliness that makes individuals turn to media characters (Rubin et al., 1985; Fabian, 1993; Ashe & McCutcheon, 2001). Rather, individuals with high social abilities tend to report strong PSI/PSR (Tsao, 1996; Cole & Leets, 1999). But shyness (a perceived lack of social competency) might hinder individuals to fulfill their need for interpersonal interactions. Vorderer and Knobloch (1996) found that PSI/PSR are highest among individuals who are shy but feel a high need for social interaction. In sum, the findings suggest that "social and parasocial interaction are complementary, perhaps because they require similar social skills" (Cohen, 2004, p. 192). Analogous, research shows that PSR develops quite similar compared to real-life relationships (Boon & Lomore, 2001; Cohen, 2004; Perse & Rubin, 1989; Rubin & McHugh, 1987).

Uses-and-gratifications-oriented research found that PSI/PSR are strongly related to motives of selective exposure in general (e.g., Rubin et al., 1985; Perse, 1990; Conway & Rubin, 1991), and to entertainment motives in particular (e.g., the anticipation of suspense; c.f. Gleich, 1997; Perse & Rubin, 1988; Rubin & Perse, 1987). The findings suggest that PSI/PSR play a crucial role in the processes of media entertainment (Maltby et al., 2002). Gleich (1997) argues that PSR serve as a contextual frame in which entertaining experiences can unfold during media exposure. Thus, PSI with media characters might be a rewarding experience in itself, but also allow for subsequent positive experiences. Research that links PSR to Affective Disposition Theory (Zillmann, 1996; Raney, this volume) shows that PSR strongly affect the formation of positive or negative dispositions towards media characters and, consequently, the experienced level of suspense (Stuke, Hartmann, & Daschmann, 2005). However, PSI/PSR might not only affect entertainment in situations of exposure, but also thereafter. Findings show, for example, that high PSI/PSR result in more intense postviewing discussions and postviewing cognitions (Fabian, 1993, Perse & Rubin, 1988; Rubin & Perse, 1987), which might add to the entertainment value of media fare (cf. Giles & Maltby, 2004).

Overall, past theoretical and empirical research on PSI/PSR has revealed various parallels between the appreciation of media personae and social behavior in real-life encounters. Moreover, the data show that media personae fill in specific and important positions within people's perceived (or construed) social networks. Observing, feeling with, thinking about, or even talking to people on the screen (or in other media entertainment products) is attractive for people for different reasons, and a variety of processes or mechanisms that link PSI/PSR and enjoyment is expectable. However, a systematic elaboration of those links requires some conceptual integration, because the theoretical heterogeneity impedes a full exploitation of past PSI/PSR studies for theorizing about entertainment.

PSI REVISITED: A THEORETICAL MODEL ON INTERPERSONAL INVOLVEMENT WITH MEDIA PERSONAE

The Benefits of a New Model

The historical review of research on PSI and parasocial relationships has revealed a number of conceptual problems that communication researchers should resolve in order to further employ the concepts successfully in entertainment studies. A process-based view of PSI that conceptualizes viewers' responses to media personae as dynamic and dependent on both media/persona and recipient characteristics is suggested as most promising resolution to the discussed shortcomings: It allows for defining cognitive, emotional, and behavioural phenomena that occur during viewers' appreciation of a media persona, and to link the dynamics of these parasocial phenomena (e.g., with respect to intensity of involvement with a persona) to conditions and consequences (e.g., media effects). In this section, we outline such a process-based model of PSI that incorporates much previous work from communication and has received substantial theoretical input from social psychology. The condensed explication of the model follows more elaborate discussions that have been published by Hartmann et al. (2004a, 2004b).

Two Levels of Involvement with Personae

The basic idea of the model is that PSI is actually a set of different (sub)processes that can occur in viewers (or users of other entertainment media; in the following, we refer to them only as "viewers"). From social psychology, a broad variety of cognitive, affective, and behavioral responses to an interaction partner is well-known, and many of them exist in PSI as well (see section on PSI processes following). What the different PSI processes have in common is the potential variation in intensity. Viewers may think very peripherally about a persona, or they may invest enormous cognitive effort in following her/his words. Their emotional reactions to a persona may be very vigorous or very weak. And behavioral activities sometimes do not occur at all, but reach surprising levels of intensity at other times.

We suggest that there are individual differences in the general intensity of parasocial responses to a persona, but that there are also strong dynamics in these responses within one course of media reception of each viewer. Cognitive activity that relates to a persona, for example, may rise and fall during reception because of certain actions the persona performs, or because of changing situational circumstances (e.g., a ringing telephone in the viewer's home), or because of changing viewer attributes (e.g., increasing tiredness). We refer to the conditions of procedural dynamics in PSI later. For the introduction of the model of PSI, the mere notion of variable intensities in PSI processes is more important, as the process of persona-oriented media consumption is conceptualized as course of dynamically changing processes that together form a parasocial interaction (pattern) between a viewer and a persona.

Although the intensity of parasocial responses toward a persona may vary with many gradations, and should therefore be modelled as a continuum, we structure our model into two prototypical levels, namely low and high levels of interpersonal involvement (cf. Rubin et al., 1985). In the case of low interpersonal involvement ("low-level PSI"), all occurring PSI processes display weak intensities. In this situation, the persona is not relevant to the viewers, as they do not devote major cognitive processing resources, emotional energy, and/or behavioral activity to the persona. If at least some PSI processes reach high levels of intensity, however, the overall parasocial appreciation of the persona is strong ("high-level PSI"), and, depending on the type of intense response processes, a specific pattern of increased involvement with the persona emerges. Those patterns of high-intensity PSI might change rapidly during the course of reception, as the underlying processes can dynamically change. This account of variable intensities in viewers' reaction to personae is rather descriptive, of course, but it allows for precise analyses of conditions and consequences of PSI within entertainment research. We return to this issue at the end of the chapter.

First Contact with a Persona: Impression Formation or Recognition

One parallel between social encounters in the real world and PSI is that the first contact with media personae (presumably) triggers the same automatic processes as the appearance of real people does. These processes serve to formulating a first impression that is available very quickly (Kanning, 1999). Attributes of the person(a) that are easy to access, for example, skin colour, clothing, and mimic expression are rapidly perceived and connected to top-down knowledge of person(a) categories (e.g., Brewer, 1988; Fiske, Lin & Neuberg, 1999). The initial limitation of the person-oriented information processing to peripheral cues, and the immediate connection to available knowledge structures, allows a first impression of a person(a) to form at high pace and with minimal cognitive efforts. Necessarily, this impression comprises only a few pieces of information about the person(a), but it feeds forward impulses for subsequent behavior that the individual should direct toward the person(a). For example, impression formation at first contact allows one to distinct obviously dangerous from harmless people, and it indicates whether a person fits to one's motivational dispositions (information about interest value and/or attractiveness). The course and quality of PSI to a persona then, partly depends on the results of the impression formation process that begins once the viewers have detected the persona and that is similar to real-world processes of person perception (cf. Babrow, O'Keefe, Swanson, Meyers, & Murphy, 1980).

As many personae appear frequently in the media, most "first contacts" with them during a viewing session do not trigger processes of impression formation, but should rather be construed as person recognition. Instead of making connections between perceived information and general category knowledge (e.g., "this is a young woman"), viewers activate specific knowledge about the persona that they have collected during past media consumption ("this is Gwyneth Paltrow"). This knowledge is stored as part of a relationship schema (Baldwin, 1992; see the section on PSI and media entertainment to follow). Accessing knowledge about the recognized person(a) generates more specific information that is fed forward to subsequent interaction behavior (e.g., it may trigger a motivation to focus on the person's physical appearance). The initial intensity and subsequent dynamics of the activated PSI processes are therefore expected to vary between personae whom viewers encounter for the first time and personae who viewers already know and recognize once they appear on the screen.

PSI Subprocesses: Cognitions, Emotions, and Behavior

The core of the proposed model is the definition of cognitive, emotional, and behavioral phenomena or processes that may occur in viewers as a response or activity that follows the impression formation/recognition and that is directed toward a persona. These processes are the actual building blocks of PSI. We present a list of twelve different processes that theoretically can occur separately from each other. Most often, however, a selection of them will be observable conjunctly. They have been derived from the literature of communication, social psychology, and media psychology. The list may not be exhaustive, as research on human interaction is a very rich field and additional (or more precisely defined) processes of how people respond to each other may be distilled from further literature analysis and applied to the reaction to media personae (for example, an alternative process structure is suggested by Konijn and Hoorn, 2005).

Building on existing research on human interaction, we model a broad variety of cognitive responses of viewers to media personae. They all relate closely to human information processing (Wicks, this volume):

- Attention allocation (Anderson, this volume): One basic cognitive process that viewers direct towards a persona is the deployment of attentional resources. Perceiving and processing what the persona looks like, what s/he is saying and doing, and maintaining the attentional focus on the persona even when the mediated presentation centers around other personae or objects are the foundation for many other PSI processes. In the case of intensified attention allocation (high-level PSI), viewers will seek actively for new information from or about the persona, whereas in the condition of low attention (low-level PSI), they will process information from the persona only incidentally and without much effort.
- Comprehension and reconstruction: Based on attentional processes, viewers may try to understand a persona's goals, attitudes, utterances etc. (Harris, this volume). If viewers find a persona relevant and important, they will invest cognitive efforts to comprehend the sense of her/his actions and statements. Mentally reconstructing the thoughts of personae displays some similarities with the concept of identification (Cohen, this volume). However, identification typically implies that one will put aside self-related cognitions, as the mind is occupied by the perspective of the persona with whom people identify. Efforts of comprehension and reconstruction do not necessarily demand one to discard self-related cognitions, and therefore they rather mirror a cognitive component of empathy (Omdahl, 1995; Zillmann, Chapter 10, this volume). In phases of low-level PSI, viewers will limit their activities to understand the persona's thoughts and actions to a minimum and rely on category-based information ("this is only a supporting actor"), whereas in phases of high-level PSI, viewers will mobilize considerable cognitive energy to understand what is in the persona's mind.
- Activation of prior media and life experience: Another part of viewers' cognitive responses to a media persona is the access of memorized contextrelevant information. Viewers may compare the situation and the actions of a persona with circumstances in which they have observed the persona in the past, or with events they have experienced themselves in their real life. Such connections between the actually perceived information about the persona and memory contents allow for the identification of behavior patterns of the persona (e.g., mimic expressions of a police investigator indicating that he has solved a puzzle), and the detection of contradictions between current and past behaviors of the persona (see the section on PSI and media entertainment following). Retrieving memorized information also allows one to make judgments at the formal level

of the media product, as viewers may refer both to knowledge about a fictional character (Konijn & Hoorn, 2005) and the actual actor who is playing the character ("character synthesis," cf. Wulff, 1996).

- Anticipatory observation: Thinking about the future of a persona, that is, his/her further appearance in the watched program or his/her fortune outside of the program, is another cognitive process that belongs to PSI. In situations of high-level PSI, viewers may anticipate the consequences of a persona's actions or words (e.g., politically incorrect statements of a talk show host) for her/his social acceptability and career; during low-level PSI, in contrast, viewers will not think extensively about the future that the persona should expect.

- Evaluations: Judgments about the persona's thoughts, utterances, and actions are another important cognitive PSI process. They may refer to moral dimensions (e.g., Zillmann, 1996), but can, for instance, also target performance issues ("this was a poor argument"), the appearance of the persona, or the truthfulness of her/his words. Evaluations contribute substantially to the image that viewers create from a persona and are key determinants of the development of parasocial relationships (see the section on PSI and media entertainment following).

- Construction of relations between persona and self: Another part of viewers' ruminations about a persona concerns the viewers' self. People compare themselves with media personae on various dimensions (e.g., Mares & Cantor, 1992), and they may look for similarities between them and the personae. Social comparisons (Festinger, 1954) with people on the screen may serve various functions, for example, in the construction of identity (Trepte, Chapter 15, this volume). Persona–self-relations are not limited to comparison processes, however. They can also refer to imagining a social group to which both the persona and the viewer belong (e.g., supporters of the same sports team) and thus create a sense of affiliation, or they might imply intended learning in that viewers wonder how their own lives could benefit from the experiences the persona communicates (e.g., in viewers of teleshopping programs, cf. Fritchie & Johnson, 2003). It is more likely that viewers establish such relations between the persona and themselves in conditions of high-level PSI than during low-level PSI, as personae in the latter case are perceived as less relevant, and establishing such relations with them does not produce much valuable information for the viewers.

One might add more items to this list of cognitive PSI processes. However, the types of thinking about a media persona discussed certainly cover a major portion of the cognitive activities that people perform during real-life interactions, so we believe that they also represent a good deal of viewers cognitions during person-oriented media reception.

According to the model, emotional PSI processes do not display as many subtypes as cognitive PSI do. We distinguish between empathic emotions (feeling with the persona, cf. Zillmann, this volume), persona-generated own emotions, and "mood contagion" phenomena (Neumann & Strack, 2000).

Empathic Reactions to media personae have been extensively investigated by Zillmann Chapter 10, (this volume). According to his affective disposition theory (Raney, Chapter 9, this volume), a broad variety of socio-emotional experiences may occur depending on viewers' moral evaluation of a persona's actions. Viewers may experience (virtually) the same emotions as a persona is expressing, if they consent morally with her/him, or emotions of the opposite valence than the feelings expressed by the persona may occur (if viewers morally disagree with the persona). We model empathy-based emotional processes as affective counterparts of the cognitive PSI process "comprehension and reconstruction" (see prior); however, Zillmann conceptualizes empathy to include both cognitive and affective processes of reproducing a persona's internal condition (Zillmann, Chapter 10, this volume). With respect to the PSI model, empathic emotional reactions represent the major portion of affective PSI processes. As personae might display a great bandwidth of different emotions, so would empathically (or counter-empathically) interacting viewers experience a broad variety of feelings.

Persona-Generated Own Emotions.   In addition to socio-emotions directly reproduced from a persona's emotions, observing a persona may cause self-related feelings (egoemotions, cf. Vorderer, 1998). For instance, a viewer who perceives great similarity between her-/himself and a persona may feel ashamed due to politically incorrect contentions of the persona, because s/he feels to be part of a group who has been discredited. A positive example would be the pride of viewers who watch their favorite sports team win a match, because they perceive themselves as part of the successful group ("basking in reflected glory," cf. Cialdini et al., 1976). Such persona-generated own emotions are the affective counterpart to the cognitive PSI process of establishing relations between persona and self (see above).

Mood Contagion, finally, is the automatic and unintended transfer of mood from one person to another (Neumann & Strack, 2000). Watching a laughing child triggers spontaneous happiness in many people, for example. Such processes may also occur between personae who are expressing a certain emotion and a viewer. Observers of a romantic episode in a love movie may acquire the same flowery condition that the personae in the movie exhibit. Cognitive dispositions, such as moral dissent, may prevent such contagion processes, as negative evaluations cause additional vigilance that allows viewers to override spontaneous emotions. If viewers do not sustain such vigilance, however, mood contagion phenomena as one subtype of emotional PSI processes are conceivable.

Similarly to cognitive PSI processes, affective PSI can display substantial procedural dynamics with respect to intensity. Regardless of the mechanism that generates emotional responses to personae (empathy, own emotions, or mood contagion), intensity of the resulting emotion(s) varies over time and depends on persona and viewer factors (see the following sections). In situations of high-level PSI, strong emotional experiences (which are, for instance, accompanied by high levels of physiological arousal) will occur repeatedly, whereas during low-level PSI, weak or next-to-zero affective experiences would be observable. Affective PSI processes may therefore lead to an "emotional roller-coaster ride" just as affective responses to a real person can do.

Behavioral PSI processes, finally, are probably the empirical starting point of all PSI theory. The idea that viewers respond to media personae similarly as to real people receives immediate support by the casual observation of individuals who shout at media personae or gesticulate vigorously in front of a TV set. Such observable reactions to media personae are, of course, accompanied by internal (affective and cognitive) PSI processes, but can be analytically and empirically separated. We distinguish three behavioural PSI processes:

Motor Activity that aims at keeping one's perceptual apparatus directed toward the personae is a basic behavioral process that is closely connected to attention allocation (see prior). Just as in interactions with real people, viewers may turn their head and move their eyes to keep (visual) contact with a moving persona. Such activity would be more energetic in the condition of high-level PSI, whereas during low-level PSI, viewers would not spend much motor energy in holding contact to the persona.

Physical Activity (Especially Mimics and Gestures) towards a persona is also a parasocial behavioral pattern that resembles real-world interaction. Smiling back to a smiling persona and setting up a romantic facial expression when a beautiful persona is appearing are typical examples of mimic responses to media figures. Frequently, also gestures toward media personae occur, such as viewers of sports broadcasts indicating a promising route to pass through the defense of the opposite team, or outraged viewers of a talk show forming obscene signs towards the personae with their fingers. Presumably, such behavior patters are automatic and not consciously controlled. However, knowing that the persona cannot take notice of such behaviors opens up degrees of freedom in social behavior, allowing one to switch off processes of self-control that would prevent social sanctions, for example, following obscene signs in real interactions. The internal (cognitive, meta-cognitive and emotional) processes that trigger mimics and gestures as responses to personae are potentially complex and deserve more theoretical and empirical research. With respect to the model of PSI, however, we define such behavior as an important conative component of PSI processes.

Verbal Utterances to a persona are the third category of behavioral PSI processes. Depending on the type of persona and situation, anecdotes of very different verbal viewer responses have been reported. For instance, people formulate expressions of love and affinity to likeable personae, or they use explicitly negative language when talking to a disliked persona. Comments, recommendations, expressions of understanding or indifference, and statements of agreement or dissent with a persona are further examples of the long list of conceivable types of verbal responses to media personae. Similarly to mimics and gestures, the question of automaticity versus planned behavior (based on knowledge of the mediation of the social situation) remains unsolved. Moreover, some viewers may just pretend that they address a persona, although their utterances actually aim at communicating a message (e.g., about their emotional condition or about their competence) to their co-viewers. Yelling at a soccer player who has not hit the goal, for instance, may be an indirect message for co-viewers that contains information about the viewer's engagement with a team and/or his competence in soccer issues. Alternatively, the shouting may even be a strategy to reassure oneself of one's competence or self-esteem that simply utilizes a media persona as an unspecific projection screen. These examples already indicate that there is much complexity in the issue of verbal viewer responses to media personae. Nevertheless, there is no doubt that verbal actions towards a persona belong to the behavioral classes of PSI processes. The frequency of verbal responses is presumably higher in the condition of high-level PSI than during low-level PSI, as viewers are likely to direct most utterances to personae they regard as relevant and important.

Intermediate Summary

So far, the explication of the process-oriented model of PSI has focused on the conceptual deconstruction of PSI, and has identified component processes that can, but must not necessarily, occur as viewer response to a media persona. A variety of cognitive, affective, and behavioral processes has been proposed, but we believe by no means that the present list is complete. Rather, it indicates the large complexity of viewers' "interaction" with media personae, as very different processes may occur synchronously. Future research on PSI may use the modelled structure of PSI to investigate the role of personae and their appreciation through the viewers in mass communication processes in more detail (see the subsequent sections of this chapter). However, to complete the explication of the PSI model, the effect of personae and viewer characteristics on the procedural dynamics of PSI patterns is addressed first.

The Role of the Persona

Without doubt, the quality and intensity of viewers' PSI largely depend on attributes of the media persona under consideration. What a persona looks like, and what he says and does affects the type of cognitive processing, emotional, and behavioural responses that occur in viewers (see, for example, Horton and Wohl's (1956) example of "The Lonesome Gal"). Very specific persona attributes may be of great relevance for individual viewers, but we identify a few factors that should be regarded as generally important for the (procedural) quality and intensity of PSI:

- Obtrusiveness and persistence: The more noticeable a persona within a media product is, the more probable are intense parasocial responses from the viewers. If a media offering centers around one protagonist and presents her/him frequently and with many close-up shots, s/he is highly obtrusive and persistent, which increases noticeability and thus the likelihood of high-level PSI. Personae who appear only briefly and are not moved into the focus of the media presentation, in contrast, are less probable to induce intense PSI.
- Addressing performance: Intensive PSI frequently emerge when personae treat viewers as if they were participating in a real social interaction, for example, by saluting or addressing viewers directly (e.g., Auter, 1992). This is the reason why news anchors are a preferred object of PSI research, as they belong to those media personae who display a variety of behaviours that they adopt from real-life interaction (e.g., Rubin et al., 1985).
- Physical appearance: Visual information about a persona is not only relevant for impression formation (see prior) but also affects subsequent PSI processes. The reason for this is that viewers possess much knowledge that enables them to infer characteristics of a persona from visual information, for example, about his/her professional occupation, style preferences, or personality traits (Todorov & Uleman, 2002). Because visual information is easy to receive and available all the time that a persona is on screen (in contrast, for example, to verbal or action-related information), we assume that visual data is in general an important persona factor in the quality, intensity, and dynamics of PSI processes.

Highlighting the importance of these three persona attributes does not imply that all other characteristics are less relevant. On the contrary, further theorizing and research should establish connections between these and other persona variables and specific PSI processes. The frequency of mood contagion phenomena (an emotional PSI process), for example, may be affected primarily by facial expressions of the persona, whereas anticipatory thinking about the persona (a cognitive PSI process) will rather rely on verbal or action-related information. As PSI patterns may change very rapidly and very often throughout the course of media exposure, the matrix of the effects of persona characteristics on PSI processes is probably extremely complex.

The Role of the Audience

In addition to persona attributes, most PSI processes depend on characteristics of the viewers. Both trait and state variables may influence the development of PSI processes and patterns. Tired viewers, for example, are more likely to limit cognitive processes to very peripheral intensities and not to reach high levels of cognitive involvement with a persona. An important viewer trait might be openness for experience (Costa & McRae, 1992), which one may assume to correlate positively with the intensity of many different types of PSI processes, as highly open viewers are more interested and capable to internalize the utterances, thoughts, and experiences of a persona than are viewers with smaller openness values. In general, most state and trait variables that can affect PSI processes relate to viewers' motivation to become involved with a media persona. Motivation for PSI refers both to the general readiness to "interact" with a persona at all, which determines the probability of high-level versus low-level PSI, and to the preference for a specific PSI pattern. For instance, one viewer might be motivated to center her/his parasocial response pattern around the perception and processing of a persona's physical attractiveness, whereas another viewer would prefer to actualize her/his admiration for a celebrity (e.g., McCutcheon, Ashe, Houran, & Maltby, 2003). A key determinant of PSI is of course the parasocial relationship that viewers already hold with a persona, as the relationship schema indicates what viewers should expect from interactions with a persona (see section on PSI and media entertainment below).

The umbrella term "motivation" incorporates viewers' contributions to "close" the social interactions that personae "offer" in very specific and sometimes unique ways. Individual and situational differences impede a more precise modelling of these contributions within this chapter, but it should be clear that viewer characteristics affect the quality, intensity, and procedural dynamics of PSI at least to the same extent as persona attributes do (Hartmann et al., 2004b).

Model Summary

Based on the history of PSI research, and adopting theories from communication, media, and social psychology, we have laid out a process-oriented model of PSI that conceptualizes viewers' responses to media personae as being composed of different cognitive, emotional, and/or behavioral processes (see Fig. 17.1). These processes follow on initial impression formation (or persona recognition), can emerge into different interaction patterns, can change dynamically within the course of media exposure, and are strongly influenced both by persona and viewer variables. Overall, viewers' appreciation of a media persona is highly complex and dynamic, and its analysis must refer to a broad variety of research on human interaction.

PSI/PSR AND MEDIA ENTERTAINMENT: SELECTIVE EXPOSURE, CONSUMPTION EXPERIENCE, AND MEDIA EFFECTS

The broad variety of phenomena that conceptually belongs to the constructs PSI and PSR suggests to model multiple connections between them and the psychology of entertainment. As personae are key components of the production, promotion and consumption of media entertainment, PSI/PSR with them are theoretically linked to enjoyment at the level of entertainment media selection, the level of entertainment media consumption, processing and experience, and the level of entertainment media effects. We discuss the role of PSI/PSR for each of theses dimensions of media entertainment in the following subsections.

PSI/PSR and Selective Exposure to Media Entertainment

Advertisements for many entertainment media—for example, Hollywood movies—highlight the personae who appear in them. Celebrities are likely to attract people, and the commercial success of many entertainment media products partly depends on the reputation of the involved personnel (e.g., Wallace, Seigerman, & Holbrook, 1993). This holds also true for

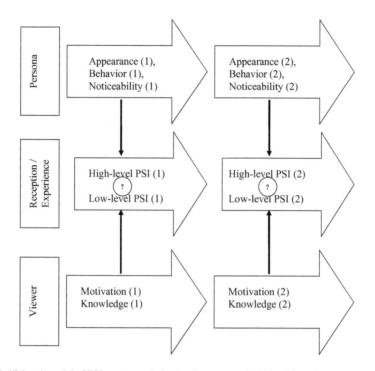

FIG. 17.1. A model of PSI processes during media consumption (simplified visualization of the original model by Hartmann et al., 2004b). Horizontal arrows indicate procedural dynamics over time (e.g., changes in a persona's appearance or behavior), vertical arrows indicate causal influences (e.g., the noticeability of a persona at time 1 influences the quality and intensity of PSI at time 1). Numbers in brackets indicate timedependence of variable (e.g., viewers' motivation for PSI maybe different at time 1 from time 2).

non-entertainment media, such as TV newscasts, which heavily rely on "anchormen" (Rubin et al., 1985).

From the perspective of media consumers, the presence of well-known and/or interesting personae in a media product is apparently an important argument in the decision process of selective exposure. People who select a specific media offering because of the expected appearance of a persona, anticipate rewarding experiences that build on PSI with that persona. Expectations that involvement with a persona will be gratifying arise from certain motivational dispositions, for example, an interest in observing physically attractive individuals, or a demand to experience closeness with successful people (Cialdini et al., 1976). Such motivational structures may be enduring (trait) or situation-bound (state) conditions and determine the selection behavior in situations of media choice.

Even more important than general motivational structures that correspond with characteristics of many personae (e.g., physical attractiveness) are parasocial relationships with a single persona. Repeated exposure to media offerings centered around one specific persona leads to a large number of PSI processes and allows the consumers to accumulate knowledge about the persona. Continued contact leads to the formation of a social relationship schema: People construct a mental representation of their relationship with the persona. This mental representation includes information on the characteristics of the persona and how the individual evaluates them, as well as information about how the individual sees him-/herself within the relationship and the "quality" of the relationship itself (Baldwin, 1992).

Most media users have created relationship schemata with many media personae and added them to their inventory of relationship schemata that concern real people from their social network (Gleich, 1997). When a media offering in which a specific persona is going to appear is available, viewers activate the mental representation of their relationship with the persona. Depending on the quality of the relationship schema—that is, how attractive do viewers find the persona, how rewarding do they evaluate their (parasocial) interaction history with him/her, how much similarities do viewers detect between themselves and the persona—-the motivation to select that media product is altered by the notice of the persona's appearance. If viewers extrapolate from their relationship schema that a new contact to the persona will be entertaining in a way that fits their situational motivation, the probability that they choose the correspond- ing media offering will increase. For example, if a viewer remembers funny experiences of (parasocially) interacting with a certain comedian and if s/he is seeking a hilarious stimulus in order to improve her/his mood (Zillmann, 1988), s/he is likely to prefer a media product showing the comedian over alternative media offerings. If, in contrast, the parasocial relation- ship schema indicates a high probability of negative interaction experiences to the viewer, for example, in the case of a disagreeable celebrity appearing in a talk show, the likelihood that s/he will select the according media product decreases.

The motivation to actualize and revive a parasocial relationship with a media persona is therefore an important determinant for selecting specific media products, and the desire to avoid (further) negative interaction experiences with a persona is a strong force that can repel viewers from watching a given program. Applying the proposed model of PSI, we can further specify the role of parasocial processes in selective exposure to media entertainment: People may anticipate that the general involvement (which includes multiple parasocial processes) with a media persona will be enjoyable in the desired way, or they may expect that one specific type of PSI with the persona will be entertaining.

For example, viewers may evaluate the presence of a familiar talk show host as "generally" gratifying, because the persona creates a positive emotional atmosphere and is likeable inde- pendently from the situational circumstances. Such a persona may help viewers to deal with very fundamental motivational dispositions, for example, to enjoy social contacts in spite of their shyness (Vorderer & Knobloch, 1996). In this case, any type of anticipated (parasocial) interactions (except for negative, disagreeing cognitions and socio-emotions) with the persona will appear to be enjoyable. Other personae may promise entertaining interaction experiences only if viewers structure their parasocial behavior in a special way. For instance, viewers of action movies might find PSI with the superhero most enjoyable if they do not activate evalu- ative or critical cognitions, but observe him empathically (Zillmann, Chapter 10, this volume) and focus their thoughts on justifying arguments for his (violent) behaviors. This way, more specific motivational dispositions may be served by the PSI processes, for example, a need for diversion from experienced helplessness and inferiority in real life (escapism, cf. Katz & Foulkes, 1962).

Because parasocial relationship schemata (may) include knowledge about "how to deal" with a persona, viewers may consider such persona-specific interaction patterns when they evaluate how pleasurable the consumption of a media product presenting the persona will be. They may, for instance, decide that remembering biographical details about one persona who is appearing in a talk show, which is required to enjoy the conversation between the host and the persona, is too demanding in their current situation of tiredness, and prefer another media product in which the personae are "easier" to (parasocially) interact with.

Relationship schemata affect media selection processes not only at the stage when viewers compare available media offerings against each other. Strong parasocial relationships (i.e., fandom) may motivate people to actively search for media products in which a favorite persona appears (Giles, 2002). For example, a person who is holding a strong parasocial binding with a certain actress will lay out her/his search pattern at a video rental store in a way that maximizes the probability to find all available movies in which this actress appears.

In sum, the motivation to select a given piece of media entertainment is strongly affected by the parasocial relationship that viewers hold to the involved personae. A variety of psychological mechanisms is in effect that link relationship schemata and anticipated qualities of PSI to viewing motivation and actual selection behavior. These mechanisms fit into general action-theoretical models of entertainment media selection (e.g., Dohle, Klimmt & Schramm, 2004; Vorderer, 1992; Vorderer, Klimmt, & Ritterfeld, 2004), as parasocial relationship schemata and anticipated (experiential qualities of) PSI represent parts of the expectations toward and evaluations of a media offering that the individual considers for selection. An analysis of decision processes in the consumption of media entertainment must therefore refer to parasocial processes both when considering the qualities of the available media products (e.g., attractiveness and addressing behavior of the appearing personae) and the characteristics (e.g., knowledge and motivational dispositions) of the viewers.

PSI and the Experience of Media Entertainment

Conceptually, the proposed notion of PSI as a set of cognitive, emotional, and/or behavioral reaction processes toward personae on the screen is very closely connected to the experience of consuming media entertainment. Thinking about, feeling with, or talking to a media persona during the reception process most often contributes to the entertainment experience. For example, watching a comedian is fun (i.e., enjoyable), because viewers process the moves and jokes of the persona (attention allocation and cognitive involvement with the persona) and, consequently, feel exhilaration.

The experience of "being entertained" may be a side-effect of parasocial processes (as portrayed in the earlier), or certain PSI (processes) may be structural components of the enjoyment itself. In this section, we discuss both types of links between PSI and entertainment experiences.

Enjoyment as Side-Effect of PSI Processes

In many cases, PSI with a media persona trigger specific experiential processes that viewers regard as enjoyment. Personae often function as key to the comprehension of complex narrative and/or social structures and, thus, enable viewers to access enjoyable experiences that arise from comprehending the mediated situation. For example, cognitive elaboration of the thoughts of the protagonist of a conspiracy thriller (e.g., comparing them to available knowledge from other conspiracy stories, or generating one's own creative ideas on the conspiracy) contributes to the epistemic curiosity (e.g., Groeben & Vorderer, 1988) about the progress of the plot, which breeds interest and suspense (Vorderer & Knobloch, 2000).

Self-related experiences of enjoyment may also occur as side-effects of PSI processes. For example, accessing memorized statements of a familiar persona that contradict her/his current utterances (i.e., a cognitive PSI subprocess) allows the viewer to unmask the persona as a liar, which consequently can trigger feelings of competence and pride (increase in selfesteem, cf. Weiner, 1985), as well as the impression of superiority compared to real or imagined other viewers of the program who have presumably not detected the contradiction in the persona's words. Such self-related emotions of pride and superiority form a major component within the entertainment experience (cf. Klimmt, 2003; Vorderer, Klimmt & Ritterfeld, 2004).

Behavioral PSI processes can launch another important mechanism of enjoyment. For instance, part of the entertainment experience in watching sports broadcasts is the increase in physiological arousal that occurs due to the observation of a dramatic game and the "atmosphere" in the stadium (Dohle, Klimmt & Schramm, 2004). Simulating typical behaviors of stadium visitors, such as cheering, singing, or egging on the athletes (the personae), helps TV viewers to enhance and maintain their physiological arousal and the exciting atmosphere, and thus boosts their entertainment experience.

The proposed model of PSI highlights the importance of viewers' knowledge about the mediated quality of their "interaction" with a persona. Although people may suspend this knowledge in certain situations and perceive the personae just as close as a real individual ("social presence", cf. Lee, 2004), awareness of the screen that separates personae and viewers sometimes allows for playful interaction behaviors (Schramm et al., 2002). Knowing that a politician in a TV talk show cannot hear viewers' comments opens up new degrees of freedom for viewers' verbal responses to her/his utterances. Even severe insults do not cause social sanctions as they would in real interpersonal interaction. Such "unleashed" social behavior provides simulated experiences of extreme or unusual interaction patterns, which may breed additional forms of enjoyment. In the above example, aggressive and insulting comments to the politician (behavioral PSI processes) foster viewers' perception of themselves as candid citizens who do not mince matters when it comes to important issues. Such self-perceptions, in turn, would be associated with positive emotions that viewers construe as enjoyment. Other situations in which viewers utilize the "intimacy at a distance" (Horton & Wohl, 1956) for purposes of social-interactive enjoyment are conceivable. For instance, viewers may act out their social curiosity about strange, ugly, or disgusting individuals by interacting with them via television, which allows them to quickly create more distance if they want to, whereas distancing would be more difficult (e.g., socially inappropriate) in a real encounter.

PSI Processes as Component of Entertainment Experiences

The second conceptual link between PSI and the entertainment experience is that PSI processes can be components or dimensions of enjoyment itself. Affective PSI processes certainly display the strongest affinity to experiential phenomena that belong to the concept of media entertainment, such as enjoyment, fun, and pleasure (Bosshart & Macconi, 1998; Vorderer, Klimmt & Ritterfeld, 2004). Feeling happy with a persona empathically (Zillmann, Chapter 10, this volume) is part of enjoyment, as well as "mood contagion" processes (Neumann & Strack, 2000) between a persona and a viewer. Another well-known example is TV viewers suffering from the fear that their favorite sports team may lose the observed game (Raney, 2003, Chapter 9, this volume).

Cognitive and behavioral responses to media personae can represent important parts of the entertainment experience, too. Sometimes cognitive involvement with the statements and thoughts of a media persona may be exactly the kind of diversion that viewers find enjoyable. This sort of diversion has been conceptualized as escapism (Henning and Vorderer, 2001; Katz & Foulkes, 1962), or as epistemic modes of pleasure (e.g., the enjoyment of successful perceptual organization, or the satisfaction of confirming one's own expectations, cf. Groeben & Vorderer, 1988). Cognitive PSI processes that lead to such distraction from real-life thinking, and rearrange viewers' cognitions in a preferred way, represent enjoyment processes by themselves and thus do not belong to the factors that only cause entertainment experiences as side-effects (see prior). If a certain way of thinking or intellectual stimulation is viewers' goal for using media entertainment, then, cognitive PSI processes form key components of the resulting consumption experience.

Behavioral PSI processes, finally, may also represent dimensions of entertainment experiences rather than originators of enjoyment. For instance, viewers who bodily participate in a boxing match they are watching (e.g., who are performing punches they regard as useful in a given fight situation while standing in front of the screen) add the fun of their own physical activity to their overall appreciation of the broadcast. Such activities require an effective comprehension of the ongoing fight, of course, and thus come with certain cognitive and affective PSI processes as well, which may also influence other components of the entertainment experience. But, analytically, the experience of one's own movement may be separated from other dimensions of the entertainment experience. In certain cases, other behavioral PSI processes, such as talking to a persona or clapping to the rhythm of a persona's song, add similar dimensions to the entertainment experience.

The discussed role of PSI in entertainment experiences has focused on examples of positive relationships between PSI and enjoyment. However, the considered connections also function in a negative way, in that PSI processes can cause experiential phenomena that diminish enjoyment (e.g., disagreeing cognitions about a persona impede exhilaration about her/his jokes) or in that PSI processes, which could be part of an entertainment experience, do not reach the required intensity or duration, which would at least prevent enjoyment to fully unfold. For instance, a talk show host could perform insufficiently with respect to addressing the viewers, and consequently, no extensive emotional PSI processes evolve, which keeps the overall appreciation of the show low. In sum, entertainment experiences rise and fall with the quality and intensity of the parasocial processes that viewers create during consumption.

PSI and Effects of Media Entertainment

One immediate effect of PSI to personae of media entertainment products is directed towards subsequent selection behavior. Virtually all theoretical accounts of media selection (e.g., Rosengren, Wenner, & Palmgreen 1985; Slater, Henry, Swaim, & Anderson, 2003; Zillmann & Bryant, 1985) highlight the importance of past media experience for future decisions on media consumption. Viewers are more probable to re-select (types of) programs that they have found interesting and enjoyable in the past, than to decide for media offerings that they have negativly evaluated before. This general notion of learning from prior experience is certainly also applicable to the effect of PSI processes toward personae in entertainment media on subsequent selection behavior. Each parasocial contact affects the relationship schema (Baldwin, 1992) that viewers maintain for the given persona, which in turn determines future decisions on selective exposure (see above). Gleich (1997) has modelled the connections between PSI and parasocial relationships as a cyclic process, which is similar to Slater et al.'s (2003) spiral model of media selection and media effects.

People apply information from PSI processes and relationship schemata concerning personae not only in actions of media selection. Some theories of media effects, for instance, the social-cognitive theory of mass communication (Bandura, 2001), suggest that people acquire knowledge from observing objects, events, and social entities in the media. Comprehension as well as adoption and performance of acquired knowledge partly depend on the characteristics of the "models" that viewers observe. If, for example, a media (entertainment) product portrays a specific behavior to provide an effective solution to a problem (e.g., the "model" receives a reward for the behavior), the probability that viewers will try out the behavior in a similar situation increases. PSI processes, such as positive affective responses to a persona, can moderate the outcome of such social learning processes. Viewers may adopt certain behaviors simply because they observed them in personae with which they experienced rewarding PSI. Moreover, they may prefer one specific action in a given situation of problem solving over conceivable alternatives, because their strong affinitiy (PSR) with a persona has enhanced the cognitive accessibility of the kind of behavior that the persona typically applies (priming). Finally, the parasocial relationship that results from PSI may affect motivational structures: For instance, perceived similarity to the protagonist of a crime drama TV series may add to the motivation to prefer morally acceptable action alternatives. Intense PSI processes may effect a broad variety of viewers' behavior, then, because people learn about facts, attitudes, and behaviors during PSI with personae (knowledge acquisition and formation of decision making heuristics: "what would persona X do in this situation?"), and their relationship schemata affect their action-related cognitive structures (priming) as well as motivational dispositions (e.g., with respect to moral reasoning).

A related type of effects of (the characteristics of) PSI processes in entertainment media is persuasion. Parasocial responses to personae might affect processes of attitude formation and attitude change that are explicitly or tacitly intended by communicators. Testimonial advertising, for example, has been practiced for a long time. Its underlying assumption is that an "image transfer" from a celebrity to a product will help to make the product appear attractive to (potential) customers (e.g., Byrne, Whitehead, & Breen, 2003). Similarly, entertainment-education programs seek to communicate credible messages about health behavior or social

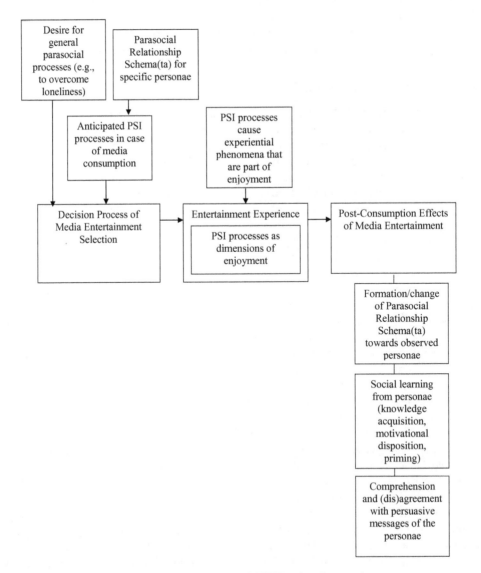

FIG. 17.2.   Theoretical links between PSI/PSR and media entertainment.

change through popular media characters (e.g., Papa et al., 2000). From the perspective of the proposed model, a set of PSI processes directed toward the appearing personae causes (or can cause) the effects desired by the communicators of testimonial advertisements and entertainment education messages. For example, an occurring persona to which viewers hold a positively evaluated parasocial relationship schema may attract attention to the advertised product and stimulate cognitive elaboration and memorization of the product characteristics that the persona talks about (cognitive PSI processes). Affective responses to the persona (e.g., mood contagion) may link positive emotions to the product, which could be revived if viewers recognize the product's logo at a supermarket. Parasocial responses to a persona can of course lead to unintended, counterproductive effects also: If a disliked persona praises a product, cognitive PSI processes such as counter-arguing and affective reactions, such as anger, may lead to the formation of negative attitudes toward the product and decrease the probability of purchase. Similarly, the individual relationship schema will lead to specific PSI processes that either facilitate the comprehension of and agreement to an entertainment-education message or, rather, cause disagreement and resistance to the suggested behavioral change. Depending on the quality and course of PSI processes, then, the persuasive effects of (entertaining) media products that rely on celebrities or other media personae can vary substantially. The proposed model of PSI as multi-faceted phenomenon may help to identify more specific mechanisms of viewers' appreciation of a persona's advocacy for a message or product.

Summary

Conceptual connections between PSI and media entertainment have been identified at the stages of selective exposure, consumption experience, and media effects. The proposed model of PSI as a multi-faceted phenomenon allows us to define specific processes that occur during media consumption and affect viewers' cognitions, emotions, and behavior both before media selection and afterward. It breaks down the important role of personae for the selection, comprehension, appreciation, and effects of media entertainment into concrete and empirically investigable processes and relationships. Figure 17.2 summarizes the discussed dimensions of PSI for the psychology of entertainment.

CONCLUSION: DIRECTIONS FOR FUTURE RESEARCH ON ENJOYMENT OF PSI/PSR

In this chapter, we have outlined a process-oriented model of PSI with media characters, and explored the relations between PSI, PSR, and media entertainment. The outcome is a large set of assumptions that may inspire future research on viewers' response to media personae. Most important is the proposition that a large number of internal processes may occur synchronously during the observation of a media persona, and that viewers' appreciation of personae is dynamic and extremely complex. In this sense, PSI and PSR are broader concepts compared to approaches such as empathy (Zillmann, Chapter 10, this volume) or identification (Cohen, Chapter 11, this volume). Clarifying the differences and overlaps between these approaches remains a primary theoretical challenge for entertainment research.

Moreover, the proposed links between PSI/PSR and entertainment should also be investigated empirically, since most existing research on PSI/PSR did not reach the suggested level of precision with respect to the definition of separable process components (e.g., empathy vs. ego-emotions or retrieving persona-related memories vs. thinking about the persona's future). Maybe the suggested model (and its relations to entertainment) is overprecise and cannot be illustrated empirically. But the differentiation of PSI processes may help to better understand many phenomena in media reception that facilitate or belong to enjoyment. This way, the application of the above considerations may support theoretical progress in entertainment theory (Bryant, 2004), as many realizations of "media enjoyment" are still in need of conceptualization (Vorderer et al., 2004).

The discussed role of PSI/PSR in entertainment holds, also, theoretical value for media effects research. Studies on entertainment-education as well as advertising research may benefit from the suggested process structure of PSI and identify aspects of persona behavior and viewer responses that are critical for the achievement of communication goals. The appreciation of personae in non-entertainment contexts, such as TV news (e.g., Rubin et al., 1985), should also be reconsidered in the light of the proposed links between PSI processes and entertainment.

All suggested lines of further research on PSI/PSR and media entertainment require methodological progress as well. Existing instruments such as the PSI-scale (Rubin et al., 1985), should be revised, as only distinct measures for different PSI processes and for PSR (schemata) allow for the transfer of the modeled viewers responses to empirical studies (Hartmann et al., 2004b). In addition, alternative instruments that do not rely on ex-post questionnaires should be developed. Online measures, such as observation, the think-aloud methodology (e.g., Shapiro, 1994), physiological and neuropsychological methods could assess more of the complexity and dynamics of cognitive, affective, and behavioral PSI processes and help to test the connections between specific persona characteristics, viewer attributes, and experiential processes related to enjoyment and entertainment media effects.

In sum, the role of viewers' response to media personae in entertainment contexts is certainly a very important, but also very complex field that is still in need of theoretical integration and elaboration. However, expanding Horton and Wohl's (1956) original ideas to the universe of modern media entertainment is an endeavor that promises to reveal substantial insights for entertainment theory and communication research in general.

REFERENCES

Ashe, D. D., & McCutcheon, L. E. (2001). Shyness, loneliness, and attitude toward celebrities. *Current Research in Social Psychology, 6*, 124–133.

Auter, P. J. (1992). TV that talks back: An experimental validation of a parasocial interaction scale. *Journal of Broadcasting and Electronic Media, 36*, 173–181.

Auter, P. J., & Palmgreen, P. (2000). Development and validation of a parasocial interaction measure: The audience-persona interaction scale. *Communication Research Reports, 17*, 79–89.

Babrow, A. S., O'Keefe, B. J., Swanson, D. L., Meyers, R. A. & Murphy, M. A. (1988). Person perception and children's impression of television and real peers. *Communication Research, 15*, 680–698.

Baldwin, M. W. (1992): Relational schemas and the processing of social information. *Psychological Bulletin, 112*, 461–484.

Bandura, A. (2001). Social cognitive theory of mass communication. *Media Psychology, 3*, 265–299.

Bente, G., Krämer, N. C., & Petersen, A. (Eds.). (2002). Virtuelle Realitäten [Virtual realities]. Göttingen: Hogrefe.

Boon, S. D., & Lomore, C. D. (2001). Admirer–celebrity relationships among young adults: Explaining perceptions of celebrity influences on identity. *Human Communication Research, 27*, 432–465.

Bosshart, L. & Macconi, I. (1998). Defining "Entertainment." *Communication Research Trends, 18* (3), 3–6.

Brewer, M. B. (1988): *A dual process model of impression formation.* In T. Srull & R. Wyer (Eds.), *Advances in social cognition* (Vol. 1, pp. 1–36). Hillsdale, NJ: Lawrence Erlbaum Associates.

Bryant, J. (2004). Critical communication challenges for the new century. *Journal of Communication, 54*, 389–401.

Byrne, A., Whitehead, M., & Breen, S. (2003). The naked truth on celebrity endorsement. *British Food Journal, 105*, 288–296.

Caughey, J. L. (1984). *Imaginery social worlds: A cultural approach.* Lincoln: University of Nebraska Press.

Caughey, J. L. (1986). *Social relations with media figures.* In G. Gumpert & R. Cathcart (Eds.), *Inter/Media. Interpersonal communication in a media world.* New York : Oxford University Press.

Cialdini, R. B., Borden, R. J., Thorne, A., Walker, M. R., Freeman, S., & Sloan, L. R. (1976). Basking in reflected glory: Three (football) field studies. *Journal of Personality and Social Psychology, 34*, 366–375.

Cohen, J. (2001). Defining identification: A theoretical look at the identification of audiences with media characters. *Mass Communication & Society, 4*, 245–264.

Cohen, J. (2003). Parasocial breakups: Measuring individual differences in responses to the dissolution of parasocial relationships. *Mass Communication and Society, 6*, 191–202.

Cohen, J. (2004). Parasocial break-up from favourite television characters: The role of attachment styles and relationship intensity. *Journal of Social and Personal Relationships, 21*, 187–202.

Cole, T., & Leets, L. (1999). Attachment styles and intimate television viewing: Insecurely forming relationships in a parasocial way. *Journal of Social and Personal Relationships, 16*, 495–511.

Conway, J. C., & Rubin, A. M. (1991). Psychological predictors of television viewing motivation. *Communication Research, 18*, 443–463.

Costa, P. T., & McCrae, R. R. (1992). *Revised NEO personality inventory (NEO PI-R) and NEO five factor inventory (NEO-FFI): Professional Manual.* Odessa, FL: Psychological Assessment Resources.

Dohle, M., Klimmt, C., & Schramm, H. (2004, May). *Rationality in media selection processes: Analyzing the match between motivations, expectations, and ex-post evaluations.* Presentation to the 54th conference of the International Communication Association (ICA), May 27–31, 2004, New Orleans.

Ellis, G. J., Streeter, S. K., & Engelbrecht, J. D. (1983). Television characters as significant others and the process of vicarious role taking. *Journal Of Family Issues, 4*, 367–384.

Fabian, T. (1993). *Fernsehen und Einsamkeit im Alter. Eine empirische Untersuchung zu parasozialer Interaktion [Television and Loneliness in elder people. An empirical investigation of parasocial interactions].* Münster: LIT.

Festinger, L. (1954): A theory of social comparison process. *Human Relations, 7*, 117–140.

Fiske, S. T., Lin, M., & Neuberg, S. L. (1999). *The continuum model. Ten years later.* In S. Chaiken & Y. Trope (Eds.), *Dual-process theories in social psychology* (pp. 231–254). New York: Guilford Press.

Flora, C. (2004). Seeing by starlight. *Psychology Today, 37* (4), 36–41.

Fritchie, L. L., & Johnson, K. K. (2003). Personal selling approaches used in television shopping. *Journal of Fashion Marketing and Management, 7*, 249–258.

Giles, D. C. (2002). Parasocial interaction: A review of the literature and a model for future research. *Media Psychology, 4*, 279–305.

Giles, D. C., & Maltby, J. (2004). The role of media figures in adolescent development: Relations between autonomy, attachment, and interest in celebrities. *Personality and Individual Differences, 36*, 813–822.

Gleich, U. (1997). *Parasoziale Interaktionen und Beziehungen von Fernsehzuschauern mit Personen auf dem Bildschirm: ein theoretischer und empirischer Beitrag zum Konzept des aktiven Rezipienten.* [Parasocial Interactions and relationships of television viewers] Landau: Verlag Empirische Pädagogik.

Gleich, U. (1999). *Parasoziale Bindungen zu Politikern?* In P. Winterhoff-Spurk & M. Jäckel (Eds.), *Politische Eliten in der Mediengesellschaft* [Political elites in media society] (pp. 151–168). München : R. Fischer.

Grant, A. E., Guthrie, K. K., & Ball-Rokeach, S. J. (1991). Television shopping. A media system dependency perspective. *Communication Research, 18*, 773–798.

Groeben, N., & Vorderer, P. (1988). *Leserpsychologie: Lesemotivation – Lektürewirkung* [The psychology of readers: Reading motivation-effects of reading]. Münster: Aschendorff.

Hartmann, T., Klimmt, C., & Vorderer, P. (2001). Avatare: Parasoziale Beziehungen zu virtuellen Akteuren [Avatars: Parasocial relationships with virtual actors]. *Medien- und Kommunikationswissenschaft, 49*, 350–368.

Hartmann, T., Schramm, H., & Klimmt, C. (2004a). *Vorbereitende Überlegungen zur theoretischen Modellierung parasozialer Interaktionen im Prozess der Medienrezeption* [Preliminary considerations on modelling parasocial interaction in the process of media consumption]. Retrieved 01/10/2004 from http://www.ijk.hmt-hannover .de/psi/.

Hartmann, T., Schramm, H., & Klimmt, C. (2004b). Personenorientierte Medienrezeption: Ein Zwei-Ebenen-Modell parasozialer Interaktionen [Person-oriented media reception: A two-level model of parasocial interactions]. *Publizistik, 49*(1), 25–47.

Henning, B., & Vorderer, P. (2001). Psychological escapism: Predicting the amount of television viewing by need for cognition. *Journal of Communication, 51*, 100–120.

Hoerner, J. (1999). *Scaling the web. A parasocial interaction scale for world wide web sites.* In D. W. Schumann & E. Thorson (Eds.), Advertising and the world wide web (pp. 135147). Mahwah, NJ: Lawrence Erlbaum Associates.

Hoffner, C. (1996). Children's wishful identification and parasocial interaction with favorite television characters. *Journal of Broadcasting and Electronic Media, 40*, 389–402.

Horton, D., & Wohl, R. (1956). Mass communication and para-social interaction: Observation on intimacy at a distance. *Psychiatry, 19*, 215–229.

Houlberg, R. (1984). Local television news audience and the para-social interaction. *Journal of Broadcasting, 28*, 423–429.

Kanning, U. P. (1999): *Die Psychologie der Personenbeurteilung* [The psychology of person judgments]. Göttingen: Hogrefe.

Katz, E., & Foulkes, D. (1962). On the use of mass media for escape: Clarification of a concept. *Public Opinion Quarterly, 26*, 377–388.

Klimmt, C. (2003). Dimensions and determinants of the enjoyment of playing digital games: A three-level model. In M. Copier & J. Raessens (Ed.), *Level Up: Digital Games Research Conference* (pp. 246–257). Utrecht: Faculty of Arts, Utrecht University.

Konijn, E. A., & Hoorn, J. F. (2005). Some like it bad: Testing a model for perceiving and experiencing fictional characters. *Media Psychology, 7* (2), 107–144.

Krotz, F. (1996). *Parasoziale Interaktion und Identität im elektronisch mediatisierten Kommunikationsraum.* In P. Vorderer (Eds.): Fernsehen als "Beziehungskiste". Parasoziale Beziehungen und Interaktionen mit TV-Personen [TV as relationship machine: Parasocial relationships and interactions with TV personae] (pp. 73–90). Opladen: Westdeutscher Verlag.

Lee, K. M. (2004). Presence, explicated. *Communication Theory, 14* (1), 27–50. Levy, M. R. (1979). Watching TV news as parasocial interaction. *Journal of Broadcasting, 27,* 68–80.

Levy, M. R.(1979). Watching TV news as para-social interaction. *Journal of Broadcasting, 27,* 68–80.

Maltby, J., Houran, J., Lange, R., Ashe, D., & McCutcheon, L. E. (2002). Thou shalt worship no other gods—unless they are celebrities: The relationship between celebrity worship and religious orientation. *Personality and Individual Differences, 32,* 1157–1172.

Mares, M. L., & Cantor, J. (1992). Elderly viewers' responses to televised portrayals of old age: Empathy and mood management vs. social comparison. *Communication Research, 19,* 459–478.

McCutcheon, L. E., Ashe, D. D., Houran, J., & Maltby, J. (2003). A cognitive profile of individuals who tend to worship celebrities. *The Journal of Psychology, 137,* 309–322.

Neumann, R., & Strack, F. (2000). "Mood contagion": The automatic transfer of mood between persons. *Journal of Personality and Social Psychology, 79,* 211–223.

Nordlund, J. (1978). Media interaction. *Communication Research, 5,* 150–175.

Omdahl, B. L. (1995). *Cognitive appraisal, emotion, and empathy.* Mahwah, NJ: Lawrence Erlbaum Associates.

Palmgreen, P., Wenner, L. A., & Rayburn, J. D. (1980). Relations between gratifications sought and obtained. A study of television news. *Communication Research, 7,* 161–192.

Papa, M. J., Singhal, A., Law, S., Pant, S., Sood, S., Rogers, E. M., & Shefner-Rogers, C. L. (2000). Entertainment-education and social change: An analysis of parasocial interaction, social learning, collective efficacy, and para-doxical communication. *Journal of Communication, 50,* 31–55.

Perse, E. M., & Rubin, A. M. (1988). Audience activity and satisfaction with favorite television soap opera. *Journalism Quarterly, 68,* 368–375.

Perse, E. M., & Rubin, R.B. (1989). Attribution in social and parasocial relationships. *Communication Research, 16,* 59–77.

Perse, M. P. (1990). Media involvement and local news effects. *Journal of Broadcasting and Electronic Media, 34,* 17–36.

Raney, A. A. (2003). *Disposition-based theories of enjoyment.* In J. Bryant, D. R. Roskos- Ewoldsen, & J. Cantor (Eds.), Communication and emotion: Essays in honor of Dolf Zillmann (pp. 61–84). Mahwah, NJ: Lawrence Erlbaum Associates.

Ritterfeld, U., Klimmt, C., Vorderer, P., & Steinhilper, L. K. (2005). The effects of a narrative audio tape on preschoolers'entertainment experience and attention. *Media Psychology, 7,* 47–72.

Rosengren, K. E., Wenner, L., & Palmgreen, P. (Eds). (1985). *Media gratifications research: Current perspectives.* Beverly Hills Sage.

Rubin, R. B., & McHugh, M. P. (1987). Development of parasocial interaction relationships. *Journal of Broadcasting and Electronic Media, 31,* 279–292.

Rubin, A. M., & Perse, E. M. (1987). Audience activity and soap opera involvement: A uses and effects investigation. *Human Communication Research, 14,* 246–292.

Rubin, R. B., & Rubin, M. (2001). *Attribution in social and parasocial relationships.* In V. Manusov & J.H. Harvey (Eds.), Attribution, communication behavior, and close relationships (pp. 320–337). Cambridge, UK: Cambridge University Press.

Rubin, A. M., & Step, M. M. (2000). Impact of motivation, attraction, and parasocial interaction on talk-radio listening. *Journal of Broadcasting and Electronic Media, 44,* 635–654.

Rubin, A. M., Perse, E. M., & Powell, R.A. (1985). Loneliness, parasocial interaction, and local television news viewing. *Human Communication Research, 12,* 155–180.

Schramm, H., Hartmann, T. & Klimmt, C. (2002). Desiderata und Perspektiven der Forschung über parasoziale Interaktionen und Beziehungen zu Medienfiguren [Desiderata and perspectives of research on parasocial interactions and relationships with media figures]. *Publizistik, 47,* 436–459.

Schramm, H., Hartmann, T., & Klimmt, C. (2004). *Parasoziale Interaktionen und Beziehungen mit Medienfiguren in interaktiven und konvergierenden Medienumgebungen. Empirische Befunde und theoretische Überlegungen.* In U. Hasebrink, L. Mikos, & E. Prommer (Eds.), *Mediennutzung in konvergierenden Medienumgebungen* [Media usage in convergent media environments] (pp. 299–320). München: Fischer.

Shapiro, M. A. (1994). Think-aloud and thought-list procedures in investigative mental processes. In A. Lang (Ed.), *Measuring psychological responses to media* (pp. 1–14). Hillsdale, NJ: Lawrence Erlbaum Associates.

Slater, M., Henry, K. L., Swaim, R. C. & Anderson, L. L. (2003). Violent media content and aggressiveness in adolescents: A downward spiral model. *Communication Research, 30* (6), 713–736.

Stuke, D., Hartmann, T., & Daschmann, G. (2005, May). Parasocial Relationships with Drivers Affect Suspense in Racing Sport Spectators. Paper presented at the annual conference of the International Communication Association (ICA), May 26–30 2005, New York.

Teichert, W. (1973). "Fernsehen" als soziales Handeln II [Watching television as social action]. *Rundfunk und Fernsehen, 21*, 356–382.

Thallmair, A., & Rössler, P. (2001). *Parasoziale Interaktion bei der Rezeption von Daily Talk shows. Eine Befragung von älteren Talk-Zuschauern.* In C. Schneiderbauer, (Eds.), Daily Talk shows unter der Lupe [Looking at daily talk shows] (p. 179–208). München: Fischer.

Todorov, A., & Uleman, J. S. (2002). Spontaneous trait inferences are bound to actors' faces: Evidence from a false recognition paradigm. *Journal of Personality and Social Psychology, 83*, 1051–1065.

Trepte, S., Zapfe, S., & Sudhoff, W. (2001). Orientierung und Problembewältigung durch TV-Talk shows: Empirische Ergebnisse und Erklärungsansätze [Orientation and coping through TV-talk shows: Empirical findings and explanations]. *Zeitschrift für Medienpsychologie, 13*, 73–84.

Tsao, J. (1996). Compensatory media use: An exploration of two paradigms. *Communication Studies, 47*, 89–109.

Turner, J. R. (1993). Interpersonal and psychological predictors of parasocial interaction with different television performers. *Communication Quarterly, 41*, 443–453.

Visscher, A. & Vorderer, P. (1998). *Parasoziale Beziehungen von Vielsehern.* In H. Willems & M. Jurga (Eds.), Inszenierungsgesellschaft [The staged society] (pp. 453–469). Wiesbaden : Westdeutscher Verlag.

Vorderer, P. (1992). *Fernsehen als Handlung: Fernsehfilmrezeption aus motivationspsychologischer Perspektive* [Watching TV as action: TV film consumption from the perspective of motivation]. Berlin: Edition Sigma.

Vorderer, P. (1996). *Picard, Brinkmann, Derrick & Co. als Freunde der Zuschauer.* In Vorderer, P. (Eds.), Fernsehen als "Beziehungskiste". Parasoziale Beziehungen und Interaktionen mit TV-Personen [TV as relationship machine: Parasocial relationships and interactions with TV personae] (pp. 153–171). Opladen: Westdeutscher Verlag.

Vorderer, P. (1998). *Unterhaltung durch Fernsehen: Welche Rolle spielen parasoziale Beziehungen zwischen Zuschauern und Fernsehakteuren?* In G. Roters, W. Klingler, & O. Zöllner (Eds.), Fernsehforschung in Deutschland. Themen, Akteure, Methoden [TV research in Germany. Issues, agents, methods] (pp. 689–707). Baden-Baden: Nomos.

Vorderer, P., & Knobloch, S. (1996). Parasoziale Beziehungen zu Serienfiguren: Ergänzung oder Ersatz? [Parasocial relationships to TV series characters: Completion or replacement?] *Medienpsychologie, 8*, 201–216.

Vorderer, P., & Knobloch, S. (2000). *Conflict and suspense in drama.* In D. Zillmann & P. Vorderer (Eds.), Media entertainment: The psychology of its appeal (pp. 59–72). Mahwah, NJ: Lawrence Erlbaum Associates.

Vorderer, P., & Klimmt, C. & Ritterfeld, U. (2004). Enjoyment: At the heart of media entertainment. *Communication Theory, 4*, 388–408.

Wallace, W. T., Seigerman, A., & Holbrook, M. B. (1993). The role of actors and actresses in the success of films: How much is a movie star worth? *Journal of Cultural Economics, 17*, 1–27.

Weiner, B. (1985). An attributional theory of achievement motivation and emotion. *Psychological Review, 92*, 548–573.

Wulff, H. J. (1992), Fernsehkommunikation als parasoziale Interaktion: Notizen zu einer interaktionistischen Fernsehtheorie [TV communication as parasocial interaction: Notes on an interactional television theory]. *Semiotische Berichte, 3*, 279–295.

Wulff, H. J. (1996). Parasozialität und Fernsehkommunikation [Parasociality and TV communication]. *Medienpsychologie, 8*, 163–181.

Zillmann, D. (1988). Mood management through communication choices. *American Behavioral Scientist, 31*, 327–340.

Zillmann, D. (1996). The psychology of suspense in dramatic exposition. In P. Vorderer, H. Wulff, & M. Friedrichsen (Eds.), *Suspense: Conceptualizations, theoretical analyses, and empirical explorations* (p. 199–231). Mahwah, NJ: Lawrence Erlbaum Associates.

Zillmann, D., & Bryant, J. (Eds.). (1985). *Selective exposure to communication.* Hillsdale, NJ: Lawrence Erlbaum Associates.

18

Why Horror Doesn't Die: The Enduring and Paradoxical Effects of Frightening Entertainment

Joanne Cantor
University of Wisconsin-Madison

Frightening entertainment has long remained a paradox and has presented researchers with many challenging questions. Among the most intriguing are: What types of images and events in the media provoke fear, and why do children often experience fear from watching fare that adults would not consider scary? How commonly do fright reactions linger for long periods of time after the entertainment experience has ended? Why do audiences experience fear while witnessing mediated events that they know are fictional? Why do long-term memories of traumatic media exposure often result in irrational behavior? And, finally, what methods are best suited to reducing emotional trauma caused by media exposure? This chapter pulls together a variety of theories and findings in an effort to grapple with these questions.

It's Not What You'd Think: Age Differences in What Frightens Children in the Media

In general, what viewers find frightening in the mass media is similar to what they find frightening in real-life. Three overarching categories of frightening stimuli have been identified as (1) violence or the threat of harm; (2) visual depictions of distortions of natural forms (such as monsters, mutants, and deformed creatures); and (3) the depiction of other people experiencing endangerment or fear (see Cantor, 2002, for a more extensive analysis). Beyond these major categories, what specifically frightens an individual is no doubt a function of his or her temperament and previous experiences, as well as co-occurring real-world events. In addition to these idiosyncratic factors, research has determined that the viewer's chronological age is an important determinant of the source and intensity of media-induced fears.

Most parents are surprised when their young children become frightened by programs or movies that adults would consider harmless (Cantor, 1998). This is because differences in cognitive development at different ages strongly affect the way children perceive and understand the world around them, mass media included. A growing body of research has examined the types of mass media stimuli and events that frighten children at different ages based on theories and findings in cognitive development. Laboratory experiments have tested rigorously controlled variations in program content and viewing conditions, using a combination of self-reports, physiological responses, the coding of facial expressions of emotion, and behavioral measures. For ethical reasons, only small excerpts of relatively mild television shows and movies are used in experiments. As a complement to these experiments, surveys and retrospective studies have investigated the responses of children who have been exposed to a particular mass media offering in their natural environment, without any researcher intervention. Although less tightly controlled than experiments, these post-hoc studies permit the exploration of responses to much more intensely frightening media fare and can chronicle effects of longer duration.

Research exploring developmental differences has identified three content characteristics that are important determinants of age differences: physical appearance, fantasy (vs. reality), and abstractness.

The Importance of Appearance

Research on cognitive development indicates that, in general, very young children react to stimuli predominantly in terms of their perceptible characteristics, and that with increasing maturity, they respond more and more to the conceptual aspects of stimuli (see Flavell, 1963; Melkman, Tversky, & Baratz, 1981). Research findings support the generalization that the impact of appearance in frightening media decreases as the child's age increases. In other words, preschool children (up to the age of about 5 years) are more likely to be frightened by something that looks scary but is actually harmless than by something that looks attractive but is actually harmful; for older elementary school children (approximately 9 to 11 years), appearance carries much less weight, relative to the behavior or destructive potential of a character, animal, or object.

This generalization is supported by a survey conducted in 1981 (Cantor & Sparks, 1984), in which parents were asked to name the programs and movies that had frightened their children the most. In this survey, parents of preschool children most often mentioned offerings with grotesque-looking characters, such as the television series *The Incredible Hulk* and the feature film *The Wizard of Oz*; parents of older elementary school children more often mentioned programs or movies that involved threats without a strong visual component, and that required imagination to comprehend. Sparks (1986) replicated this study, using children's self-reports rather than parents' observations, and obtained similar findings. Both surveys included controls for possible differences in exposure patterns in the different age groups.

A study supporting a similar conclusion explored children's reactions to excerpts from *The Incredible Hulk* (Sparks & Cantor, 1986). [Although this program, about a man who transforms into a grotesque superhero to perform good deeds, was not intended to be scary, Cantor and Sparks' (1984) survey reported that it was named by 40% of the parents of preschoolers as a show that had scared their child.] When children were shown a shortened episode of the *Hulk* program and were asked how they had felt during different scenes, preschool children reported the most fear after the attractive, mild-mannered hero had transformed into the monstrous-looking Hulk. Older elementary school children, in contrast, reported the least fear at this time, because they understood that the Hulk was really the benevolent hero in another physical form, and that he was using his superhuman powers to rescue a character who was in danger. Preschool children's unexpectedly intense reactions to this program seem to have been partially due to their over-response to the visual image of the Hulk character and their inability to look beyond his appearance and appreciate his benevolent behavior.

Another study (Hoffner & Cantor, 1985) tested the effect of appearance more directly, by creating a story in four versions, so that a major character was either attractive and grandmotherly looking or ugly and grotesque. The character's appearance was factorially varied with her behavior—she was depicted as behaving either kindly or cruelly. In judging how nice or mean the character was and in predicting what she would do in the subsequent scene, preschool children were more influenced than older children (6–7 and 9–10 years) by the character's looks and less influenced by her kind or cruel behavior. As the age of the child increased, the character's looks became less important and her behavior carried increasing weight. A follow-up experiment revealed that in the absence of information about the character's behavior, children in all age groups engaged in physical appearance stereotyping, that is, they thought that the ugly woman would be mean and the attractive woman would be nice.

Responses to Fantasy Content

Research shows that the ability to distinguish between reality and fantasy develops only gradually throughout childhood (see Flavell, 1963; Kelly, 1981; Morison & Gardner, 1978). Consistent with this slow development, research shows that as children mature cognitively, they become less responsive to fantastic dangers and more responsive to realistic threats depicted in the media. In Cantor and Sparks' (1984) survey of what had frightened children, the parent's tendency to name fantasy offerings, depicting events that could not possibly occur in the real world, decreased as the child's age increased, and the tendency to mention fictional offerings, depicting events that could possibly occur, increased. Sparks (1986) replicated these findings using children's self-reports. Further support comes from a study of children's fright responses to television news (Cantor & Nathanson, 1996). A survey of parents of children in kindergarten, second, fourth, and sixth grades showed that the percentage of children frightened by fantasy programs decreased as the child's grade increased, whereas the percentage frightened by news stories increased with age.

Responses to Abstract Threats

Theories and findings in cognitive development show that the ability to think abstractly emerges relatively late in cognitive development (e.g., Flavell, 1963). Consistent with this emerging ability, as children mature, they become frightened by media depictions involving increasingly abstract concepts. Data supporting this generalization come from a survey of children's responses to the television movie *The Day After* (Cantor, Wilson, & Hoffner, 1986). Although many people were concerned about young children's reactions to this movie, which depicted the devastation of a Kansas community by a nuclear attack (Schofield & Pavelchak, 1985), developmental considerations regarding the ability to think abstractly led to the prediction that the youngest children would be the least affected by it. In a random telephone survey of parents, conducted the night after the broadcast of this movie, children under 12 were reportedly much less disturbed by the film than were teenagers, and parents were the most disturbed. The very youngest children were the least frightened. The findings seem to be due to the fact that the emotional impact of the film comes from the contemplation of the potential annihilation of the earth as we know it—a concept that is beyond the grasp of a young child. The visual depictions of injury in the movie were quite mild compared to the enormity of the consequences implied by the plot.

Three major conclusions emerge from the studies of developmental differences in media-induced fears: (1) The importance of appearance in fear-evoking stimuli decreases as the age of the child increases; (2) As children mature they become less likely to be frightened fantastic, impossible events in media programs, and more likely to be frightened by realistic depictions; and (3) As children mature they are more likely to be frightened by media depictions of abstract threats.

"You Are Not Alone"—The Surprising Prevalence of Media-Induced Trauma

Although experimental studies can provide information about age differences, they do not present an indication of how common and how intense fright reactions to media are in the general public. Research interest in the prevalence of fright reactions to mass media goes back as far as Herbert Blumer (1933), who reported that 93% of the children he questioned said they had been frightened or horrified by a motion picture. Although sporadic attention was paid to the media as a source of children's fears in the succeeding several decades (see Cantor, 2002), research attention began to focus on this issue more prominently in the 1980s, after the release of several blockbuster frightening films in the 1970s. As anecdotal reports of intense emotional responses to such popular films as *Jaws* and *The Exorcist* proliferated in the press, public attention became more focused on the phenomenon. The uproar over children's reactions to especially intense scenes in the 1984 movies *Indiana Jones and the Temple of Doom* and *Gremlins* prompted the Motion Picture Association of America to add "PG-13" to its rating system, in an attempt to caution parents that a film might be inappropriate for children under the age of 13 (Zoglin, 1984). In addition, as the number of cable channels expanded, most films produced for theatrical distribution, no matter how violent or bizarre, eventually ended up on television, and thus became accessible to large numbers of children, often without their parents' knowledge.

Correlational studies show that amount of television exposure is related to both anxieties and sleep disorders. A survey of third through eighth graders revealed that as the number of hours of television viewing per day increased, so did the prevalence of symptoms of psychological trauma, such as anxiety, depression, and post-traumatic stress (Singer, Slovak, Frierson, & York, 1998). Moreover, a survey of parents of public school children in kindergarten through fourth grade revealed that the amount of children's television viewing (especially television viewing at bedtime), and having a television in their own bedroom, were significantly related to sleep disturbances (Owens, Maxim, McGuinn, Nobile, Msall, & Alario, 1999). Although these survey data cannot rule out the alternative explanation that children experiencing trauma or sleep difficulties are more likely to turn to television for distraction, they are consistent with the conclusion that exposure to frightening and disturbing images on television con-tributes to a child's level of stress and anxiety. Indeed, 9% of the parents in the study by Owens et al. (1999) reported that their child experienced TV-induced nightmares at least once a week.

An experimental study suggests that witnessing scary media presentations may also lead children to avoid engaging in activities related to the events depicted (Cantor & Omdahl, 1991). In this study, kindergarten through sixth-grade children who were exposed to drama-tized depictions of a deadly house fire from *Little House on the Prairie* showed increased self-reports of worry about similar events in their own lives. Moreover, they also expressed less interest in learning to build a fire in a fireplace than children who were not shown the episode. Similarly, children who saw a scene involving a drowning expressed more con-cerns about water accidents and were less willing to learn canoeing than children who had not watched that scene. Although the duration of these effects was not measured, the ef-fects were undoubtedly short-lived, especially because debriefings were employed and safety guidelines were taught so that no child would experience long-term distress (Cantor & Omdahl, 1999).

There is an increasing body of evidence that the fear induced by mass media exposure endures well beyond the time of viewing, with sometimes intense and debilitating effects (Cantor, 1998). In a random survey by Johnson (1980), 40% of the adults interviewed said that they had seen a motion picture that had disturbed them "a great deal." The median length of the reported disturbance was three days. Respondents also reported on the type, intensity, and duration of symptoms such as nervousness, depression, fear of specific things, and recurring thoughts and images. Based on these reports, Johnson judged that 48% of these respondents (19% of the total sample) had experienced, for at least two days, a "significant stress reaction" as the result of watching a movie.

Retrospective studies of adults' detailed memories of having been frightened by a television show or movie provide more evidence of the severity and duration of media-induced fear (Harrison & Cantor, 1999; Hoekstra, Harris, & Helmick, 1999). In these studies, involving samples of undergraduates from three universities, the presence of vivid memories of enduring media-induced fear was nearly universal. All of the participants in one study (Hoekstra et al., 1999) reported such an incident. In the other study (Harrison and Cantor, 1999), 90% of the participants reported an intense fear reaction to something in the media, in spite of the fact that the respondents could receive full extra credit for participating in the study if they simply said "no," (meaning they never had such an experience), and thereby avoid writing a paper and filling out a 3-page questionnaire.

Both studies revealed a variety of intense reactions, including generalized anxieties, specific fears, unwanted recurring thoughts, and disturbances in eating and sleeping. Moreover, Harrison and Cantor (1999) reported these fears to be long-lasting: One-third of those who reported having been frightened said that the fear effects had lasted more than a year. Indeed, more than one-fourth of the respondents said that the emotional impact of the program or movie (viewed an average of six years earlier) was still with them at the time of reporting.

Recent research has explored in more detail the nature of long-term reactions to specific movies by conducting a content analysis of papers written by more than 500 university students over a three-year period (Cantor, 2004a). Students wrote papers describing an intense experience of media-induced fright, and although they were given the option of writing about someone else, 93% wrote about something they had experienced themselves. Perhaps surprisingly, the overwhelming majority of students (91%) wrote about entertainment fare (fiction or fantasy) as opposed to reality presentations (the news or a documentary), and eight of the ten most frequently cited presentations involved the supernatural. The 91 papers written about the four presentations cited most frequently (the movies *Poltergeist, Jaws, The Blair Witch Project,* and *Scream*) were content-analyzed for indications of symptoms that spilled over into the viewer's subsequent life.

Of the students who wrote about these popular movies, 46% indicated that the movie had affected their bedtime behavior (for example, inducing sleep disturbances) and 75% reported effects on waking life (for example, causing the viewer to experience anxiety in real-life situations related to the movie). Only 12% of the writers of these papers failed to mention any effects that endured beyond the movie itself. The effects on waking life were particularly striking, including difficulty swimming after *Jaws,* uneasiness around clowns, televisions displaying "snow," and trees outside windows after *Poltergeist,* avoidance of camping and the woods following *The Blair Witch Project,* and anxiety when home alone after *Scream.* Approximately one-third of the writers reported effects that were still ongoing at the time they wrote their papers, and this was true even for the movies *Poltergeist* and *Jaws,* for which the median age at viewing was 7 years. Typical ongoing effects of *Jaws* involved anxiety when swimming in lakes and pools (where there are no sharks!) as well as in the ocean, and many viewers of *Poltergeist* reported an ongoing dread of clowns, televisions, or trees, even though they were well aware, as adults, that these things would not harm them.

Explaining the Prevalence of Lingering Effects

Why do scary movies have such powerful effects on the ongoing lives of people who see them for entertainment? Why, in fact, are they frightening at all, given that they are viewed on a screen, in a situation where the viewer is in no objective danger? A number of factors seem to contribute to viewers' fears.

Young Children's Vulnerabilities

Developmental considerations can explain some of these reactions, at least those that occur in young children. For children who are especially responsive to the physical appearance of characters, it stands to reason that the hideous visual images in a movie like *Poltergeist* would alarm them. Moreover, for those who have not yet grasped the distinction between fantasy and reality, it is no wonder that they should worry about clown dolls, trees, or demons in the television attacking them after seeing it happen to other children in that movie.

These developmental considerations explain some of the problems young children have with frightening movies. However, they do not help in understanding the reactions of children over the age of eight, nor do they have any bearing on the enduring reactions of adults.

Why Fictitious Events Are Scary

Once viewers come to know that fictional movies are scripted by a screenwriter for the purpose of entertaining us, they still become scared while watching for a variety of reasons (see Cantor, 2002, 2004a, for more extensive discussions of these arguments). Certain visual images, such as attacking animals and physical deformities, automatically arouse fear, although adults often can moderate their responses to them when they see them in movies. Humans are also naturally inclined to empathize with the emotions of protagonists, especially those that they like and admire. Therefore, if the protagonists in a movie are intensely afraid or are threatened with harm, viewers often feel fear, too. Viewers have also been said to adopt what has been called "the willing suspension of disbelief" in order to enjoy a more intense experience of a scary movie. Moreover, accomplished filmmakers include such features as suspense, surprise, and scary music, which are designed to increase viewers' fear.

These factors help explain why adults become frightened while watching fictional movies, but not necessarily why they continue to exhibit feelings of fear after the movie has ended. Even if viewers care about the killer's victims while watching *Scream*, once the movie is over, they should be reminded that it was only a movie, and they should no longer be worrying about the killer. But viewers often do continue to feel anxious after such movies, and with good reason. Even though they know that that specific killer never lived and that those murders never took place, the story vividly reminds them of real threats that (although improbable) do exist. Realistic fiction, although the product of an artist's imagination, leads people to believe that it is based on something that actually happened to someone. The details have likely been changed, but because it is plausible, realistic fiction can have profound effects on the way individuals view their own world.

However, the fact that realistic fiction is based on real happenings does not explain the paradox that the overwhelming majority of retrospective reports of intense fright to scary media are based on dramatic fiction rather than the news or documentaries. If a story's basis in reality were the determining factor, the news would hold a more prominent position in viewers' memories of scary media. Yet, whether or not something actually happened does not seem to be as important a determinant of the fear response as the emotional character of the event. Research shows that the more sensational an event, the more likely viewers are to remember it, and the greater their tendency to overestimate its probability of occurrence (see Lichtenstein, Slovic, Fischoff, Layman, & Combs, 1978; Tversky & Kahneman, 1973).

Although viewers may not be worried about the specific masked killer in the movie *Scream* (who is only an actor playing a part), those vivid, sensational depictions of stalking, terror, and gore remind them of their own vulnerability, and may have a stronger effect than the more mundane stories of murder that they hear about in the daily news. And as for the vulnerability to shark attacks, although viewers know that Jaws was killed (and that, in fact, he was only a mechanical monster), witnessing his bloody attacks on helpless victims makes them intensely and memorably aware that there is the possibility of a shark attack in the ocean. This seems to change viewers' perception of their own safety when at the beach much more profoundly than the brief, matter-of-fact reports of shark attacks that they typically hear about in the press.

The Supernatural: A Gray Area Between Fantasy and Fiction

Why are supernatural events—events that have no objective reality—such prominent sources of media-induced fear? The supernatural is a difficult category to define. Children by the age of eight seem to understand basic distinctions between fantasy and reality. They know that the Wicked Witch of the West (an evil character with a pointed black hat, who rides a broom) is just a make-believe movie character. However, young adults seem to be less certain about whether or not someone like the Blair Witch could exist. For many people, stories of witchcraft, demonic possession, and alien intruders are not easy to dismiss as impossible.

Our society and culture seem to reinforce the ambiguity about supernatural events. People who call themselves witches exist in today's society (see, for example, Walker & Jung, 2003). Stories of demonic possession are reported in the press, and religious exorcisms are still being performed (Cuneo, 2001). Reports of unidentified flying objects are also common, and scientists hold out the possibility that we will someday discover intelligent life on other planets. Many of these events are reported not only in fiction, but in so-called reality programs like *Unsolved Mysteries*, and on talk shows in which self-proclaimed victims of supernatural intervention tell their stories.

The Enduring Influence of Emotional Memory

The factors discussed above—young children's cognitive immaturity, the ability of fiction to sensitize viewers to real dangers, and the ambiguity about threats from supernatural forces—explain some of the long-lasting effects of scary movies. But they do not account for a good portion of the lingering effects. Why is it that almost all of the respondents in the study by Cantor (2004a) who had difficulty swimming in the ocean, also report anxiety in lakes or pools, where they know sharks cannot be found? Why do many adults who saw *Poltergeist* as a child continue to feel dread when exposed to clown dolls that they know cannot harm them? Why are these people's bodily reactions behaving in a way that seems inconsistent with their conscious thoughts of being objectively safe? Recent research in neuroscience on the neurophysiology of fear, seems relevant here.

In *The Emotional Brain*, Joseph LeDoux (1996), brings together current knowledge of the brain mechanisms involved in emotion. To simplify his analysis greatly, there are two brain memory systems that work in parallel in the fear response. Explicit, conscious memories of a fear-inducing event are mediated by a system involving the *hippocampus*, whereas implicit, not necessarily conscious emotional memories are mediated by an area involving the *amygdala* (LeDoux 1996). In a frightening situation, the amygdala responds more quickly, even before the cause of alarm has reached the individual's state of awareness, and orchestrates more automatic responses, such as tensed muscles, blood pressure and heart rate changes, and the release of adrenaline into the bloodstream. These reactions, which contribute to the way our bodies feel when we are afraid, are part of the "fight-or-flight" response, that prepares us to defend ourselves from harm.

Citing studies involving species ranging from laboratory animals to humans, LeDoux explains the process of fear conditioning: A laboratory rat exhibits a fearful reaction upon receiving an electric shock. If the shock is paired with the sound of a tone, that tone comes to elicit the fear reaction, even when the shock does not accompany it. Similarly, if a man has a serious, traumatic automobile accident during which the horn of his car gets stuck, he is likely to experience bodily reactions associated with fear in future situations when hearing the sound of a horn. The horn may, in fact, remind him of the accident, and he may consciously associate his feelings with that event. However, over time, he may forget about the association of the horn with the accident, but still have physiological responses associated with fear whenever he hears a horn sound. In these cases, the implicit (nonconscious) emotional memory system has been activated to create the bodily experience of emotion, even after the conscious memories have faded. Other contextual features of the accident, which may never have been consciously associated with it, may also trigger the implicit emotional memory—a particular make of car, a certain type of intersection, or any other detail that was prominent at the time of the accident.

According to LeDoux, evolution favors the survival of animals (including humans) that can quickly identify stimuli that are life-threatening and that immediately take defensive action. In addition, the emotional memory system makes sure that memories of things that have endangered us in the past are extremely accurate, so that whenever we encounter similar things even years later, we will be prepared to act quickly again. Because of this, implicit fear memories are especially enduring. LeDoux's research shows that although conscious memories of fearful situations are not always correct and are quite malleable over time, implicit fear memories are highly resistant to change. In fact, LeDoux calls them "indelible":

> Unconscious fear memories established through the amygdala appear to be indelibly burned into the brain. They are probably with us for life. This is often very useful, especially in a stable, unchanging world, since we don't want to have to learn about the same kinds of dangers over and over again. But the downside is that sometimes the things that are imprinted in the amygdala's circuits are maladaptive. In these instances, we pay dearly for the incredible efficiencies of the fear system (LeDoux, 1996, p. 252).

What may be happening with these lingering effects of movies is similar to LeDoux's descriptions of fear conditioning. If a person experienced intense fear while watching *Jaws*, the implicit fear reactions, for example, the heart-rate increases, blood pressure changes, and muscle tension, became conditioned to the image of the shark, to the notion of swimming, to the musical score—most likely to a combination of the stimuli in the movie. Later, any of these stimuli—or even thoughts of these stimuli—trigger these unconscious reactions, even after the conscious mind has gotten past the problem. Similarly, people who were traumatized while watching *Poltergeist* experienced fear conditioning to images of clowns and other situations from the film. In spite of now knowing that harm from these agents is impossible, these people experience bodily reactions and anxious feelings in response to related images.

It should be noted that LeDoux was not referring to viewers of frightening movies in his analysis of the enduring effects of fear, but rather to people who suffer from phobias, panic attacks, and post-traumatic stress as the result of events they have experienced in their own lives. However, the similarity between the experience of these frightened movie viewers and people with phobic reactions is quite remarkable. Although there are a few published cases in the psychiatric literature in which people were hospitalized as the result of movie-induced trauma (Buzzuto, 1975; Mathai, 1983; Simons & Silveira, 1994), the examples reported here would not likely rise to the level of a clinical diagnosis. It is clear, however, that fright reactions to movies have influenced the lives of many people in strong and enduring ways, often with disruptive effects on some aspects of their lives.

Getting Over It: Strategies for Alleviating Media-Induced Fear

The choice of an effective coping strategy is not always obvious. Research shows that there are consistent age differences in the strategies that typically work in alleviating media-induced fear, based on differing levels of cognitive development.

Developmental differences in children's information-processing abilities yield differences in the effectiveness of strategies to prevent or reduce their media-induced fears (Cantor & Wilson, 1988). The findings of research on coping strategies can be summarized as follows: In general, preschool children benefit more from "noncognitive" than from "cognitive" strategies; both cognitive and noncognitive strategies may be effective for older elementary school children, although this age group tends to prefer cognitive strategies.

Noncognitive Strategies

Noncognitive (or "nonverbal") strategies are those that do not involve the processing of verbal information and that appear to be relatively automatic. The most heavily tested noncognitive strategy is desensitization, or gradual exposure to threatening images in a nonthreatening context. This strategy has been shown to be effective for both preschool and older elementary school children. Studies have used prior exposure to filmed snakes, live lizards, still photographs of worms, and rubber replicas of spiders to reduce the emotional impact of frightening movie scenes involving similar creatures (Weiss, Imrich, & Wilson, 1993; Wilson, 1987, 1989a; Wilson & Cantor, 1987). In addition, fear reactions to the Hulk character in *The Incredible Hulk* were shown to be reduced by exposure to footage of Lou Ferrigno, the actor who plays the character, having his make-up applied so that he gradually took on the menacing appearance of the character (Cantor, Sparks, & Hoffner, 1988). None of these experiments revealed developmental differences in the effectiveness of desensitization.

Other noncognitive strategies involve physical activities, such as clinging to a loved one or an attachment object, having something to eat or drink, or leaving the situation and becoming involved in another activity. Although these techniques can be used by viewers of all ages, there is reason to believe they are more effective for younger than for older children. First, it has been argued that the effectiveness of such techniques is likely to diminish as the infant's tendency to grasp and suck objects for comfort decreases (Bowlby, 1973). Second, it seems likely that the effectiveness of such techniques is partially attributable to distraction, and distraction techniques should be more effective in younger children, who have greater difficulty allocating cognitive processing to two simultaneous activities (e.g., Manis, Keating, & Morison, 1980).

Children seem to be intuitively aware that physical techniques work better for younger than for older children. In a study asking children to evaluate the effectiveness of various strategies for coping with media-induced fright, preschool children's ratings of "holding onto a blanket or a toy" and "getting something to eat or drink" were significantly higher than those of older elementary school children (Wilson, Hoffner, & Cantor, 1987). Similarly, Harrison and Cantor's (1999) retrospective study showed that the percentage of respondents who reported having used a "behavioral" (noncognitive) coping strategy to deal with their media-induced fear declined as the respondent's age at exposure to the frightening fare increased.

Another noncognitive strategy that has been shown to have more appeal and more effectiveness for younger than for older children is covering one's eyes during frightening portions of a presentation. In an experiment by Wilson (1989b), when covering their eyes was suggested as an option, younger children used this strategy more often than older children. Moreover, the suggestion of this option reduced the fear of younger children, but actually increased the fear of older children. Wilson noted that the older children recognized the limited effectiveness of covering their eyes (while still being exposed to the audio features of the program) and that they may have reacted by feeling *less* in control, and therefore more vulnerable, when this strategy was offered to them.

Cognitive Strategies

In contrast to noncognitive strategies, cognitive (or "verbal") strategies involve verbal information that is used to cast the threat in a different light. These strategies involve relatively complex cognitive operations, and research consistently finds such strategies to be more effective for older than for younger children.

When dealing with fantasy depictions, the most typical cognitive strategy seems to be to provide an explanation focusing on the unreality of the situation. This strategy should be especially difficult for preschool children, who do not have a full grasp of the implications of the fantasy-reality distinction. In an experiment by Cantor and Wilson (1984), older elementary school children who were told to remember that what they were seeing in *The Wizard of Oz* was not real showed less fear than their classmates who received no instructions. The same instructions did not help preschoolers, however. A study by Wilson and Weiss (1991) showed similar developmental differences in the effectiveness of reality-related strategies.

Children's beliefs about the effectiveness of focusing on the unreality of a media offering have been shown to be consistent with these experimental findings. In the study of perceptions of fear-reducing techniques (Wilson et al., 1987), preschool children's ranking of the effectiveness of "tell yourself it's not real" was significantly lower than that of older elementary school children. In contrast to both preschool and elementary school children, who apparently view this strategy accurately, parents do not seem to appreciate the inadequacy of this technique for young children. Eighty percent of the parents of both the preschool and elementary school children who participated in another study (Wilson & Cantor, 1987), reported that they employed a "tell them it's not real" coping strategy to reduce their child's media-induced fear.

To reduce fright reactions to media depictions involving realistic threats, the most prevalent cognitive strategy seems to be to provide an explanation that minimizes the perceived severity of the depicted danger. This type of strategy is not only more effective with older children than with younger ones, in certain situations it has been shown to have a fear-enhancing rather than anxiety-reducing effect with younger children. In an experiment involving the frightening snake-pit scene from *Raiders of the Lost Ark* (Wilson & Cantor, 1987), children were first exposed to an educational film involving the presence or absence of reassuring information about snakes (including, for example, the statement that most snakes are not poisonous). Although this information tended to reduce the fear of older elementary school children while watching the snake-pit scene, kindergarten and first-grade children seem to have only partially understood the information, responding to the word "poisonous" more intensely than to the word "not." For them, negative emotional reactions were more prevalent if they had heard the supposedly reassuring information than if they had not heard it.

Data also indicate that older children use cognitive coping strategies more frequently than preschool children do. In the survey of reactions to *The Day After* (Cantor et al., 1986), parents' reports that their child had discussed the movie with them after viewing it increased with the age of the child. In a laboratory experiment involving exposure to a scary scene (Hoffner & Cantor, 1990), significantly more 9- to 11-year-olds than 5- to 7-year-olds reported that they had spontaneously employed cognitive coping strategies (thinking about the expected happy outcome or thinking about the fact that what was happening was not real). Similarly, Harrison and Cantor's (1999) retrospective study showed that the tendency to employ a cognitive strategy to cope with media-induced fear increased with the respondent's age at the time of the incident.

Studies have also shown that the effectiveness of cognitive strategies for young children can be improved by providing visual demonstrations of verbal explanations (Cantor, Sparks, & Hoffner, 1988), and by encouraging repeated rehearsal of simplified, reassuring information (Wilson, 1987). In addition, research has explored some of the specific reasons for the inability of young children to profit from verbal explanations, such as those involving relative quantifiers (e.g., "some are dangerous, but most are not," Badzinski, Cantor, & Hoffner, 1989) and probabilistic terms (e.g., "this probably will not happen to you," Hoffner, Cantor, & Badzinski, 1990). It is clear from these studies that it is a challenging task to explain away threats that have induced fear in a child, particularly when there is a strong perceptual component to the threatening stimulus, and when the reassurance can only be partial or probabilistic, rather than absolute (see Cantor & Hoffner, 1990).

Expressive communication as a coping strategy. Research on coping with emotional distress produced by situations other than media exposure suggests that certain ways of communicating about frightening media may also be helpful. Although mulling over thoughts of a distressing experience may not reduce distress and can exacerbate negative feelings (see Sapolsky, Stocking, & Zillmann, 1977), cognitive therapy, one of most widely studied psychological interventions for anxiety disorders (Deacon & Abramowitz, 2004) is based on the notion that individuals may gain control over their emotions through talking disturbing situations over with a caring listener. Indeed, a study by Cloven and Roloff (1993) suggests that just *thinking about* talking a problem over with someone makes thoughts become better suited to solving the problem, whether or not the conversation ever ends up taking place. Perhaps this is because talking with another person involves an attempt to see a problem from their perspective, and allows the individual to perceive the situation more clearly.

More directly to the point here, a good deal of research has been conducted on the therapeutic value of *writing about* past frightening experiences. In his book *Opening Up: The Healing Power of Expressing Emotions,* John W. Pennebaker (1997) provides evidence of the physical as well as the psychological benefits of writing about traumatic events that have happened to the individual in the past. These benefits have been observed in measures of medical visits and immune function as well as reports of psychological well-being (Lepore & Smyth, 2002). Although young children are often unable to talk about their feelings and certainly are not equipped to write about them, many art therapists have reported that children can reduce their anxieties by drawing pictures of what frightens them in conjunction with interaction with a therapist or other care-giver (Horovitz, 1983; Roje, 1995).

In an effort to help young children and their parents cope with frightening images on television and in the movies, Cantor (2004b) wrote a children's story book. *Teddy's TV Troubles* is an illustrated picture book about a little bear who has been frightened by something on TV. After recognizing that words don't work to calm his fears, he and his mother go through a series of calming activities —such as drawing a picture of what scared him and making it look less scary—that help him cope with his feelings. The book is intended to promote the type of parent-child interaction that helps young children cope with their fearful feelings.

Summary and Conclusions

Psychological research on the processes involved in the experience of frightening media has begun to illuminate many issues involved in this paradoxical form of entertainment. Research in cognitive development helps explain why children's reactions vary markedly at different ages and why effective coping strategies differ as well. Surveys and retrospective studies demonstrate the near universality of the experience of intense fright from entertainment. And findings in fields ranging from cognitive psychology to neurophysiology help explain why many frightened movie viewers behave in irrational ways for long periods of time after viewing something they originally chose as an entertaining diversion.

REFERENCES

Badzinski, D., Cantor, J., & Hoffner, C. (1989). Children's understanding of quantifiers. *Child Study Journal, 19*, 241–258.

Blumer, H. (1933). *Movies and conduct*. New York: Macmillan.

Bowlby, J. (1973). *Separation: Anxiety and anger*. New York: Basic Books.

Buzzuto, J. C. (1975). Cinematic neurosis following *The Exorcist*. *Journal of Nervous and Mental Disease, 161*, 43–48.

Cantor, J. (1998). *"Mommy, I'm scared": How TV and movies frighten children and what we can do to protect them*. San Diego, CA: Harcourt.

Cantor, J. (2002). Fright reactions to mass media. In J. Bryant & D. Zillmann (Eds.), *Media effects: Advances in theory and research* (2d ed., pp. 287–306). Mahwah, NJ: Lawrence Erlbaum Associates.

Cantor, J. (2004a). "I'll never have a clown in my house": Why movie horror lives on. *Poetics Today, 25*, 283–304.

Cantor, J. (2004b). *Teddy's TV troubles*. Madison, WI: Goblin Fern Press.

Cantor, J., & Hoffner, C. (1990). Children's fear reactions to a televised film as a function of perceived immediacy of depicted threat. *Journal of Broadcasting & Electronic Media, 34*, 421–442.

Cantor, J., & Nathanson, A. (1996). Children's fright reactions to television news. *Journal of Communication, 46*(4), 139–152.

Cantor, J., & Omdahl, B. (1991). Effects of fictional media depictions of realistic threats on children's emotional responses, expectations, worries, and liking for related activities. *Communication Monographs, 58*, 384–401.

Cantor, J., & Omdahl, B. (1999). Children's acceptance of safety guidelines after exposure to televised dramas depicting accidents. *Western Journal of Communication, 63*, 1–15.

Cantor, J., & Sparks, G. G. (1984). Children's fear responses to mass media: Testing some Piagetian predictions. *Journal of Communication, 34*(2), 90–103.

Cantor, J., Sparks, G. G., & Hoffner, C. (1988). Calming children's television fears: Mr. Rogers vs. the Incredible Hulk. *Journal of Broadcasting & Electronic Media, 32*, 271–188.

Cantor, J., & Wilson, B. J. (1984). Modifying fear responses to mass media in preschool and elementary school children. *Journal of Broadcasting, 28*, 431–443.

Cantor, J., & Wilson, B. J. (1988). Helping children cope with frightening media presentations. *Current Psychology: Research & Reviews, 7*, 58–75.

Cantor, J., Wilson, B. J., & Hoffner, C. (1986). Emotional responses to a televised nuclear holocaust film. *Communication Research, 13*, 257–277.

Cloven, D. H., & Roloff, M. E. (1993). Sense-making activities and interpersonal conflict II: The effects of communicative intentions on internal dialogue. *Western Journal of Communication, 57*, 309–329.

Cuneo, M. W. (2001). *American exorcism: Expelling demons in the land of plenty*. New York: Doubleday.

Deacon, B. J., & Abramowitz, J. S. (2004). Cognitive and behavioral treatments for anxiety disorders. A review of meta-analytic findings. *Journal of Clinical Psychology, 60*, 429–441.

Flavell, J. (1963). *The developmental psychology of Jean Piaget*. New York: Van Nostrand.

Harrison, K., & Cantor, J. (1999). Tales from the screen: Enduring fright reactions to scary media. *Media Psychology, 1*(2), 97–116.

Hoekstra, S. J., Harris, R. J., & Helmick, A. L. (1999). Autobiographical memories about the experience of seeing frightening movies in childhood. *Media Psychology, 1*(2), 117–140.

Hoffner, C., & Cantor, J. (1985). Developmental differences in responses to a television character's appearance and behavior. *Developmental Psychology, 21*, 1065–1074.

Hoffner, C., & Cantor, J. (1990). Forewarning of a threat and prior knowledge of outcome: Effects on children's emotional responses to a film sequence. *Human Communication Research, 16*, 323–354.

Hoffner, C., Cantor, J., & Badzinski, D. M. (1990). Children's understanding of adverbs denoting degree of likelihood. *Journal of Child Language, 17*, 217–231.

Horovitz, E. G. (1983). Preschool aged children: When art therapy becomes the modality of choice. *Arts in Psychotherapy, 10*, 23–32.

Johnson, B. R. (1980). General occurrence of stressful reactions to commercial motion pictures and elements in films subjectively identified as stressors. *Psychological Reports, 47*, 775–786.

Kelly, H. (1981). Reasoning about realities: Children's evaluations of television and books. In H. Kelly & H. Gardner (Eds.), *Viewing children through television* (pp. 59–71). San Francisco: Jossey-Bass.

LeDoux, J. (1996). *The emotional brain: The mysterious underpinnings of emotional life.* New York: Simon & Schuster.

Lichtenstein, S., Slovic, P., Fischoff, B., Layman, M., & Combs, B. (1978). Judged frequency of lethal events. *Journal of Experimental Psychology: Learning and Memory, 4*, 551–578.

Lepore, S. J., & Smyth, J. M. (2002). *The writing cure: How expressive writing promotes health and emotional well-being.* Washington, DC: American Psychological Association Books.

Manis, F. R., Keating, D. P., & Morison, F. J. (1980). Developmental differences in the allocation of processing capacity. *Journal of Experimental Child Psychology, 29*, 156–169.

Mathai, J. (1983). An acute anxiety state in an adolescent precipitated by viewing a horror movie. *Journal of Adolescence, 6*, 197–200.

Melkman, R., Tversky, B., & Baratz, D. (1981). Developmental trends in the use of perceptual and conceptual attributes in grouping, clustering and retrieval. *Journal of Experimental Child Psychology, 31*, 470–486.

Morison, P., & Gardner, H. (1978). Dragons and dinosaurs: The child's capacity to differentiate fantasy from reality. *Child Development, 49*, 642–648.

Owens, J., Maxim, R., McGuinn, M., Nobile, C., Msall, M., & Alario, A. (1999). Television-viewing habits and sleep disturbance in school children. *Pediatrics, 104*(3), 552, e 27. http://pediatrics.aappublications.org/cgi/content/full/104/3/e27

Pennebaker, J. W. (1997). *Opening up: The healing power of expressing emotions.* New York: Guilford Press.

Roje, J. (1995). LA '94 earthquake in the eyes of children: Art therapy with elementary school children who were victims of disaster. *Art Therapy, 12*, 237–243.

Sapolsky, B. S., Stocking, S. H., & Zillmann, D. (1977). Immediate vs. delayed retaliation in male and female adults. *Psychological Reports, 41*, 197–198.

Schofield, J., & Pavelchak, M. (1985). "The Day After": The impact of a media event. *American Psychologist, 40*, 542–548.

Simons, D., & Silveira, W. R. (1994). Post-traumatic stress disorder in children after television programmes. *British Medical Journal, 308*, 389–390.

Singer, M. I., Slovak, K., Frierson, T., & York, P. (1998). Viewing preferences, symptoms of psychological trauma, and violent behaviors among children who watch television. *Journal of the American Academy of Child and Adolescent Psychiatry, 37*(10), 1041–1048.

Sparks, G. G. (1986). Developmental differences in children's reports of fear induced by the mass media. *Child Study Journal, 16*, 55–66.

Sparks, G. G., & Cantor, J. (1986). Developmental differences in fright responses to a television program depicting a character transformation. *Journal of Broadcasting and Electronic Media, 30*, 309–323.

Tversky, A., & Kahneman, D. (1973). Availability: A heuristic for judging frequency and probability. *Cognitive Psychology, 5*, 207–232.

Walker, W., & Jung, F. (Eds., 2003) *Witches' Voice* (web site). http://www.witchvox.com/twv/meet.html

Weiss, A. J., Imrich, D. J., & Wilson, B. J. (1993). Prior exposure to creatures from a horror film: Live versus photographic representations. *Human Communication Research, 20*, 41–66.

Wilson, B. J. (1987). Reducing children's emotional reactions to mass media through rehearsed explanation and exposure to a replica of a fear object. *Human Communication Research, 14*, 3–26.

Wilson, B. J. (1989a). Desensitizing children's emotional reactions to the mass media. *Communication Research, 16*, 723–745.

Wilson, B. J. (1989b). The effects of two control strategies on children's emotional reactions to a frightening movie scene. *Journal of Broadcasting & Electronic Media, 33*, 397–418.

Wilson, B. J., & Cantor, J. (1987). Reducing children's fear reactions to mass media: Effects of visual exposure and verbal explanation. In M. McLaughlin, (ed.), *Communication Yearbook 10* (pp. 553–573). Beverly Hills, CA: Sage.

Wilson, B. J., Hoffner, C., & Cantor, J. (1987). Children's perceptions of the effectiveness of techniques to reduce fear from mass media. *Journal of Applied Developmental Psychology, 8*, 39–52.

Wilson, B. J., & Weiss, A. J. (1991). The effects of two reality explanations on children's reactions to a frightening movie scene. *Communication Monographs, 58*, 307–326.

Zoglin, R. (1984, June 25). Gremlins in the rating system. *Time*, p. 78.

19

Personality

Mary Beth Oliver, Jinhee Kim, and Meghan S. Sanders
Penn State University

The variety and volume of entertainment offerings is arguably greater now than at any other time in history—multiplex cinemas present numerous films simultaneously, cable television includes hundreds of channels, and video and DVD rentals and purchases make entertainment choices virtually limitless. Although some critics would characterize this variety of entertainment as a "glut," we believe that this diversity of entertaining diversions also reflects the diversity of the audience members and their variety of interests and tastes. For instance, whereas some people are diehard science fiction fans, other people take great pleasure in sentimental, "weepy," romance films.

Individuals' interests in different genres or types of entertainment obviously reflect a variety of variables. Some variables, such as mood or viewing situations, imply interests that vary within individuals at different points in time. Other variables, though, are more enduring, including demographic variables such as cohort, ethnic identity, or education, that predict entertainment preferences. Among these variables that are more long-lived, or stable, the personality characteristics, traits, or enduring dispositions of the viewer are perhaps among the most crucial determinants, as these variables arguably transcend demographic boundaries and may further play important roles in affecting transitory states such as emotion or mood.

Definitions of *personality* are diverse, with different approaches stressing a variety of explanations for the elements that best describe the nature of and explanations for enduring individual differences (see Funder, 2001, for an overview of differing perspectives). For the purposes of this chapter, we employ the term "personality" broadly, using it to refer to "individuals' characteristic patterns of thought, emotion, and behavior" (Funder, 2001, p. 198) that demonstrate *relative* stability and consistency within individuals over time and across situations (see Oliver, 2002). Included in our definition of personality are such constructs as traits, temperaments, and dispositions. With this definition in mind, this chapter examines the role of personality in the entertainment experience, considering not only how personality predicts entertainment preferences, but also how it functions as an important moderator in predicting effects. Finally, we conclude this chapter by considering the utility of exploring the role of media in affecting or shaping personality.

SELECTION AND ENJOYMENT

Given that most conceptualizations of personality understand it to be an enduring characteristic, it follows that the preponderance of media research that has employed personality has utilized it as an independent variable in predicting selection of media content and preference for (or enjoyment of) media offerings. Although selection and enjoyment are likely strongly correlated, these two variables are conceptually distinct. That is, it is possible for a person to select content without necessarily enjoying it, or to enjoy content that was not deliberately selected. Consequently, some researchers have explored entertainment selection and the reasons for such selection *per se*, apart from viewers' reactions to or enjoyment of such content.

Use and Selection

Uses and Gratifications. Individuals' uses of media content and their motivations for media selection have been explored extensively from uses and gratifications approaches, examining users' expected utility of media in fulfilling various needs (Palmgreen, Wenner, & Rosengren, 1985; Rubin, 2002). Research from this perspective has identified a number of common motivations for media use, including passing the time, habit, companionship, relaxation, information, and escape, among others (see, for example, Rubin, 1983). Although it is likely that viewers' motivations for viewing entertainment are more ritualistic than instrumental (Rubin, 2002), the specific types of "ritualistic" motivations are undoubtedly varied, and the motivations for viewing different genres (e.g., game shows vs. soap operas) are likely to be diverse as well. Importantly, too, even when viewers may choose to view identical entertainment offerings, their reasons for doing so may differ considerably. As a result of this variation, many scholars have considered the importance of personality in predicting individuals' uses and gratifications for media use *per se*, suggesting that these motivations may ultimately play important roles in moderating any potential effects that result from consumption.

For example, Weaver (2003) examined five different television use motives among participants who had been categorized according to three personality dimensions: psychoticism, neuroticism, and extroversion (Eysenck, Eysenck, & Barrett, 1985). Neurotic personality types reported significantly higher scores on motivations related to passing the time, companionship, relaxation, and stimulation, whereas extroverted personality types reported significantly lower scores on companionship motivations. Weaver interpreted these results as consistent with the idea that individuals who are shy, isolated, and emotional (neurotic personalities) display high levels of affiliation for viewing television, whereas individuals who are outgoing and sociable (extroverted personalities) appear to perceive that television is a poor substitute for actual, interpersonal interaction.

Sherry (2001) also examined individuals' motivations for media use, though he used measures of "temperament" as predictors, suggesting that media-use motivations may reflect neurophysiological differences. His results revealed that higher media-use motivations overall were associated with higher levels of rigidity, with low task orientation, and with a more negative general mood. Sherry interpreted these findings as consistent with the notion that biologically based individual differences may play a powerful role in viewers' media behaviors (see also Sherry, 2004).

Mood Management. Similar to uses and gratifications, research from a mood-management perspective also explores the processes that influence individuals' media selections (Zillmann, 1985, 1988a, 1988b, 2000; Zillmann & Bryant, 1985). However, mood management differs in several important respects from uses and gratifications. First, whereas uses and gratifications research examines a variety of media-use motivations, mood management focuses specifically on the ways that individuals use the media as a means of regulating their moods. In brief, mood-management perspectives generally suggest that people use media offerings to intensify or prolong positive states, and/or to diminish or terminate negative states. Given that entertainment is often designed to elicit a variety of emotional responses among viewers (e.g., humor, thrills, tranquility), mood management is particularly applicable to entertainment specifically, though the theory can be applied to instructional or news-related content as well. A second difference between uses and gratifications and mood management is that the latter does not assume that individuals are necessarily cognizant of their motivations for media use, nor that they are always willing to divulge their motivations when they do happen to be known (Zillmann, 1985; Zillmann & Bryant, 1985). As a result, studies from a mood-management perspective typically employ experimental designs where mood or affect is manipulated, and assessment of media selection is made via behavioral indicators.

The importance of mood regulation in viewers' media selections has been demonstrated in a wide variety of settings and for a host of different types of content, including comedy (Zillmann, Hezel, & Medoff, 1980), music (Knobloch & Zillmann, 2002), news (Biswas, Riffe, & Zillmann, 1994), and game shows (Bryant & Zillmann, 1984), among others. Interestingly, however, the importance of personality or other individual differences in moderating mood-management behaviors has not been frequently examined. This lack of attention may reflect the idea that mood management may be largely non-specific to any particular type of individual—that is, the maintenance of positive moods may be a universal motivation. Alternatively, mood management studies may infrequently explore personality as a result of the experimental methodologies that are typically employed. Namely, random assignment to experimental conditions may simply allow for individual differences to "come out in the wash," with variations due to personality or other traits treated as error variance or "noise."

Although personality has yet to play an important role in the majority of existing mood-management studies, some findings suggest that future research may benefit from a greater exploration of the role of personality and trait measures in predicting not only the initial mood states that are subsequently regulated by media exposure, but also the types of content that different individuals may find useful in achieving these ends (Zillmann, 2000). For example, Mares and Cantor (1992) explored media preferences among a sample of elderly participants who had been categorized as either "lonely" or "non-lonely" on the basis of a pretest. Although the non-lonely participants showed a preference for media entertainment featuring uplifting or positive portrayals of older people, lonely participants showed a preference for more negative portrayals, evidencing increases in positive affect after viewing a brief film featuring an elderly character depicted as forlorn and neglected. Similar results have been obtained in terms of musical choices. For example, Knobloch and Zillmann (2003) found that participants who reported satisfaction with their romantic-relationship status opted to listen to music classified as love-celebrating rather than love-lamenting, whereas the opposite was true for participants who reported dissatisfaction with their romantic-relationship status (see also Gibson, Aust, & Zillmann, 2000).

Although both of these examples focus on individual differences that are arguably more akin to states rather than traits, the importance of these differences in moderating media exposure suggests considerable variation in response to media content that may best serve mood-management ends for different people. That is, people who are particularly empathic, hostile, extroverted, or optimistic may respond very differently to the same media depiction. Whereas an empathic person may find joy and hopefulness from viewing a sad film, a more hard-hearted person may react with boredom or disdain. Similarly, whereas a more rebellious viewer may respond with excitement and enthusiasm to a violent horror film, a more emotionally sensitive viewer may react with disgust and horror. Consequently, understanding how individuals use media to regulate their moods would undoubtedly benefit from a clearer understanding of the ways that viewers respond to different media offerings.

Enjoyment

As implied by our discussion of media selection, individuals' enjoyment of media content is undoubtedly influenced by the personality characteristics of the viewer. Although researchers have explored the role of personality in predicting a variety of genres—including reality-based programming (Nabi, Biely, Morgan, & Stitt, 2003), sporting events (McDaniel, 2003), and talk shows (Rubin, Haridakis, & Eyal, 2003), among others—curiously, the preponderance of research on personality and enjoyment has tended to explore entertainment that is thought to have harmful effects on viewers (e.g., media violence) or that generally evokes emotional reactions that most people would consider decidedly unpleasant (e.g., horror films, sad films). Perhaps this focus is a reflection of the idea that some genres that are associated explicitly with positive affect (e.g., comedy) are so universally enjoyed that there is little variation to be explained. Conversely, the focus on morbid or potentially harmful media may reflect a tendency to try to find explanatory mechanisms for media preferences that, on the face of it, appear odd, curious, or paradoxical. Regardless of the explanation for this research focus, studies on violent entertainment, pornography (or erotica), and sadness-evoking genres underline the importance of personality in predicting individuals' media preferences.

Media Violence.   In terms of violent entertainment, research generally supports the notion that enjoyment is highest among viewers who possess characteristics associated with aggressive, hostile, or calloused dispositions. For example, traits such as masculinity, aggressiveness, psychoticism, and machiavellianism have been shown to be positively correlated with greater enjoyment of violent entertainment and with greater interest in viewing aggression and harm to others (Aluja-Fabregat, 2000; Aluja-Fabregat & Torrubia-Beltri, 1998; Bushman, 1995; Tamborini, Stiff, & Zillmann, 1987; Weaver, 2000). Similar results have been obtained in terms of preferences for music that some people consider "violent" or "aggressive." For example, Schwartz and Fouts (2003) reported that among adolescents in their sample, greater liking of "hard" music (e.g., tough, loud, wild, rock) than "light" music was associated with higher scores on measures of tough-mindedness, forcefulness, pessimism, and discontentedness, and lower scores on measures of impulse control and concern for others. Hansen and Hansen (2000) reported comparable trends in their review of research on personality and musical tastes, noting that fans of hard rock and heavy metal tend to score higher on machiavellianism, tough-mindedness, and machismo, and that fans of punk rock tend to score lower on acceptance of authority (Bleich, Zillmann, & Weaver, 1991; Hansen & Hansen, 1991; Knobloch & Mundorf, 2003; Robinson, Weaver, & Zillmann, 1996). In contrast, Carpentier, Knobloch, and Zillmann (2003) found that selective exposure to "defiant" rock and rap music was associated with higher levels of disinhibition and proactive rebelliousness (e.g., excitement-seeking), but was unrelated to hostility and reactive rebelliousness (e.g., retaliatory behavior). As a consequence, these authors suggested that consumption of some types of "hard" music may be better described as "fun-seeking than hostility-inspired" (p. 1653), though they did entertain the possibility that more extreme forms of defiant music may hold appeal for angry or hostile listeners.

Although the idea that "violent" personality types prefer "violent" entertainment is intuitively appealing, the observation of this relationship does not explain *why* this relationship exists. To date, a variety of explanations have been considered. For example, to the extent that one's media preferences are an extension of one's identity, conspicuous consumption of identifiable genres (e.g., horror, death metal music, etc.) may serve to "advertise" to others one's personality or character (Carpentier et al., 2003; Zillmann & Bhatia, 1989). Alternatively, violent entertainment may be more meaningful to individuals with more aggressive or hostile traits, as such media offerings likely present scenarios, characters, or situations that are familiar and therefore more relevant (Hoffner & Cantor, 1991). In contrast, individuals who harbor antagonistic dispositions may actually be less likely than more sensitive individuals to perceive violent entertainment as gruesome or disturbing, thereby allowing them to derive enjoyment from some forms of entertainment (e.g., slasher films, thrillers) that other viewers would find too unsettling. Finally, research on disposition theory and enjoyment of media entertainment might suggest that people with more hostile dispositions may be more likely to see the victims in violent entertainment as more "deserving" of their victimization, thereby increasing the experience of gratification that accrues when "justice is restored," however that justice may be defined by the individual viewer (Oliver, 1993a; Oliver & Armstrong, 1995; Raney, 2003; Raney & Bryant, 2002; Zillmann, 1991; Zillmann & Cantor, 1977). Of course, these various explanations are speculative at this point, suggesting that further research is needed to more fully understand the role of personality in predicting preferences for media violence.

Pornography. As with enjoyment of violent entertainment, research on sexually explicit materials generally suggests that individuals who are more permissive about their own sexuality report greater interest in and/or enjoyment of this type of media fare than do less permissive individuals. For example, Zuckerman and Litle (1986) reported that sensation seeking (and particularly disinhibition) was positively associated with more frequent viewing of X-rated films among both male and female participants. Similarly, Lopez and George (1995, Study 1) reported that erotophilia was positively associated with liking of erotic photographs, and that among male participants, was also associated with longer viewing times (see also Becker & Byrne, 1985). Finally, Bogaert (1996) highlighted the important methodological implications of individual differences in his study concerning volunteer bias. Namely, among the all-male sample of participants in his study, individuals who volunteered to participate in a study involving the viewing of a sexually-explicit film scored higher than did non-volunteers on measures of sensation-seeking, erotophilia, phychoticism, hypermasculinity, and delinquency. Additionally, volunteers reported more frequent and varied sexual experiences, including exposure to pornography.

Whereas the studies discussed above appear to imply that enjoyment of pornography largely reflects an interest in and openness to sexually exciting entertainment *per se*, research that has differentiated between different types of sexually explicit materials has tended to report more nuanced patterns in the extent to which personality variables predict liking or enjoyment. For example, Bogaert (2001) reported that whereas higher levels of dominance and antisocial tendencies (machiavellianism, psychoticism, hypermasculinity) were associated with greater likelihood of viewing sexually-explicit films featuring violence, children, or women with "insatiable" sexual appetites, these variables were unrelated to likelihood of viewing "erotic" films (i.e., devoid of aggression or unconventional sexual activities). Indeed, the only individual-difference variable associated with likelihood of viewing erotica was erotophilia. Barnes, Malamuth, and Check (1984) reported similar results in their study of males' responses to sexually explicit audiotapes. Namely, among males who scored higher on psychoticism, self-reported and physiological indicators of sexual arousal were higher for stories featuring rape depictions than for non-violent stories, whereas the opposite was true among males who scored lower on psychoticism.

As with our former discussion of enjoyment of violence, the specific explanations for *why* personality would play important roles in the enjoyment of sexually explicit media are not conclusive. For example, the aforementioned roles of relevance, similarity, and affective dispositions may apply to this genre as well. In addition, it is important to note that although our discussion of personality and enjoyment has "treated" personality as an independent variable, it is also plausible that consumption of sexually-explicit or violent entertainment may affect or shape dispositions such as hostility, rebelliousness, or dominance. We further explore this interpretation of personality as a dependent variable in the concluding section of this chapter.

Sadness-Evoking Entertainment.   Although entertainment such as sad films, mournful love songs, and tragedies differ considerably from genres such as horror films, crime dramas, or violent pornography, sad or tragic entertainment has gained notice among media scholars, in part because of the paradox it presents. Namely, why would viewers elect to consume entertainment that is designed to evoke (and presumably successfully evokes) emotions with negative hedonic valence? As a result of this paradox, numerous explanations have been entertained, including the role that personality or dispositions play in predicting enjoyment of this theoretically-curious entertainment fare.

At first glance, it might seem that enjoyment of sadness-evoking entertainment would be most evident among viewers who would be *least* likely to respond with sadness. After all, viewers who are most susceptible to feeling blue from watching tear-jerkers or listening to sad songs should seemingly find this type of entertainment particularly distasteful. In contrast, though, research generally suggests that dispositions implying *greater* affective responsiveness are associated with attraction to these types of genres. For example, research has reported that enjoyment of sad films is associated with higher levels of empathy (Oliver, 1993b), communal (feminine) gender-role self perceptions (Oliver, Sargent, & Weaver, 1998), and loneliness (Mares & Cantor, 1992). Further, de Wied, Zillmann, and Ordman (1994) reported that participants who scored high on measures of empathy were more likely than were participants who scored low to report *both* higher levels of distress *and* enjoyment in response to scenes from a sad film.

Although this existing research suggests that personality characteristics that predict empathic distress to sad entertainment also predict enjoyment, the reasons why distress and enjoyment would be positively correlated are still largely speculative at this point. For example, de Wied et al. (1994) discussed the idea that through excitation transfer (Bryant & Miron, 2003; Zillmann, 1971), the distress evoked from viewing beloved characters suffer ultimately leads to heightened gratification when satisfying resolutions are presented at a film's conclusion. From this perspective, individual-difference variables that serve to heighten empathic distress should contribute to increased enjoyment, provided that satisfying resolutions are indeed forthcoming. Alternatively, Goldenberg, Pyszczynski, Johnson, Greenberg, and Solomon (1999) suggested terror management as an explanatory mechanism for enjoyment of tragedy. These authors argued that individuals who are reminded about their own mortality may find tragic entertainment more meaningful and comforting, as such fare allows individuals to confront their fears in a safe and non-threatening environment. Although mortality salience is generally conceptualized to be a state rather than a trait, future research may benefit from exploring how chronic accessibility of death-related thoughts may be related to coping behaviors, including media consumption.

In addition to suggesting that sadness-inducing entertainment may aid in gratification and coping, other researchers have explored individuals' responses to the experience of negative affect *per se*. For example, Oliver (1993b) suggested that individuals who tend to have positive appraisals of negative emotions (i.e., positive "meta-emotions," see Mayer & Gaschke, 1988) should be most likely to derive enjoyment from viewing sad films. Although pointing out that some people "like to feel sad" begs the question as to *why* they like to feel sad, some researchers have suggested that feeling sad in response to others' suffering allows individuals to feel that they have behaved appropriately (i.e., empathically) to others' misfortunes (Feagin, 1983; Mills, 1993). Alternatively, other researchers have suggested that individual differences may account for variations in the experience of emotion, with some people being more receptive to experiencing a wide range of feelings and being more reflective about their emotional states (see Salovey, Mayer, Goldman, Turvey, & Palfai, 1995). For example, Maio and Esses (2001) conceptualized the "need for affect" as a trait reflecting "the general motivation of people to approach or avoid situations and activities that are emotion inducing for themselves and others" (p. 585). Further, these authors reported that individuals who scored higher on a measure of need for affect reported greater preference for viewing "emotional" rather than "unemotional" films.

Clearly, the list of possible explanations for the enjoyment of sad or tragic entertainment is incomplete and exploratory, calling for greater research attention. Additionally, research in this area needs to take special care to avoid tautological or non-falsifiable explanations (e.g., individuals enjoy sad films because sadness is enjoyed; viewing sad films must result in positive states because individuals choose to view them). Consequently, future research that more closely examines the role of personality or other individual-difference measures that predict and explain the experience of and enjoyment of sadness or other negative affective reactions to entertainment stands to provide much insight into what might be reasonably understood to be a paradoxical diversion.

PERSONALITY AS A MODERATOR OF MEDIA EFFECTS

Given the importance of personality in predicting media selection and enjoyment, it is not surprising that researchers have further explored the role that personality plays in moderating effects of media on viewers. Research from this perspective has typically employed associative-priming models as an interpretive framework (Berkowitz & Rogers, 1986; Jo & Berkowitz, 1994), suggesting that many individual differences reflect variations in the structure and strength of cognitive associations and the frequency and ease with which different cognitions are activated. As a result of these variations in knowledge structures, media content may serve to prime or activate certain cognitions in some viewers, whereas for other viewers the lack of relevant or strong cognitive associations may result in little to no priming.

The importance of individual differences in moderating media effects has been explored most extensively in the context of media violence. For example, Bushman (1995) reported that exposure to violent films resulted in greater increases in aggressive affect and behaviors among individuals who scored higher on trait aggressiveness than among individuals who scored lower on this characteristic. Anderson and Dill (2000, Study 1) found similar results in their study of video-game violence. Namely, self-reported exposure to violent video games was associated with greater aggressive behavior, but only among males scoring higher on measures of aggressive personality. Additional research examining alternative trait measures has revealed similar findings. For example, Scharrer (2001) reported that males' exposure to a violent television program resulted in increases in self-reported aggression/hostility, but only among the participants who scored higher on measures of hypermasculinity. Likewise, Zillmann and Weaver (1997) found that exposure to films featuring gratuitous violence (e.g., *Death Warrant*, *Total Recall*) resulted in greater acceptance and perceived effectiveness of violent conflict resolution, but only among male respondents who scored higher on psychoticism.

Additional research on sexually explicit materials has shown similar patterns in the moderating role of personality. For example, McKenzie-Mohr and Zanna (1990) found that gender-schematic males (i.e., males scoring high on masculinity and low on femininity) who had viewed an erotic film were more likely to display sexually suggestive mannerisms in a subsequent interaction with a woman, whereas aschematic males were largely unaffected by the pornography exposure. Although the erotic film employed in their study was relatively devoid of aggressive behavior, additional research suggests that personality variables can further interact with the *type* of erotic material in question to produce differential effects. For example, Malamuth and Check (1985) found that exposure to physically *aggressive pornography* in which the woman was depicted as aroused resulted in higher estimates of the percent of women who enjoy being raped or forced into having sex, but only among males who reported a high likelihood of rape prior to exposure (see also Bogaert, Woodard, & Hafer, 1999).

Together, the results of the role of personality and dispositions as moderating variables in media effects paint a rather disturbing portrait of media influence, and particularly when combined with scholarship concerning media preference. That is, personality variables that have been shown to predict selection and enjoyment of entertainment such as violence or sexually aggressive content also appear to set the stage for increased likelihood of susceptibility to harmful effects (e.g., aggression, inappropriate sexual behaviors, etc.). This pessimistic interpretation is similar to Slater, Henry, Swaim, and Anderson's (2003) "downward spiral model" of media violence, in which exposure to violent portrayals is predicted to increase aggressiveness, and aggressiveness, in turn, is predicted to increase selection of media violence. Situated in the context of personality or disposition, one might further predict that not only would aggressive dispositions result in increased exposure to media violence, but that such exposure would result in more pronounced effects on aggressive than non-aggressive viewers, further accelerating the "speed" at which the spiral may progress. Given the complexity of such a model, future research would benefit from exploring the mutual influence of not only effects and selection, but also the role that media may play in cultivating the very dispositions or personality characteristics that appear to amplify both enjoyment and susceptibility to influence.

PERSONALITY AS A DEPENDENT VARIABLE

This chapter began with a recognition that *personality* is conceptualized as enduring and stable, and therefore, unlike transitory states such as arousal or mood, is typically employed by media scholars as a predictor or moderator of media-related behaviors and responses. In this final section, we explore the utility of expanding research on media and personality to consider how personality may be conceptualized as a dependent variable, with media behaviors serving to create or shape individuals' traits, dispositions, or more enduring characteristics.

Media scholars' infrequent exploration of personality as a dependent variable may reflect, in part, the way that personality is often conceptualized. For example, conceptualizations of personality that emphasize biological or genetic origins or that characterize personality as a small number of global traits (e.g., the "Big Five") often downplay environmental or cultural influences (e.g., Eysenck & Eysenck, 1985; McCrae et al., 2000; Zuckerman, 2004). In contrast, social-cognitive conceptualizations frequently acknowledge cultural and social influences, allowing for the relationship between cognitive, emotional, and behavioral elements to strengthen or diminish as a function of environmental changes (Bandura, 2002; Funder, 2001; Mischel & Shoda, 1998). In addition to "opening the door" for media exposure to play a potential role in shaping personality, social-cognitive conceptualizations of personality also appear to fit nicely with many existing media effects theories that have examined the influence of media on individuals' cognitions, affect, and behavior.

The effect of media violence on viewers' aggression is one area of research that is particularly relevant to the notion of personality as a dependent variable. For example, Anderson and Bushman's (2002) discussion of the general aggression model (GAM) integrates numerous theories employed by media scholars to examine the effects of media violence, including script theory, excitation transfer, social learning, and priming (see also Anderson & Huesmann, 2003). In brief, GAM conceptualizes aggressive actions as resulting from the combination of situational and person factors that contribute to internal states (affect, cognition, arousal). These states are then appraised by the individual, with decisions applied (deliberately or automatically), and behavioral outcomes then resulting. The importance of personality to this model is two-fold. First, person factors are largely conceptualized as enduring traits or dispositions reflecting scripts, knowledge structures, and schemas that prepare a person to aggress (or not). As these authors argued, "In a very real sense, personality is the sum of a person's knowledge structures" (p. 35). In addition, personality is also conceptualized not only as a moderator and predictor, but also a reflection of past experiences, including media violence. As an example, Anderson and Bushman discussed the potential long-term effects of playing violent video games, arguing that cumulative exposure can result in "the creation and automatization of ... aggression-related knowledge structures [that serve to] change the individual's personality" (p. 42).

In addition to increasing aggression, this same type of model may be usefully employed to understand alternative effects of media violence such as those suggested by cultivation (Gerbner, Gross, Morgan, Signorielli, & Shanahan, 2002). For example, Shrum's (Shrum, 1995, 2002; Shrum & O'Guinn, 1993) research concerning the explanatory mechanisms associated with cultivation has revealed evidence that media provide heavy viewers with frequent and vivid exemplars that are readily accessible and that are therefore influential in individuals' judgments and perceptions. As such, the accessibility of such constructs, and particularly their chronic accessibility (Bargh & Pratto, 1986), may be characterized as representing fundamental aspects of a person's knowledge structure. Although changes in knowledge structures resulting from media exposure might be thought to represent changes only in attitudes or beliefs, the constellation of similar attitudes, behavioral tendencies, and beliefs may also be conceptualized as the foundation of personality characteristics such as authoritarianism, punitiveness, or "mean world" outlooks (see, for example, Shanahan, 1995).

A final example that illustrates personality as a dependent variable concerns desensitization. A host of studies have demonstrated that exposure to media violence and to violent pornography can result in lower levels of arousal to victims' suffering, to greater levels of callousness, and to a general lack of empathy (Linz, Donnerstein, & Adams, 1989; Linz, Donnerstein, & Penrod, 1988; Thomas, Horton, Lippincott, & Drabman, 1977; Zillmann, 1989; Zillmann & Bryant, 1982; Zillmann & Weaver, 1989). Although studies have typically examined such effects in relatively short durations of time (e.g., hours, days, or weeks), the more enduring cognitive and emotional changes that desensitization may reflect suggest that media portrayals could play a role in the formation of related personality characteristics such as callousness or deviance.

The aforementioned examples of how media researchers may usefully employ personality as a dependent variable are far from exhaustive. Consequently, the reader should understand them to be but a sample of the variety of ways that scholars may usefully examine the cumulative and long-term effects of media on viewers' more enduring and stable dispositions. This suggestion for future research is not meant to imply that such inquiries will be easily conducted. Indeed, the exploration of the influence of cumulative exposure is wrought with a host of methodological challenges. Nevertheless, explorations of the role of personality beyond that of predictor or moderator hold promise of demonstrating the important and powerful role of media in affecting viewers in ways that go well beyond the immediate viewing situation.

CONCLUDING COMMENTS

At the beginning of this chapter we noted that the variety of entertainment offerings is a reflection of the diversity of the viewing audience and their tastes and dispositions. The importance of personality to the viewer–entertainment relationship is pervasive, predicting viewers' exposure, affecting their enjoyment, and moderating their responses. Further, if our arguments concerning the conceptualization of personality as a dependent variable are viable, the importance of including individual differences in models of media influence become imperative.

With this call for greater attention to personality in mind, we hope that future research attempts to address the many questions and gaps that remain in the literature. For example, explanations for *why* personality predicts entertainment preferences are arguably speculative at this point, calling for greater study of the specific mechanisms that account for these associations. Similarly, examining the mutually reinforcing role of media selection and media effects on viewers will likely yield a greater understanding of how personality moderates media influence. And finally, attention to how existing theories of media effects may profitably conceptualize their dependent variables as dispositions or enduring knowledge structures stands to broaden the scope of media research and to highlight the idea that entertainment may be even more pervasive than we currently realize.

REFERENCES

Aluja-Fabregat, A. (2000). Personality and curiosity about TV and films violence in adolescents. *Personality and Individual Differences, 29*, 379–392.

Aluja-Fabregat, A., & Torrubia-Beltri, R. (1998). Viewing of mass media violence, perception of violence, personality and academic achievement. *Personality and Individual Differences, 25*, 973–989.

Anderson, C. A., & Bushman, B. J. (2002). Human aggression. *Annual Review of Psychology, 53*, 27–51.

Anderson, C. A., & Dill, K. E. (2000). Video games and aggressive thoughts, feelings, and behavior in the laboratory and in life. *Journal of Personality and Social Psychology, 78*, 772–790.

Anderson, C. A., & Huesmann, L. R. (2003). Human aggression: A social-cognitive view. In M. A. Hogg & J. Cooper (Eds.), *Handbook of Social Psychology* (pp. 296–323). London: Sage.

Bandura, A. (2002). Social cognitive theory of mass communication. In J. Bryant & D. Zillmann (Eds.), *Media effects: Advances in theory and research* (2nd ed., pp. 121–153). Mahwah, NJ: Lawrence Erlbaum Associates.

Bargh, J. A., & Pratto, F. (1986). Individual construct accessibility and perceptual selection. *Journal of Experimental Social Psychology, 22*, 293–311.

Barnes, G. E., Malamuth, N. M., & Check, J. V. (1984). Psychoticism and sexual arousal to rape depictions. *Personality and Individual Differences, 5*, 273–279.

Becker, M. A., & Byrne, D. (1985). Self-regulated exposure to erotica, recall errors, and subjective reactions as a function of erotophobia and Type A coronary-prone behavior. *Journal of Personality and Social Psychology, 48*, 760–767.

Berkowitz, L., & Rogers, K. H. (1986). A priming effect analysis of media influences. In J. Bryant & D. Zillmann (Eds.), *Perspectives on media effects* (pp. 57–82). Hillsdale, NJ: Lawrence Erlbaum Associates.

Biswas, R., Riffe, D., & Zillmann, D. (1994). Mood influence on the appeal of bad news. *Journalism Quarterly, 71*, 689–696.

Bleich, S., Zillmann, D., & Weaver, J. (1991). Enjoyment and consumption of defiant rock music as a function of adolescent rebelliousness. *Journal of Broadcasting & Electronic Media, 35*, 351–366.

Bogaert, A. F. (1996). Volunteer bias in human sexuality research: Evidence for both sexuality and personality differences in males. *Archives of Sexual Behavior, 25*, 125–140.

Bogaert, A. F. (2001). Personality, individual differences, and preferences for the sexual media. *Archives of Sexual Behavior, 30*, 29–53.

Bogaert, A. F., Woodard, U., & Hafer, C. L. (1999). Intellectual ability and reactions to pornography. *Journal of Sex Research, 36*, 283–291.

Bryant, J., & Miron, D. (2003). Excitation-transfer theory and three-factor theory of emotion. In J. Bryant, D. Roskos-Ewoldsen, & J. Cantor (Eds.), *Communication and emotion: Essays in honor of Dolf Zillmann* (pp. 31–59). Mahwah, NJ: Lawrence Erlbaum Associates.

Bryant, J., & Zillmann, D. (1984). Using television to alleviate boredom and stress: Selective exposure as a function of induced excitational states. *Journal of Broadcasting, 28*, 1–20.

Bushman, B. J. (1995). Moderating role of trait aggressiveness in the effects of violent media on aggression. *Journal of Personality and Social Psychology, 69*, 950–960.

Carpentier, F. D., Knobloch, S., & Zillmann, D. (2003). Rock, rap, and rebellion: Comparisons of traits predicting selective exposure to defiant music. *Personality and Individual Differences, 35*, 1643–1655.

de Wied, M., Zillmann, D., & Ordman, V. (1994). The role of empathic distress in the enjoyment of cinematic tragedy. *Poetics, 23*, 91–106.

Eysenck, H. J., & Eysenck, M. W. (1985). *Personality and individual differences: A natural science approach.* New York: Plenum.

Eysenck, S. B. G., Eysenck, H. J., & Barrett, P. (1985). A revised version of the psychoticism scale. *Personality and Individual Differences, 6*, 21–29.

Feagin, S. L. (1983). The pleasures of tragedy. *American Philosophical Quarterly, 20*, 95–104.

Funder, D. C. (2001). Personality. *Annual Review of Psychology, 52*, 197–221.

Gerbner, G., Gross, L., Morgan, M., Signorielli, N., & Shanahan, J. (2002). Growing up with television: Cultivation processes. In J. Bryant & D. Zillmann (Eds.), *Media effects: Advances in theory and research* (2nd ed., pp. 43–67). Mahwah, NJ: Lawrence Erlbaum Associates.

Gibson, R., Aust, C. F., & Zillmann, D. (2000). Loneliness of adolescents and their choice and enjoyment of love-celebrating versus love-lamenting popular music. *Empirical Studies of the Arts, 18*, 43–48.

Goldenberg, J. L., Pyszczynski, T., Johnson, K. D., Greenberg, J., & Solomon, S. (1999). The appeal of tragedy: A terror management perspective. *Media Psychology, 1*, 313–329.

Hansen, C. H., & Hansen, R. D. (1991). Constructing personality and social reality through music: Individual differences among fans of punk and heavy metal music. *Journal of Broadcasting & Electronic Media, 35*, 335–350.

Hansen, C. H., & Hansen, R. D. (2000). Music and music videos. In D. Zillmann & P. Vorderer (Eds.), *Media entertainment: The psychology of its appeal* (pp. 175–213). Mahwah, NJ: Lawrence Erlbaum Associates.

Hoffner, C., & Cantor, J. (1991). Perceiving and responding to mass media characters. In J. Bryant & D. Zillmann (Eds.), *Responding to the screen: Reception and reaction processes* (pp. 63–101). Hillsdale, NJ: Lawrence Erlbaum Associates.

Jo, E., & Berkowitz, L. (1994). A priming effect analysis of media influences: An update. In J. Bryant & D. Zillmann (Eds.), *Media effects: Advances in theory and research* (pp. 43–60). Hillsdale, NJ: Lawrence Erlbaum Associates.

Knobloch, S., & Mundorf, N. (2003). Communication and emotion in the context of music and music television. In J. Bryant, D. Roskos-Ewoldsen, & J. Cantor (Eds.), *Communication and emotion: Essays in honor of Dolf Zillmann* (pp. 491–509). Mahwah, NJ: Lawrence Erlbaum Associates.

Knobloch, S., & Zillmann, D. (2002). Mood management via the digital jukebox. *Journal of Communication, 52*, 351–366.

Knobloch, S., & Zillmann, D. (2003). Appeal of love themes in popular music. *Psychological Reports, 93*, 653–658.

Linz, D., Donnerstein, E., & Adams, S. M. (1989). Physiological desensitization and judgments about female victims of violence. *Human Communication Research, 15*, 509–522.

Linz, D., Donnerstein, E., & Penrod, S. (1988). Effects of long-term exposure to violent and sexually degrading depictions of women. *Journal of Personality and Social Psychology, 55*, 758–768.

Lopez, P. A., & George, W. H. (1995). Men's enjoyment of explicit erotica: Effects of person-specific attitudes and gender-specific norms. *Journal of Sex Research, 32*, 275–288.

Maio, G. R., & Esses, V. M. (2001). The need for affect: Individual differences in the motivation to approach or avoid emotions. *Journal of Personality, 69*, 583–615.

Malamuth, N. M., & Check, J. V. (1985). The effects of aggressive pornography on beliefs in rape myths: Individual differences. *Journal of Research in Personality, 19*, 299–320.

Mares, M. L., & Cantor, J. (1992). Elderly viewers' responses to televised portrayals of old age: Empathy and mood management versus social comparison. *Communication Research, 19*, 459–478.

Mayer, J. D., & Gaschke, Y. N. (1988). The experience and meta-experience of mood. *Journal of Personality and Social Psychology, 55*, 102–111.

McCrae, R. R., Costa, P. T., Ostendorf, F., Angleitner, A., Hrebickova, M., Avia, M. D., Sanz, J., Sanchez-Bernardos, M. L., Kusdil, M. E., Woodfield, R., Saunders, P. R., & Smith, P. B. (2000). Nature over nurture: Temperament, personality, and life span development. *Journal of Personality and Social Psychology, 78*, 173–186.

McDaniel, S. R. (2003). Reconsidering the relationship between sensation seeking and audience preferences for viewing televised sports. *Journal of Sport Management, 17*, 13–36.

McKenzie-Mohr, D., & Zanna, M. P. (1990). Treating women as sexual objects: Look to the (gender schematic) male who has viewed pornography. *Personality and Social Psychology Bulletin, 16*, 296–308.

Mills, J. (1993). The appeal of tragedy: An attitude interpretation. *Basic and Applied Social Psychology, 14*, 255–271.

Mischel, W., & Shoda, Y. (1998). Reconciling processing dynamics and personality dispositions. *Annual Review of Psychology, 49*, 229–258.

Nabi, R. L., Biely, E. N., Morgan, S. J., & Stitt, C. R. (2003). Reality-based television programming and the psychology of its appeal. *Media Psychology, 5*, 303–330.

Oliver, M. B. (1993a). Adolescents' enjoyment of graphic horror: Effects of viewers' attitudes and portrayals of victim. *Communication Research, 20*, 30–50.

Oliver, M. B. (1993b). Exploring the paradox of the enjoyment of sad films. *Human Communication Research, 19*, 315–342.

Oliver, M. B. (2002). Individual differences in media effects. In J. Bryant & D. Zillmann (Eds.), *Media effects* (2nd ed., pp. 507–524). Mahwah, NJ: Lawrence Erlbaum Associates.

Oliver, M. B., & Armstrong, G. B. (1995). Predictors of viewing and enjoyment of reality-based and fictional crime shows. *Journalism & Mass Communication Quarterly, 72*, 559–570.

Oliver, M. B., Sargent, S. L., & Weaver, J. B. (1998). The impact of sex and gender role self-perception on affective reactions to different types of film. *Sex Roles, 38*, 45–62.

Palmgreen, P. C., Wenner, L. A., & Rosengren, K. E. (1985). *Uses and gratifications research: The past ten years.* Beverly Hills, CA: Sage.

Raney, A. A. (2003). Dispositon-based theories of enjoyment. In J. Bryant, D. Roskos-Ewoldsen, & J. Cantor (Eds.), *Communication and emotion: Essays in honor of Dolf Zillmann* (pp. 61–84). Mahwah, NJ: Lawrence Erlbaum Associates.

Raney, A. A., & Bryant, J. (2002). Moral judgment and crime drama: An integrated theory of enjoyment. *Journal of Communication, 52*, 402–415.

Robinson, T. O., Weaver, J. B., & Zillmann, D. (1996). Exploring the relation between personality and the appreciation of rock music. *Psychological Reports, 78*, 259–269.

Rubin, A. M. (1983). Television uses and gratifications: The interactions of viewing patterns and motivations. *Journal of Broadcasting, 27*, 37–51.

Rubin, A. M. (2002). The uses-and-gratifications perspective of media effects. In J. Bryant & D. Zillmann (Eds.), *Media effects: Advances in theory and research* (2nd ed., pp. 525–548). Mahwah, NJ: Lawrence Erlbaum Associates.

Rubin, A. M., Haridakis, P. M., & Eyal, K. (2003). Viewer aggression and attraction to television talk shows. *Media Psychology, 5*, 331–362.

Salovey, P., Mayer, J. D., Goldman, S. L., Turvey, C., & Palfai, T. P. (1995). Emotional attention, clarity, and repair: Exploring emotional intelligence using the Trait Meta-Mood Scale. In J. W. Pennebaker (Ed.), *Emotion, disclosure, & health* (pp. 125–154). Washington, DC: American Psychological Association.

Scharrer, E. (2001). Men, muscles, and machismo: The relationship between television violence exposure and aggression and hostility in the presence of hypermasculinity. *Media Psychology, 3*, 159–188.

Schwartz, K. D., & Fouts, G. T. (2003). Music preferences, personality style, and developmental issues of adolescents. *Journal of Youth and Adolescence, 32*, 205–213.

Shanahan, J. (1995). Television viewing and adolescent authoritarianism. *Journal of Adolescence, 18*, 271–288.

Sherry, J. L. (2001). Toward an etiology of media use motivations: The role of temperament in media use. *Communication Monographs, 68*, 274–288.

Sherry, J. L. (2004). Media effects theory and the nature/nurture debate: A historical overview and directions for future research. *Media Psychology, 6*, 83–109.

Shrum, L. J. (1995). Assessing the social influence of television: A social cognition perspective on cultivation effects. *Communication Research, 22*, 402–429.

Shrum, L. J. (2002). Media consumption and perceptions of social reality: Effects and underlying processes. In J. Bryant & D. Zillmann (Eds.), *Media effects: Advances in theory and research* (2nd ed., pp. 69–95). Mahwah, NJ: Lawrence Erlbaum Associates.

Shrum, L. J., & O'Guinn, T. C. (1993). Processes and effects in the construction of social-reality: Construct accessibility as an explanatory variable. *Communication Research, 20*, 436–471.

Slater, M. D., Henry, K. L., Swaim, R. C., & Anderson, L. L. (2003). Violent media content and aggressiveness in adolescents: A downward spiral model. *Communication Research, 30*, 713–736.

Tamborini, R., Stiff, J., & Zillmann, D. (1987). Preference for graphic horror featuring male versus female victimization: Personality and past film viewing experiences. *Human Communication Research, 13*, 529–552.

Thomas, M. H., Horton, R. W., Lippincott, E. C., & Drabman, R. S. (1977). Desensitization to portrayals of real-life aggression as a function of television violence. *Journal of Personality and Social Psychology, 35*, 450–458.

Weaver, J. B., III. (2000). Personality and entertainment preferences. In D. Zillmann & P. Vorderer (Eds.), *Media entertainment: The psychology of its appeal* (pp. 235–248). Mahwah, NJ: Lawrence Erlbaum Associates.

Weaver, J. B., III. (2003). Individual differences in television viewing motives. *Personality and Individual Differences, 35*, 1427–1437.

Zillmann, D. (1971). Excitation transfer in communication-mediated aggressive behavior. *Journal of Experimental Social Psychology*, 419–434.

Zillmann, D. (1985). The experimental exploration of gratifications from media entertainment. In K. E. Rosengren, L. A. Wenner, & P. Palmgreen (Eds.), *Media gratifications research: Current perspectives* (pp. 225–239). Beverly Hills, CA: Sage.

Zillmann, D. (1988a). Mood management through communication choices. *American Behavioral Scientist, 31*, 327–340.

Zillmann, D. (1988b). Mood management: Using entertainment to full advantage. In L. Donohew, H. E. Sypher, & E. T. Higgins (Eds.), *Communication, social cognition, and affect*. Hillsdale, NJ: Lawrence Erlbaum Associates.

Zillmann, D. (1989). Effects of prolonged consumption of pornography. In D. Zillmann & J. Bryant (Eds.), *Pornography: Research advances and policy considerations* (pp. 127–157). Hillsdale, NJ: Lawrence Erlbaum Associates.

Zillmann, D. (1991). Empathy: Affect from bearing witness to the emotions of others. In J. Bryant & D. Zillmann (Eds.), *Responding to the screen: Reception and reaction processes* (pp. 135–167). Hillsdale, NJ: Lawrence Erlbaum Associates.

Zillmann, D. (2000). Mood management in the context of selective exposure theory. In M. E. Roloff (Ed.), *Communication Yearbook* (Vol. 23, pp. 103–123). Thousand Oaks, CA: Sage.

Zillmann, D., & Bhatia, A. (1989). Effects of associating with musical genres on heterosexual attraction. *Communication Research, 16*, 263–288.

Zillmann, D., & Bryant, J. (1982). Pornography, sexual callousness, and the trivialization of rape. *Journal of Communication, 32*, 10–21.

Zillmann, D., & Bryant, J. (1985). Affect, mood, and emotion as determinants of selective exposure. In D. Zillmann & J. Bryant (Eds.), *Selective exposure to communication* (pp. 157–190). Hillsdale, NJ: Lawrence Erlbaum Associates.

Zillmann, D., & Cantor, J. R. (1977). Affective responses to the emotions of a protagonist. *Journal of Experimental Social Psychology, 13*, 155–165.

Zillmann, D., Hezel, R. T., & Medoff, N. J. (1980). The effect of affective states on selective exposure to televised entertainment fare. *Journal of Applied Social Psychology, 10*, 323–339.

Zillmann, D., & Weaver, J. B. (1989). Pornography and men's sexual callousness toward women. In D. Zillmann & J. Bryant (Eds.), *Pornography: Research advances and policy considerations* (pp. 95–125). Hillsdale, NJ: Lawrence Erlbaum Associates.

Zillmann, D., & Weaver, J. B. (1997). Psychoticism in the effect of prolonged exposure to gratuitous media violence on the acceptance of violence as a preferred means of conflict resolution. *Personality and Individual Differences, 22*, 613–627.

Zuckerman, M. (2004). The shaping of personality: Genes, environments, and chance encounters. *Journal of Personality Assessment, 82*, 11–22.

Zuckerman, M., & Litle, P. (1986). Personality and curiosity about morbid and sexual events. *Personality and Individual Differences, 7*, 49–56.

20

Emotion and Cognition in Entertainment

Dorina Miron
University of Alabama

DRIVEN BY PLEASURE

Have you enjoyed or dreamed about operating the "ultimate driving machine," a BMW, or a perfectionist Mercedes? The human brain is like a Mercedes plant, where a curious visitor can see perpetual work in progress: Various models are being put together as they move on, at a steady pace, down the assembly line. Major operations are on display, while others, probably messy, are hidden in order to keep the brand, or the human species, safely competitive.

Layers of Brain History at Work

The advantage of the human brain over a Mercedes plant is that the brain museum is the production line: Each one of us carries a reptilian brain, on top of which there is an old-mammalian brain or the limbic system, physically covered by the neo-mammalian brain or the neocortex. In this "triune brain" (MacLean, 1990), the reptilian or instinctual brain is the locus of basic motor plans and basic emotions, such as seeking, fear, aggression, and sexuality, that have driven all individual animals throughout history; the old-mammalian or emotional brain adds sophistication to fear and anger in conjunction with the new social emotions it elaborates, and which are associated with life in groups of increasingly complex structure; finally, the neo-mammalian or rational brain is the locus of logic or cause–effect reasoning that cuts across time, linking the present with the past through learning and with the future through learning-based anticipation and planning.

How did this layered structure come about? Humans were and are relatively small animals—just think about the size of dinosaurs. In the inter-species race for survival, humans succeeded by taking a leap in natural selection. Instead of outgrowing the largest animals, they overpowered the other species by accelerating adjustment to and control over the environment. From the slow-paced species-level selection and genetic encoding of effective survival features, what was to become the human species shifted to massive individual lifetime learning. The human-defining cortical functions that specialize in learning, anticipating, and planning emerged as regulators and enhancers of the basic animal functions performed by the old- and neo-mammalian layers of the brain. The neocortex derives from individual and social experience optimization patterns for subsequent behaviors. In addition to relying on genes, humans massively draw on memory-stored knowledge to shape/direct their actions. Originally captive within the environments in which they were born, humans have progressed toward adjusting those environments through their actions, for survival purposes.

Moved by Good and Bad Feelings

For animals as well as humans the lack of external life-sustaining resources (e.g., food, water, heat) *feels* bad because it leads to depletion of internal resources and death. On the contrary, the use/consumption of such resources feels good because it recharges the body, keeps it functioning, and maintains life. Needs for various life-sustaining resources automatically drive or consciously motivate actions, which can be broadly classified as approach, or seeking of good things and withdrawal, or avoidance of bad things. Aggressive feelings and behaviors were historically triggered by scarcity of resources and competition for them. The performance of sexual activity feels good because it maintains and increases populations, and enhances their ability to compete with other species for resources and survival. Such *feelings* are "neurophysiological homologies that exist across all mammalian species" (Panksepp, 1998, p. 10) and "have an important role in controlling behavior, especially conditionally" (p. 13). The enjoyment/pleasure of consummatory behaviors has emerged as a natural mechanism for rewarding and reinforcing successful survival behaviors, whereas unpleasant feelings punish and discourage behaviors that jeopardize survival. This polar system of good–bad feelings has perpetuated the human species by making it more adaptable to fast environmental changes that occur within the timeframe of an individual human life. The fact that pleasure is the oldest and still key function regulating human behavior points to the importance of hedonic science and positions entertainment theory as a study of voluntary, cognitively based, hedonic pursuits.

Emotions

The Self-Reference Feature of Emotions.   Emotional states are "subjectively experienced feelings" that "arise from material events (at the neural level) that mediate and modulate the deep instinctual nature of many human and animal action tendencies" (Panksepp, 1998, p. 14). Emotions were made possible by the emergence in higher animals of "neurally based self-representation systems" (p. 14). Emotional experiences result from interactions between the neural networks that support different feelings and those that support self-representation. The role of emotions is to provide self-referential value-coding of world events. Such coding guides individuals' adaptive behaviors, which are more appropriate in atypical or changing situations and have superior survival efficiency as compared to the genetically coded responses to the typical environment, which are stereotypical[1].

Distinguishing Emotions.   According to Valenstein (1973), stimulus-bound activations of different neural configurations are subjectively experienced as distinguishable states that prompt specific and coherent behavioral patterns. Although "there are no unambiguous 'centers' or loci for discrete emotions in the brain," and "everything ultimately emerges from the

---

[1] All individuals of a species respond to the same or similar environmental conditions through similar behavioral patterns.

interaction of many systems" (Panksepp, 1998, p. 147), each emotion is associated with certain specific and indispensable circuits. Therefore, from a neurological perspective, the primary criterion for classifying emotions is not the type of stimuli that triggers a specific emotion, but the functionally specialized neural network that is activated, causes the subjective perception of a distinct internal state, and works as an operating system that elicits specific response behaviors.

Emotion classifications vary depending on the taxonomist's interests. According to Panksepp (1998), "there are at least seven innate [basic] emotional systems ingrained within the mammalian brain" (p. 47): fear, anger, sorrow, anticipatory eagerness, play, sexual lust, and maternal nurturance. The same author distinguished between "primary emotions" (e.g., anger, sorrow, fear, joy) that are supported by neural networks located at the lowest/oldest levels of the brain and newer emotions (e.g., jealousy, depression, desire, compassion, surprise, disgust) that are supported by neural networks located in the limbic system (old-mammalian brain) that elaborates social emotions. As human societies around the globe vary in terms of natural environment and resources, organization, level of economic development, cultural heritage, and surely many other factors, it is reasonable to expect that the newest human-only emotions present greater variance in their neural mapping according to the geographic location and social status of the people who experience them.

Emotions Under the Heavy Lid of Culture. Because the newer "social" emotions vary more across cultures and subcultures, their communication requires more feature identification and comparative analysis—that is, more rational activity—than does the sharing of older, "primary"/"universal" emotions. From an entertainment perspective, this difference has created a production and distribution bias/preference for the heavy use of old primary emotions as a prerequisite of commercial success. Newer emotions would be more interesting/attractive to the public due to their novelty and variability of individual experiences, but they would sell to smaller subcultural markets. Thus the economic pressure for product popularity maintains cultural coherence and species cohesiveness at the cost of slow innovation. For example, we will have more first-person shooter videogames to choose from rather than more types of videogames or new types of entertainment. Ultimately, the conservative mainstreaming effect of the popular culture market counterbalances the diverging effect of the self-reference function of emotions, which produces within-species emotional and behavioral diversity that might lead to incommunicability among individuals.

Emotion and Cognition as Necessary Functions for Individual Coherence. If we consider internal processes, the self-reference feature of emotions makes possible the integration of internal states caused by environmental conditions with response behaviors. This integration is achieved through massive interconnectedness in the human brain, "whose every part can find an access pathway to any other part" (Panksepp, 1998, p. 70). Emotive circuits have "reciprocal interactions with brain mechanisms that elaborate higher decision-making processes and consciousness" (p. 150). The result of those ascending interactions is that no emotion occurs without a thought, and many thoughts can evoke emotions. In addition, emotive circuits have interactions with lower brain areas that support automatic physiological functions and govern movement. The outcome of descending interaction is that, "there is no emotion without a physiological or behavioral consequence, and many of the resulting bodily changes can also regulate the tone of emotional systems in a feedback manner" (p. 27). Overall, the activity of a human brain has two major directions: the hypothalamic-limbic axis or the "stream of feeling," which is involved in visceral information processing, and the thalamic-neocortical axis or the "stream of thought" (p. 58), which is involved in somatic information processes. Both streams "converge on basic sensory-motor programs of basal ganglia to generate behavior in which both somatic and visceral processes are blended and yield coherent behavior output" (p. 62). The mediation role of the emotive circuits, which involve self-referencing, ensures and optimizes an individual's self-serving responses to the environment.

Emotion, Cognition, and Context Appropriateness.

The inextricable linkage between emotions and rational activities supported by the neocortex serves a survival purpose by enabling not only self-referential but also situation-relevant decision making. According to the network theory of affect, the arousal of an affective state spreads activation throughout a network of cognitive associations linked to that emotion (Bower, 1981; Bower & Cohen, 1982; Clark & Isen, 1982; Isen, 1984). As a result, "material that is associatively linked to the current mood[2] is more likely to be activated, recalled, and used in various constructive cognitive tasks, leading to a marked mood congruency in constructive associations, evaluations, and judgments" (Forgas, 1999, p. 591).

In philogenesis, emotions preceded rationality. The older/primary emotions, served by neural networks situated deeper in the brain architecture, trigger primitive (automatic and simple) response actions (e.g., fight-or-flight when faced with life-threatening danger). There is little that reason (located in the neocortex) can do about such emotions. However, there is a lot that the neocortex can do when it comes to handling newer/social emotions. Consequently, collaborative decision making and social action strategizing can greatly benefit from rationality. The less we fight about and shoot around social problems, the more we can limit impulsive reactions like flight-or-fight or destruction and benefit from the creativity of the top layer of our brain. This scientific observation is encouraging for movie producers who are willing to engage their audiences in social debate, because it promises more rational involvement from viewers if producers drop the carnage scenes.

Closed and Open Emotional-Cognitive Processes.

Speaking of creativity, animal behavior programs are mostly "closed" (Panksepp, 1998, p. 62), that is, environmental events can hardly change them. Animals generally respond through simple successive actions that address different survival needs in a genetically encoded hierarchy of priorities. The development of the neocortex in the human species shifted the balance in favor of open response programs that are much more sensitive to environmental circumstances (e.g., language, agriculture, politics). In addition, cortical capacity allows for concurrent assessment of various and sometimes conflicting feelings, and makes possible novel responses to increasingly complex situations.

The philogenetic growth of the neocortex relative to the rest of the human body corresponds to an expansion of human choice and invention. However, the enduring role of emotions as triggers of decisions and behaviors should be expected to maintain the fundamental prosurvival direction of rational (cortex designed) human action. On the other hand, the new cortical activities should be expected to follow the course set by the older conservative (life preserving) emotional functions. The openness of newer social emotions and particularly the openness of cortical behavioral decisions are the bread and butter of entertainment. Fresh emotions will always be available, waiting to be explored. On the other hand, the blooming repertoire of novel human behaviors is virtually unlimited, unfolding before our eyes.

Side Effect of Human Creativity: Growing Complexity of the Environment.

The openness of cortical activity and the creativity of human behavior have unfortunately raised an enormous new problem. Humans no longer behave as mere respondents to nature. A growing population of successful self-referential survival-motivated humans has dramatically enriched

---

[2]Moods are weaker but more persistent emotional states, reflecting smaller and slower physiological changes in a human body caused by changes in environmental conditions.

and diversified human habitats. The complexity of man-manipulated environments has, in turn, expanded the range of human needs, emotions, decisions, choices, and possible actions. Humans' emotional-rational system for integrating information about the environment, internal states, and behaviors is overwhelmed. The natural self-referential evaluation allowed by emotions worked well with early humans because their life circumstances were little differentiated and their range of decisions and actions was small. Cohesion in one's life and within a social group was feasible. Now an individual has to spend a lot more time and effort to figure out what is going on in the environment and inside his or her body, establish a personally relevant hierarchy of problems to address, consider more factors and response options in decision-making, and master decision implementation technologies. This situation is conducive to less physical action and more rational activity for humans, and an increasing number of machines that take over human work. Thus, the winning species in the surviving race has paradoxically made both individual and social life harder, that is, emotionally conflicting and overloaded, and more taxing in terms of rational operations. We sweat less working for a living, but have more headaches; individuals live longer, but the species is in greater danger of conflicts of interests and self-destruction. Good news for entertainment though: The human need and market demand for recreation have registered a phenomenal growth.

THE RACE FOR PLEASURES

Pleasure as Related to Basic Survival Functions

In physiological terms, pleasure is often defined as activation of the limbic areas of the brain (Campbell, 1973). The neurotransmitters involved are the opioid and dopamine systems (Hoebel, 1988; Wise, 1982). The nerve fibers that serve pleasure seeking are inextricably interwoven with fibers that control physiological functions that are indispensable for individual and species survival, such as heart beat, breath, blood pressure, and sexual excitation. Those functions can activate pleasure networks, or be activated by neural constellations in the pleasure areas. Thus the limbic system emerged as a coordinator of the basic survival functions: Activities that sustain the species get to be performed because they reward individuals with pleasures; activities that are detrimental to the species are punished and discouraged through displeasure. This explains how humans' basic (default) goal in life came to be the maximization of pleasure and minimization of displeasure. What do humans do to maximize pleasure?

The Potentials of Our Five Senses For Cognitively Enhanced Pleasures

The five senses are the interface between the environment and an individual's autonomic system, which is the part of the human brain that looks after basic survival and reproduction processes and ensures that these functions are performed automatically. The pleasure provided by the autonomic activity in general and sensorial activity in particular (i.e., the stimulation of neurons located in the pleasure areas by inputs from the five senses) has survival value because it motivates us to stay connected to the world and adjust to it, that is, respond in ways that minimize discomfort and maximize pleasure (Greenfield, 2000). How have sensorial pleasures been exploited for entertainment?

Smell. Our use of the smell and touch senses are largely similar to that of animals in general: Unhealthy dirty places, spoiled food, and dead bodies smell bad, and life-sustaining fresh food, clean environments, and mating partners smell good. In addition, humans developed through learning ways to artificially reduce the inconvenience of bad odors and enhance olfactory pleasure: They developed fragrances that increase environmental comfort, or enhance appeal and merchandise attractiveness. Various pleasure-seeking activities emphasize fragrances both as inherent components (e.g., fine cuisine, wine tasting) and as artificially produced odors (e.g., perfumes, incenses, fragrant candles).

Touch. Feeling the temperature, texture, weight, and movement is generally associated with danger and desirability assessment in the case of objects that are hard to identify using safer non-contact senses. With animals, repeated smell and touch experiences result in learned associations between olfactory and tactile qualities and good or bad things, thus facilitating recognition of danger or opportunity. With humans, touching is highly relevant to sexual behavior, clothing, and activities that require a lot of contact with (manipulation of) objects. Both smell and touch can be pleasure enhancers, but they rarely suffice to support an entertainment activity without the involvement of other senses.

Taste. The situation is different with taste. Cuisine arts and the restaurant industry cater for humans' gustative pleasures. The cognitive components added to the palate sensations produced by the food and beverages we seek and consume are fragrance discrimination, recipe knowledge and cooking skills, knowledge and skills for producing/processing the plants, animals, and minerals used as food ingredients, food and beverage market knowledge, *connoisseur* proficiency in the *gourmet* culture, aesthetic appreciation of the visual component of food and table displays and of the auditory and visual environment, as well as social knowledge of people who enjoy similar culinary and enological pleasures and who might enhance individual consummatory pleasures by sharing/communicating about them.

Hearing. Similarly, our auditory pleasure is served by the art of music, the acquisition of singing skills, the knowledge and skills involved in the making, playing, and marketing of musical instruments, the knowledge of music history needed to fully enjoy music from old times and distant geographic locations, the knowledge and skills of blending music with visual and performance arts, and the social knowledge and skills for socializing and sharing auditory pleasures with other *aficionados*.

Food is something concrete and shared by all humans and animals. Cooking art and wine making/tasting are therefore very popular cultural assets. Unlike food, music is immaterial. It depends on signing abilities that vary across individuals, instrument playing skills that take effort to learn, and the availability of instruments that may be hard to make and expensive. These features have for a long time restricted the enjoyment of music—particularly music played on sophisticated instruments and music that required a lot of learning—to members of the elite/leisure class. Media technologies that created popular/mass entertainment have considerably removed this inequality of access to musical culture in the Western societies, more so for consummatory purposes than for production purposes. The most recent developments in computer technologies have completely eliminated the quality and capacity problems related to the archiving of audio materials and their global distribution. The consummatory enjoyment of music currently rivals that of food and beverages. The massification of these entertainments under market pressure unfortunately trades quality for quantity. The every-day experiences of frozen lean cuisine and background music have low hedonic potential.

Sight. Visual discrimination is probably the most unobtrusive and informative/useful of the five human senses because it enables individuals to process a wide range of danger and opportunity cues. The exploitation of sight for pleasure peaks in the fine arts and the performance arts, which have historically built up a huge amount of knowledge and skills on both the consumer side (for enjoyment purposes) and the producer side (for business/work purposes). The visual and performance arts assisted by media technologies have increased visual complexity, speed of change, and artificiality in the human environment, which has put pressure on the neocortex to develop rational analytical skills for abstraction/symbol processing and meaning assessment in order to cope with the information overload. Consequently, visual enjoyment has become more "intellectual," referential, dependent on personal history, and restricted by cultural trends, operating as idiosyncratic choice in a rather stereotypical environment.

This brief comparison of our senses' potentials for artificially enhanced pleasures, developed through learning and the creation of objects that add to the sensorial stimulation available in the natural environment, shows that humans have made considerable progress in self-stimulation. Compared to animals, for which the automatically regulated seeking behavior enhances access to environmental resources for survival purposes, humans are less needy, particularly in the Western world, where for most people survival is no longer an issue. They use much of their sensorial capacity for pleasure enhancement, which is served by specialized leisure, entertainment, and auxiliary/support industries. History shows that the combination of different types of sensorial stimuli and the development of associated knowledge supported by cortical processes have been two practically inexhaustible avenues for hedonic enhancement.

The Potentials of Emotions for Cognitively Enhanced Pleasures

While looking for opportunities to use cognitive processes for the purpose of enhancing pleasure through entertainment activities, our focus is on the biunivocal relationship between emotions and cognition: Emotions powerfully modify critical appraisal and memory processes and vice versa (Christianson, 1992). For exemplification, we will address primary emotions[3] that are highly relevant to entertainment. According to Panksepp (1998), "each basic emotional system can energize a number of distinct behavioral options" (p. 283).

Seeking as the Fundamental Emotion. Humans as pleasure seekers are in constant search of stimulation. When the activation of a pleasure area decreases, nerve impulses are sent to the motor centers that control the muscles involved in exploratory behavior, until the individual finds a new source of sensory stimulation (i.e., a new source of temporary pleasure). This scheme is to be regarded as the most fundamental and basic neural mechanism of behavior (Campbell, 1973). Panksepp (1998) defined seeking as arousal in the lateral hypothalamic area and display of "intense interest," "engaged curiosity," "eager participation," and "foraging behavior" (pp. 147, 149). Stimulation in major sites in the lateral hypothalamus yields "frantically energized" behavior, whereas stimulation in other sites like the media septal area and locus coeruleus produces "slow and methodical" seeking behavior (p. 149). Environment exploration is associated with a theta rhythm that is typical for encoding information, that is, "translating recent experiences into long-term memories" (p. 129).

In general, the orientation in the seeking process is provided by need states (e.g., thirst, hunger, thermal imbalance, sexual needs) that produce a perceptual bias manifested as increased sensitivity to environmental cues signaling resources that would satisfy the currently most urgent need. In humans, the neural networks that support seeking behavior interact with "higher brain systems such as the frontal cortex and hippocampus that generate plans by mediating higher-order temporal and special information processing" (Panksepp, 1998, p. 151). The hippocampus has strong downward connections with the hypothalamus via the descending

---

[3] According to Panksepp (1998), the primary emotions are seeking, fear, anger, sorrow, joy, social bonding, separation distress, play, and emotions related to sexuality, hunger, and thirst.

fiber bundle of the fornix, and these connections may convey cues from spatial maps to foraging impulses (Vanderwolf, 1992).

Although seeking is clear and evident in young children (Izard & Buechler, 1979), it rarely has intense outward expression in adults, who have learned the social convention of being cool and collected. Nevertheless, seeking is a basic and constant function in our lives as it energizes and motivates the body for interactions with the environment, which are crucial for survival. The sensorial stimulations resulting from such interactions, emotionally (self-referentially) coded as "bad" or "good," are gradually linked to their sources through repeated experience, which means that we learn causal links between our behaviors, objects in the environment, and our internal feelings occasioned by those objects. Such causal learning leads to the development of neural networks that include areas in the brain that perform the analysis of sensorial stimuli, pleasure areas that support self-referential evaluation, and cortical areas that contribute context assessment (location, time, order, cause) and produce decisions for behavioral optimization of the non-automatic responses.

Once a pleasure experienced through interaction with a specific object is "learned" (i.e., its complete causal linkage is stored in memory), that piece of knowledge motivates further seeking of the same object or of similar objects. The person's sensorial filtering system has been enriched with an additional criterion for selecting sources of pleasure.

Based on these neurophysiological considerations, "we can be certain that the seeking system does interact with higher brain circuits that mediate each animal's ability to anticipate rewards" (Panksepp, 1998, p. 147). In contrast to the behaviorist view that external reward conditions behavior, current neuropsychology emphasizes the role of the internal/neural reward mechanism that links spontaneous sensorial stimulation to limbic/pleasure processes and cortical/decisional processes: "Rewards in the world are meaningless unless animals can search them out" (p. 151).

"The seeking system is commonly tonically engaged rather than phasically active" (Panksepp, 1998, p. 149), but, of course, there is variability in this respect. "The level of activity in this system may be related to the personality dimensions of positive emotionality, sensation seeking, and other measures of appetitive engagement with the world[4]" (p. 149). More outgoing, avid seekers of hedonically positive stimulation are overall happier people. This variability is highly relevant to the entertainment industry as it defines the market. The market value of entertainment is maximized by the most compulsive seekers of pleasures.

Unlike most other emotions, a seeking state is rewarding in itself, because it energizes the body for action and makes it feel good in order to sustain the pursuit. Thus seeking is a basic form of self-stimulation. The intrinsic pleasure of seeking is indifferent to the type of object stimulation being pursued at each point in time, and the effort involved in seeking is regularly rewarded by various consummatory pleasures, all of which reinforce the seeking behavior. A potential problem for the entertainment industry is that consummation causes a transient inhibition of appetitive arousal, and seeking is resumed only when consummatory pleasure ends or declines. This pleasure inertia is a safety feature that conserves the energy people invest in seeking. As long as they enjoy a specific pleasure, they won't be seeking other sources of pleasure.

When the current pleasure declines or ends, the renewed pursuit of stimulations/pleasures is greatly facilitated by learning from personal or other people's pleasurable experiences. Such learning is the prerequisite of rational optimization of consumer behavior, which involves systematic comparison between the pleasure provided by a current or targeted activity and

---

[4]Testing conducted with Spilberger's 60-question State Train Personality Inventory (STPI), which is used to evaluate tonic and phasic anxiety, curiosity, and anger, showed a high positive correlation between tonic and phasic curiosity, but no such correlation for anger and anxiety (Spilberger, 1975).

potential pleasures obtainable through alternative activities. In the process of learning about pleasure sources, neocortical networks develop downward projections to the sensorial processing and pleasure areas associated with various entertainment activities. The activation of such broad neural constellations is experienced as "deeper"/stronger and more robust/reliable emotions and pleasures. On the other hand, extensive neural networks are easier to access, because they can be activated not only bottom-up (i.e., pleasurable sensorial stimuli evoke related knowledge) but also top-down (i.e., knowledge activated in the working memory evokes related sensations and pleasures). Consequently, pleasure activities that involve learning cease to be simply "here and now," they are to a large extent "trans-temporal." This feature points to the practically infinite possibility of producing and enjoying culturally enriched entertainments.

Fear and Anxiety. Because the ability to detect life-threatening situations is maximally important for survival, it evolved as a genetically ingrained function, protected from the vagaries of individual learning. "Evolution created several coherently operating neural systems that help orchestrate and coordinate perceptual, behavioral, and physiological changes that promote survival in the face of danger" (Panksepp, 1998, p. 206).

The executive system of fear is concentrated in the lateral and central zones of the amygdala, the anterior and medial hypothalamus, and specific areas in the periaqueductal gray of the midbrain (Panksepp, 1990). From there, the system projects into specific autonomic and behavioral output components of the lower brain stem and spinal cord, which control the physiological symptoms of fear (increased blood pressure, startle response, elimination, and perspiration) (Panksepp, 1998). Conditioned fears access the fear system at the central nucleus of the amygdala (Aggleton, 1992). The amygdala is also instrumental in the overall integration of fear responses (Panksepp, 1998). Many anxieties are associated with the arousal of the pituitary adrenal stress responses (Puglisi-Allegra & Oliverio, 1990).

The intensity dimension to fear combines with two temporal dimensions—speed of arousal and its duration—to produce two main varieties of fear: dread/terror, when the circuit is precipitously aroused, and arousal reaches high levels for a relatively short interval of time; and chronic anxiety, when arousal is milder and more sustained (Panksepp, 1998).

Fear is an aversive state of the nervous system. It is experienced as a combination of worry and tension, and is accompanied by specific forms of autonomic and behavioral arousal (vigilance, increased heart rate, sweating, increased muscle tension, fidgeting) that prepare the organism for response behavior that is energy-intensive (fight-or-flight).

Because fear is a sure-fire emotion, easily triggered by stimuli that can be manipulated with high precision, it has been increasingly used in entertainment to enhance consumer excitation, which is a strong enjoyment factor. This pleasure increasing strategy is based on excitation transfer (Zillmann, 1971). The procedure uses the physiological lag of excitation states triggered by different stimuli to build up excitement through a rapid succession of frightful and/or horrific events. Cumulative fear-derived excitement fuels the enjoyment of whatever desirable events are included in the entertainment experience (particularly a happy ending) and/or the appreciation of peripheral aesthetic features of the entertainment object or experience.

"Higher cortical processes are not necessary for the activation of learned fears, although those processes refine the types of perceptions that can instigate fear" (Panksepp, 1998, p. 206). Adding analytical complexity (cortical activities) to the fear-based entertainment experience expands the activated neural networks and slows down their turnover, which deepens and stabilizes the experienced emotion, moving it from the dread/terror range (typical of quick-paced action movies) to the anxiety range (typical of psychological drama).

Humans appear to be prewired for fear of dark and high places, sudden sounds and movements, snakes and spiders, and approaching strangers, especially those with angry faces (Gray, 1987). The dangers associated with social life are all learned, and a wide range of them are learned only vicariously and, mostly, from news and entertainment. As noted by Gerbner (1969, 1970, 1971, 1972) in his cultivation theory and by Zillmann (1999) in his exemplification theory, the disproportionate frequency in the media of life-threatening natural and social situations (e.g., accumulation of tornados, earthquakes, floods, car accidents, plane crashes, crimes, and wars from all parts of the world in the daily newscast and/or every other day's movie) develops in the audience an erroneous perception of the world as a place that is much more dangerous than it is in reality. Violence-loaded entertainment thus thwarts the original adaptive role of dramatic play, cultivating in the public insecurity and anxiety (labeled by Gerbner the mean world syndrome), and offering mostly anti-social coping solutions (destructive action schemata).

Anger and Rage. Aggression is behavior that causes physical harm or destruction of others. It is fueled by anger, which develops from irritations and frustrations related to threats to one's freedom of action, obstructed access to resources, or "reward and expectation mismatches" (Panksepp, 1998, p. 198). Anger is supported by subcortical circuits (shared by humans and other mammals) that "run from medial areas of the amygdala, through discrete zones of the hypothalamus, and down into the periaqueductal gray of the midbrain" (Panksepp, 1998, p. 187). Animals and humans that have constitutional low brain serotonin activity are more prone to aggression and impulsive acting out of other emotions (Coccaro, 1996), and males are generally more aggressive than females because of the activational effects of testosterone on their brains (Ferris & Grisso, 1996).

Anger as an emotional state is experienced as a "fiery mental storm" associated with increased heart rate, blood pressure, body temperature, muscular tonus, and the "desire to strike out at the offending agent" (Panksepp, 1998, p. 191). Anger tends to automatically trigger simple physical action scripts (Christianson, 1992). The narrow scope of anger-related cognitions (focus on self-interest) tends to cause self-serving biases in any subsidiary cortical activities (attribution, decision-making). Aggressive actions triggered by anger are instrumental in dissipating uncomfortable rage states while carrying out predatory activities (eliminating competition for life-sustaining resources), delivering vengeance/retribution, or achieving sexual goals (facilitating sexual competition and/or mating).

As anger states and aggressive behaviors are generally occasioned by competition for environmental resources, they are related to the seeking behavior. Unsuccessful seeking, particularly in a highly competitive environment, is conducive to emotional frustration and fierce fighting behavior (for survival purposes). As fighting with competitors involves personal risks, it activates fear and self-protection behaviors. The spiraling emotions of anger and fear exacerbate into rage, which is extreme anger, with further reduced cognition (diminished risk analysis), and almost completely uncensored aggressive behavior. Anger and rage are supported by somewhat different neural circuits: The two circuits are fairly clearly segregated in the amygdala, with fear being more lateral and rage more medial. But at high levels of arousal, both circuits may be concurrently aroused (Panksepp, 1998). The shift from the activation of fear pathways to the activation of rage pathways reflects in the behavioral shift from flight to fight responses.

In addition to the passionate heat of rage, aggression may play out as hatred, which is a "more calculated, behaviorally constrained, and affectively colder" emotion (Panksepp, 1998, p. 191), under stronger cortical control, which may increase the effectiveness of the aggressive actions taken to resolve the emotional discomfort.

An important characteristic of aggressive behavior is self-reinforcement. Successful aggression is accompanied by an increased level of testosterone, which is likely to facilitate future assertive behavior (Booth & Mazur, 1998; Dabbs, Dabbs, & Mazur, 2002; Kreutz, Rose, & Jennings, 1972; Mazur & Booth, 1998; Mazur & Lamb, 1980). Aggression-intensive entertainments (e.g., sports and games) that develop skills for such behaviors in real life, maintain in habitual players high levels of testosterone that establish aggression as a dominant pattern in their behavior.

Cortical learning is relevant to the construction of expectations and the costs/benefits analysis that mediate anger. It is also relevant to the acquisition of knowledge about social status, dominance hierarchy (access rights), and typically human strategies and instruments of aggression, from deferred individual gratification to coalition building for group aggression, which may range from gang skirmishes to international wars. The benefits of such knowledge are both a more accurate cause–effect mapping of frustrating scenarios and a wider inventory of response options. Both increase rational control over anger and improve behavioral efficiency by reducing personal effort and risks, or by eliminating useless interpersonal fighting (social loss).

The problem with learned controls is that the neural circuits that support anger and aggressive behavior "are hierarchically arranged so that higher functions are dependent on the integrity of lower ones," which means that "the most recently evolved controls continue to depend critically on the nature of preexisting emotional circuit functions" (Panksepp, 1998, pp. 187, 190). Cool detachment may be the pride of human race, but is hard to achieve under high competition stress and frustration that tend to trigger automatic aggressive responses ingrained at low levels in the brain.

Entertainment that features anger and aggression could be expected to provide a cool and detached learning experience to the public, and to enable the audience to consider more effective and/or efficient non-aggressive responses. The problem is that the pace of aggression-based entertainment (e.g., action movies, first-person shooter games) is so fast that it leaves little time to the entertainment user to do optimization thinking. On the other hand, the producers' race for excitement is based on escalation of anger and aggressive behavior, not the resolution of negative emotional states by rational means. The audience is captive to a fictional environment packed with stimuli that elicit frustrations and anger, and is deprived of the possibility to respond through personal action. Such an entertainment environment dominated by the anger/aggression content increases the accessibility of anger states and aggressive behavioral scripts in real life situations. This risk is amplified by the fact that entertainment scripts also develop in the public's memory a repertoire of spectacular but socially maladaptive behaviors. Such undesirable learning effects of aggression-laden entertainment may be mitigated by an abundant use of play features that yield pleasurable states and undercut excitation buildup, which makes consumers of violent entertainment prone to aggressive behaviors immediately after exposure to intense fictional violence.

Individual Versus Social Emotions. Seeking, fear, and anger position a human being as distinct from the natural and social environment, in other words, they emphasize singularity, or individuality. Seeking as the fundamental emotion that makes people move/act in order to survive is intrinsically pleasurable in order to be self-sustainable. But fear and anger are hedonically negative states, and only the behaviors that resolve them are enjoyable for the person who experiences such emotions. This means that a lonely individual against his/her environment has to cope with a great deal of negative feelings. The hedonic balance is redressed by the major primary "social emotions," called so because they are experienced together with other humans. Playfulness promotes social interaction for learning purposes, enabling the sharing and development of social knowledge. Sexual desire, pair bonding, and offspring nurturing are crucial for species survival. The social bond (mutual dependence and support) ensures the convergent and constructive evolution of the human species. Social emotions have to provide pleasure rewards to individuals in exchange for their altruistic efforts that benefit others, and primarily the human species. This observation provides a theoretical explanation for the entertainment producers' empirically derived wisdom of building their business on human/social interactions, which are inherently dramatic as they constantly swing between diversity-related conflicts and similitude-based bonding.

Play.   Roughhouse play is ontologically the earliest and one of the most stable and intense sources of joy in human life. Its evolutionary roots may go back to "an era predating the divergence of mammalian and avian lines more than a hundred million years ago" (Panksepp, 1998, p. 282). Play in general reflects genetically ingrained ludic impulses of the nervous system supported by the "parafascicular and posterior thalamic nuclei" (p. 281) and projections to the "vestibular, cerebellar, and basal ganglia systems that control movement" (p. 291). The features of the playful emotional state are lightness, joy, and flow. Roughhousing behavior is "a flurry of dynamic, carefree rambunctiousness," "play solicitations," "vigorous interaction," and rapid "role reversals" (p. 283).

The playfulness of the human species, extended beyond childhood throughout adult life, is due to a large extent to the massive involvement of cognitive activities that "add a great deal of diversity to our playful behaviors" (Panksepp, 1998, p. 287) and create age-appropriate forms of play.

According to Slade and Wolf (1994), human play can be rough-and-tumble play, exploratory/sensorimotor play, relational/functional play (all three shared with other species), and constructive play, dramatic/symbolic play, and games-with-rules play (which are typical for humans and involve cortical elaborations). Specifically human play options (e.g., sports, games, performing arts, linguistic activities such as puns, joking, and verbal jibes) represent the species' developments of the higher animals' relational/functional play, but they often include rough-and-tumble and exploratory/sensory-motor elements (e.g., sports, action movies).

The ontological role of play, for all species, is learning. Play facilitates the acquisition and constant testing of physical skills (e.g., self-defense and aggression), social skills (e.g., cooperating, competing, resolving conflicts, courting, parenting), and cognitive skills (e.g., identifying sources of danger and opportunities, anticipating responses, problem solving, decision making). Play allows young individuals to be effectively assimilated into the structures of their society. Because of its learning function, play has an inverted U-shaped developmental function, increasing during the early juvenile period, remaining stable through youth, and diminishing as individuals reach adulthood (Barrett & Bateson, 1978).

Neuroscientists in search of the "fountain of youth" or the "ludic cocktail"—that is, pharmaceuticals that may enhance and maintain playfulness and the associated exhilaration—found that psychostimulants such as amphetamines, which invigorate exploratory activities, markedly reduce rather than enhance play behaviors (Beatty, Dodge, Dodge, White, & Panksepp, 1982). This suggests that the involvement of very complex and diverse neural networks in different types of play makes playfulness a pretty robust emotional function, not open to chemically induced excesses. Nevertheless, people can enhance play by creating environments that elicit play behaviors. The entertainment industries explore this avenue.

Although pleasurable, playfulness is not completely risk-free. "In the midst of play, an animal may gradually reach a point where true anger, fear, separation distress, or sexuality is aroused" and the player begins to process the situation in "more realistic and unidimensional emotional terms" (Panksepp, 1998, p. 283). Such shifts result in the loss of the play-related feelings of safety and joy. If we want play objects/events to perform their function (deliver learning and joy), it makes sense to hold the entertainment industries responsible for possible elicitation of fear and anger that spoil the fun and may trigger anti-social behaviors rather than the acquisition of useful behaviors.

The volatility of playfulness (risk to escalate into aggressive behavior) is due to the players' dependence on voluntary cooperation and observance of rules by all parties involved. A child may get overly excited, mishandle a toy, and hit a playmate, and a parent may get irritated and slap an unruly child. On the other hand, many adult sports and games, assumed to have evolved as institutionalized ways to dissipate the dominance drive and the "aggressive energies that might otherwise cause chaos in peaceful societies" (Panksepp, 1998, p. 286), can degenerate into dangerous fighting. Although real fighting and play differ in terms of behavioral structure "rules" and level of physical exertion (Pellis, 1988), professional players do "mean business," and that shuts down the play circuits in their brains. At high levels of violence among players, the play circuits in the supporters' brains shut down too, and the anger and rage circuits become activated. That mental state prompts the audience to get involved in actual fighting with other audience members, referees, players, and/or order-keeping personnel. Thus games may turn from entertainment into real drama or tragedy. This fragility of play emphasizes the importance and desirability of adding cortical activities to games (e.g., complex rules and such tasks as risk assessment, attribution, self-control, decision making). Extensive involvement of the neocortex in play has the potential to enhance pleasure and safety through top-down control.

Sexual Love, Pair Bonding, and Offspring Nurturance. "Sex is not essential for the bodily survival of any individual member of a species, 'merely' for the survival of the species itself" (Panksepp, 1998, p. 228). Intense orgasmic pleasure has evolved to keep individuals constantly interested in an activity that does not otherwise benefit them directly. Feelings of lust "diminish only with age, stress, and illness" (p. 226).

Sex-related emotions have been a gold mine for entertainment producers for various reasons: They are universally shared, elicit high excitation, and have high dramatic potential due to intrinsic differences among the sexes. The biological male-female differences cause distinct self-referential valuation of environmental features and events, leading to often conflicting behavioral tendencies. Sex differences originate in the ontogenetic process of sexualization governed by hormones (testosterone and two closely related metabolic products, estrogen and dihydrotestosterone) that start controlling brain and body development while the baby is still in the womb and finalize sexual differentiation during puberty.

If we consider "male" and "female" as possible directions of development for both brain and body, the outcome may be one of four possible combinations: two common configurations (male body and brain, female body and brain) and two less frequent, androgynous configurations (male body and female brain, and female body and male brain). The actual sexual pattern of each individual depends on the timing and intensity of hormonal signals, and real patterns are never pure cases. The resulting sexual behavior at the species level is diverse, reflecting "biochemically determined gradients of brain and body masculinization and feminization" (Panksepp, 1998, p. 232), and leading to largely unpredictable socio-sexual dynamics that can make interesting entertainment stories.

In addition to this fuzzy sexual determination, a broad stable gender difference in goal orientation (due to anatomical reproductive differentiation) has a great potential for socio-sexual conflict: "Females are seeking companions who are powerful and willing to invest resources on their behalf, whereas males are swayed by youth and beauty" (Panksepp, 1998, p. 226). Aggravating this goal dissimilarity, evolution has built into the brain "the potential for social devotion and deception, which can serve to maximize reproductive success" (Panksepp, 1998, p. 226). Although sex-related deceptiveness is a feature of both sexes, it greatly enhances the dramatic potential of female pursuit of extra-sexual goals, making it one of the perennial topics in entertainment.

On the other hand, "male sexuality and aggression/assertiveness interact to a substantial extent in subcortical areas of the brain" (Panksepp, 1998, p. 229). This makes male sex-related violence and violence-related sexuality two major themes in entertainment. Temporal lobe areas (where aggression circuitry is concentrated) are more active in males, while cingulated areas (where nurturance and other social emotional circuitries are concentrated) are more active in females (Gur, et. al., 1995). Consequently, "males are more aggressive and power-oriented, while females are more nurturant and socially motivated" (Panksepp, 1998, p. 230). These facts are eternal fodder for entertainment and have produced the domineering, power-hungry, and belligerent male types, and the motherly, overly social, and gossipy female types.

Pleasurable erotic feelings in both sexes are critical in sustaining sexual activities. But orgasm rewards primarily males for their intense energy investment. Because egg fertilization is unrelated to female orgasm, and female sexual appetite is hormone-bound, females experience orgasmic pleasure less frequently and predictably. Neuroscientists speculated that female orgasm "is presently emerging in an evolutionary sense" to "help females identify males who have the right characteristics for social bonding and hence are likely to support the woman's future needs" (Panksepp, 1998, p. 244). This orgasm-related difference between sexes generated a few more topics for entertainment: female frigidity, orgasm faking by females, and orgasm obsession in males.

Unlike animal sexuality, human sexuality "has become strongly dissociated from immediate reproductive concerns" (Panksepp, 1998, p. 228) and more pleasure-oriented. The changing and often conflicting motivations and behaviors of people seeking recreational sex or marriage (pair bonding and child raising) have gained prominence in entertainment since societal norms have become more permissive and legitimized women's pursuit of no-strings-attached sexual gratification. In addition to the classical male Don Juan type and female professional sex provider, modern culture has given new luster to formerly minor stereotypes, the gigolo and the lustful vamp.

Offspring nurturance and sexual motivation "are partially independent but also intertwined in the brain" (Panksepp, 1998, p. 227). This biological setup enables parents to assume nurturing roles if needed, not automatically. With mammalian species, mothers are the primary caregivers because they are equipped to breastfeed the infants. Male hit-and-run tactics in sexual behavior are more likely when a single individual (the mother) can easily rear offspring successfully to reproductive age (Barash & Lipton, 2001). Male fidelity (long-term attachment) and the amount of male investment in pair bonding and offspring care vary vastly across species: Some species make no lasting bond, while others remain paired for life (Carter, DeVries, & Getz, 1995). The degree and forms of male involvement in offspring rearing vary more in the human species than in other mammalian species because of the enormous diversity of circumstances in different societies (e.g., wealth, age, education, health, social and cultural environment), which results in different needs for father contribution. Human males can be "trained to exhibit a high level of nurturance, but their care is rarely as natural or as intense a motive as it is for the mother" (Panksepp, 1998, p. 246). The negotiation of, bickering about, and fun or disaster of paternal activities related to childcare has constantly been a productive subject area in entertainment, reflecting cultural stereotypes as well as trends in cultural norms.

"Females generally exhibit greater hemispheric coordination, since their right and left lobes are integrated more extensively via the larger fiber connections of the corpus callosum" (Panksepp, 1998, p. 235). Consequently, females tend to use both hemispheres in speech while males tend to use only the left side of their brains (Shaywitz et al., 1995). This generally makes male speech more "rational," and female speech more "incomprehensible" to males as it shifts between rational and emotional arguments. The communication difficulties between the sexes are often played out in dramatic arts at this linguistic level.

To close this exemplification of male–female differences and their potential for entertainment, one more observation is in place. Sex "is not just a peripheral bodily need but a brain need that has profound consequences for each species. Thus it is not surprising that it is 'highly politicized' (Panksepp, 1998, p. 228) in the human species through the institution of marriage and the historic battle for equal rights waged by the physically weaker, less domineering (power-oriented), and less aggressive sex. It seems an irony of fate that women's rights movement be due to a biological accident: Tomboyishness in females was massively promoted by maternal injections of diethylstilbestrol, an estrogenic hormone that was given to pregnant mothers during the second trimester to prevent miscarriages in the 1940s and 1950s (Ehrhardt et al., 1985). The girls born in the 1940s were in their twenties in the 1960s. It is plausible that they had enough brain maleness and power-drive in their generation to mount aggressive action and shape a more equitable code of gender rights. That societal change also points to the sweeping role that cortical/rational activities related to sex can play.

According to Panksepp (1998), "the neural programs for sexuality are much more 'open' to higher mental influence in humans than in other species" (p. 239). This may be due to the biological male–female differences and the partial anatomo-physiological dissociation between the lust and the nurturing neural systems. Both differentiations make necessary behavioral adjustments that require more extensive and sophisticated analysis of circumstances, options, and decision-making. Although cortical activity can provide overriding principles of cultural control, the subcortical emotional circuits have preemptive power and "may still be decisive in the sexual quality of individual lives—the ability to sustain receptivity and potency and to have experiences of intimacy and pleasure" (Panksepp, 1998, p. 245). In spite of increasing education and cultural sophistication, the conflicting dynamics between lower-brain sexual drives and rational decision making still wreck one in every two marriages (Panksepp, 1998).

As far as entertainment is concerned, the turbulence of sexual and marital life is very "productive" because it maintains throughout people's lives a strong need to learn about sexual and family relations, preferably in a vicarious and playful way. Such social learning from entertainment fare certainly presents the risks of stereotype acquisition at the individual level and stereotype reinforcement at the societal level. The more "mainstream" the sex and family content of the media is, for commercial reasons, the stronger the stereotyping becomes. Another inconvenience is that the fun bias of entertainment media provides inaccurate rosy and caricatured pictures that cultivate in the audience maladaptive expectations. Young people's acting on unrealistic expectations in their real sexual and family life are prone to chronic dissatisfaction and cognitive dissonance. The currently hyper-sexed and conflict-ridden pop culture is unlikely to debilitate our societies or species, though: The winners of social encounters typically exhibit elevations in circulating testosterone, while losers exhibit declines (Booth & Mazur, 1998; Dabbs, Dabbs, & Mazur, 2002; Kreutz, Rose, & Jennings, 1972; Mazur & Booth, 1998; Mazur & Lamb, 1980). This genetically coded physiological adaptation reinforces social dominance behaviors and reproductive behaviors among social winners, which optimally serves the survival at the species level. In addition, maternal stress sets in motion internal neurochemical changes that tend to leave the brains of male offspring in their primordial female-like condition (Panksepp, 1998, p. 237), causing androgynism. Consequently, at times of "high social and environmental stress, increased levels of homosexuality may be adaptive by limiting reproduction that could be wasteful (Ward, 1984). Also, a focus on pleasure rather than a productive life among women manifested as consumption of opioids tends to demasculinize their male fetuses (Johnston, Payne, & Gilmore, 1992). Moreover, opiate addicts who consume strong drugs such as heroin, "report feeling an orgasmic rush, with a warm erotic feeling centered in the abdomen, when the drug hits their system" (Panksepp, 1998, p. 243), so they will have their surrogate sexual pleasure without jeopardizing the gene pool of our species. These considerations may be regarded as cynical and might convey a wrong drug-permissive message, but the severe negative effects of addiction that punish pleasure-greedy humans at the individual level (to be addressed in the next section of this chapter) should function as a deterrent, particularly if they are accurately depicted in the media and persistently publicized.

The Social Bond Versus Loneliness, Separation Anxiety, Panic, and Grief.   The "subtle feeling of social presence is almost undetectable, until it is gone" (Panksepp. 1998, p. 261). We feel normal and comfortable as long as we have the social support systems within which we were born. But the loss, especially unexpected, of people we have invested genetic effort in (i.e., our children) or those who have substantially supported us (i.e., parents, friends), plunge us into sorrow or grief. The latter "verges on panic in its most intense and precipitous forms" (p. 261). Milder and more persistent forms are experienced as separation anxiety, which is accompanied by "feelings of weakness and depressive lassitude" (p. 212), urges to cry, tightness in the chest, and the feeling of having a lump in the throat. The evolutionary role of this psychic pain is to teach humans the importance of the social support roles.

The neural system for separation anxiety emerged from the more primitive distress mechanisms that mediate feelings related to basic survival functions such as hunger, pain, and coldness, and activates circuits of primitive audio-vocal communication (Panksepp, 1998). The panic system involves the midbrain, the medial diencephalons (especially the dorsomedial thalamus), the ventral septal area, the preoptic area, and sites in the bed nucleus of the stria terminalis that play an important role in sexual and maternal behaviors. Crying is supported by a system of neurons situated at lower levels in the brain than the system that supports laughter (Black, 1982).

The neurotransmitters that are important in the control of social emotions, the elaboration of social attachments and the various forms of human love, both nurturant and erotic, are the endogenous opioids oxytocin and prolactin (Panksepp, Nelson, & Bekkedal, 1997). Without brain opioids, an individual tends to feel "psychologically weaker" (Panksepp, 1998, p. 285) and be more prone to experience separation distress.

In entertainment, the use of separation anxiety, panic, and grief associated with the loss of a loved person is similar to the use of fear and anger: It serves the purpose of dramatic buildup of excitation for the enjoyment of other pleasant features of the entertainment experience. Because the neural network of separation anxiety is broad and involves low-level circuits that trigger automatic responses, it is very reliable for producing vicarious distress in the audience. In comedy, the whole range of negative emotions caused by threats to or breakdowns in social bonding may be caricatured as neediness or clinginess, and ridiculed for their uncontrollable behavioral manifestations (e.g., crying, ineffective despondency).

Particularly relevant to entertainment are the "remarkable similarities between the dynamics of opiate addiction and social dependence" (Panksepp, 1998, p. 261). Under morphine, individuals experience "heightened social confidence, a feeling of psychological strength that emerges from the neurochemical correlates of social bonding" (p. 285). This accounts for the crucial role of social context in addictions, experienced as companionship (comfortable gregariousness) and "nurturing" environment (teaching gambling or drug consumption techniques).

As far as cognition is concerned, it is interesting to note the contrast between the relative closeness of the sorrow/panic system and the openness of the play system. This difference is related to the distinct utilities of the two systems: Sorrow and panic deliver mostly species-relevant knowledge whose acquisition could not be left to chance, whereas play is crucial for individual learning and offers a wide range of behavioral schemata from which individuals can choose. Entertainment belongs to the play class of human activities and has the potential to greatly expand knowledge about available behavioral options. In the evolution of the human brain, crying, which is linked to separation distress, preceded the ability to laugh, which is linked to play. Both laughter and crying are social bonding functions that help to discriminate friends and family from strangers (Panksepp, 1998, p. 288). The newer play/laughter function draws heavily on the creative potential of the neocortex and offers a wider range of manipulations in entertainment as compared to the few basic schemata that trigger crying. But laughter cues in entertainment may be problematic to the extent they require complex, culture-/learning-based cortical analysis: Some people may not invest enough thinking to decode humor and may not "get it."

Learning and Hedonic Escalation

Techniques for Hedonic Enhancement.

The autonomic system energizes and monitors two types of functions: routine maintenance processes, for which resource deployment is minimized, and processes that support emergency activities, for which resource deployment needs to be maximized (e.g., fight-or-flight response, sexual intercourse).

Autonomic activities surface in consciousness typically in relation to deprivation (pain in case of deficiency, and pleasure, in case of remedied deficiency—e.g., pleasure of movement after immobility, pleasure of breathing fresh air after stifling in a crowded room). However, it is also possible to experience pleasure during energy-intensive activities (e.g., runner's high). The role of pleasure in such situations is to supplement resource investment beyond the automatic deployment encoded for the base level of that function.

Autonomic pleasure (associated with the five senses and basic activities such as eating, drinking, physical activity, and sexual activity) can be enhanced by sequencing deprivation and excess, repeating the sequence, simultaneously applying the schema to several autonomic processes, and further synergizing by means of combinations with other non-autonomic stimulations. The additional pleasure does not come without costs, though: Both deprivation and excesses throw the neatly coordinated autonomic system out of balance and reduce its functional efficiency and/or effectiveness (cause resource dissipation/waste). If practiced long term, they may lead to physical exhaustion and illnesses. Fortunately, some natural protection mechanisms have evolved to balance human greed for pleasure and to promote homeostasis (system stability).

Natural Protections Against Hedonic Excesses

Adaptation. The most basic safety mechanism is adaptation, that is, the decay of excitation at the level of peripheral receptors (Campbell, 1973). The pleasure seeker's strategies for circumventing adaptation are changing the sources of stimulation and searching for new sources. Each novel stimulus triggers fewer associations and is more readily displaced. The turnover rate of processing novel stimuli is faster, and that keeps a pleasure seeker bombarded by his/her senses, experiencing intense pleasure (Greenfield, 2000).

Hedonic Reversal. A second natural defense against stimulation excesses is hedonic reversal. At too high speed of change and excessive stimulus density, the processing capacity is overwhelmed and the hedonic quality of the experience is spontaneously reversed from pleasure to displeasure. "Pleasure shades into fear when the stimulation is just *too* fast and *too* novel" (Greenfield, 2000, p. 113). Pleasure seekers will therefore try to optimize, rather than maximize, arousal. Zillmann's (1988; Bryant & Zillmann, 1984) mood-management theory and Apter's (1994) reversal theory of enjoyment of violence endorse the notion that pleasure is a curvilinear function of arousal, with displeasure occurring at too high, or too low, levels of stimulation.

Another strategy for avoiding hedonic reversal is "sustaining pure pleasure" (Greenfield, 2000) by adding a "protective frame" (Apter, 1994, p. 9) that undercuts the fear reaction. Examples of "protected" experiences are sports, which exploit environmental dangers and control risk primarily through equipment (e.g., skydiving with a parachute); games, which exploit dangers associated with personal interactions and control risk primarily through rules and equipment; and spectacles, which exploit all possible dangers and limit risk through environment artificiality (controllability) and the indirect (vicarious) nature of the experience that gives the spectator a choice between empathy and detachment.

Habituation. A third natural defense against excessive stimulation is habituation, a "safety cognition" that develops through repeated stimulation experiences. If an unpleasant (potentially dangerous) stimulus has not involved aggravation (in terms of harmful effects) in prior experiences, then the brain will decide not to pay attention simply because there is nothing to worry about (Campbell, 1973). Pleasure seekers' strategy to overcome this natural defense is beating memory on its own ground by increasing stimulus intensity (which heightens arousal and strengthens the memory trace) and by repetition (which develops chronic accessibility). A typical example is that of a teenager playing the same music over and over again, and turning up the volume higher and higher: This keeps giving him/her pleasure, but it also prevents the neighbors from habituating. At the market level, the anti-habituation battle is fought season by season with new products and trends launched by the entertainment industries.

Choice, Learning, and Hedonic Optimization

A human being has about 10^6 genes that contain the "programmed" survival activity of the species. For activities that are not autonomic, a mechanism of choice (individual latitude) has developed to handle the increasing richness of pleasure opportunities historically produced by social life. Our choices are instrumented by a system of 10^{11} neurons with possibilities of connecting through 10^{15} synapses (Greenfield, 2000). We can exercise choice (respond in nonautomatic ways) only in situations of which we are conscious. The more arousing a situation and the more frequently it is repeated, the more extensive is the constellation of neurons it activates in our brains, and the deeper our consciousness of it.

Each activity has a hedonic component, which means that the neural constellation activated during the experience of that activity includes neurons located in the pleasure areas. Repeated activation through similar experiences builds meaning into our experience of the world by means of quasi-permanent connections among neurons (preferred pathways) that personalize our brain (Greenfield, 2000). The most sophisticated part of memory is supported by the prefrontal cortex that individualizes a brain by means of time and space referencing, which is essential for an individual's history. The contextualization of experiences makes possible the recognition and choice of situations (rather than isolated stimuli) that provide similar pleasures. Theoretically, this can serve both hedonic optimization and homeostasis. Practically, this linkage makes us prisoners of our gradually stabilizing (closing) universe of pleasures. Responsible for this unfortunate tendency is the natural decay of our memory (i.e., deactivation of unused links), which is an efficiency bias that pushes us toward extreme and easy (sensorial) sources of pleasure and makes us disregard other activities that have either lower hedonic potential or have higher hedonic potential but from which it is more difficult to derive pleasure.

What happens when pleasure more intense than that naturally provided by ordinary life experiences becomes available? Experiments with intracranial self-stimulation revealed compulsive pleasure-seeking behavior that preempts other activity. The most dramatic effects are related to recreational drug consumption. "The analysis of subjective responses to psychostimulants such as cocaine and amphetamines suggests that an energized psychic state accompanies arousal of the seeking system" (Panksepp, 1998, p. 149). These drugs induce a highly energized state of psychic power, increase engagement with the world, and enhance an individual's ability to pursue various goal-directed activities. Frantic activity artificially sustained by drugs eventually causes serotonin depletion that aggravates behavioral problems[5]: Affected individuals may become behaviorally disinhibited, "more aggressive, hypersexual, and generally exhibit more motivational/emotional energy" (p. 141). Because the activities involved in such behaviors are generally controlled by neural networks that are situated lower in the brain and are activated, largely, automatically, cortical control becomes increasingly unlikely. Therefore, such drug-sustained behavior usually leads to progressive decline of personal health through energy depletion, and to the deterioration of social relations through impulsivity (weakened logic and diminished self-censorship).

"Bad" Entertainments: Ways to Overcome Displeasure

The relationship between pleasure and displeasure is critically important in the process of hedonic maximization or escalation. Let us consider an isolated experience of simultaneous feelings of pain and pleasure. A trivial pain will attenuate through adaptation and, if not completely eliminated from consciousness, will tend to be habituated. If the pain is nontrivial, it will trigger what has traditionally been called the instinct of self-preservation or fear and avoidance of pain. The outcome will be avoidance behavior that will momentarily prevail over our general tendency to pursue pleasure. Thus the activity that caused simultaneous pleasure and nontrivial pain will be discontinued. If the same activity is repeated, even nontrivial pain will tend to be habituated if it does not aggravate or develop alarming effects. The brain will cease to pay attention to those components of the stimulation that cause stable pain. The neural constellation corresponding to the experience of that activity will lose pain connections, which will shift the individual's behavioral priority away from self-preservation toward pleasure maximization, "liberating" him/her for enjoyment (e.g., football players' insensitivity to severe bodily injuries while happily engaged in their games; moviegoers' enjoyment of violent or horror movies; gamblers' disregard of financial loss and deterioration of family relations). One problem associated with this type of practices is the amount of harm an individual can incur without being aware, because the harm is eliminated from his/her consciousness through habituation. Another problem is the projection of the individual's personal harm (incurred for pleasure) onto the social structure to which he/she belongs, through modified perception, values, and behavior (e.g., tolerance to drug consumption, risky, violent, and antisocial behaviors, promiscuous relations, bad language).

A more sophisticated exploitation of habituation for the purpose of pleasure maximization is that in which pleasure is not co-present in reality but merely imagined, or evoked from memory. Such imagination may naturally develop through repetition of specific sequences of painful and pleasurable activities: During the unpleasant activity, the brain may anticipate the pleasure expected to follow. This "fore-pleasure" (Bousfield, 1926/1999, p. 84) is enjoyed, and the pain of the current activity is habituated. This model would cover pleasures associated with "perverse" practices such as sadomasochism. The problems associated with these practices based on anticipation and memory are more serious. At the individual level, the activation of neural pathways through imagination mobilizes top–down control (Hobson, 1994) by the neocortex. Cortically controlled pleasure-seeking behaviors are hedonically more efficient than haphazard trial-and-error behavioral patterns because the coordination among brain and body activities tends to produce a coherent and stable set of hedonic practices at the individual level. Although perverse practices are naturally avoided and socially discouraged because they

---

[5]One function of serotonin is to "sustain stability in perceptual and cognitive channels." A mild reduction in brain serotonin activity loosens up rational control and inertia, which allows for the generation of new insights and ideas. A sustained reduction of serotonin might lead to "chaotic feelings and perceptions, contributing to feelings of discoherence and mania" (Panksepp, 1998, p. 142).

consistently cause nontrivial harm, their adopters tend to put pressure on the social system to accept and legitimize such activities (remove restrictions and sanctions).

A different class of hedonic behaviors, originally described for drug consumption, includes practices that combine initial lack of pleasure with subsequent co-presence of pain and pleasure. Even repeated experiences with such activities by individuals in isolation would have a negative hedonic balance that would discourage adoption. But individuals do not live in isolation, and social/peer pressure plays a major role in cultivating such practices.

Hirsch, Conforti, and Graney (1998) revamped Becker's (1953) three-stage process theory of drug consumption, applicable to this entire class. The theory posits that in order for individuals to experience pleasure, they need to develop a conception of the means (drug, sexual practice, etc.) as a source of pleasure. The construction of that concept (the means–pleasure neural linkage in memory) involves learning the technique (in the case of drugs, quantities and administration procedures), experiencing and recognizing the effects, and learning to enjoy the effects (i.e., habituating to initially unpleasant feelings and anticipating subsequent pleasures). This learning process requires initiation by more "advanced" practitioners and encouragement to overcome the pain during the habituation process.

This technique of pleasure escalation that requires "initiation" is the most aggressive. At the individual level, it may be physically harmful. At the societal level, curious, outgoing, and gregarious people (as most youth are) have abundant opportunities to experiment. If the trials prove effective, very intense sensorial stimulation floods consciousness with fleeting "pure pleasures" (Greenfield, 2000, p. 116), and the high turnover of small neural constellations precludes the development of links to higher levels of the brain that support contextualization and control functions. Very simple sensorial stimulation produces standard neural activation that makes group members feel more "alike" and "together." This feeling is conducive to uncontrolled herd behavior. In the public arena, cohesive groups of pleasure seekers develop that tend to proselytize and propagandize for their preferences. Larger groups gain market and political power and begin to shape social life from values to consumption patterns and distribution of wealth.

If this train of thought seems far-fetched, and the danger of pleasure seeking appears to be exaggerated, let us consider the felicitous situation of people who's work is pleasure. Here is what a West Point graduate has to say: "I've been around the world, I've jumped out of planes, jumped out of helicopters, got to play with explosives [...]. If what I'm doing is an adventure, if I'm having a good time, I'll stick with it. If it's not fun anymore, I'll get out" (Lipsky, 2003, pp. 67–68). This young man's victims and his parents are not as happy, and he may not be lucky to get bored and quit. How much does a career of utmost excitement cost a person and society?

ACKNOWLEDGEMENTS

I wish to express my special gratitude to Jaak Panksepp, whose treatise on *Affective Neuroscience* was a valuable source of neurological information for this chapter and suggested a framework for an integrative assessment of entertainment as cognitive exploitation of emotions for pleasure enhancement.

REFERENCES

Aggleton, J. P. (Ed.). (1992). *The amygdala: Neurobiological aspects of emotion, memory, and mental dysfunction.* New York: Wiley.
Apter, M. J. (1994, October). *Why we enjoy media violence: A reversal theory approach.* Paper presented at the International Conference on Violence in the Media, St. John's University, New York, NY.

Barash, D. P., & Lipton, J. E. (2001). *The myth of monogamy.* New York: Freeman.

Barrett, P., & Bateson, P. (1978). The development of play in cate. *Behaviour, 66,* 106–120.

Beatty, W. W., Dodge, A. M., Dodge, L. J., White, K., & Panksepp, J. (1982). Psychomotor stimulants, social deprivation, and play in juvenile rats. *Pharmacology, Biochemistry, and Behavior, 16,* 417–422.

Becker, H. S. (1953). Becoming a marijuana user. *American Journal of Sociology, 59,* 235–242.

Black, D. (1982). Pathological laughter: A review of the literature. *Journal of Nervous and Mental Disorders, 170,* 67–71.

Booth, A., & Mazur, A. (1998). Old issues and new perspectives on testosterone research. *Behavioral and Brain Sciences, 21,* 386–390.

Bousfield, P. (1999). *Pleasure and pain: A theory of the energic foundation of feeling.* London: Routledge. (Original work published 1926)

Bower, G. H. (1981). Mood and memory. *American Psychologist, 36,* 129–148.

Bower, G. H., & Cohen, P. R. (1982). Emotional influences in memory and thinking: Data and theory. In M. S. Clark & S. T. Fiske (Eds.), *Affect and cognition* (pp. 291–332). Hillsdale, NJ: Lawrence Erlbaum Associates.

Bryant, J., & Zillmann, D. (1984). Using television to alleviate boredom and stress: Selective exposure as a function of induced excitational states. *Journal of Broadcasting, 28,* 1–20.

Campbell, H. J. (1973). *The pleasure areas: A new theory of behavior.* New York: Delacorte.

Carter, S. C., DeVries, A. C., & Getz, L. L. (1995). Physiological substrates of mammalian monogamy: The prairie vole model. *Neuroscience and Biobehavioral Reviews, 19,* 303–314.

Christianson, S.-A. (Ed.). (1992). *The handbook of emotion and memory: Research and theory.* Hillsdale, NJ: Lawrence Erlbaum. Associates.

Clark, M. S., & Isen, A. M. (1982). Towards understanding the relationship between feeling states and social behavior. In A. H. Hastorf & A. M. Isen (Eds.), *Cognitive social psychology* (pp. 73–108). New York: Elsevier.

Coccaro, E. F. (1996). Neurotransmitter correlates of impulsive aggression in humans. *Annals of the New York Academy of Sciences, 794,* 121–135.

Dabbs, J. M., Dabbs, M. G., & Mazur, A. (2002). Heroes, rogues, and lovers: Testosterone and behavior. *Contemporary Psychology, 47,* 275–276.

Ehrhardt, A. A., Meyer-Bahlburg, H. F. L., Rosen, R. L., Feldman, J. F., Veridiano, N. P., Zimmerman, I., & McEwen, B. S. (1985). Sexual orientation after prenatal exposure to exogenous estrogen. *Archives of Sexual Behavior, 14,* 57–78.

Ferris, C. F., & Grisso, T. (Eds.). (1996). Understanding aggressive behavior in children. Special issue of *Annals of the New York Academy of Sciences, 794.*

Forgas, J. P. (1999). Network theories and beyond. In T. Dalgleish & M. J. Power (Eds.), *Handbook of cognition and emotion.* New York: Wiley.

Gerbner, G. (1969). Toward "Cultural Indicators": the analysis of mass mediated message systems. *Communication Review, 17*(2), 137–148.

Gerbner, G. (1970). Cultural indicators: The case of violence in television drama. *Annals of the American Academy of Political and Social Science, 388,* 69–81.

Gerbner, G. (1971). Violence in television drama: Trends and symbolic functions. In G. A. Comstock & E. A. Rubinstein (Eds.), *Television and social behavior. Vol. 1. Content and control.* Washington: Government Printing Office, pp. 28–187.

Gerbner, G. (1972). Communication and social environment. *Scientific American, 227*(3), 152–160.

Gray, J. A. (1987). *The psychology of fear and stress.* New York: Cambridge University Press.

Greenfield, S. (2000). *The private life of the brain: Emotions, consciousness, and the secret of the self.* New York: Wiley.

Gur, R. C., Mozley, L. H., Mozley, P. D., Resnick, S. M., Karp, J. S., Alavi, A., Arnold, S. E., & Gur, R. E. (1995). Sex differences in regional cerebral glucose metabolism during a resting state. *Science, 267,* 528–531.

Hirsch, M. L., Conforti, R. W., & Graney, C. J. (1998). The use of marijuana for pleasure: A replication of Howard S. Becker's study of marijuana use. In J. A. Inciardi & K. McElrath (Eds.), *The American drug scene: An anthology* (2nd ed., pp. 27–35). Los Angeles, CA: Roxbury.

Hobson, J. A. (1994). *The chemistry of conscious states: How the brain changes its mind.* New York: Little, Brown & Co.

Hoebel, B. G. (1988). Neuroscience and motivation: Pathways and peptides that define motivational systems. In R. C. Atkinson, R. J. Herrenstein, G. Lindzey, & R. D. Luce (Eds.), *Steven's handbook of experimental psychology* (pp. 547–626). New York: Wiley.

Isen, A. M. (1984). Toward understanding the role of affect in cognition. In R. S. Wyer & T. K. Srull (Eds.), *The handbook of social cognition* (Vol. 3, pp. 179–236). Hillsdale, NJ: Lawrence Erlbaum Associates.

Izard, C. E., & Buechler, S. (1979). Emotion expressions and personality integration in infancy. In C. E. Izard (Ed.), *Emotions in personality and psychopathology* (pp. 447–472). New York: Plenum.

Johnston, H. M., Payne, A. P., & Gilmore, D. P. (1992). Perinatal exposure to morphine affects adult sexual behavior of the male golden hamster. *Pharmacology Biochemistry and Behavior, 42*, 41–44.

Kreutz, L. E., Rose, R. M., & Jennings, J. R. (1972). Suppression of plasma testosterone levels and psychological stress. *Archives of General Psychiatry, 26*, 479–483.

Lipsky, D. (2003). *Absolutely American: Four years at West Point*. Boston, MA: Houghton Mifflin.

MacLean, P. (1990). *The triune brain in evolution*. New York: Plenum Press.

Mazur, A., & Lamb, T. A. (1980). Testosterone, status, and mood in human males. *Hormones and Behavior, 14:* 236–246.

Mazur, A., & Booth, A. (1998). Testosterone and dominance in men. *Behavioral and Brain Sciences, 21*, 353–363.

Panksepp, J. (1990). The psychoneurology of fear: Evolutionary perspectives and the role of animal models in understanding anxieties. In G. D. Burrows, M. Roth, & R. Noyes (Eds.), *Handbook of anxiety. Vol. 3. The neurobiology of anxiety* (pp. 3–58). Amsterdam: Elsevier.

Panksepp, J. (1998). *Affective neuroscience: the foundations of human and animal emotions*. New York: Oxford University Press.

Panksepp, J., Nelson, E., & Bekkedal, M. (1997). Brain systems for the mediation of social separation-distress and social-reward: Evolutionary antecedents and neuropeptide intermediaries. *Annals of the New York Academy of Sciences, 807*, 78–100.

Pellis, S. M. (1988). Agonistic versus amicable targets of attack and defense: Consequences for the origin, function, and descriptive classification of play-fighting. *Aggressive Behavior, 14*, 85–104.

Puglisi-Allegra, S., & Oliverio, A. (Eds.). (1990). *Psychology of stress*. Dordrecht: Kluwer.

Shaywitz, B. A., Shaywitz, S. E., Pugh, K. R., Constable, R. T., Skudlarski, P., Fulbright, R. K., Bronen, R. A., Fletcher, J. M., Shankweller, D. P., Katz, L., & Gore, J. C. (1995). Sex differences in the functional organization of the brain for language. *Nature, 373*, 607–609.

Slade, A., & Wolf, D. P. (Eds.) (1994). *Children at play*. New York: Oxford University Press.

Valenstein, E. (1973). *Brain control*. New York: Wiley.

Vanderwolf, C. H. (1992). Hippocampal activity, olfaction, and sniffing: An olfactory input to the dentate gyrus. *Brain Research, 593*, 197–208.

Ward, I. L. (1984). The prenatal stress syndrome: Current status. *Psychoneuroendocrinology, 9*, 3–11.

Wise, R. A. (1982). Neuroleptics and operant behavior: The anhedonia hypothesis. *Behavioral Brain Science, 5*, 39–87.

Zillmann, D. (1971). Excitation transfer in communication-mediated aggressive behavior. *Journal of Experimental Social Psychology, 7*, 419–434.

Zillmann, D. (1988). Mood management: Using entertainment to full advantage. In L. Donohew, H. E. Sypher, & E. T. Higgins (Eds.), *Communication, social cognition, and affect* (pp. 147–171). Hillsdale, NJ: Lawrence Erlbaum Associates.

Zillmann, D. (1999). Exemplification theory: Judging the whole by some of its parts. *Media Psychology, 1*, 69–94.

APPLICATION OF PSYCHOLOGICAL THEORIES AND MODELS TO ENTERTAINMENT THEORY

21

Sensation Seeking in Entertainment

Marvin Zuckerman
University of Delaware

"Novelty is always the condition of enjoyment"

—(Freud, 1922/1955, p. 35)

As with many of Freud's pithy observations, the above one is only partly true. For most persons maximal enjoyment of sensory experiences lies somewhere between familiarity and novelty. The balance of preference is in part a function of the trait of sensation seeking defined as: " ...the seeking of varied, novel, complex and intense sensations and experiences, and the willingness to take risks for the sake of such experience." (Zuckerman, 1994, p. 27). The last part of the definition, willingness to take ...risks, is not relevant in this chapter, most of which concerns purely sensory or vicarious experience that entails no risks.

The concept of sensation seeking arose from experiments in sensory deprivation (Zuckerman, 1969). Subjects in these experiments were put in a sound-proof room in total darkness with further restrictions on touch and movement. Naturally the stress increased with time in these conditions. But a control condition, in which the same subjects were confined in the same room but without much variety in stimulation, was found to be equally stressful (Zuckerman, et al., 1966). Subjects had taken an early form of the Sensation Seeking Scale (SSS), (Zuckerman, Kolin, Price, & Zoob, 1964). Subjects scoring high on this SSS showed more increasing behavioral restlessness over time in both conditions compared with low scorers.

LABORATORY STUDIES

Novelty

The primary elements of sensation hypothesized to be positively motivational for high sensation seekers (novelty, intensity, complexity) have been investigated both in the laboratory and in

367

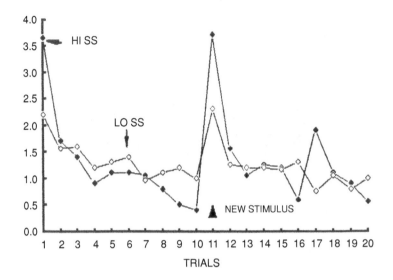

FIG. 21.1. Skin conductance responses (SCRs) to a simple (trials 1–10) and a complex (trials 11–20) visual stimulus. Data from Neary & Zuckerman (1976). Figure from "The psychophysiology of sensation seeking," by M. Zuckerman, 1990, *Journal of Personality, 58*, Fig. 1, p. 322. Copyright 1990 by Duke University Press. Reprinted by permission.

prefererences in televison, film, art, music, and humor. Laboratory studies have often used psychophysiological measures of arousal, interest, or fear. The orienting reflex (OR) is a measure of arousal and interest related to attention to a stimulus. It can be triggered by any novel object appearing in a perceptual field. The reaction habituates with each stimulus repetition after the first presentation. We may continue to be aware of the stimulus but attention to it and arousal by it diminish with each repetition. Apart from novelty, the OR may be enhanced by the intensity of the stimulus or the emotional association of the stimulus content.

Neary and Zuckerman (1976) presented subjects with a simple visual stimulus, consisting of a rectangle of light, measuring the skin conductance response (SCR) to the stimulus. The stimulus was then repeated nine more times at randomly determined but short intervals. After the tenth presentation a new stimulus, consisting of a complex colored design, was presented without warning and then this stimulus was repeated nine times. The SCRs to the 20 stimulus presentations are shown in Figure 21.1.

High sensation seekers, as indicated by scores on the General SSS, had a stronger OR (SCR) to the first stimulus presentation than low sensation seekers, but on the subsequent presentations of the same stimulus they dropped to the habituated levels of the lows. On the presentation of a new (novel) stimulus the high sensation seekers' ORs again exceeded those of the low sensation seekers, but as with the previous stimulus there were no differences in SCR on subsequent presentations.

Some subsequent studies were able to replicate this result, but others were not. Smith, Perlstein, Davidson, and Michael (1986) replicated the results using auditory stimuli (tones). The high sensation seekers had a stronger SCR to the first presentation of a tone, but differences between highs and lows disappeared on subsequent presentation.

Smith et al. also used a series of auditorily presented words, and another series of visually presented objects or activities using slides and video tapes. Each category of stimuli had a subset of stimuli of neutral content (e.g., a landscape or clock) or one loaded by content that was of greater interest to sensation seekers (e.g., boxing, mountain climbing). Both high and low sensation seekers had a larger SCR for loaded than for neutral words but the difference

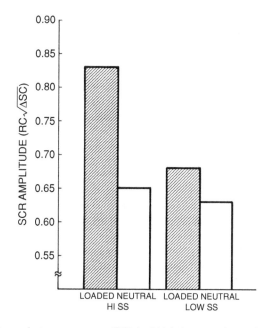

FIG. 21.2. Skin conductance responses (SCRs) of high low sensation seekers to neutral and loaded words on first presentations. From "Sensation seeking: Differential effects of relevant, novel stimulation on electrodermal activity," by B. D. Smith et al., 1986, *Personality and Individual Differences, 7*, Fig. 4, p. 449. Copyright 1986 by Pergamon Press. Reprinted by permission.

between the two categories was signficantly larger for the high than for the low sensation seekers (Fig. 21.2). This interaction effect was particularly significant for the initial presentation of each type of word. The response to the videotapes showed a similar type of interaction. The OR reaction to novelty in the high sensation seeker is enhanced by content related to their interests.

INTENSITY

Smith and his colleagues investigated the effect of the intensity of stimulus content on the OR (Smith, Davidson, Smith, Goldstein, & Perlstein, 1989). They used tape recorded words classified as low, medium, or high in intensity of sexual and aggressive content. Figure 21.3 shows the SCRs of high and low sensation seekers on first presentations of words classified by intensity. Amplitude of SCRs increased as a function of intensity for both high and low sensation seekers but the increase was greater for the high sensation seekers. In fact there was no difference between the groups in response to the low intensity words, some difference to the medium intensity, and a marked difference in response to the high intensity words.

The problem in using the SCR as a measure of OR, or indication of interest, is that the SCR cannot differentiate between OR and defensive reflex (DR) or startle reflex (SR). DRs are associated with negative emotional responses, such as fear or shock. Heart rate (HR) reactions offer a way of differentiating ORs from DRs because they are biphasic, that is, they can show a decelerating or accelerating pattern in response to a stimulus. The former is characteristic of an OR and is elicited in response to first presentations of low-to-moderate intensities of stimuli. The latter (acceleration) is characteristic of a DR (or a SR) and is elicited by high intensity stimuli.

Orlebeke and Feij (1979) used an auditory stimulus of high moderate intensity (80-dB) and measured change in HR in the ten seconds following stimulus presentation. The subjects were

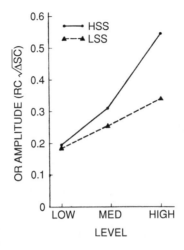

FIG. 21.3. Skin conductance responses (SCRs) of high and low sensation seekers to low-, medium-, and high-intensity sexual and aggressive words on first presentations. From "Sensation seeking and arousal: Effects of strong stimulation on electrodermal activation and memory task performance," by B. D. Smith et al., 1989, *Personality and Individual Differences, 10,* Fig. 1, p. 674. Copyright 1989 by Pergamon Press. Reprinted by permission.

high or low scorers on the disinhibiton (Dis) subscale of the SSS. Figure 21.4 shows the HR changes from prestimulus levels over the first three trials. Actually major differences between the groups were found on the first exposure and by the third exposure responses were minimal (habituation) in both groups. The low Dis group showed a marked HR acceleratory pattern (DR or SR) whereas the high Dis group showed a HR deceleratory response (OR).

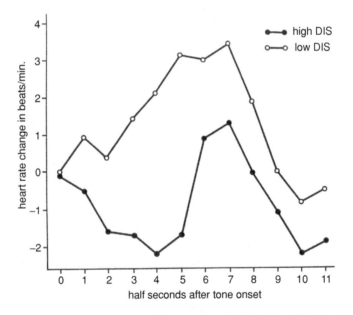

FIG. 21.4. Heart rate response of high and low scorers on the disinhibition (Dis) subscale of the Sensation Seeking Scale (SSS) averaged over the first three presentations of an 80dB tone. From "The orienting reflex as a personality correlate," by J. F. Orlebeke & J. A. Feij, 1979. In H. D. Kimmel, E. H. van Olst, & J. F. Orlebeke (Eds.), *The orienting reflex in humans,* Fig 33.1, p. 579. Hillsdale, NJ: Lawrence Erlbaum Associates. Copyright 1979 by Lawrence Erlbaum Associates. Reprinted by permission.

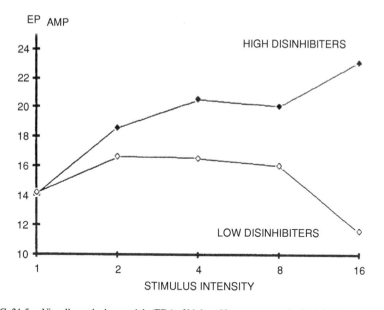

FIG. 21.5.   Visually evoked potentials (EPs) of high and low scorers on the Disinhibition subscale of the Sensation Seeking Scale (SSS) as a function of stimulus intensity. From "Sensation seeking and cortical augmenting-reducing," by M. Zuckerman et al., 1974, *Psychophysiology, 11,* p. 539. Copyright 1974 by the Society for Psychophysiological Research. Reprinted by permission.

Zuckerman, Simons, and Como (1988) varied the stimulus intensities as well as comparing high and low Dis subjects. At the lowest intensity of 50dB most subjects showed HR deceleration, but the high Dis showed a stronger OR than the low Dis subjects on the first presentation only. At the highest intensity of 95dB most subjects showed an acceleratory pattern but the low Dis subjects showed a stronger acceleration (DR) than the high Dis subjects.

These experiments show a novelty-intensity interaction with sensation seeking since the group differences only occur on the first or second stimulus presentations. The cortical evoked potential (EP) also varies as a function of intensity, but is less subject to habituation than electrodermal and heart rate responses.

Buchsbaum and Silverman (1968, also Buchsbaum, 1971) developed the method called EP augmenting-reducing (A-R). The EP method presents different intensities of stimuli, and measures the amplitudes of an early EP component in response to the different stimulus intensities. An augmenting pattern is defined by increasing EP amplitudes as a direct function of stimulus intensity. A reducing pattern is defined by little intensity dependence of EP on stimulus intensity, or an actual decrease in intensity at the highest intensities of stimulation. The slope of the intensity-amplitude relationship is sometimes used as the measure of A-R. This measure tends to be normally distributed.

Zuckerman, Murtaugh, and Siegel (1974) related the visually evoked A-R to sensation seeking. A high correlation ($r = .59$) was found between augmenting and the disinhibiton (Dis) scale of the SSS. Figure 21.5 contrasts the A-R reactions of those high and low on the Dis scale. The high Dis subjects showed increasing EP amplitudes with increases in stimulus (flashing lights) intensities, whereas the low Dis subjects show little or no increase in EPs over the first 4 intensities and a significant decrease in response to the highest intensity o light.

Various investigators extended the A-R paradigm to auditory evoked potentials. Figure 21.6 shows the results of a study by Zuckerman, Simons, and Como (1988). High Dis subjects showed an augmenting pattern and low Dis subjects showed a reducing pattern. The results have been well-replicated, particularly for the auditory EP (Zuckerman, 1990)

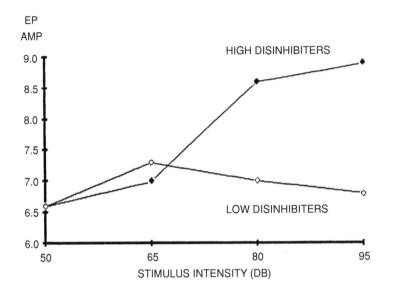

FIG. 21.6.   Auditory evoked potentials (EPs) of high and low scorers on the Disinhibition subscale
of the Sensation Seeking Scale (SSS) as a function of stimulus intensity for the short interstimulus
intervals (2 seconds). From "Sensation seeking and stimulus intensity as modulators of cortical,
cardiovascular, and electrodermal response: A cross-modality study," by M. Zuckerman et al.,
1988, *Personality and Individual Differences, 9*, Fig. 6, p. 368. Copyright 1988 by Pergamon Press.
Reprinted by permission.

We believe that EP augmenting signifies a cortical capacity to process and respond to
the higher intensity ranges of stimulation, whereas reducing indicates a protective mechanism dampening cortical response at high intensities. The auditory A-R pattern is particularly relevant to music preferences as will be discussed in a later section of this
chapter.

Complexity

Zuckerman, Neary, and Brustman (1970) found that high sensation seekers scored higher than
low sensation seekers on preferences for complex designs on the Welsh (1959) Figure Preference test. In a subsequent study the design preferences of high and low sensation seekers
were compared (Zuckerman, Bone, Neary, Mangelsdorf, & Brustman, 1972). The designs
preferred more by high sensation seekers are shown in Figure 21.7 and those preferred relatively more by low sensation seekers are found in Figure 21.8. Highs prefer designs that are
complex, assymmetrical, and suggestive of movement. Lows prefer designs that are simple
and symmetrical. These results have obvious relevance to art preferences discussed in a later
section.

MEDIA PREFERENCES

"Entertainment" generally refers to passive, vicarious types of activities, rather than active
types, for instance engagement in sports, social interactions, aggression, and sex. Unlike active
engagement, there is no risk in vicarious activity. The main categories to be discussed here are
television, film, music, art, humor, and vacation preferences.

FIG. 21.7. Designs liked more by high sensation seekers. From "What is the sensation seeker? Personality trait and experience correlates of the Sensation Seeking Scales," by M. Zuckerman et al., 1972, *Journal of Consulting and Clinical Psychology, 39*, p. 317. Copyright by American Psychological Association, 1972. Reprinted by permission.

Television and Film

Television and film have a greater capacity to increase involvement and arousal than simple auditory or textual materials because of the involvement of two senses and the immediacy of the experiences. Television has been labeled as the "cooler" medium because the limited screen size (until recently) provided a more detached experience than the large screen film viewed in darkness. Historically there has been more restriction on the content of television. Sexual and very aggressive "horror/slasher" shows were confined to the movie screen. Now, however, these films are shown on television although the more explicit sexual and bloody parts are usually edited out or cut in duration. Although simulated sex has become commonplace in film, there seems to be much more concern about aggression and sadism because of the debate over the role of such portrayals as causes for actual aggressive behaviors. Are aggressive persons simply drawn to these themes, or do the films, television programs, and video-games act as instigators of violent behavior? Is the aggressive trait the only explanation for preferences or can the simple need for arousal and novel stimulation (sensation seeking) explain the taste for violence in the media?

Schierman and Rowland (1985) asked college students to rate their typical participation in various types of entertainments. In both men and women, SSS scores correlated positively with reading "X"-rated magazines and in the women with going to "X"-rated (sexually explicit) movies. The attendance of sexually explicit movies by high sensation seekers was also found in other studies (Brown, Ruder, Ruder & Young, 1974; Zuckerman & Litle, 1986). However, in the men interests were not confined to TV and movies; positive correlations were also found with news magazines, non-fiction books, and TV news reports. Negative correlations were

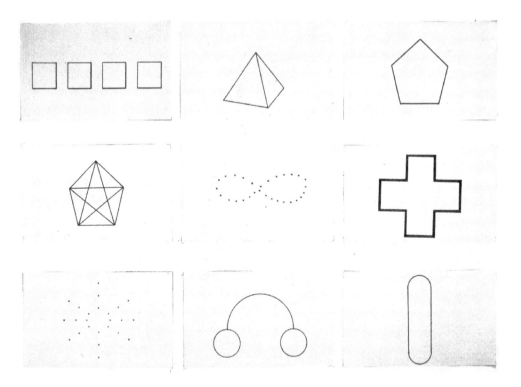

FIG. 21.8. Designs liked more by low sensation seekers. From "What is the sensation seeker? Personality trait and experience correlates of the Sensation Seeking Scales," by M. Zuckerman et al., 1972. *Journal of Consulting and Clinical Psychology, 39,* p. 318. Copyright by American Psychological Association, 1972. Reprinted by permission.

found with musical movies and romantic fiction. College women preferred active experience, pubs, lounges, and night-clubs with or without entertainment, and attending rock concerts. Low sensation seeking women liked the theater.

Rowland, Fouts, and Heatherton (1989) found that high sensation seekers watched less TV than medium- and low groups overall, and particularly on Friday afternoons and evenings, the pub and party time at their university. However Perse (1996), using an older non-college sample, found no relationship between TV viewing time and sensation seeking. High sensation seeking was related positively to preferences for action-adventure and music programs, and negatively to preferences for game-shows and news programs. Aluja-Fabregat and Torrubia-Beltri (1998) also found a preference for violent action films related to sensation seeking, particularly on the disinhibition subscale.

In an actual laboratory situation with a forced choice among programs, but with permission to switch channels whenever they liked, both male and female high sensation seekers preferred the action film to the others (Schierman & Rowland, 1985). But the most striking finding was the high positive correlations between the number of channel switches and SSS scores (.69 for men and .54 for women).

This last finding is an example of the need for variety in high sensation seekers. Variety is the need for change in stimulation, even if the stimulation is not completely novel. The variety of drugs used (Zuckerman, 1983, 1994), and the variety of sexual partners (Zuckerman, Tushup, & Finner, 1976) are also related to sensation seeking. Habituation to marriage works against monogamous fidelity.

High sensation seekers tend to leave the TV on while engaging in other activities, reading, eating, or even "cuddling" with a boy- or girlfriend (Perse, 1996; Rowland, et al., 1989). TV is used as a secondary source of stimulation, with shifting attention between it and the primary activity. Attention is alternated between the TV and the homework task in order to maintain an optimal level of arousal.

Perse (1996) examined the reasons for channel switching by high sensation seekers. They reported changing to avoid commercials, watch more than one show at a time, see what they are missing on another channel, and simply because they "get bored."

The studies by Smith and his colleagues on the effect of content of stimuli on the OR arousal showed that intense sexual and aggressive content produced more initial arousal in high sensation seekers and tend to maintain their arousal over more exposures than other types of stimuli (Smith et al., 1989). If such stimuli are more arousing for high sensation seekers they should be attracted to X-rated (pornographic) and aggressive-sadistic horror films. If such films were offered on TV, high sensation seekers say they would watch them and lows say they would avoid them (Rowland et al., 1989). Zuckerman and Litle (1986) found significant correlations between sensation seeking and scales expressing general curiosity about sexual events and curiosity about morbid events, as well as actual attendance at horror films of the "slasher" variety and X-rated sexual films. Aluja-Faberegat (2000) replicated the results on violence in media preferences using a young adolescent sample (13–14 year olds), adapting the general curiosity about morbid events scale for Spanish adolescents, and asking about attendance at violent films.

Both of the last two studies also included the Eysenck personality questionnaire (EPQ, Eysenck, Eysenck, & Barrett, 1985) and both found that the psychoticism (P) scale related to curiosity about violence in media and viewing of violent TV and movies in the same way as did sensation seeking. Psychoticism correlates highly with sensation seeking, particularly disinhibition and boredom susceptibility, and P is the best marker for the sensation seeking/impulsivity factor in factor analyses including both scales (Zuckerman, Kuhlman, & Camac, 1988; Zuckerman, Kuhlman, Thornquist, & Kiers, 1991). Extraversion and neuroticism scales showed little relationship to these variables. In the Spanish study, the combination of high P and high sensation seeking (disinhibition) accounted for the major variance in morbid curiosity. Bruggerman and Barry (2002) exposed high and low P scorers to 10 presentations of a violent video segment alternating with a comic video segment and recorded skin conductance level (SCL) continuously. The high P group preferred the violent presentations to the comic ones and perceived the violent video as more enjoyable and comical (!) than the the low P group. The high P subjects had a higher SCL on first exposures to the violence video, which habituated rapidly with repetition. The low P subjects showed a sensitzation (increase in SCL), rather than habituation, to the second presentation, which then habituated on subsequent trials.

Litle (1986, see description in Zuckerman, 1994) showed subjects a 20 minute segment of a gory horror movie, Friday the Thirteenth, and measured SCL continuously during the showing. SCRs varied with the content of the scenes, peaking during parts involving assaults or discovery of dead bodies, and subsiding between such scenes. Until almost the end of the movie, there were no differences between the high and low sensation seekers in their reactions to the film scenes. During the climactic scene the hero decapitates the mad killer and his head goes flying off with blood spurting everywhere. At this point there was an extreme peak in the SCRs of the low sensation seekers but little response in the high sensation seekers. Apparently the highs had been habituating during the film, whereas the lows were becoming more and more sensitized to the violence. Alternatively the low sensation seekers may have simply been more fear reactive to this highly intense image.

Given their negative emotional reactions to this type of film, it is easy to understand why low sensation seekers avoid it. But why do high sensation seekers enjoy them? Johnston

(1995) analyzed their self-reported sources of enjoyment. Four correlated factors were found: (1) sadism with admitted enjoyment of the bloody suffering of the victim; (2) thrills through vicarious emotion, or enjoyment of the suspense and fear engendered by identification with the victim; (3) self-enhancement by pride in the ability to control fear; and (4) compensation for problems, attempting to escape from negative feelings of loneliness, anger, and problems at home. The first two of these were both associated with high thrill and adventure seeking, but sadistic motivation was associated with low empathy, whereas thrill motivation correlated with high empathy. This is consistent with the differential identification with an aggressor in the former and with the victim in the latter.

Slater (2003) found that both sensation seeking and aggression predicted use of violent media by 14 year old adolescents. Frequency of internet use and gender were also predictive. Boys are more attracted to this type of media content than girls. Aluja-Fabregat and Torrubia-Beltri (1998), using a similar age group, found that boys watched violent cartoons and action oriented and violent films more than girls, and enjoyed them more. They also found violence on TV as more funny and thrilling than girls. Zuckerman and Litle (1986), using college age subjects, found that males had higher horror film attendence and more curiosity about morbid events in general than females. Men also scored higher on sensation seeking and the P scale, proven to be predictors of consumption of violent and horror films.

Harris et al. (2000) reported the reactions of males and females who went to horror movies on a date or in a group. During the movie more men felt amused and entertained while more women reported feeling jumpy or disgusted and even yelling or screaming in some parts. Women reported holding onto their date and more men reported feeling "sexually turned on." Arousal from fear may potentiate sexual arousal and increase bonding. A study of couples showed that shared participation in novel and arousing activities increased the experienced quality of their relationships (Aron, Norman, Aron, McKenna, & Heyman, 2000). Boredom was a mediating factor, reduced by joint participation in novel and arousing activities and negatively related to relationship quality. Assortative dating and mating is found for sensation seeking, that is, highs date and marry other highs, and lows are similarly attracted (Zuckerman, 1994). Preferences in entertainment may be a factor in attraction and mating among sensation seekers.

Judging from advertisements, the concept of sensation seeking has penetrated the advertising industry. Ads for particular products, such as automobiles, seem to be targeted at high sensation seekers. Sensation seeking theory and research have been used to design anti-drug ads (Donohew, Lorch, and Palmgreen, 1991). The ads encourage them to quit using drugs or not to start using them. The ads designed for high sensation seekers incorporate the qualities of stimulation shown to attract, arouse, and get their attention: novelty, change, complexity, intensity, uncertainty, incongruity, affective connotations, color, and reward value. Ads designed for low sensation seekers are the opposite in these qualities. Donohew et al. found that an anti-marijuana message with high sensation value was more effective with high sensation seekers, and one with low sensation value worked better with low sensation seekers in creating an intention to call a "hot-line" mentioned in the ad.

Lorch, Palmgreen, Donohew, Helm, and Baer (1994) found that the context of the TV program in which the ad was embedded affected attention paid to the ad. They used two comedies and two dramas of which one of each genre had a high sensation value and the other had a low sensation value. High sensation seekers paid greater attention to the ad embedded in the high sensation value programs, whereas low sensation seekers paid more attention to the ads in the low sensation value programs. These effects were even stronger in those involved in recent drug use.

Stephenson and Palmgreen (2001) found that the sensation value of anti-drug ads increased cognitive, narrative, and sensory processing, but that personal involvement in marijuana use tended to reduce cognitive processing, particularly in high sensation seekers. However, increased sensation value of the message increased cognitive processing in all subjects and therefore could counteract the influence of personal drug involvement. Novelty of the message was a much stronger predictor of attentiveness for high than for low sensation seekers (Palmgreen, Stephenson, Everett, Baseheart, & Francies, 2002).

These studies show the usefulness of considering the sensation seeking characteristics of the viewer in designing media messages to discourage drug use or any other risky activity.

Music

Music is a mode of popular entertainment that may also vary along the dimensions of intensity, novelty, and complexity that determine the preferences of high and low sensation seekers. Zuckerman et al. (1966) used music to counteract the effects of sensory deprivation as a control for an extreme deprivation situation. They offered the college subjects a choice of classical, jazz, or low-key popular music. The high sensation seekers tended to choose either jazz or classical music while the low sensation seekers chose the bland popular music. This and other studies performed about that time used a limited range of musical styles.

Litle and Zuckerman (1986) devised a questionnaire with a broad range of styles and examples of popular performers for each. Subjects are asked to rate their like or dislike for the styles. Sensation seeking correlated positively with a liking for rock music of all types, and negatively with movie sound-track music, typically a bland type. The experience seeking subscale of the SSS correlated with a broader selection of styles including folk, classical, hard rock, and soft rock music.

Dollinger (1993) used the Litle-Zuckerman musical preference scale (LZMP) along with Costa and McCrae's (1985) NEO-PI scale, which contains a facet scale of extraversion for excitement seeking (ExS). ExS correlated positively with a liking for hard rock, and negatively with gospel music. Openness to experience, a scale most closely related to experience seeking in the SSS, correlated with a variety of music style preferences: classical, jazz, rhythm and blues, new age, reggae, and folk–ethnic.

Rawlings, Hodge, Sherr, and Dempsey (1995) correlated the LZMP with Eysenck's EPQ and impulsivity scales. The P scale correlated positively with a liking for hard rock, and negatively with a liking for soft popular music.

The LZMP was constructed in the mid-1980's, and did not include types of music and groups that came later. Carpentier, Knoblock, and Zillman (2003) included modern rock and hip hop songs from the Adult Top-40 Billboard list and classified them into two groups: Defiant and non-defiant music. Defiant music included songs from rock and hip hop selected for their "edgy" sound, and social and politically defiant lyrics. Non-defiant songs were selected for their melodious nature and absence of defiant lyrics. Personality measures included scales for negativism, rebelliousness, hostility, and the disinhibition subscale of the SSS. Preference was based on self-selected exposure durations to samples of each type of music. High disinhibition and proactive rebelliousness predicted increased listening of defiant music, but high hostility and reactive rebelliousness did not.

Music preferences vary with generations and the most popular styles change. Stratton and Zalanowski (1997) compared moods associated with different types of music using three groups: a younger college student, and two middle-aged groups, a college faculty and staff and a non-college group. Nearly all of the college students reported listening to rock compared to only 61–64% of the older groups. Classical music was occasionally listened to by about half of the college and faculty groups but only by a fifth of the non-college group. "Easy listening" was common among the older groups but less commonly listened to by the college students.

The affect scale used was the trait form of the Multiple Affect Adjective Check List-Revised (MAACL-R, Zuckerman & Lubin, 1985), which includes scales for anxiety, depression, hostility, positive, and sensation seeking (surgency) affects. In the college student group the listening time for rock music correlated *positively* with sensation seeking affect, but also with anxiety and depression, and negatively with positive affect. Among the faculty sensation seeking affect correlated *negatively* with listening time to rock, and positively to listening time to country music. Listening time for classical music correlated positively with depression and hostility. Correlations in the non-college group were nearly all non-significant. The relationships between sensation seeking affect and rock were opposite in direction in the student and faculty groups. It is not clear if this is a change with age or due to generational differences.

Young college students are clearly attracted to "heavy metal" types of rock. One of the characteristics of this music is an exaggerated bass sound transmitted at high intensities through huge loudspeakers. As a fan of rock once told me: "you listen with your whole body, not just your ears." A study of college students examined this factor, controlling for the musical selections by playing the same selection in normal or enhanced bass modes (McCown, Keiser, Mulhearn, & Williamson, 1997). The EPQ was used to assess personality. A preference for the enhanced bass was demonstrated by those scoring high on psychoticism and extraversion, although the psychoticism relationship was much stronger. Strong gender differences were also found; males preferred the high bass choice.

Art and Photography

The still image is usually less compelling than the moving image. To some extent, tastes in art reflect familiarity, so we would expect pleasant realistic art to be more liked and preferred to abstract or surreal art. While sensation seekers might be expected to go along to some extent with this bias toward the familiar, as in realistic art, their attraction to novelty and emotionally arousing, even if unpleasant, themes may make them more ready to accept modern art and unpleasant themes in art and photographs.

Rawlings, Barrantes i Vidal, and Furnham (2000) tested the relationships between music and art preferences and sensation seeking in two samples, one of Spanish and the other of English students. The relationships between preferences in the two media, music and art, were explored in factor analyses. A major factor in both samples contrasts liking for violent-abstract paintings and hard rock music with dislike, or low liking, of neutral-realistic painting and easy-listening type music.

As in studies discussed in the previous section, sensation seeking, particularly disinhibition (Dis), correlated with liking of hard rock. The SSS subscale of experience seeking (ES) correlated with a broader range of music, including hard rock, electronic, jazz, and techno in both samples. Preferences for painting styles correlated primarily with ES and Dis subscales of the SSS. ES correlated positively with liking for violent-abstract art and Dis correlated positively with liking of erotic abstract art. Both ES and Dis correlated negatively with neutral-realistic art in both countries. The preferences related to sensation seeking involve both style (abstract) and emotional value of content (violent or sexual). It should be noted that abstract styles included are not completely non-representational but are reality distorted.

Zuckerman, Ulrich, and McLaughlin (1993) restricted their stimuli to 19th century nature paintings. The paintings were rated for their complexity and tension values. Complexity was positively related to liking and tension was negatively related to liking among both high and low sensation seekers, but tension interacted with sensation seeking in their effect on liking. There was no difference between high and low sensation seekers on liking for low tension paintings, but the high sensation seekers liked the medium and high tension paintings relatively more regardless of style. Factor analysis resulted in five style categories of paintings: (1) a hazy, semi-abstract category, mostly Turner paintings; (2) realistic turbulent scenes; (3) romantic, fantasylike landscapes; (4) placid realistic pastoral landscapes; and (5) early expressionist, such as Van Gogh. Most subjects liked the more realistic paintings in categories 3 and 4 more than the less representational semi-abstract paintings in categories 1 and 5. However, the highs liked the paintings in the expressionist category (5) relatively more than the low sensation seekers.

Surreal art tends to be in a realistic style, portraying actual objects or persons but juxtaposing them in incongruous, ambiguous ways, in the manner of dream images. The novelty of these images would be expected to appeal to sensation seekers. Furnham and Avison (1997) found that sensation seeking correlated negatively with preference ratings of realistic non-surreal, and positively with ratings of surreal art.

Furnham and Walker (2001a) explored the personality and art preference relationship using abstract, pop art, and representational paintings. They used form VI of the SSS which contains only subscales for experience or intention to engage in thrill and adventure seeking or disinhibition activities (Zuckerman, 1984a). Only the intention scales were used. Both TAS and Dis activity intentions correlated with liking for abstract and pop art, but TAS also correlated with representational art. Openness to experience from the NEO also correlated with liking for all 3 art styles, whereas conscientiousness only correlated with liking of conventional realistic style art. Art education influenced the liking for abstract art and exposure to art in museums, but only exposure correlated with liking for Pop art. Personality variables added signficant variance to liking for abstract art based on education and exposure, but personality variables alone accounted for significant variance in ratings of Pop art.

In another study, Furnham and Walker (2001b) found few relationships between painting preferences and art education or visits to galleries, although these variables influence familiarity with the different styles: abstract, representational, pop-art, and Japanese traditional art. Using the SSS form VI they found that the TAS intention scale correlated with preferences for abstract art in galleries or homes, and both Dis and TAS correlated with a preference for Pop-art as something the subjects would hang in their own homes. Conscientiousness from the NEO and conservatism correlated negatively with liking for Pop-art, and the latter correlated negatively with abstract and Japanese art as well. The unconventionality of Pop-art probably accounts for the dislike by conservatives and the liking of it by sensation seekers, who are mostly on the liberal end of the political spectrum.

Thus far, I have discussed style of paintings, but another body of research looks at the content in terms of affective tone or pleasantness–unpleasantness of its content. Previously, I discussed sensation seekers liking of horror movies with their highly unpleasant portrayals of gore and death. Other studies have been done using static images in photographs and paintings rated for pleasantness or unpleasantness.

Zaleski (1984) used emotionally positive photos, including scenes of celebration and "mild love-making," and negative images including scenes of torture, hanging, and corpses. Subjects were asked to pick out the picicture they liked the most. Eighty percent of low sensation seeking subjects chose a positive picture as their first choice with the remainder divided between neutral (10%) and negative (10%) pictures. In contrast, nearly 50% of the high sensation seekers chose a negative picture as their favorite and only 38% chose a positive picture.

Rawlings (2003) used representational and abstract and pleasant and unpleasant paintings and photographs as stimuli. Sensation seeking correlated positively with a liking for unpleasant paintings and photographs and negatively with a liking for pleasant photographs, regardless of style. The P scale of the EPQ showed the same pattern of results. The ES subscale of the SSS and the openness scale of the NEO showed the same pattern, except that they also correlated with a liking for pleasant abstract art. In this case of these subscales, the style had some influence.

Rawlings and Bastion (2002) also looked at the interaction between affective content and style in paintings in relation to impulsive sensation seeking, as measured by that subscale of the Zuckerman-Kuhlman Personality Questionnaire (Zuckerman, 2002). This form is divided into two subscales: sensation seeking (SS) and impulsivity (Imp). SS was related to preferences for erotic, violent, and unpleasant art in the abstract form, but not in realistic form. Imp was not related to any kind of art preference.

Negative images have the capacity to elicit fear vicariously, through identification with the victims in the images. In the previously discussed study by Litle (1986), a particularly gory image in a film elicited a stronger skin conductance reaction in low than in high sensation seekers. Lissek and Powers (2003) used the startle-blink method for measuring the fear reactivity to positive, neutral, and threatening images. This method measures the amount of potentiation of the unconditioned startle eye blink (EMG) response (UCR) to an unconditioned stimulus (UCS)—a loud noise—by the presentation of a picture just before the UCS. High and low sensation seekers did not differ on the basal UCR without the pictorial stimuli, but low sensation seekers had larger potentiation of startle responses by threatening pictures compared to positive pictures, whereas the high sensation seekers showed weak startle responses to both types of pictures.

There were no differences between high and low sensation seeking groups in trait anxiety, or in state anxiety after electrode attachment but before the experiment started. The low sensation seeker apparently shows negative physiological arousal in reaction to stimuli regarded as unpleasant accounting for their dislike of such stimuli. The absence of such reactions in the high sensation seeker accounts for their liking and even preferences for such stimuli.

Humor

Analyses of TV preferences by Potts, Dedmon, and Halford (1996) showed that sensation seeking correlated with a liking for stand-up comedy performance shows but not with situation comedies. Perse (1996) did not find correlations with situation comedy or comedy and variety shows. However, use of a situation sense of humor task, which asks subjects to rate how funny they would find a situation in everyday life, such as a waiter accidentally spilling a drink on oneself, correlated highly with the total SSS, form V, in American and German samples (Deckers & Ruch, 1992). Lourey and McLachlan (2003) used the same humor scale with Arnett's (1994) inventory of sensation seeking (AISS). High sensation seekers on both the intensity and novelty subscales of the AISS perceived more situations as funny and believed that they would give strong laughing expression to such situations.

Ruch (1988) studied the types of humor that most appeal to high or low sensation seekers. High sensation seekers like nonsense humor based on incongruity or absurdity (e.g., animals behaving or talking like humans) and sexual humor. Low sensation seekers tend to find nonsense and sexual humor aversive, but like non-sexual humor in which incongruity is resolved by the "punch-line." The preferences of high sensation seekers for unresolved incongruity may be related to their liking of surreal art.

Internet

Armstrong, Phillips, and Saling (2000) examined the personality of subjects in relation to internet usage. Ten percent of their subjects came from an "internet addiction" group. The Dis scale of the SSS was used, along with other personality measures. Dis did not predict scores on an internet problem scale. Low self-esteem did predict such scores. Unlike most media forms, which are often pursued in the company of others, interactions on the internet are usually a solitary activity and may be a substitute for real human interactions. Slater (2003) looked at the use of the internet for video games and for violence-oriented site use. Sensation seeking (using a two-item scale) was not related to the use of the internet for games, but was weakly related, along with aggression and alienation, to its use for violent stimulation. These personality variables did not account for much variance and were probably only significant because of the very large sample size.

Vacations

Vacations are the time when people are free to seek their favored types of outdoor or indoor activities without the constraints of job activities. Not surprisingly, a group of tourists choosing to go on overland travel holidays in Africa scored higher on the total, and the thrill and adventure seeking, and experience seeking subscales of the SSS form V (Gilchrist, Povey. Dickinson, & Povey, 1995). Eachus (2004) examined holiday preferences in an unselected sample. He used the brief sensation seeking scale (BSSS, Hoyle, Stephenson, Palmgreen, Lorch, & Donohew, 2002) consisting of only 2 items for each of the 4 factors in the regular SSS. His holiday preference scale contains four subscales: (1) adventure preference, such as hiking and camping; (2) beach preference, including sun, sand, and nightlife; (3) cultural preference, including visits to museums and art galleries and; (4) indulgent preference, including luxury cruises and health spas where one is waited on and pampered. the total BSSs correlated positively with adventure and beach preference and not at all with cultural and indulgent preferences. Experience seeking (ES) correlated most highly with adventure preference and disinhibition (Dis) with beach preference. ES also correlated positively with culture preference and negatively with indulgent and beach preferences.

Fighting as Recreation

Strange as it may seem to us academic types, some groups in the male population see fighting as a form of recreation and look forward to the Saturday night bar brawl. The disinhibition subscale, in particular, is related to physical aggression (Joireman, Anderson, & Strathman, 2003). Joireman et al. asked subjects about their desire to engage in a physical fight with large cushioned mallets and found that individuals scoring high on Dis, BS and TAS expressed a desire to engage in such combat. Naturally there were also gender difference, but sensation seeking predicted combativeness over and above gender. Even verbal aggression, as measured by the desire to participate in an argument, was predicted by sensation seeking beyond gender differences.

Many of those who attend sporting events seem to enjoy letting out their aggression verbally toward the opponent team. This can even take the form of physical aggression in the form of throwing objects at the enemy players or their fans. In Europe there are organized fans who go to games with the explicit idea of starting brawls with the opposing fans. Sometimes the brawls start in the street outside of pubs before the game actually begins. Male spectators at hockey games in Finland and Canada were asked about their motivation to see fights among the players and their actual personal participation in brawls with other fans (Mustonen, Arms, & Russell, 1996). Sensation seeking was among the predictors of involvement in fighting and motivation to witness fights among the players.

BIOSOCIAL BASES OF SENSATION SEEKING

A levels approach to personality says that personality and its behavioral expressions are based on successively deeper biological levels of analysis (Zuckerman, 2005). Underlying the behavioral expressions, entertainment preferences in this chapter are physiological mechanisms like orienting, defensive, and startle reflexes, and cortical augmenting or reducing in relation to intensity of stimulation. Underlying these are biochemical traits, characteristic reactivities of neurotransmitters and hormones. At the next lower level are the neuronal systems which are mediated by specific neurotransmitters and the receptors which are keyed to them. Finally we have the genes that create these biological structures and their products. A complete analysis of traits must define connections between phenomena at all levels.

A complete discussion of the psychobiology of sensation seeking is beyond the goals in this chapter. The reader must be referred elsewhere for details (Zuckerman, 1984b, 1991, 1994, 1995, 2003, 2005, in press). Starting from the bottom level, genetics, biometric twin studies have shown a substantial heritability for the trait of sensation seeking (about 60%) (Fulker, Eysenck, & Zuckerman, 1980; Hur and Bouchard, 1997; Koopmans, Boomsma, Heath, & Lorenz, 1995). A gene for the dopamine receptor D4 (DRD4) has been related to the personality trait of "novelty seeking," which is highly correlated with sensation seeking (Ebstein et al., 1996; Prolo & Licinio, 2002). The form of this gene associated with novelty seeking is also related to orientation to novel stimuli in newborn infants and less distress in reaction to such stimuli (Ebstein & Auerbach, 2002).

The neurotransmitter dopamine is implicated in sensation seeking by these genetic results and other findings. The enzyme monoamine oxidase (MAO-type B), which is a catabolic regulator with a preferential link to the monoamine dopamine, is low in the platelets of high sensation seekers suggesting that dopamine reactivity is high in sensation seekers (Zuckerman, 1994 for a summary). In animals, dopamine is related to exploration and approach to novel stimuli, whereas another monoamine, serotonin, is related to inhibition of approach and exploration (Zuckerman, 1984b). Dopamine is released in the neurons of the mesolimbic dopamine pathway when a rat explores a novel stimulus. Nuclei in this system are responsive to electrical stimulation or drugs, which produce intrinsic reward affects. Evidence in animals and humans suggests low serotonergic reactivity in high sensation seekers. Moderate arousal is characteristic of response to novelty, but excessive arousal is associated with fear and inhibition of behavior. Arousal is partially mediated by a noradrenergic system which stimulates the cortex and, in a descending pathway, activates the peripheral sympathetic nervous sytem. Low levels of norepinephrine have been found in the CSF of high sensation seekers. The hormone testosterone is positively related to sensation seeking and to assertive and impulsive behavior in humans and other species.

The proposed model for sensation seeking involves the balance between approach, avoidance, and arousal as mediated by the balance between dopamine, serotonin, and norepinephrine, and moderated by the enzymes and hormones regulating these systems (Zuckerman, 1994, 1995, 1996). The optimal levels of stimulation for high and low sensation seekers, which affect their sensory and emotional preferences, may ultimately depend on these genetically regulated biological traits as well as the possibilities for entertainment provided in their particular cultures.

SUMMARY

The validity of the definition of sensation seeking as "the seeking of varied, novel, complex, and intense experiences ..." has been tested in various laboratory studies. High sensation seekers have strong physiological ORs to novel stimuli, but quickly habituate when stimuli are repeated. Stimuli with content of interest to them, or with emotionally intense meanings, potentiate the OR of high sensation seekers and prolong habituation. Low sensation seekers have weaker ORs to novel stimuli of moderate intensity but stronger DRs to stimuli of high intensity. The high sensation seekers can more effectively process high intensity stimuli as shown by augmentation of the cortical EP in response to such stimulation. Preferences for complexity and novelty are shown in reactions to simple designs.

These basic reactions are translated into preferences for more complex media stimuli. High sensation seekers prefer live entertainment like parties, clubs, and rock concerts, but also like and watch films with action, violence, and explicit sex (high intensity). Low sensation seekers prefer quiet media fare such as game shows, situation comedies, and romantically themed films and TV shows. Gender differences follow along similar lines, because males are generally higher in sensation seeking.

Given control of the remote, high sensation seekers show more switching of TV channels (variety seeking), and they often use TV as a background for other activities. They demonstrate high "boredom susceptibility" (one of the 4 subtraits of sensation seeking).

Reactions to what is "unpleasant content" in films (e.g., horror films), art, and photographic images tend to be more positive in high than in low sensation seekers. High sensation seekers respond positively to the arousal value, lows to the valence (pleasant vs. unpleasant) of the stimuli. Low sensation seekers show stronger physiological arousal in augmented startle reflexes to unpleasant stimuli.

The demonstrated media preferences of high sensation seekers have been used to design anti-drug ads that engage their attention and are more effective in getting the message to them. The ads are even more effective when embedded in the kind of media programs that they like.

Tastes in music are affected by peer popularity but also by personality. High sensation seekers among the younger generations like rock music, particularly "hard rock," and dislike bland, less intense, "easy-listening" types of popular music. The high sensation seekers respond positively to the intense, augmented bass of rock music. They also like the "defiant" lyrics of some rock and hip-hop music. But one subtype of sensation seeking, experience seeking, and the related openness to experience scale of the NEO, are related to a broader variety of styles including classical, jazz, blues, and folk music.

In art, high sensation seekers have a relatively greater liking than lows for expressionistic, abstract, surreal, Pop art, and art with erotic or violent and unpleasant themes. Low sensation seekers have a relatively greater preference for representational art with pleasant content. As with music, experience seekers have a broader range of preference than disinhibiters.

In humor, high sensation seekers find more humor in everyday life situations and they like humor in cartoons and jokes that involve incongruity, nonsense, and sexual content. Interet use is not related to sensation seeking, but there is a weak association with use of violence oriented sites. Sensation seeking influences vacation preferences. Sensation seekers prefer adventurous travel to unfamiliar locations or to beach resorts with lively night life over cultural travel or cruises. The subscale of experience seeking is an exception in that they enjoy cultural and abjure beach vacations. A subgroup of the (primarily) male population enjoy their sensation seeking in fighting, either through direct participation or watching others do it at sports events.

Sensation seeking and its expressions in entertainment preferences are some product of environmental and biological influences. The latter affects the different types of physiological reactivities underlying preferences in stimulation. Genetic and biochemical influences operate on the trait.

Schneirla (1959) formulated a postulate to describe the motivating effect of intensity of stimulation:

"For all organisms in early ontogenetic stages, low intensities of stimulation tend to evoke approach reactions, high intensities withdrawal reactions with reference to the source" (Schneirla, 1959, p. 3).

Adding novelty to intensity, I have postulated a basic personality dimension involving individual differences in reactions to these qualities of stimulation and given it the label of "sensation seeking." The construct and its operational definitions in the form of trait tests have found some usefulness in both scientific and applied contexts.

REFERENCES

Aluja-Fabregat, A. (2000). Personality and curiosity about TV and films violence in adolescents. *Personality and Individual Diffferences, 29,* 379–372.

Aluja-Fabregat, A., & Torrubia-Beltri, R. (1998). Viewing of mass media violence, personality, and academic achievement. *Personality and Individual Differences, 25,* 973–989.

Armstrong, L., Phillips, J. G., & Saling, L. L. (2000). Potential determinants of heavier internet usage. *International Journal of Human Computer Studies, 53,* 537–550.

Arnett, J. (1994). Sensation seeking a new conceptualization and a new scale. *Personality and Individual Differences, 16,* 289–296.

Aron, A., Norman, C. C., Aron, E. N., McKenna, C., & Heyman, R. E. (2000). *Journal of Personality and Social Psychology, 78,* 273–284.

Brown, L. T., Ruder, V. G., Ruder, J. H., & Young, S. D. (1974). Stimulation seeking and the Change Seeker Index. *Journal of Consulting and Clinical Psychology, 42,* 311.

Bruggemann, J. M., & Barry, R. J. (2002). Eysenck's P as a modulator of affective and electrodermal responses to violent and comic film. *Personality and Individual Differences, 32,* 1029–1048.

Buchsbaum, M. S. (1971). Neural events and the psychophysical law. *Science, 172,* 502.

Buchsbaum, M. S., & Silverman, J. (1968). Stimulus intensity control and the cortical evoked response. *Psychosomatic Medicine, 30,* 12–22.

Carpenter, F. D., Knoblock, S., & Zillman, D. (2003). Rock, rap, and rebellion: Comparisons of traits predicting selective exposure to defiant music. *Personality and Individual Differences, 35,* 1643–1655.

Costa, P. T. Jr., & McCrae, R. R. (1985). *The NEO Personality Inventory Manual.* Odessa, FL: Psychological Assessment Resources.

Deckers, L., & Ruch, W. (1992). Sensation seeking and the Situational Humour Response Questionnaire (SHRQ): Its relationship in German and American samples. *Personality and Individual Differences, 13,* 1051–1054.

Dollinger, S. J. (1993). Personality and music preference. Extraversion and excitement seeking or openness to experience? *Psychology of Music, 21,* 73–77.

Donohew, L., Lorch, E., & Palmgreen, P. (1991). Sensation seeking and targeting of televised anti-drug PSA's. In L. Donohew, H. E. Snyder, & W. Bullenski (Eds.), *Persuasive communication and drug abuse prevention* (pp. 209–226). Hillsdale, NJ: Lawrence Associates Erlbaum.

Eachus, P. (2004). Using the Brief Sensation Seeking Scale to predict holiday preferences. *Personality and Individual Differences, 36,* 141–153.

Ebstein, R. P., & Auerbach, J. G. (2002). Dopamine D4 receptor and serotonin transporter promoter polymorphisms and temperament in early childhood. In J. Benjamin, R. P. Ebstein, & R. H. Belmaker (Eds.), *Molecular genetics and the human personality* (pp. 137–149). Washington, DC: American Psychiatric Publishing.

Ebstein, R. P., Novick, O., Umansky, R., Priel, B., Osher, Y., Blaine, D., Bennett, E. R., Nemanov, L., Katz, M., & Belmaker, R. H. (1996). Dopamine D4 receptor (D4DR) exon III polymorphism associated with the human personality trait of novelty seeking. *Nature Genetics, 12,* 78–80.

Eysenck, S. B. G., Eysenck, H. J., & Barrett, P. (1985). A revised version of the psychoticism scale. *Personality and Individual Differences, 6,* 21–29.

Freud, S. (1922/1955). Beyond the pleasure principle. In J. Strachey (Ed.), *Collected works of Sigmund Freud, Volume 18.* London: Hogarth Press.

Fulker, D. W., Eysenck, S. B. G., & Zuckerman, M. (1980). A genetic and environmental analysis of sensation seeking. *Journal of Research in Personality, 14,* 261–281.

Furnham, A., & Avison, M. (1997). Personality and preference for surreal paintings. *Personality and Individual Differences, 23,* 923–935.

Furnham, A., & Walker, J. (2001a). Personality and judgments of abstract, pop art, and represenational paintings. *European Journal of Personality, 15,* 57–72.

Furnham, A., & Walker, J. (2001b). The influence of personality traits, previous experiences of art, and demographic variables on artistic preferences. *Personality and Individual Differences, 31,* 997–1017.

Gilchrist, H. Povey, R., Dickinson, A., & Povey, R. (1995). The sensation selective scale: its use in a study of the character of people choosing 'adventure holidays'. *Personality and Individual Differences, 19,* 513–516.

Harris, R. J., Hoekstra, S. J., Scott, C. L., Sanborn, F. W., Karafa, J. A., & Brandenberg, J. D. (2000). Young men's and women's different autobiographical memories of the experience of seeing frightening movies on a date. *Media Psychology, 2*, 245–268.

Hoyle, R. H., Stephenson, M. T., Palmgreen, P., Lorch, E. P., & Donohew, R. L. (2002). Reliability and validity of a brief measure of sensation seeking. *Personality and Individual Differences, 32*, 401–414.

Hur, Y. M., & Bouchard, T. J. Jr. (1997). The genetic correlation between impulsivity and sensation seeking traits. *Behavior Genetics, 27*, 455–463.

Johnston, D. D. (1995). Adolescents' motivations for viewing graphic horror. *Human Communication Research, 21*, 522–552.

Joireman, J., Anderson, J., & Strathman, A. (2003). The aggression paradox: Understanding links among aggression, sensation seeking, and the consideration of future consequences. *Journal of Personality and Social Psychology, 84*, 1287–1302.

Koopmans, J. R., Boomsma, D. I., Heath, A. C., & Lorenz, J. P. D. (1995). A multivariate genetic analysis of sensation seeking. *Behavior Genetics, 25*, 349–356.

Lissek, S., & Powers, A. S. (2003). Sensation seeking and startle modulation by physically threatening images. *Biological Psychology, 63*, 179–197.

Litle, P. A., (1986). *Effects of a stressful movie and music on mood and physiological arousal in relation to sensation seeking.* Doctoral dissertation, University of Delaware. August, 1986.

Litle, P. A., & Zuckerman, M. (1986). Sensation seeking and music preferences. *Personality and Individual Differences, 7*, 575–577.

Lorch, E. P., Palmgreen, P., Donohew, L., Helm, D., & Baer, S. A. (1994). Program context, sensation seeking, and attention to televised anti-drug public service announcements. *Human Communication Research, 20*, 390–412.

Lourey, E., & McLachlan, A. (2003). Elements of sensation seeking and their relationship with two aspects of humour appreciation-perceived funniness and overt expression. *Personality and Individual Differences, 35*, 277–287.

McCown, W., Keiser, R., Mulhearn, S., & Williamson, D., (1997). The role of personality and gender in preference for exaggerated bass in music. *Personality and Individual Differences, 23*, 543–547.

Mustonen, A., Arms, R. L., & Russel, G. W. (1996). Predictors of sports spectators' proclivity for riotous behaviour in Finland and Canada. *Personality and Individual Differences, 21*, 519–525.

Neary, R. S., & Zuckerman, M. (1976). Sensation seeking, trait and state anxiety, and the electrodermal orienting reflex. *Psychophysiology, 13*, 205–211.

Orlebeke, J. F., & Feij, J. A. (1979). The orienting reflex as a personality correlate. In H. D. Kimmel, E. H. van Olst & J. F. Orlebeke (Eds.), *The orienting reflex in humans* (pp. 567–585). Hillsdale, NJ: Lawrence Erlbaum Associates.

Palmgreen, P., Stephenson, M. T., Everett, M. W., Baseheart, J. R., & Francies, R. (2002). Perceived message sensation value (PMSV) and the dimensions and validation of a MPSV scale. *Health Communications, 14*, 403–428.

Perse, E. M. (1996). Sensation seeking and the use of television for arousal. *Communication Reports, 9*, 37–48.

Potts, R., Dedmon, A., & Halford, J. (1996). Sensation seeking, television viewing motives and home television viewing patterns. *Personality and Individual Differences, 21*, 1081–1084.

Prolo, P., & Licinio, J. (2002). DRD4 and novelty seeking. In J. Benjamin, R. P. Ebstein, & R. H. Belmaker (Eds.), *Molecular genetics and the human personality* (pp. 91–107). Washington, DC: American Psychiatric Publishing.

Rawlings, D. (2003). Personality correlates of liking for 'unpleasant' paintings and photographs. *Personality and Individual Differences, 34*, 395–410.

Rawlings, D., Barrantes i Vidal, N., & Furnham, A. (2000). Personality and aesthetic preferences in Spain and England: Two studies relating sensation seeking and openness to experience to liking for paintings and music. *European Journal of Personality, 14*, 553–576.

Rawlings, D., & Bastion, B. (2002). Painting preference and personality. *Empirical Study of the Arts, 20*, 177–193.

Rawlings, D., Hodge, M., Sherr, D., & Dempsey, A. (1995). Toughmindedness and preference for musical excerpts, categories, and triads. *Psychology of Music, 23*, 63–80.

Rowland, G. L., Fouts, G., & Heatherton, T. (1989). Television viewing and sensation seeking: Uses, preferences and attitudes. *Personality and Individual Differences, 10*, 1003–1006.

Ruch, W. (1988). Sensation seeking and the enjoyment of structure and content of humor: Stability of findings across four samples. *Personality and Individual Differences, 9*, 861–871.

Schierman, M. J., & Rowland, G. L. (1985). Sensation seeking and selection of entertainment. *Personality and Individual Differences, 5*, 599–603.

Schneirla, T. C. (1959). An evolutionary and developmental theory of biphasic processes underlying approach and withdrawal. In M. J. Jones (Ed.), *Nebraska Symposium on motivation* (vol. 7). Lincoln, NE: University of Nebraska Press.

Slater, M. D. (2003). Alienation, aggression, and sensation seeking as predictors of adolescent use of violent film, computer, and website content. *Journal of Communication, 53*, 105–121.

Smith, B. D., Davidson, R. A., Smith, D. L., Goldstein, H., & Perlstein, W. (1989). Sensation seeking and arousal: Effects of strong stimulation on electrodermal activation and memory task performance. *Personality and Individual Differences, 10,* 671–679.

Smith, B. D., Perlstein, W. M., Davidson, R. A., & Michael, K. (1986). Sensation seeking: Differential effects of relevant novel stimulation on electrodermal activity. *Personality and Individual Differences, 4,* 445–452.

Stephenson, M. T., & Palmgreen, P. (2001). Sensation seeking, perceived message sensation value, personal involvement and processing of anti-marijuana PSAs. *Communication Monographs, 68,* 49–71.

Stratton, V. N., & Salanowski, A. H. (1997). The relationship between various characteristic moods and most commonly listened to types of music. *Journal of Music Therapy, 34,* 129–140.

Welsh, G. S. (1959). *Preliminary manual for the Welsh Figure Preference Test.* Palo Alto: CA: Consulting Psychologists Press.

Zaleski, Z. (1984). Sensation seeking and preference for emotional visual stimuli. *Personality and Individual Differences, 5,* 609–611.

Zuckerman, M. (1969). Theoretical formulations: I. In J. P. Zubek (Ed.), *Sensation seeking: Fifteen years of research* (pp. 407–432). New York: Appleton.

Zuckerman, M. (1983). Sensation seeking: The initial motive for drug abuse. In E. H. Gotheil, K. A. Druley, & H. M. Waxman (Eds.), *Etiological aspects of alcohol and drug abuse* (pp. 202–220). Springfield, IL: Charles C. Thomas.

Zuckerman, M. (1984a). Experience and desire: A new format for sensation seeking scales. *Journal of Behavioral Assessment, 6,* 101–114.

Zuckerman, M. (1984b). Sensation seeking: A comparative approach to a human trait. *Behavioral and Brain Sciences, 7,* 413–434; 453–471.

Zuckerman, M. (1990). The psychophysiology of sensation seeking. *Journal of Personality, 58,* 313–345.

Zuckerman, M. (1991). *Psychobiology of personality.* Cambridge, UK: Cambridge University Press.

Zuckerman, M. (1994). *Behavioral expressions and biosocial bases of sensation seeking.* New York: Cambridge University Press.

Zuckerman, M. (1995). Good and bad humors: Biochemical bases of personality and its disorders. *Psychological Science, 6,* 325–332.

Zuckerman, M. (1996). The psychobiological model for Impulsive Unsocialized Sensation Seeking: A comparative approach. *Neuropsychobiology, 34,* 125–129.

Zuckerman, M. (2002). Zuckerman-Kuhlman Personality Questionnaire (ZKPQ): An alternative five-factorial model. In B. DeRaad & M. Perugini (Eds.), *Big five assessment* (pp. 377–396). Seattle, WA: Hogrefe & Huber.

Zuckerman, M. (2003). Biological bases of personality. In T. Millon & M. J. Lerner (Eds.), *Handbook of psychology: Vol. 5: Personality and social psychology* (pp. 85–116). Hoboken, NJ: Wiley.

Zuckerman, M. (2005). *Psychobiology of personality, second edition, revised and updated.* New York: Cambridge University Press.

Zuckerman, M. (in press). Biosocial bases of sensation seeking. In T. Canli (Ed.), *Biology of personality and individual differences.* New York: Guilford.

Zuckerman, M., Bone, R. N., Neary, R., Mangelsdorf, D., & Brustman, B. (1972). What is the sensation seeker? Personality trait and experience correlates of the Sensation Seeking Scales. *Journal of Consulting and Clinical Psychology, 39,* 308–321.

Zuckerman, M., Kolin, E. A., Price, L., & Zoob, I. (1964). Development of a sensation seeking scale. *Journal of Consulting Psychology, 28,* 477–482.

Zuckerman, M., Kuhlman, D. M., & Camac, C. (1988). What lies beyond E and N? Factor analyses of scales believed to measure basic dimensions of personality. *Journal of Personality and Social Psychology, 54,* 96–107.

Zuckerman, M., Kuhlman, D., Thornquist, M., & Kiers, H. (1991). Five (or three) robust questionnaire scale factors of personality without culture. *Personality and Individual Differences, 12,* 929–941.

Zuckerman, M., & Litle, P. (1986). Personality and curiosity about morbid and sexual events. *Personality and Individual Differences, 7,* 49–56.

Zuckerman, M., & Lubin, B. (1985). *Manual for the Multiple Affect Adjective Check-List Revised (MAACL-R).* San Diego, CA: Educational and Industrial Testing Service.

Zuckerman, M., Murtaugh, T. T., & Siegel, J. (1974). Sensation seeking and cortical augmenting-reducing. *Psychophysiology, 11,* 535–542.

Zuckerman, M., Neary, R. S., & Brustman, B. A. (1970). Sensation seeking correlates in experience (smoking, drugs, alcohol, "hallucinations" and sex) and preference for complexity (designs). *Proceedings of the 78th Annual Convention of the American Psychological Association,* pp. 317–318. Washington, DC: American Psychological Association.

Zuckerman, M., Persky, H., Hopkins, T. R., Murtaugh, T., Basu, G. K., & Schilling, M. (1966). Comparison of stress effects of perceptual and social isolation. *Archives of General Psychiatry, 14,* 356–365.

Zuckerman, M., Simons, R. F., & Como, P. G. (1988). Sensation seeking and stimulus intensity as modulators of cortical, cardiovascular, and electrodermal response: A cross-modality study. *Personality and Individual Differences, 9*, 361–372.

Zuckerman, M., Tushup, R., & Finner, S. (1976). Sexual attitudes and experience: Attitude and personality correlations and changes produced by a course in sexuality. *Journal of Consulting and Clinical Psychology, 44*, 7–19.

Zuckerman, M., Ulrich, R. S., & McLaughlin, J. (1983). Sensation seeking and reactions to nature paintings. *Personality and Individual Differences, 15*, 563–576.

22

(Subjective) Well-Being

Margrit Schreier
University of Bremen

Psychological research on well-being has burgeoned since the 1960s. In this context a number of factors have been identified that have a decisive impact on how we feel about ourselves and life in general: love, friendships, a fulfilling job, all are among the things in life that immediately come to mind and that have indeed been confirmed as contributing to our sense of well-being. But apart from such large-scale factors as one's partner or one's work, small, everyday matters also affect how we feel, and, among these, leisure time satisfaction has been shown to be significantly related to well-being (Argyle, 1987, Chap. 4). Leisure time, nowadays, is often media time—which raises the question that will be at issue in this chapter: Is there a relationship between our media use, especially our use of entertaining media fare, and our sense of well-being? To answer this question, I will begin by giving an outline of some major concepts and results from research on well-being in psychology, distinguishing between different aspects or forms of well-being: subjective well-being, psychological well-being, and health. In the following section, the relationship between each of these forms of well-being and entertainment is examined, with a focus on entertainment as presented in the audiovisual media, notably television.

WELL-BEING: CONCEPTS AND EMPIRICAL FINDINGS

Although the majority of research in psychology on well-being has been inductive, with a focus on assessments of happiness and life satisfaction in the population and on determining antecedents and consequences of well-being, two broad theoretical strands can be distinguished (Keyes & Waterman, 2003; Ryan & Deci, 2001; Waterman, 1993). The first of these has alternatively been termed the 'hedonic' view, or 'emotional well-being'; in this tradition, well-being is conceptualized as an affectively pleasant state ("pleasure and pain" approaches

389

according to Diener, 1984). Proponents of the second tradition, the 'eudaimonic' view, also called 'psychological well-being,' claim that well-being cannot—and should not be—reduced to pleasure, that well-being comprises living in accordance with one's inner self, one's 'demon' (daimon); the focus here is on living a meaningful life and on achieving self-realization as a fully functioning person ("telic" approaches in Diener, 1984). Some authors have proposed additional components of well-being, such as social as distinct from psychological well-being (Keyes & Waterman, 2003), health is sometimes included as a sub-dimension (Argyle & Martin, 1991), and other conceptualizations rest upon how well-being can be achieved, as opposed to what it comprises (Diener, 1984). The large majority of present approaches to well-being, however, can be captured by the distinction between hedonic and eudaimonic traditions. In the following, both traditions will be described in more detail.

The Hedonic Approach

Structure of Subjective Well-Being.   Within the hedonic approach, three subcomponents of subjective (emotional) well-being (SWB) have been distinguished: positive affect, negative affect, and life satisfaction (on the hedonic view see Kahneman, Diener & Schwarz, 1999; Strack, Argyle, & Schwarz, 1991). Measures of positive and negative affect tap—as the name indicates—the affective dimension of well-being, whereas measurement of life satisfaction involves a predominantly cognitive judgement. Life satisfaction has alternatively been assessed overall or separately for various life domains such as marriage, friendship, work, health, or leisure (Argyle & Martin, 1991).

A person's well-being will be higher the more positive affect and the less negative affect that person reports over a long period of time, and the higher his/her life satisfaction in different domains. In this, positive and negative affect have been shown to be independent (Bradburn, 1969): Presence of positive affect does not automatically imply absence of negative affect, and vice versa. How often someone feels cheerful, happy, satisfied is thus unrelated to how often that person feels low and hopeless over the life course. The two can be considered independent, however, only to the extent that average affect over a longer period is assessed. Diener has presented evidence that, in the short term, positive and negative affect are mutually exclusive (at least in relation to the same life domain: Diener, Sandvik & Pavot, 1991). Nor are they independent if affect is assessed in terms of intensity: If a person tends to have intense feelings, she will feel intense pleasure as well as intense pain; conversely, if a person's feelings usually remain superficial, they will do so with respect to both the good and the bad things in life. There continues to be some debate about whether it is intensity or frequency of affect that contributes most to overall emotional well-being. Who will be happier—he who experiences a succession of small, but frequent pleasures, or she who has felt supremely, deeply happy once, an experience that never again had its like? Diener et al. (1991) also point out that it is the frequency, rather than the intensity, that makes the difference. In fact, intense happiness may even be counterproductive to overall well-being by setting a standard against which all smaller, less intense pleasures pale.

Measurement.   There exists a large number of instruments for assessing emotional well-being and life satisfaction, ranging from single-item measures to multiple-scale inventories (Diener, 1984; Mayring, 2003; on measurement issues concerning well-being see also Larsen & Fredricksen, 1999). Single-item measures are usually based on the self-anchoring ladder developed by Cantril (1965). Respondents are asked to rate their overall life at present by marking one rung of a nine-rung ladder, ranging from "best life for you" to "worst life for you." While easy to apply, they usually score below multi-item scales on reliability, and they do not, of course, allow for simultaneously assessing the multiple dimensions of emotional well-being. With multi-item scales, respondents are usually asked to rate themselves on a number of indicators for positive and negative affect, such as: "During the last 30 days, how much of the time did you feel cheerful; in good spirits; extremely happy; calm and peaceful; satisfied; and full of life?" (for positive affect) and "During the last 30 days, how much of the time did you feel so sad nothing could cheer you up; nervous; restless or fidgety; hopeless; that everything was an effort; worthless?" (for negative affect), response options ranging from "all the time" to "none of the time" (Keyes & Waterman, 2003, p. 482). Multi-item scales include, for instance, Bradburn's affect balance scale or the happiness measures developed by Fordyce (Diener, 1984). Multi-scale inventories go beyond assessing positive or negative affect; they usually comprise both the various aspects of emotional well-being and related concepts, such as health concerns and overall lifestyle (for instance Dupuy's general well-being schedule; Fordyce's self-description inventory: Diener, 1984). More recent multi-scale instruments as they have been developed in a health and quality of life related context extend beyond the hedonic approach, comprising dimensions such as social well-being, autonomy, and the like (Mayring, 2003; see following). The majority of instruments (excepting the single-item measures) has achieved satisfactory to good measures on reliability, whereas validity, according to Mayring, has not always been satisfactorily assessed.

Antecedents and Correlates of Subjective Well-Being. A major question in research on SWB relates to antecedents: Can it be deliberately affected, can we learn to increase our own happiness? This issue also has important consequences for the relationship between entertainment and well-being: The media will be able to affect well-being only to the extent that this does not depend on factors external to the self and beyond our control.

Socio-demographic factors such as income, age, ethnicity, gender, social class, and education are among those over which we have little (such as income) to no (gender, ethnicity, age) control. Overall, these factors are indeed related to SWB, but with correlations around 0.10–0.15 the relationship is rather weak (Argyle, 1999; Keyes & Waterman, 2003). For income, education, and social class, the relationship is as expected: Persons with higher education and greater income, who are higher up on the social ladder, also tend to score higher on SWB. Most likely, however, this is not a case of money or knowledge directly increasing happiness. Rather, the three factors are closely linked and work together so as to provide opportunities for activities that, in their turn, correlate with higher well-being (such as satisfactory leisure time activities, see following). This is probably also the mechanism underlying ethnicity-related differences in SWB, with Caucasians reporting greater happiness than, for instance, African Americans and members of other ethnic groups. With age and gender, the relationship to well-being is somewhat more complex. With age, there appears to exist a small positive correlation, that is, older people rate themselves as somewhat happier. Ryff, however, has shown that this applies only to some (eudaimonic) dimensions of well-being, such as autonomy and mastery, but not to others (Ryff, 1995; Ryff & Keyes, 1995; see following). For gender, the results have been somewhat mixed, but overall men and women do not differ on SWB (Nolen-Hoeksema, & Rusting, 1999).

Another group of factors that cannot easily be changed at will comprises personality traits (Diener & Lucas, 1999). Positive correlations have been established between SBW and extraversion, optimism, self-esteem, and self-efficacy; SWB has also been shown to correlate negatively with neuroticism.

The correlation between SWB and extraversion may well be mediated by the engagement in social activities and by the readiness to enter into social relationships: Extraverts will be more likely to participate in social relationships, and such participation in turn is positively correlated with SWB. This applies to a wide variety of relationships. Marriage is most strongly related to SWB, with those who are married rating themselves higher on well-being, marriage being perceived as a source of emotional and social support. Having close friends, relatives, helpful neighbors, friendly co-workers—in most general terms, others with whom to share activities and experiences—also significantly contributes to SWB.

Other factors that are related to SWB include, for instance, religion and employment. Significant relationships have also been demonstrated between SWB and health as well as life events, although in both cases it is the subjective perception that correlates with SWB, not necessarily the 'objective' event or state of a person's health.

The Eudaimonic Approach

Eudaimonic Conceptualizations of Well-Being. The eudaimonic approach arose out of a fundamental dissatisfaction with the hedonic tradition, criticizing the latter on account of its essentially inductive procedure, accompanied by a neglect of important aspects of positive psychological functioning (see for instance Ryff, 1989; Ryan & Deci, 2000; Waterman, 1993). Hedonic pleasure, say these authors, is only an insufficient translation of the Greek word for happiness, 'eudaimonia.' Rather, 'eudaimonia' refers to an inner self, a 'daimon,' and a happy life against this background would be a life in which the person strives to live in accordance with that inner self. Succeeding in this may be a source of pleasure; yet at other times realizing one's inner potential may only be possible at the expense of foregoing an immediate pleasurable experience.

Drawing upon growth-models put forward in developmental psychology (such as Erickson's stages of psychosocial development), clinical psychology in the humanistic tradition (for instance Maslow's concept of self-actualization or Roger's conceptualization of the 'fully functioning person'), and selected approaches from the mental health literature (such as Jahoda's positive criteria of mental health), Ryff (1989) proposed a six-dimensional conceptualization of psychological (as distinguished from emotional) well-being (PWB). These six dimensions comprise self-acceptance in the sense of holding positive attitudes toward one's self (self-acceptance), the ability to enter into positive relationships with others (positive relations), drawing upon personal standards in evaluating oneself (autonomy), selecting, controlling, and changing one's environment (environmental mastery), having a sense of purpose and direction in life (purpose in life), and the ability to remain open to experience and to continue to grow throughout life (personal growth). Confirmatory factor analysis supported this six-dimensional structure, with a single second-order factor corresponding to the overall construct of PWB (for instance Ryff, 1989; Ryff & Keyes, 1995).

Another conceptualization of well-being within the eudaimonic approach is self-determination theory as put forward by Ryan and Deci (2000). Ryan and Deci focus on the assumption that the fulfilment of one's needs is essential for well-being, that is, their concern is less with the definition of well-being than with the conditions that foster well-being in a variety of contexts (Ryan & Deci, 2001). They postulate the three basic needs of autonomy, competence, and relatedness; well-being is conceptualized in terms of psychological growth, integrity, life-satisfaction, psychological health, the experience of vitality, and congruence. Unlike PWB, according to Ryff, that is considered as distinct from SWB, self-determination theory thus comprises aspects of both subjective and psychological well-being, although the growth-related aspects predominate.

A concept of well-being that focuses on social aspects has been proposed by Keyes (1998). She distinguishes between the felt degree of belonging (social integration), the evaluation of one's contribution to society (social contribution), the sense that the social world constitutes a well-structured and coherent whole (social coherence), the belief in the potential and the generally positive direction of social development (social actualization), and the trust in the essential goodness of others (social acceptance). Confirmatory factor analysis showed that this five-dimensional structure did, indeed, provide the best fit.

Interrelations Between Types of Well-Being. Results on the relationship between psychological, social, and subjective well-being are mixed (for an overview, see Keyes & Waterman, 2003, p. 481; for a debate between the proponents of the hedonic and the eudaimonic approach see Diener, Sapyta, & Suh, 1998; Ryff & Burton, 1998). Social and psychological well-being (as conceptualized by Ryff and collaborators) are regarded as separate and as distinct from subjective well-being. Keyes could indeed show that social and psychological well-being constitute different concepts (1996). By exploratory factor analysis she also demonstrated that social well-being differs from the positive dimensions of SWB (happiness and satisfaction).

Ryff and Keyes (1995) also explored the relationship between the six dimensions of PWB and the most common measures of SWB, that is, positive affect, negative affect, and life satisfaction. The strongest correlations emerged between SWB and self-acceptance and environmental mastery (strongly positive for life satisfaction; positive, although somewhat weaker, for positive affect; strongly negative for negative affect, such as depression). The remaining dimensions of PWB showed weaker or mixed relationships with SWB. The dimensions of PWB as suggested by Ryff thus clearly tap aspects of well-being that lie outside the scope of SWB.

In counterdistinction to Ryff and Keyes, Ryan and Deci in their theory of self-determination assume that SWB constitutes one indicator among others of PWB in the eudaimonic sense (2001); they thus place SWB alongside, for instance, integrity or vitality. The exact relationship between these indicators, according to Ryan and Deci, is not absolute and fixed, but varies according to context. In some contexts, such as engaging in a pleasantly challenging task, an increase in SWB will go along with an increase in, for instance, vitality. In other contexts— when, for instance, the same task is superimposed, not freely chosen—SWB, but not necessarily one's sense of vitality, will increase in the case of success (Nix, Ryan, Manly, & Deci, 1999). The results obtained by Waterman (1993) also support the assumption that the relationship between hedonic and eudaimonic aspects of well-being is context-dependent. He asked the participants in two studies to list five activities of importance and to then rate these on six opportunities these activities afford for satisfaction and on hedonic enjoyment and personal expressiveness, the latter constituting a measure of PWB. There emerged a substantial correlation between measures of SWB and PWB. At the same time, the activities clearly differed in the extent to which they provided opportunities to experience the two kinds of well-being. Personal expressiveness, the eudaimonic type of well-being, was more strongly associated with higher levels of skills and the feeling of being challenged by a task, whereas feelings of hedonic enjoyment more frequently went along with a sense of relaxation, happiness, and losing track of time.

In sum, although conceptually distinct, the three aspects of well-being also appear to be linked. Social and psychological well-being share a basic growth orientation as well as the assumption that developing into a fully functioning person requires the ability of entering into positive relationships with others. With eudaimonic (comprising both psychological and social aspects) and hedonistic concepts, the relationship appears to be such that some activities, experiences, and situations afford an opportunity for both types of well-being; others may further our growth, which need not necessarily be a hedonically pleasant experience; and yet others may be very pleasant in hedonic terms, without offering any growth potential. Media usage, especially entertainment oriented usage, will most likely be of the hedonically pleasant type that offers little potential for growth. The application of this distinction to media use and entertainment will be explored in more detail below.

Antecedents and Correlates of PWB.   Because there exists no such clear-cut and consensual definition for the eudaimonic as for the hedonic type of well-being, results are more difficult to summarize and will therefore necessarily be selective, focusing on the results obtained by Ryff and collaborators (for a more general overview see Keyes & Waterman, 2003; Ryan & Deci, 2001).

Among socio-demographic variables, gender, age, and socioeconomic status have been explored for their relation to PWB. For gender, Ryff and Keyes (1995) showed that women score higher on positive relationships to others and tend to score higher on personal growth; for the remaining aspects of psychological well-being, no consistent gender differences emerged. Age, or rather, more generally, the development of psychological well-being over the life course, has been of special concern to Ryff and her collaborators (Ryff & Marshall, 1999). In their studies they have consistently found a differential pattern for the various components of PWB: With age, environmental mastery and autonomy increase and purpose in life and personal growth decrease, whereas self-acceptance and positive relations with others show no reliable variation (Ryff, 1995; Ryff & Keyes, 1995). Socio-economic status also stands in a consistent relationship to PWB, with persons with lower socio-economic status scoring lower on self-acceptance, purpose, mastery, and personal growth (Ryff, Magee, Kling & Wing, 1999). For all relationships between socio-demographic factors and dimensions of PWB, Ryff assumes that these are mediated by social comparisons, appraisal of feedback received by others, attributional processes, and the subjective importance attached to life experiences (Ryff, 1995). In addition, Schmutte and Ryff (1997) have explored the relationship between the dimensions of PWB and a number of personality factors. The results are quite complex, yielding differential patterns of correlations between factors and dimensions. Considering the straightforward relationship between SWB and extraversion and neuroticism (with positive correlations between positive affect, life satisfaction, and extraversion), this lends further support to the assumption that PWB and SWB are conceptually distinct.

Health and Well-Being

It appears intuitively plausible that health and well-being should be closely related, including both physical and mental health, and subjective as well as psychological well-being. To a large extent, the assumed positive relationship could indeed be empirically confirmed.

Self-rated physical health and SWB have been shown to correlate (r around .3: Argyle, 1987, Chap. 10; Diener, 1984; Okun, Stock, Haring, & Witter, 1984). The correlation decreases, however, if the effect, of additional factors such as leisure time activities are taken into account, and it is also considerably lower if physicians' ratings are used as an 'objective' assessment of health (Diener, 1984; Ryan & Deci, 2001). Nevertheless, the positive mood states that go along with SWB have been shown to facilitate immune functioning, leading in turn to improved health, and especially with older people, higher SWB has played a part in delaying the onset of disability (Keyes & Waterman, 2003). Research on the relationship between physical health and PWB has been more scarce, but findings so far have yielded evidence of a positive relationship in this area as well. In particular, vitality and positive relationships with others have been demonstrated to have a positive effect on physical health (Ryan & Deci, 2001).

The dividing line between mental health and SWB has not always been entirely clear; negative affect as one dimension of mental health has, for instance, sometimes been assessed with scales measuring depression (Diener, 1984). Where a distinction has been made, both subjective and psychological well-being have displayed substantial negative correlations with depression (Keyes & Waterman, 2003). Low SWB and PWB, especially in terms of positive relations with others, have in addition been shown to go along with an increased risk of depression and suicide.

Overall, both hedonic and eudaimonic measures of well-being are thus positively related to physical as well as mental health, with the ability to enter and maintain relations to others playing a particularly important role among the eudaimonic dimensions.

FROM ENTERTAINMENT TO WELL-BEING?

In the following section, the focus will be on exploring the relationship between media use, especially entertainment, and well-being. The large majority of relevant research so far has concentrated on media use in relation to SWB. In order to systematize the highly diverse findings in this area, two further distinctions will be made. The first draws upon the subdivision between positive and negative affect within the concept of SWB (the cognitive dimension of life satisfaction can be neglected in this context), the second takes up the distinction between the reception experience itself and the (longer-lasting) effects of that experience. In sum, four areas of the interrelation between media use and SWB will be summarized: The relationship between the two during the reception experience, comprising both positive and negative affect, and the effects of media use upon SWB, again distinguishing between positive and negative affect. Subsequently, the relationship between media use and PWB will be explored, with a focus on positive effects, followed by a summary of findings on media use and health.

Subjective Well-Being and Entertainment

The Reception Experience.   In a sense, to be entertained is to be in a good mood, in a state of positive affect—entertainment thus seems to be almost synonymous with a state of well-being (Bryant & Miron, 2002). Research on experiential states while watching television, however, presents a somewhat divided picture (Rubin, 1994). On the one hand, television viewing has been shown to go along with a state of drowsiness and relaxation, somewhat similar to daydreaming (Hills & Argyle, 1998; Kubey & Csikszentmihalyi, 1990). On the other hand, the viewing experience has repeatedly been described in terms of arousal. According to affective-disposition theory, the recipient will evaluate the (fictional) characters on moral grounds, and on this basis develop positive or negative feelings towards them; these feelings in their turn are assumed to shape the hopes and fears with respect to the characters and the outcome of the story (Raney, 2003; Chapter 9, this volume). Such hopes and fears, before the hopes are finally fulfilled and the fears averted, are accompanied by what can be considerable states of arousal (Zillmann, 1991).

Arousal, however, need not in itself be pleasurable and involve positive affect. How arousal during media reception is experienced, will depend on a number of factors, such as personality characteristics (high sensation seekers, for instance, have been shown to enjoy higher levels of arousal: Oliver, 2002), genre (Raney, 2003), the motives underlying the reception experience (Rubin, 1994; for mood-management theory cf. following), and, not least of all, whether all harm has in fact been averted from the 'good' protagonists and the 'bad' ones have undergone their just punishment. Excitation-transfer theory (Bryant & Miron, 2003; Chapter 13, this volume) posits that the arousal experienced throughout the reception process is transformed, on the resolution of the situation or plot that caused the arousal in the first place, into feelings of relief and gratification; and it is further assumed that these feelings of relief will be proportionate to the intensity of the arousal previously experienced.

According to the present state of research on entertainment, the television viewing experience can thus, in sum, be either relaxing or arousing (Rubin, 1994), presumably depending on genre, content, and other factors. Whereas a state of relaxation will most likely be pleasurable in and of itself, thus contributing to SWB, this claim cannot necessarily be made for arousal. Excitation-transfer theory provides an explanation why arousal should be sought in the first place—for the sake of the relief experienced upon the resolution of the respective situation—and thus explicitly acknowledges that, at least at times, the degree of arousal felt during reception may extend beyond the pleasurable. In addition, the theory ties the experience of arousal as positive to a specific type of outcome: that which permits a resolution and the subsequent transformation of the arousal into relief. If such a transformation is not possible, and the arousal level remains high following the reception experience, the resulting affect may even be negative. The state of arousal experienced during reception therefore may, but need not, be pleasurable and contribute to SWB.

Undeniably, however, viewers also seek out reception experiences that do not allow for the subsequent relief of arousal and tension: 'Tearjerkers' and drama are an important case in point, and whoever watches a sports event invariably runs the risk that it will be the other team that wins. Moreover, recipients seem willing to experience a whole range of emotions that can hardly be described as straightforwardly pleasurable, such as sadness or even fear (Oliver, 2003; Vorderer, 2003; on fear reactions as a long-term effect cf. following). Media reception thus involves, apart from high states of arousal, experiential states that in and of themselves would seem to increase negative affect and thereby, in fact, lower SWB.

In counterdistinction to the first impression of 'entertainment' as being inherently pleasurable, the reception experience must thus count as ambivalent: It can be pleasurable and relaxing or pleasurably exciting, but it can also involve unpleasantly high degrees of arousal that are not subsequently resolved as well as other emotions, such as sadness, that are usually associated with negative affect.

The Effects of Entertainment. The paradox of recipients seeking out experiential states that are not as such pleasurable is in principle easily resolved by looking beyond the experience itself to its effects: Do they contribute to recipients' well-being? To examine this question, the following research traditions will be drawn upon: mood-management theory, uses-and-gratifications research, and research on fear reactions. Effects going beyond the affective experience that results from watching television will be covered in the context of PWB.

Mood-management theory, in positing that the choice of media fare is hedonically motivated, is very much in line with the view that entertainment contributes to subjective well-being (Oliver, 2003; Zillmann, 1988; Chapter 14, this volume). It is assumed that entertainment is selected in such a way as to regulate arousal, with persons in a state of arousal being more likely to opt for programs that will have a calming effect, and persons who are understimulated and feeling bored more likely to prefer stimulating types of entertainment. With respect to mood, the assumption is that recipients in a negative affective state will prefer programs that are likely to improve their mood, whereas persons who are experiencing a positive affective state at the time of program selection will choose media fare that helps them preserve this good mood.

Overall, empirical evidence confirms these assumptions (Oliver, 2003). There have been some exceptions, however, such as the finding that media users sometimes select music so as to deliberately intensify negative moods (Vorderer, 2003). In addition, comparisons of television viewing to other leisure time activities in terms of their effect on SWB have shown that the positive mood that accompanies viewing apparently does not extend far beyond the actual reception experience (Hills & Argyle, 1995; Kubey & Csikszentmihalyi, 1990, Chapter 7). Moreover, mood-management theory cannot, at least not in its original form, resolve the paradox that recipients clearly do select to view material that is likely to induce negative affective states, such as sad films (Vorderer, 2003). Oliver (2000; 2003) has introduced the concept of meta-emotions in order to reconstruct the enjoyment that apparently results from the experience of negative affect induced by the media. Although intriguing, this solution is nevertheless not altogether convincing. It salvages the basic assumption that the selection of media fare is hedonically motivated—if this is the case, then seeking out negative emotions must on another level engender positive affect. Yet, what if there is, other than hedonic motivation, underlying program and media choice?

Uses-and-gratifications research seeks to explain media effects in terms of motives and functions, that is, the motives underlying media use are assumed to correspond to effects obtained (Katz, Blumler & Gurevitch, 1974; Rubin, 2002). In the context of subjective well-being, two typologies appear especially pertinent: the distinction between ritualized and instrumental media use (Rubin, 1984) and the identification of a cluster of mood-management and social compensation motives for television viewing (see for instance Finn & Gorr, 1988). With ritualized use, the focus is on the medium to the neglect of the content. Ritualized media use serves the general purpose of diversion when there is little else to do, it imposes a certain time structure, and thus prevents negative affect from occurring and thereby contributes to subjective well-being (see also Huston et al., 1992, Chapter 1; Kubey & Csikszentmihalyi, 1990, Chapter 5). In instrumental media use, on the other hand, specific content is purposefully selected because it corresponds to the interests and goals of the media user (Rubin, 2002). Finn and Gorr's (1988) identification of mood management and companionship motives likewise points to SWB as an immediate effect associated with television viewing. Mood management has already been addressed. Social compensation motives are in part identical with the motives underlying ritualized media use (such as habit, or passing the time). In addition, social compensation comprises enjoyment of the parasocial relations afforded by television viewing, that is, the coming to know and responding to television characters as though they were actual persons, even friends (Horton & Wohl, 1956; Chapter 17, this volume). Although the enjoyment of parasocial interactions and relations by no means necessarily entails a lack of actual social relations (Vorderer & Knobloch, 1996), the prevalence of the social compensation motive has been shown to correlate with shyness, loneliness, and neuroticism, whereas persons high in extraversion have rejected the idea that the interaction with television characters can in any way substitute for 'real' interactions with actual persons (Finn & Gorr, 1988; Weaver, 2000). To the extent that television succeeds in providing a sense of companionship for those who watch with a social motive, it can again be assumed that gratification of this need for companionship will ward off negative affect and thus increase SWB. To the extent that parasocial interactions with characters on the screen are actively enjoyed, positive affect will also result (on TV viewing as a social occasion, see following).

So far the focus has been on positive effects of television viewing. Depending on genre and type of viewer, however, negative effects may also occur. This does not refer to negative feelings such as sadness while viewing drama, that is, negative feelings that are deliberately sought out and, potentially, on some level provide paradoxical enjoyment (cf. prior). Rather, the focus here is on negative affect as an unintentional effect of media reception. A large body of findings testifies, for instance, to the occurrence of fear reactions in both children and adults to media fare that in some cases have been reported to last well beyond the actual reception experience, thus substantially decreasing SWB even over an extended period of time (Cantor & Mares, 2001; Cantor, 2002; 2004; Chapter 18, this volume).

In sum, media reception has been shown to potentially increase SWB, both by increasing positive affect via mood management and by warding off negative affect by affording a means of diversion, of structuring time, and of providing parasocial 'company.' As for inducing positive mood, however, it remains unclear for how long this positive affective state is maintained following the reception experience. Moreover, media reception does not invariably induce a positive mood. Depending on genre, viewing can also result in negative affect, intentionally as well as unintentionally.

Psychological Well-Being and Entertainment

Summarizing research on PWB and media use is a difficult enterprise on two accounts. First, as outlined previously, there does not as yet exist a unified conceptualization of PWB; second, research on media reception and well-being has for the most part focused on SWB to the neglect of PWB. Nevertheless, a few consistent findings have emerged, concerning the use of entertainment in identity construction and media reception as a social occasion. On the negative side, there has been the fear that the (frequent) use of television may be detrimental to personal growth. In this, PWB will throughout be considered a potential effect of entertainment.

Positive Effects of Entertainment on PWB.

Entertainment has to a large extent been conceptualized as fictional. Fiction, in turn, has usually been considered as sharply distinct from everyday life, and interrelating the two (e.g., by adopting a point of view expressed by a fictional character) as dysfunctional. Only recently has it been recognized that recipients readily adopt information encountered in fictional contexts and that entertainment and fiction may in fact carry powerful persuasive messages (Green, Brock, & Strange, 2002; Prentice & Gerrig, 1999; see following for health-related information). In this vein, recipients have frequently been shown to make use of (fictional) entertainment in identity construction, with fictional characters serving as role models and suggesting ways of how to deal with personal problems (e.g., Barker, 1997; Hoffner & Cantor, 1991; Pitta, 1999; Rios, 2003). Incidentally, this type of research has largely concentrated on soaps and on female viewers. Other entertainment formats, such as talk shows, have also been demonstrated to provide social orientation and to play a part in identity construction (Bente & Feist, 2000; Trepte, Zapfe, & Sudhoff, 2001).

Social companionship has emerged as one of the major motives underlying media reception (Rubin, 2002; Weaver, 2000). Apart from affording an opportunity for parasocial interaction (as discussed in the last section), media reception frequently occurs in the context of actual interactions with others. Comparisons of leisure time activities on SWB have shown that the enjoyment of companionship constitutes a common dimension, be it sports activities or watching television (Hills & Argyle, 1998). While viewing together used to be more frequent in the past, families still turn television viewing into a social occasion (Alexander, 2001; Brown & Hayes, 2001). In addition, entertainment, even when not consumed together, still provides an opportunity for subsequent communication (Press, 1990; Sutter, 2002). The proliferation of virtual reading groups on the internet that have followed the reading circles of the 19th century (Long, 2003) bear witness to the strength of this motive for communication about media fare. Although not necessarily enabling the individual to establish positive relationships to others, media reception nevertheless provides for a number of occasions during which such relationships may arise.

Negative Effects of Entertainment on PWB.

On the one hand, media reception can thus contribute to PWB by supporting identity construction and thereby the ability to form a positive attitude toward oneself, and by enabling occasions for establishing positive relations with others. On the other hand, however, there has arisen the fear that entertainment, especially television entertainment, may prevent personal growth and thus, in fact, be detrimental to PWB. In their study of experiential states during television reception, Kubey and Csikszentmihalyi (1990) found television viewing, although pleasantly relaxing and drowsy, to also be devoid of challenges and skills, that is, failing to provide any opportunities for growth. Moreover, in particular heavy, ritualized television viewing has frequently been shown to correlate with a low activity profile, low values on extraversion, and high values on neuroticism (Hills & Argyle, 1998; Kubey and Csikszentmihalyi, 1990; Rubin, 2002). These findings suggest a differential relation between entertainment and PWB for heavy as opposed to light television viewing. Light-to-moderate television viewing based on instrumental motives can potentially contribute to PWB in the manner described. Heavy, ritualized television viewing, on the other hand, appears to be detrimental to PWB, just filling time and substituting for a variety of other activities. Incidentally, SWB in the sense of positive mood as an effect of entertainment has also been shown to decrease with heavy viewing (Kubey and Csikszentmihalyi, 1990).

Health and Entertainment

A relationship between health and entertainment has been established along a number of dimensions. On the positive side, this concerns the potential of entertainment for stress reduction and for health education ('entertainment education'); on the negative side, the relationship between, especially, television viewing, body image, eating disorders, and obesity has been at issue.

Positive Effects of Entertainment on Health. A first positive effect of entertainment on health relates to the potential of positive moods for stress reduction. In particular the response to comedy has been shown to reduce stress (Berk, Tan, Fry, et al., 1989) and to thus improve immune functioning (Berk, Tan, Napier, & Eby, 1989; overview in King, 2003).

Since the late 1960s, there have also been an increasing number of attempts to influence the health-related behavior of recipients by embedding relevant information in fictional entertainment formats, thus making use of the potential of fiction to carry powerful persuasive messages (see prior). This has either taken the form of incorporating health messages into already existing entertainment programs (e.g., embedded health messages, Brown & Walsh-Childers, 2002) or devising entire serials so as to transport health-related information in an entertaining fashion (Singhal & Rogers, 1999). Both strategies are based on Bandura's social learning theory (e.g., Bandura, 2002), which provides the viewers with attractive role models that offer a high involvement potential. The assumption is that recipients will learn new information, become more sensitive toward the issues involved, and change their behavior more readily, the higher their cognitive and emotional involvement during reception (Lampert, 2003).

Health messages have, for instance, included information on coronary heart disease as part of the medical TV series *Medisch Centrum West* (Bouman, Maas, & Kok, 1998), the Harvard School of Medical Health Campaign featuring messages against drunk driving in a number of entertainment formats (DeJong & Winsten, 1989), or pro-condom use messages embedded in the situation comedy *Friends* (Brown & Walsh-Childers, 2002). While a systematic evaluation of the effect of such embedded health messages is lacking, there is some indication that they do have the desired effect. The messages about coronary heart disease in *Medisch Centrum West,* for instance, were remembered after reception. Likewise, Gassmann, Vorderer, and Wirth (2003) could show that attitudes toward organ donation were more positive and readiness to donate increased after viewing part of an episode from a German medical TV drama where such a positive attitude was advocated. The duration of these effects, however, is still unclear, and it cannot be ruled out that it may only be short-term.

The entertainment education approach, in the sense of using entire serials in order to sensitize the recipients to specific health issues, has been especially popular and successful in third world countries. Here, where radio is the most widely used medium, the focus has been on the production of radio soap operas. Popular examples include the Tanzanian radio soap *Twende na Wakati (Let's go with the times)* that combines Aids-related information with information on family planning and the use of contraceptives. In an extensive evaluation study, Rogers et al. (1999) showed that listening to the soap increased the recipients' sense of self-efficacy, led to an increase in communication about the topics mentioned, and even went along with changes in behavior, such as an increasing use of condoms. Similarly encouraging results were obtained for other radio soaps produced in accordance with the entertainment education approach (overview in Lampert, 2003).

Negative Effects of Entertainment on Health. While entertainment education and embedded health messages thus offer promising venues for positively affecting the health of the recipients, television viewing especially has also been suspected of having a detrimental effect on viewers' health (overview in Brown & Walsh-Childers, 2002).

In the first place, there has been concern about the media in general transporting an overly thin body ideal, which may lead to body dissatisfaction and subsequently contribute to the emergence of eating disorders, especially in women. Indeed, the models depicted in women's magazines have grown progressively thinner over the years (The Media and Eating Disorders, 2001), and there exists evidence linking the exposure to the thin-body ideal to body dissatisfaction (for instance Anderson, Huston, Schmitt, Linebarger, & Wright, 2001; Harrison, 2001). The relationship between exposure to the ideal and eating disorders is, however, less clear-cut (e.g., Harrison & Cantor, 1997). Tiggemann (2003) presents evidence to the effect that the relationship is not a direct one, but mediated by self-esteem and the internalization of the ideal.

A second area of concern centers around the relationship between television and obesity. While watching television, the viewers are physically inactive. It is therefore hardly surprising that amount of television viewing and body weight are positively correlated. 'Heavy' viewers, for instance, who watch at least three hours a day are twice as likely to be obese than 'light' viewers, who restrict themselves to one hour a day or less (Tucker & Bagwell, 1991; overview in Brown & Walsh-Childers, 2002; see also Huston., 1992, Chapter 4). Moreover, the same mechanism may be at work here that in the context of entertainment education has been put to a beneficial use. Both advertising and fictional formats carry a multitude of nutritional messages. Signorielli and Lears (1992) have demonstrated that the majority of these messages are of an unhealthy kind: non-nutritious foods high in starch and sugar dominate, and the foods are primarily consumed not in order to still hunger, but in order to satisfy social and emotional needs. Not surprisingly, these messages, too, are readily learned by the viewers. Signorielli and Lears found evidence that the amount of television viewing predicts bad eating habits in children and that, the more television is watched, the lower the child viewers' nutritional knowledge.

Apart from the relationship between the amount of television viewing and obesity, the relationship between health and (television) entertainment thus seems to depend largely on content. Entertainment formats can carry any type of information, be it beneficial or detrimental to health, and, in accordance with social learning theory, contribute to the spreading of that information and thereby in the long run affect behavior. This content dependence applies equally to the potential of entertainment for stress reduction: With comedy, positive mood and stress reduction result; with unpleasantly high levels of arousal or fear viewing may even increase stress and, thus, negatively affect health.

CONCLUSIONS

In sum, (television) entertainment can have quite powerful effects on our sense of well-being. It can put us in a good mood, help us relax after a stressful day at work, distract us from our problems, or just help us fill time when there is nothing better to do—and thus make us feel good, at least for the time being. But the power of entertainment also works the other way around: If we pick the wrong program, what we see may make us feel tense, or sad, or even provide us with images of bloodshed and war that are not easily shaken off. A more differentiated picture emerges when these research results on entertainment and subjective well-being are placed in the context of findings on psychological well-being. Whether entertaining media fare will not just make us feel good for the duration of the reception experience, but will moreover contribute to our feeling good about the kinds of persons we are, seems to depend to some degree on the uses to which we put the media. If we use television primarily in a ritualized manner, collapsing in front of the television screen almost every evening, no matter what is on, just in order to pass the time and thus foregoing other activities and opportunities to interact with others—if, in short, we turn into the proverbial couch potato, this will decrease our sense of psychological well-being. Moreover, positive affect resulting from television use will become increasingly short-lived, and any positive effects of good mood on our immune system may well be offset by the detrimental effects of obesity. If, on the other hand, we use entertainment more sparingly and instrumentally, as one leisure time activity among others, we will be able to profit fully from the positive effects of entertainment on mood, enjoy co-viewing and subsequent communication with others, and use media fare to help us orient ourselves within the increasingly complex social realities surrounding us. Entertainment, in short, can positively affect both our subjective and our psychological sense of well-being—but only as long as we remain aware that there exists a world beyond the screen.

REFERENCES

Alexander, A. (2001). The meaning of television in the American family. In J. Bryant & J. A. Bryant (Eds.), *Television and the American family* (pp. 273–288). Mahwah, NJ: Lawrence Erlbaum Associates.

Anderson, D. R., Huston, A. C., Schmitt, K. L., Linebarger, D. L., & Wright, J. C. (2001). Early childhood television viewing and adolescent behaviour. *Monograph of the Society for Research in Child Development, 66* (1, Serial No. 264).

Argyle, M. (1987). *The psychology of happiness.* London: Methuen.

Argyle, M. (1999). Causes and correlates of happiness. In D. Kahneman, D. Diener, & N. Schwarz (Eds.), *Well-being. The foundations of hedonic psychology* (pp. 353–373). New York: Russell Sage Foundation.

Argyle, M., & Martin, M. (1991). The psychological causes of happiness. In F. Strack, M. Argyle, & N. Schwarz (Eds.), *Subjective well-being. An interdisciplinary perspective* (pp. 77–99). Oxford etc.: Pergamon Press.

Bandura, A. (2002). Social cognitive theory of mass communication. In J. Bryant & D. Zillmann (Eds.), *Media effects. Advances in theory and research* (pp. 121–154). Mahwah, NJ: Lawrence Erlbaum Associates.

Barker, C. (1997). Television and the reflexive project of the self: soaps, teenage talk, and hybrid identities. *British Journal of Sociology, 48*(4), 611–628.

Bente, G., & Feist, A. (2000). Affect-talk and its kin. In D. Zillmann & P. Vorderer (Eds.), *Media entertainment. The psychology of its appeal* (pp. 113–134). Mahwah, NJ: Lawrence Erlbaum Associates.

Berk, L. S., Tan, S. A., Fry, W. F., Napier, B. J., Lee, J. W., Hubbard, R. W., Lewis, J. E., & Eby, W. C. (1989). Neuroendocrine and stress hormone changes during mirthful laughter. *American Journal of the Medical Sciences, 298*, 390–396.

Berk, L. S., Tan, S. A., Napier, B. J., & Eby, W. C. (1989). Eustress of mirthful laughter modifies natural killer cell activity. *Clinical Research, 37*, 115A.

Bouman, M., Maas, L., & Kok, G. (1998). Health education in television entertainment—Medisch Centrum West: a Dutch drama serial. *Health Education Research, 13*(4), 503–518.

Bradburn, N. M. (1969). *The structure of psychological well-being.* Chicago: Aldine.

Brown, D., & Hayes, T. (2001). Family attitudes towards television. In J. Bryant & J. A. Bryant (Eds.), *Television and the American family* (pp. 111–135). Mahwah, NJ: Lawrence Erlbaum Associates.

Brown, J. D., & Walsh-Childers, K. (2002). Effects of media on personal and public health. In J. Bryant & D. Zillmann (Eds.), *Media effects. Advances in theory and research* (pp. 453–488). Mahwah, NJ: Lawrence Erlbaum Associates.

Bryant, J., & Miron, D. (2002). Entertainment as media effect. In J. Bryant & D. Zillmann (Eds.), *Media effects. Advances in theory and research* (pp. 549–582). Mahwah, NJ: Lawrence Erlbaum Associates.

Bryant, J., & Miron, D. (2003). Excitation-transfer theory and three-factor theory of emotion. In J. Bryant, D. Roskos-Ewoldsen, & J. Cantor (Eds.), *Communication and emotion. Essays in honor of Dolf Zillmann* (pp. 31–60). Mahwah, NJ: Lawrence Erlbaum Associates.

Cantor, J. (2002). Fright reactions to mass media. In J. Bryant & D. Zillmann (Eds.), *Media effects. Advances in theory and research* (pp. 287–308). Mahwah, NJ: Lawrence Erlbaum Associates.

Cantor, J. (2004). "I'll never have a clown in my house!"—why movie horror lives on. *Poetics Today, 25*(2), 283–304.

Cantor, J., & Mares, L.-M. (2001). Effects of television on child and family emotional well-being. In J. Bryant & J. A. Bryant (Eds.), *Television and the American family* (pp. 317–332). Mahwah, NJ: Lawrence Erlbaum Associates.

Cantril, H. (1965). *The pattern of human concerns.* New Brunswick, NJ: Rutgers University Press.

DeJong, W., & Winsten, J. A. (1989). *Recommendations for future mass media campaigns to prevent preteen and adolescent substance abuse* (Special Report). Cambridge, MA: Harvard School of Public Health Center for Health Communication.

Diener, E. D. (1984). Subjective well-being. *Psychological Bulletin, 95*(3), 542–575.

Diener, E. D., & Lucas, L. E. (1999). Personality and subjective well-being. In D. Kahneman, D. Diener, & N. Schwarz (Eds.), *Well-being. The foundations of hedonic psychology* (pp. 213–229). New York: Russell Sage Foundation.

Diener, E. D., & Lucas, L. E. (2000). Subjective emotional well-being. In M. Lewis & J. M. Haviland (Eds.), *Handbook of Emotions* (2nd ed., pp. 325–337). New York: Guilford.

Diener, E. D., Sandvik, E., & Pavot, W. (1991). Happiness is the frequency, not the intensity, of positive versus negative affect. In F. Strack, M. Argyle, & N. Schwarz (Eds.), *Subjective well-being. An interdisciplinary perspective* (pp. 119–139). Oxford etc.: Pergamon Press.

Diener, E., Sapyta, J. J., & Suh, E. (1998). Subjective well-being is essential to well-being. *Psychological Inquiry, 9*, 33–37.

Finn, S., & Gorr, M. B. (1988). Social isolation and social support as correlates of television viewing motivations. *Communication Research, 15*, 135–158.

Gassmann, C., Vorderer, P., & Wirth, W. (2003). Ein Herz für die Schwarzwaldklinik? Zur Persuasionswirkung fiktionaler Fernsehunterhaltung am Beispiel der Organspende-Bereitschaft [Persuasive effects of fictional television entertainment, using the example of organ donation]. *Medien- und Kommunikationswissenschaft, 51*(3–4), 478–498.

Green, M. C., Brock, T. C., & Strange, J. J. (2002). In the mind's eye. Transportation-imagery model of narrative persuasion. In M. C. Green, J. J. Strange, & T. C. Brock (Eds.), *Narrative impact. Social and cognitive foundations* (pp. 315–342). Mahwah: Lawrence Erlbaum Associates.

Harrison, K. (2001). Ourselves, our bodies: thin-ideal media, self-discrepancies, and eating disorder symptomatology in adolescents. *Journal of Social and Clinical Psychology, 20*(3), 289–323.

Harrison, K., & Cantor, J. (1997). The relationship between media consumption and eating disorders. *Journal of Communication, 47*(1), 40–67.

Hills, P., & Argyle, M. (1995). Positive moods derived from leisure and their relationship to happiness and personality. *Personality and Individual Differences, 25*, 523–535.

Hills, P., & Argyle, M. (1998). Positive moods derived from leisure and their relationship to happiness and personality. *Personality and Individual Differences, 25*(3), 523–537.

Hoffner, C., & Cantor, J. (1991). Perceiving and responding to mass media characters. In J. Bryant & D. Zillmann (Eds.), *Responding to the screen: Reception and reaction processes* (pp. 63–101). Hillsdale, NJ: Lawrence Erlbaum Associates.

Horton, D., & Wohl, R. R. (1956). Mass-communication and para-social interaction. *Psychiatry, 19*, 215–229.

Huston, A. C. (1992). *Big world, small screen. The role of television in American society.* Lincoln, London: University of Nebraska Press.

Kahneman, D., Diener, E., & Schwarz, N. (Eds.) (1999). *Well-being. The foundations of hedonic psychology.* New York: Russell Sage Foundation.

Katz, E., Blumler, J. G., & Gurevitch, M. (1974). Utilization of mass communication by the individual. In J. G. Blumler & E. Katz (Eds.), *The uses of mass communications: Current perspectives on gratifications research* (pp. 19–32). Beverly Hills, CA: Sage.

Keyes, C. L. M. (1996). Social functioning and social well-being: studies of the social nature of personal wellness (Doctoral Dissertation, University of Wisconsin, 1995). *Dissertation Abstracts International: Section B: Sciences and Engineering, 56*(12–B).

Keyes, C. L. M. (1998). Social well-being. *Social Psychology Quarterly, 61*, 121–140.

Keyes, C. L. M. & Waterman, M. B. (2003). Dimensions of well-being and mental health in adulthood. In M. H. Bornstein, L. Davidson, C. L. M. Keyes, & K. A. Moore (Eds.), *Well-being. Positive development across the life-course* (pp. 477–497). Mahwah, NJ: Lawrence Erlbaum Associates.

King, C. M. (2003). Humor and mirth. In J. Bryant, D. Roskos-Ewoldsen, & J. Cantor (Eds.), *Communication and emotion. Essays in honor of Dolf Zillmann* (pp. 349–378). Mahwah, NJ: Lawrence Erlbaum Associates.

Kubey, R., & Csikszentmihalyi, M. (1990). *Television and the quality of life.* Hillsdale, NJ: Lawrence Erlbaum Associates.

Lampert, C. (2003). Gesundheitsfoerderung durch Unterhaltung? Zum Potential des Entertainment-Education-Ansatzes fuer die Foerderung des Gesundheitsbewusstseins [Better health through entertainment? The potential of entertainment-education for improving sensitivity to own health.]. *Medien- und Kommunikationswissenschaft, 51*(3–4), 461–477.

Larsen, R. J., & Fredrickson, B. L. (1999). Measurement issues in emotion research. In D. Kahneman, E. Diener, & N. Schwarz (Eds.), *Well-being: the foundations of hedonic psychology* (pp. 40–60). New York: Russell Sage Foundation.

Long, E. (2003). *Book clubs: Women and the uses of reading in everyday life*. Chicago: University of Chicago Press.

Mayring, P. (2003). Diagnostik gesundheitlicher Ressourcen und Risiken. Gesundheit und Wohlbefinden [Diagnosing health resources and risks. Health and well-being.]. In M. Jerusalem & H. Weber (Eds.), *Psychologische Gesundheitsförderung. Diagnostik und Prävention* (pp. 1–15). Göttingen etc.: Hogrefe.

Nix, G., Ryan, R. M., Manly, J. B. & Deci, E. L. (1999). Revitalization through self-regulation: the effects of autonomous versus controlled motivation on happiness and vitality. *Journal of Experimental Social Psychology, 35*, 266–284.

Nolen-Hoeksema, S., & Rusting, C. L. (1999). Gender differences in well-being. In D. Kahneman, D. Diener, & N. Schwarz (Eds.), *Well-being. The foundations of hedonic psychology* (pp. 330–350). New York: Russell Sage Foundation.

Okun, M. A., Stock, W. A., Haring, M. J., & Witter, R. A. (1984). The social activity/subjective well-being relation: a quantitative synthesis. *Research on Aging, 6*, 45–65.

Oliver, M. B. (2000). The respondent gender gap. In D. Zillmann & P. Vorderer (Eds.), *Media entertainment. The psychology of its appeal* (pp. 215–234). Mahwah, NJ: Lawrence Erlbaum Associates.

Oliver, M. B. (2002). Individual differences in media effects. In J. Bryant & D. Zillmann (Eds.), *Media effects. Advances in theory and research* (pp. 507–524). Mahwah, NJ: Lawrence Erlbaum Associates.

Oliver, M. B. (2003). Mood management and selective exposure. In J. Bryant, D. Roskos-Ewoldsen, & J. Cantor (Eds.), *Communication and emotion. Essays in honor of Dolf Zillmann* (pp. 85–106). Mahwah, NJ: Lawrence Erlbaum Associates.

Pitta, P. (1999). Family myths and the TV media: history, impact, and new directions. In L. L. Schwartz (Ed.), *Psychology and the media: a second look* (pp. 125–145). Washington, DC: APA.

Prentice, D. A., & Gerrig, R. (1999). Exploring the boundary between fiction and reality. In S. Chaiken & Y. Trope (Eds.), *Dual-process theories in social psychology* (pp. 529–546). New York: Guilford.

Press, A. (1990). Gender, class, and the female viewer. Women's responses to 'Dynasty.' In M. E. Brown (Ed.), *Television and women's culture. The politics of the popular* (pp. 158–180). Thousand Oaks, CA: Sage.

Raney, R. R. (2003). Disposition-based theories of enjoyment. In J. Bryant, D. Roskos-Ewoldsen, & J. Cantor (Eds.), *Communication and emotion. Essays in honor of Dolf Zillmann* (pp. 61–84). Mahwah, NJ: Lawrence Erlbaum Associates.

Rogers, E. M., Vaughan, P. W., Swalehe, R. M. A., Rao, N., Svenkerud, P., & Sood, S. (1999). Effects of an entertainment-education radio soap opera on family planning behaviour in Tanzania. *Studies in Family Planning, 30*(3), 193–211.

Rios, D. I. (2003). U.S. Latino audiences of "telenovelas". *Journal of Latinos and Education, 2*(1), 59–65.

Rubin, A. M. (1984). Ritualized and instrumental television viewing. *Journal of Communication, 34*, 67–77.

Rubin, A. M. (1994). Media uses and effects: A uses-and-gratifications perspective. In J. Bryant & D. Zillmann (Eds.), *Media effects. Advances in theory and research* (pp. 417–436). Hillsdale, NJ: Lawrence Erlbaum Associates.

Rubin, A. M. (2002). The uses-and-gratifications perspective of media effects. In J. Bryant & D. Zillmann (Eds.), *Media effects. Advances in theory and research* (pp. 525–548). Mahwah, NJ: Lawrence Erlbaum Associates.

Ryan, R. M. (2001). On happiness and human potentials: a review of research on hedonic and eudaimonic well-being. *Annual Review of Psychology, 52*, 141–166.

Ryan, R. M., & Deci, E. L. (2000). Self-determination theory and the facilitation of intrinsic motivation, social development, and well-being. *American Psychologist, 55*, 68–78.

Ryan, R. M., & Deci, E. L. (2001). On happiness and human potentials: a review of research on hedonic and eudaimonic well-being. *Annual Review of Psychology, 52*, 141–166.

Ryff, C. D. (1989). Happiness is everything, or is it? Explorations on the meaning of psychological well-being. *Journal of Personality and Social Psychology, 57*(6), 1069–1081.

Ryff, C. D. (1995). Psychological well-being in adult life. *Current Directions in Psychological Science, 4*, 99–104.

Ryff, C. D., & Keyes, C. L. M. (1995). The structure of psychological well-being revisited. *Journal of Personality and Social Psychology, 69*(4), 719–727.

Ryff, C. D., & Marshall, V. W. (Eds.) (1999). *The self and society in ageing processes*. New York: Springer.

Ryff, C. D., & Singer, B. (1998). The contours of positive human health. *Psychological Inquiry, 9*(1), 1–28.

Ryff, C. D., Magee, W. J., Kling, K. C., & Wing, E. H. (1999). Forging macro-micro linkages in the study of psychological well-being. In C. D. Ryff & V. W. Marshall (Eds.), *The self and society in ageing processes* (pp. 247–278). New York: Springer.

Schmutte, P. S., & Ryff, C. D. (1997). Personality and well-being: reexamining methods and meanings. *Journal of Personality and Social Psychology, 73*(3), 549–559.

Signorielli, N., & Lears, M. (1992). Television and children's concepts of nutrition: unhealthy messages. *Health Communication, 4*(4), 245–257.

Singhal, A., & Rogers, E. M. (1999). *Entertainment-education. A communication strategy for social change.* Mahwah, NJ: Lawrence Erlbaum Associates.

Strack, F., Argyle, M., & Schwarz, N. (Eds.) (1991). *Subjective well-being. An interdisciplinary perspective.* Oxford etc.: Pergamon Press.

Sutter, T. (2002). Anschlusskommunikation und die kommunikative Verarbeitung von Medienangeboten [Communication following reception and the communicative processing of media offers.]. In N. Groeben & B. Hurrelmann (Eds.), *Lesekompetenz. Bedingungen, Dimensionen, Funktionen* (pp. 80–105). Weinheim, Muenchen: Juventa.

The Media and Eating Disorders (2001). *Rader programs.* Retrieved September 8, 2004, from: http://www.raderprograms.com/media/htm

Tiggemann, M. (2003). Media exposure, body dissatisfaction, and disordered eating: television and magazines are not the same! *European Eating Disorders Review, 11*, 418–430.

Trepte, S., Zapfe, S., & Sudhoff, W. (2001). Watching daily talkshows for orientation and problem solving. Empirical results and explanatory approaches. *Zeitschrift fuer Medienpsychologie, 13*(2), 73–84.

Tucker, L. A., & Bagwell, M. (1991). Television viewing and obesity in adult females. *American Journal of Public Health, 81*(7), 908–911.

Vorderer, P. (2003). Entertainment theory. In J. Bryant, D. Roskos-Ewoldsen, & J. Cantor (Eds.), *Communication and emotion. Essays in honor of Dolf Zillmann* (pp. 131–154). Mahwah, NJ: Lawrence Erlbaum Associates.

Vorderer, P., & Knobloch, S. (1996). Parasocial relationships with characters from a TV series: supplement or functional alternative? *Medienpsychologie, 8*(3), 201–216.

Waterman, A. S. (1993). Two conceptions of happiness: contrasts of personal expressiveness (eudaimonia) and hedonic enjoyment. *Journal of Personality and Social Psychology, 64*(4), 678–691.

Weaver, J. B. (2000). Personality and entertainment preferences. In D. Zillmann & P. Vorderer (Eds.), *Media entertainment. The psychology of its appeal* (pp. 235–248). Mahwah, NJ: Lawrence Erlbaum Associates.

Zillmann, D. (1988). Mood management. Using entertainment to full advantage. In L. Donohew, H. E. Sypher, & E. T. Higgins (Eds.), *Communication, social cognition, and affect* (pp. 147–171). Hillsdale, NJ: Lawrence Erlbaum Associates.

Zillmann, D. (1991). Television viewing and physiological arousal. In J. Bryant & D. Zillmann (Eds.), *Responding to the screen: Reception and reaction processes* (pp. 103–133). Hillsdale, NJ: Lawrence Erlbaum Associates.

23

Catharsis as a Moral Form of Entertainment[1]

Brigitte Scheele and Fletcher DuBois
University of Cologne

THE PROBLEM: ENTERTAINMENT AS STIMULATION INSTEAD OF AS CATHARSIS?

It would be helpful at the outset to give a brief explanation of just what (media) psychology understands as catharsis, but unfortunately that is not possible, because of the manifold meanings that are currently associated with the use of that term. In some cases, they reflect very divergent, and sometimes even diametrically opposed, conceptions that have been developed in the field of psychology during the last century. Looking to Aristotle and the concept of catharsis in his theory of tragedy would seem to suggest itself as a starting point, and indeed, authors often reference Aristotle's work as a source for understanding their own (later) conceptions of that term. Simply going back to Aristotle, however, will not offer any real solution to the problem either, because a one-and-only Aristotelian catharsis does not exist. This is the result of Aristotle not given (or our primary sources not reporting) a sufficiently complete and clear explication of what he meant by catharsis. The result that there is no philological consensus as to what Aristotle wanted to be understood—and what should not be understood—by catharsis. Thus, the history of reconstruction has lead to positions that, in the most extreme case, are diametrically opposed. According to one, catharsis means a psychic cleansing (purgation), and to the other, a spiritual purification.

Now the lack of clarity in the 'original' explication of catharsis would seem to require an effort on the part of the researcher to explain their own concept of catharsis at the beginning of empirical research on the topic. This expectation is, however, seldom fulfilled. Working out the central features of one's own conception is more often than not neglected, and so one is not in a position to explicitly make comparisons with, and draw distinctions between, one's own approach and Aristotelian or current approaches. All this results in what appears, not just initially, to be a highly contradictory basis for discussion, in which one can contrast—at the highest level of abstraction—two theoretical perspectives. The action-related perspective maintains—expressed in a metaphorical manner—that catharsis is "alive and kicking" while the media reception-related perspective comes up with the opposite—that it is empirically "as dead as a doornail."

However, after the theoretical modeling has been inspected, particularly in a comparative manner, one sees that both perspectives, as presented in the discussion up till now, clearly fall short. The action-related perspective has not researched catharsis as an effect of media reception in the realm of aesthetics, but rather as something completely other, namely, the motivation-reducing effect due to action that achieves its goal. This research tradition is not concerned, for example, with the potential cathartic effect from viewing a melodramatic film. Rather the focus is on the relief one feels when the opportunity (to take action) presents itself to "get back at" someone who has been a source of frustration, whereas the reception-oriented perspective does concern itself with how a media presentation is processed. It uses, however, only one of the philologically reconstructed polar concepts—that of purgation—to model its catharsis hypothesis. The second pole—that of purification—has not yet been perceived as relevant, and certainly not theoretically delineated and empirically tested.

In addition, within the purgation perspective, the so-called "aggression-catharsis" has been the main focus of research thus far. This has been the case even though Aristotle did not explicitly limit the possibility of catharsis to anger and aggression, but rather (probably) included all emotions that might in some manner be burdensome. Because it is currently not at all clear to what extent the previous results concerning anger and aggression can be generalized, it follows that one can, by no means, speak of a refutation of the Aristotelian catharsis hypothesis even with regard to the purgation pole.

Therefore, it is our contribution's primary goal to promote the (re)gaining of the (possibly already almost lost) concept of purification, and to use it in modeling media entertainment. This, however, implies that entertainment be understood not only as the experience of enjoyment, suspense or melancholy, but also to include more complex processes in which emotional, cognitive, and motivational dimensions (can) connect in creative tension (cf. Klimmt & Vorderer, 2004; Maill & Kuiken, 2002; Vorderer, 2004). This (more) inclusive conception of entertainment is not only to be preferred for methodological reasons (as regards to what extent the subject matter is included in the concept). It is, above all, preferable for anthropological reasons, because it corresponds to the model of human being as a reflective, ethically sensitive subject, which is what is required to combine the aesthetic and the moral. Just this combination is contained in the concept of purification, which, therefore, presents a value that should be striven for—including the anthropological guiding idea of the reflective, moral subject. It is the task of this article to work out, both theoretically and empirically, this guiding idea with regard to the main focus of catharsis, purification.

At this point an experimental psychologist schooled in theoretical matters will be horrified to note that we are laying our foundation with a value judgment as scientific statement. In mainstream scientific discourse, this has been considered to be something that is simply not done. The reader may well feel not only within their rights but also duty bound to literally put down this article, that is, stop reading it. To motivate as many readers as possible to keep going, we will briefly give reasons for why we in fact have gone against the postulate of value neutrality and how this does not present inadmissible irrationality. Our rationale rests both on the fact that such object-theoretical value judgments are unavoidable in research practice and that they are, contrary to the critical rationalists' contention in the wake of Max Weber, quite open to rational critique and legitimation.

This is unavoidable in so far as the model of what it means to be human, as well as the methodology, contain implicit values, which in turn crucially influence what constitutes the object under consideration. For example, in our case the subject model of the ethically

reflective individual implies a preference for the purification over the purgation approach. The experimental method, however, (as will be shown) leads to primarily testing retributive actions, which are ethically highly problematic. Opposed to this, the catharsis concept contains a positive valuation. As is the case generally when transferring terms used in everyday language to scientific concepts, this valuation can not be eliminated, but rather plays a part in constituting the essential components of the scientific object. Take "creativity" as an example: that term is understood as a constructive human(e) competence with the product-criteria of novelty and effectiveness. Those criteria would surely be fulfilled by the planning and execution of the attack on September 11th 2001, yet in this case we refer to it as criminal energy and not as creative, because the positive value implications of that (scientific) concept are not fulfilled. Denying such object-theoretical implications of value only desensitizes one toward value shifts, as happened by subsuming retributive action under the concept of catharsis (see following). Instead, one should and can explicate and legitimate aspects of value as precisely as possible. Such legitimating can be provided, for example, by using a means-ends argumentation. This (re)presents an analysis of a mixture of descriptive and prescriptive statements that opens up an empirical critique of normative propositions and their, at least relative, (with respect to overarching value statements) justification (cf. Groeben & Scheele, 1977; Groeben, 1999).

The two main lines of questioning in which our analysis will engage thus become evident. They attempt, precisely, at combining value statements with empirical-descriptive ones. First, how does one explicate and legitimate the purification concept as a higher ethical value in relation to the catharsis research that exists up until now? Second, is (and if so, how is) such an ethically worthy conceptualization of catharsis connected to the concept of entertainment? What follows will be organized such that in Part 2 we will explicate the two focal points of the elliptical catharsis construct, namely purgation and purification. Using this as a basis, we will summarize the state of empirical research concerning the purgation concept in Part 3. In Part 4 we will present a reconstruction of empirical results in order to regain the purification concept. As a conclusion in Part 5, in making a case for not reducing the catharis to its purgative half, we will sketch out possible connections between aesthetics—or rather entertainment—and the ethical domain.

CATHARSIS AS AN ELLIPTICAL CONSTRUCT: THE FOCAL POINTS OF PURGATION AND PURIFICATION

With the term 'catharsis' Aristotle described the effect of emotional relief ensuing from the reception of tragedy, but also of music. From this point on, for more than 2000 years, it has been hotly debated whether he meant a medical-organismic or an epistemological process (cf. Flashar, 1984). Possibly one can argue indirectly that "Aristotle's general strong opposition to physiological reductionism gives us a great deal of support" (Nussbaum, 2001, p. 390) for the epistemological process. Naturally, one would prefer to deduce it directly from Aristotle's conception of catharsis. To do so one could turn to the two central terms Phobos and Eleos that, according to Aristotle, should be seen as decisively responsible for the cathartic effect. However, this does not provide clarity either, because Phobos and Eleos can be translated both as Furcht und Mitleid (Fear and Pity) and as Schauder und Jammer (Shuddering and Mourning). Thus "Pity and Fear" (Nussbaum, 2001) are to be seen as the necessary conditions for the development of moral consciousness and humanity, while Shuddering and Mourning more likely can count as developmentally inconsequential. This means that already the exegesis of Aristotle's text divides itself into the two interpretative approaches, namely the concept of purgation versus. the concept of purification (cf. Luserke, 1991).

These two strands are important to us because the object characteristics of purification versus purgation, which by now have been worked out in the philosophical-philological debate about the adequate interpretation of catharsis (cf. for the history of interpretation among others, Gründer, 1968/1991; Langholf, 1990), present the central starting point for the psychological modeling of the process of reception.

The purgation position understands catharsis to be a "hedonistic" principle (Schadewaldt, 1955). This means that the vicarious experience of Phobos and Eleos leads to an emotional discharge and so to a reduction of excessive and psychologically oppressive emotional excitation. The cathartic effect appears accordingly as emotional relief, which is subjectively experienced as well-being after being freed from the experience of aversive tension. This condition is achieved most readily by essentially non-intentional, automatic-mechanical processes of discharge, which—dependent upon the intensity—are outwardly expressed in comparatively spontaneous behavior and somatic reactions. Thus, the focus of this approach is on the congruence of subjective feeling and behavior. Practically, this means: Pent up anger can be expelled, for instance through reactions of anger, and unexpressed grief through grief reactions, and so on. These variations of "aggression-" and "grief-" catharsis are prototypical examples of a "somatic-emotional-"catharsis (in contradistinction to "cognitive-emotional" as in Nichols & Zax, 1977, p. 8). Well-being that has been impaired by emotions is restored when the cognitive control function is temporarily deactivated. This is the principle involved in certain cathartic methods in therapy (cf. primal scream therapy: Janov, 1970; an overview: Möller, 1981), or in suggested possibilities of identification in the reception of fictional literature (cf. Hanly, 1986; Scheff, 1983) as well as in religious stimulation (cf. Scheff, 1983). The cathartic process undertaken as (purgative) healing is, therefore, marked by the (re)gaining of emotional-behavioral integration under greatest possible exclusion of cognitive control (cf. Scheele, 2001; Scheff & Bushnell, 1984).

As opposed to this, the purification position models catharsis as an "ethical" principle (Pohlenz, 1956). So it is maintained, and usually also demanded, that the reception of the tragedy causes a change in the individual toward higher moral standards, and not, as in the purgation position, simply relief from excessive, that is, oppressive, emotional excitation (cf. the theater as a moral institution: Schiller, 1801; or ironically as correctional institution, Bernays, 1858/1970). As a consequence, the cathartic effect should be evident in a qualitatively higher moral consciousness after the performance than before (Lessing, according to Gründer 1968/1991). What is necessary (but, however, not sufficient) for this to happen is vicarious suffering and compassion for the protagonists on the part of the audience (cf. Nussbaum, 2003). Accordingly, the purification effect consists of a change that is subjectively experienced as positive and which is motivationally determined by the "vital need for a (stronger) realization of one's own better (in the sense of more humane) capabilities and possibilities" (Scheele, 1999, p. 20). The purification process (the "clearing up": Nussbaum, 2001, p. 390) is therefore to be seen as a (increasing) congruence between emotion and cognition, that is an integration of emotion and cognition. This is achieved through a "dual cognitive-affective process" (Bohart, 1980, p.192; cf. Nichols & Efran, 1985). Thus, what we have here is not an exclusively cognitive procedure (i.e., not a discernment totally free of emotion—as is often quite falsely assumed). It is much more the case that on the emotional level the cathartic effect entails the subjective feeling of having a stronger ego-identity. On the level of motivation, the cathartic effect entails hoping to constructively resolve certain moral conflicts (be they self-referential or regarding the world), whereas on the cognitive level it implies a clearer view, at least, than before of what is problematic and why (Scheele, 1999, 2001). Thus the purification effect does not manifest itself (as does the purgation effect) primarily in a release of emotional tension but, rather, among other things, in an increase in the tension that is experienced as being positive. This leads to an orientation within processing which is emotionally involved

TABLE 23.1
Purging vs. Purification: Central Features

Cathartic process

Integration of emotion and behavior via abreaction/acting out	Integration of emotion and cognition via restructuring/transformation
-experience oriented (absorbing a-reflexivity)	-processing oriented (involved reflexivity)
-direct resolution of tension	-delayed resolution of tension
-intensity manifests itself somatically	–

Cathartic effect

-immediate relaxation of tension	-mediate relaxation of tension
-mediate healing i.e. restoration of emotional well-being	-immediate healing i.e. strengthening of psychosocial motivation
(homeostatic model)	(developmental model)

and yet reflective, such that the behavioral component has no constitutive function. Table 23.1 presents an overview and comparison of the process- and effect-characteristics that have been discussed above (for an earlier version cf. Scheele, 2001, p. 205).

What both poles have in common is a positive evaluation of the cathartic process and its effects. Therefore, it would seem to make sense and be justified to take the two different conceptions as two polar focal points of a single, integrative catharsis construct—which considering the medial presentation of the ongoing processes, can be called "symbolic catharsis." Accordingly, this symbolic catharsis can be understood as a kind of elliptical construct with two polar focal points. Out of these, two kernels of hypothetical thinking in relation to the reception of fictional works become evident: The purification pole leads to the assumption of an effect consisting of a positive change, especially regarding psycho-social motivation, which comes about through vicarious suffering for and with the characters. The purgation pole maintains, instead, a hedonistic effect and the mostly automatic discharge of an oppressive excess of emotionality through vicarious and pleasurable experience (Scheele, 2001, p. 205). Which of these assumptions have been empirically tested and to what extent, and in what manner they can be considered valid, is what we will seek to ascertain in an abbreviated summary of the relevant research.

EMPIRICAL EMPHASIS ON PURGATIVE ACTIONS

The path from the philosophical–philological catharsis conception to empirical psychological research has led to the fact that, of the two conceptual focal points, research is done practically only on the purgation pole. This research is not primarily concerned with the reception of media presentations, but instead with the efficacy of a person's own actions, above all, of an aggressive kind. Thus, the subject matter has been considerably restricted. In relation to the original concept of catharsis, this can be judged as a distorting misperception of the initial problem. Yet, interestingly enough, this has gone mostly unnoticed. The reason for this lack of awareness of the problem might be due to the fact that conceptually reductionist steps, undertaken in order to focus on this point of view, have been perceived as methodologically appropriate, even necessary.

Historically, the first step toward this shift of focus was to include the catharsis concept in the development of the psychoanalytic procedure of therapy. In their conception of the so-called cathartic method (Breuer & Freud, 1893/1955) replaced theater-catharsis with therapy-catharsis—this means catharsis taken out of the reception-aesthetic context and placed into the real functional world of the subject (who is to undergo therapy). The reason for this lies in the fact that for psychoanalysis all artistic contents are—due to the maintained parallels between them and dream content as well as dream work—only relevant as functional analogies to general human processes and structures (cf. Groeben, 1972; Leuzinger, 1997). This general functional approach was, at the same time, connected to the concentration on the purgation pole since catharsis as a therapeutical method was, with the help of hypnosis, to achieve, above all, two main effects. The first was *"bringing clearly to light the memory of the event by which it was provoked"* (Breuer & Freud, 1893/1955, p. 6; italics in the original). The second was *"allowing its strangulated affect to find a way out through speech"* (p. 17; italics in the original). Interestingly enough, Freud relatively quickly tried to bring in purification aspects as the result of his therapeutic experience. He did this by modifying the therapeutic process toward an integrated emotional and cognitive reinterpretation of the relevant conflicts, thus gradually giving up the method of hypnosis. This "cognitive turn" was not, however, taken into account in the second step of reduction.

This second step consisted of the transfer of the psychoanalytic concept of catharsis to the behavioristic concept of the frustration-aggression hypothesis, promoted by the so-called Yale group (Dollard, Doob, Miller, Mowrer, & Sears, 1939). The goal of this research group was to make the current approaches and results from various disciplines about the relationship of frustration and aggression fruitful for building empirical-psychological theory. They were particularly interested in reconstructing Freud's early results into an empirically testable form. Their efforts led to a wealth of hypotheses about the actual increase and decrease of aggression. Their work has been highly influential for subsequent research even up to today (cf. Baron & Richardson, 1994; Mummendey, 1983). Here catharsis was conceptualized in contrast to inhibition construct. They proposed "that the inhibition of any act of aggression is a frustration which increases the instigation to aggression. Conversely, *the occurrence of any act of aggression is assumed to reduce the instigation to aggression.* In psychoanalytical terminology, such a release is called '*catharsis*' (Dollard et al., 1939, p. 50; italics in the original). Catharsis was thus defined as a temporary reduction of motivation for further aggressive acts. This reduction is achieved through behaving aggressively. This behavior as the precondition for the cathartic effect could be overt, but also covert behavior, for instance, can occur in the form of "fantasy or dream or even a well thought-out plan of revenge" (p. 10). This does not mean however, that they postulate two qualitatively different catharsis processes. Quite to the contrary—conforming to the methodological position of behaviorism—the overt aggressive behavior is seen as the prototypical manifestation of the cathartic process. All the other opportunities for achieving actual deactivation of aggression, including internal behavior, are conceptually modelled on it.

Accordingly there were only two areas for catharsis to be employed within the aggression research. The first, conceptionally decisive one is "direct aggression," which refers to actual behavior that is carried out with the intention to harm; and apart from that, all other behavior with aggression-cathartic effect from indirect (aggressive) activities such as physically strenuous activities, uses of fantasy, the reception of aggressive jokes, up to *"the reception of vicariously aggressive ways of behaving in films (or other media)"* (Mummendey, 1983, p. 399, italics in the original). Through the thus-connected experimental testability, the shift in meaning from the former reception-oriented to the current action-oriented catharsis concept seems to be so unavoidable that after a brief time this behavioral, "turned around" catharsis conception was understood to be and promulgated as the classical modeling linked to Freud and, through him, to Aristotle (e.g., by Baron & Richardson, 1994, p. 24; similarly the introductory texts by, among others, Bierhoff & Wagner, 1998; Heckhausen, 1989; Herkner, 1991; Mummendey & Otten, 2002; Schneider & Schmalt, 1994; Selg, Mees, & Berg, 1997).

As a third step, however, a further limitation resulted from the association of catharsis with the very successful aggression research (à la Dollard et al.): reducing catharsis to the field of so-called hostile aggression. This limitation manifests itself in catharsis having been hardly—or not at all—researched explicitly in other possible reception-psychological areas of emotional-motivational stress, such as "disappointment," "guilt," "shame," "fear," "cowardice," "humiliation," "vanity," "jealousy," "envy," and "ill-will" (Scheele, 2001). Instead, in the second half of the 20th century, catharsis is limited to the reduction of hostile aggression motivation as a purgation effect. It is researched, above all, in relation to action, but in some cases also related to reception (for the development of this branch of the psychology of reception, see section 3 following; for more details: Scheele, 1999). Within the framework of the action theoretical elaboration, catharsis is determined to be the deactivating effect after successfully carrying out aggressive action (cf. Kornadt, 1974; Kornadt & Zumkley, 1992). In the corresponding experimental paradigm, the experimental subject, who has been initially frustrated, is given the opportunity to act aggressively and with retaliation. The findings suggest that a deactivation of aggression motivation is only successful to the extent that the represented aggression's goal is achieved to one's subjective satisfaction (Zumkley, 1978). It is also shown that (retributive) non-specific, for instance, motoric, acting out—as propagated by the (popularized) drive theory as a general purgation cure (Lorenz, 1963)—fails to deactivate hostile aggression motivation (cf. Bushman, 2002; Peper, 1981; Zillmann, Bryant, & Sopolsky, 1979; overviews: Geen & Quanty, 1977; Heckhausen, 1989; Mummendey, 1983).

The value perspectives contained in these results show most forcibly the distance to the original catharsis concept. The action psychological approach implies an antisocial tendency, as catharsis is postulated and validated as the motivational final state of revenge and retaliation that is conscious, wanted, done on one's own account, carried out and, in the end, successful. This is necessarily connected with an anti-emancipatory tendency since this action-related modeling suggests catharsis as a 'solution' to individual psychological stress which, according to Kohlberg's model of moral development, would not go beyond the conventional stage (cf. Colby & Kohlberg, 1978). Thus, more humane coping strategies for retribution motivation are eradicated or, at least, not actively included in the modeling. Beyond this the action-theoretical conceptualization of purgative effect (called catharsis) has led in the fourth and last step to an extensive inflation of the catharsis concept—thereby draining it of meaning. The central result is that after successful retributive action, the aggression motivation declines. This simply represents a specification of content of the generally valid motivation process. After achieving the goal, each action motivation is (at first) reduced. When we have finished writing this article our motivation to continue to work on it, will, reasonably enough, decrease. However, no one would talk here, in this context, about a writing-catharsis.

Thus, the experimental testing of the catharsis conception has all-in-all—and for the most part this has gone unnoticed—on the one hand led to a drastic narrowing of the concept to the purgation pole and thereby, once again, to retributive action. On the other hand, this not only proves to have highly problematic value implications, but finally also leads to an inflation of the concept and the depletion of its meaning.

EMPIRICAL-THEORETICAL APPROACHES TO REGAIN THE PERSPECTIVE OF PURIFICATION

Although for the most part in empirical-psychological research, the action-theoretical focus on the catharsis concept as purgation has been dominant, there have also been other research deployments and results. These not only clarify the limits of the purgation concepts, but also show the possibilities of modeling the purification concept. An important basis for doing this is the research on "symbolic catharsis," that is the reception perspective on the purgation concept. This began in the 1960s and was refined by Bandura's research on "observational learning" (cf. Bandura, 1969).

The Reception Perspective: Purgation as Symbolic Catharsis?

The starting point for the reception-psychological testing of the purgation thesis is a study by Feshbach (1961), in which subjects who had been initially frustrated were shown a ten-minute film with an aggressive theme (a boxing match), as compared to a neutral film. The following measuring of aggression motivation revealed a reduction in the tendency to aggression after the reception of the film with the aggressive content (but not after viewing the neutral film). This result, of "symbolic purgation," maintains at least this one central dimension of the classical catharsis concept through the perspective of reception. However, replications of the Feshbach study using improved methods (treatment check on the frustration of the subjects, a more valid difference between the films shown, and a more valid measurement of aggression including the measurement of the base line) were not able to confirm this effect of symbolic catharsis (cf. Bergler & Six, 1979; Lukesch & Schauf, 1990; Mummendey, 1983). Quite to the contrary, the research done within Bandura's approach, regarding observational learning in particular, has brought about findings which rather speak for vicarious learning—also of aggression—than for a purgative reduction. Must we therefore postulate stimulation instead of catharsis?!

This one-dimensional contrasting of catharsis and stimulation represents the—in principle inadmissible—reduction to the one identical dimension, namely the reception perspective, whereas on other dimensions there are essential differences. In these essentials, central aspects of the catharsis concept, albeit different ones, are missing. Therefore, the status of the data only seems to be contradictory and incomplete regarding a global catharsis conceptualization (including the purgation postulate). So, for example, carrying forth the Yale tradition (and the Feshbach design), Berkowitz maintains that with symbolic catharsis as well (i.e., investigating the reception dimension), the subjects must be frustrated before the vicarious experience in order to be able to speak of the presence and (potential) abreaction of an aggression motivation. Here, however, there is, as in the action, theoretical reduction, a parallel distortion of the original actual catharsis concept that is about the effect of "pity and fear" experienced through media. Therefore, with regard to the purification conception, the studies of the group around Berkowitz (cf. Berkowitz, 1984; Jo & Berkowitz, 1994; similarly Geen & Thomas, 1986) are not very meaningful on a theoretical level. At best, they are of some heuristic value for the differentiation of the purification concept (see following).

In opposition to the atorementioned, Bandura's experimental paradigm does not include the "uncathartic" preliminary condition of frustrating the viewers. Primarily, Bandura researched the learning of observed (aggressive) ways of acting, and not the reduction in motivation through such observation. The reason for this might be found in Bandura's well-known opposition to the behaviorist theories about performance and its reinforcement as necessary (but, however, not sufficient) conditions for learning. He showed their irrelevance for learning in the "acquisition"-mode (cf. Bandura, 1969). The condition of observation of learned behaviors, which has been shown to be sufficient instead, does however continue, at least initially, the behaviorist tradition of concentrating on observable behavior. In contrast, what is central to the purification catharsis is the complex cognitive-emotional-motivational processing of what has been (medially) observed.

Therefore Bandura's paradigm is not immediately relevant for the (purification) catharsis conception. Nevertheless the social-cognitive aspects of processing included here provide a foundation for the modeling of more complex internal processes. Thus his approach is (more than Berkowitz's position) suited for the theoretical explication and integration of the incoherent data which, since the 1980s, has been attempted in order to structure the research field. Along with the inclusion of ontogenetic developmental aspects, the connection with script and schema modeling from the psychology of information have played a central role. These turn Bandura's observational learning approach into an overarching theoretical framework within which, in principle, short-, midrange and long-term medial effects can be modeled in their reciprocal interaction with situational and personality variables (Rule & Ferguson, 1986; cf. Eron & Huesmann, 1980; Groeben & Vorderer, 1988; Huesmann, 1986; Vorderer, 1992).

This integrative theoretical modeling with regard to symbolic catharsis refers to an individual's ability to learn through (receptive) observation, and not only through direct experience based on one's own actions (as the conditioning theories maintain e.g., Bandura, 1974). The content of what is learned can be practically anything which admits of being learned—from simple behavior to speech, to emotionality, right up to moral rules—and not least to the control of one's own and other individuals' actions. Bandura differentiates various effects, each of which he assumes to have different underlying mechanisms. Among these, the acquisition of new behavior is to be named, as well as inhibitory and disinhibitory effects, effects of response facilitation, environmental enhancement, and arousal effects (cf. e.g., Bandura, 1976, 1986, 1994, 2000). Also, the focus placed on aggressive ways of behaving has attracted particular theoretical and empirical attention. Given that media offerings have been proven to present aggressive and violent contents (see the overview of analyses of violence profiles by Gleich, 2004), the basic assumption of Bandura's learning theory implies the thesis of learning aggression through observation, that is an anti-catharsis (Bandura, 1973; Berkowitz, 1993; Feshbach, 1989; Goranson, 1970; Huesmann & Malamuth, 1986; on the enormous range of modeling influences and effects from simple response mimicry to acquiring highly complex rules, moral attitudes etc.: see Bandura, 1986).

To demonstrate the contrast with the purgation thesis (as one half of the conceptually divided catharsis thesis) it is less important that, in deed, new aggressive behaviors can be learned and memorized through observation. This effect is stronger the more similarity there is between the observed scene and the real setting of the learner and the more the aggressive behavior is presented as successful (see overview of results by Baron & Richardson, 1994; Bergler & Six; 1979; Berkowitz, 1984, 1993; Eron, 1986; Geen & Quanty, 1977; Geen & Thomas, 1986; Mummendey, 1983). More central is the stimulation effect, or disinhibiting effect, respectively, namely that vicariously experienced violence raises the aggression potential (in relation to the already existing behavior repertoire) of the observer and certainly does not reduce it. This empirical evidence against the purgation thesis is also not weakened by the fact that an inhibiting effect has been shown to be caused by vicarious punishment. This means that if the observed model is punished, this leads to an increase in the aggression inhibition on the part of the observer, and thus to a suppression of the readiness for aggression, but not to a reduction of instigation to aggression (Kornadt, 1982). Therefore, the proof of inhibition effects caused by punishment can not be made valid for the purgation thesis.

All this means that within the reception perspective the stimulation thesis has received clearly more empirical evidence than the purgation thesis. This is particularly the case when certain factors apply to the presentation: First, when the aggression is shown as positive or pleasurable from the perspective of the aggressor—thus as hedonistic and for its own sake or done to serve a good cause (cf. Kunczik, 1982, p. 6)—when it seems to be a morally justifiable and (probably) successful instrument (cf. a.o. Bandura, 1989; Geen & Thomas, 1986; Goranson, 1970; Huesmann, 1986; Rule & Ferguson, 1986). Furthermore, this is the case when the aggression is experienced as unjustified from the perspective of the victim, so that feelings of and motivation for revenge are brought about (cf. Hartmann, 1969). Accordingly, the hypothesis of hedonistic purgation (through pleasurable vicarious experience) can be considered as largely falsified for the area of hostile aggression. This is exactly what is referred to in the statement quoted in the beginning, that the catharsis hypothesis has been finally falsified. However, this can only apply to the purgation of aggression motivation through vicarious experience.

Having said this, the potential for integration in Bandura's approach, which has been mentioned previously, is not yet depleted. From the beginning Bandura has, it is true, employed symbolic models in his research—however, the concept of symbolic control contains, above and beyond this, aspects that can be made fruitful for the elaboration of an unreduced conception of catharsis. One question among others would be the following: To what extent can a symbolic generalization, for instance of the experience of justice, deviate from one's own context of action (e.g., regarding other persons, actions, etc.), and still present a counterbalance to possibly existing tendencies toward aggression? Such questions, in principle, point out highly complex internal processes, which Bandura has brought together under the term "abstract modeling." This describes the process in which "judgmental skills and generalizable rules are being learned by observation" (Bandura, 1986, p. 100).

We do not have to discuss here whether this possibly presents an over extension, and thereby a depletion, of observational learning as a concept of vicarious experience through observation. Far more important is that we should use this perspective with its complex, cognitive, emotional, and motivational processes in order to extract out of the present empirical results heuristic starting points for a conceptualization of the purification pole.

Starting Points for an Elaboration of the Purification Pole

For this heuristic, how we have differentiated aspects of the purgation pole up until now will serve as the theoretical background for the explication of the psychological inner structure that is to be postulated for the purification pole. At the highest level of abstraction, this concerns the transformation of a hostile aggression readiness to a socially positive orientation and motivation. Such self-critical overcoming of desires for revenge, retaliation, and more diffuse destructive impulses (cf., for the different functional qualities of aggression: Berkowitz, 1993; Eckensberger & Emminghaus, 1982; Feshbach, 1964; Kornadt, 1982) should be achieved "through and in pity and fear" (Nussbaum, 2001, p. 393) as the result of media reception.

Generally speaking, what is at stake here is the moral response to what has been presented through the media. With regard to the person, the more advanced the moral development is, the greater is the likelihood that there will be a moral response (Blasi, 1983; Herzog, 1991; Montada, 2002). With regard to the situation of the content that is presented as well as the presentation's form, it is primarily the stimulation experiments of the group around Berkowitz that offer indirect indications when one focuses on the conditions that were analyzed as a contrast to the stimulation conditions. Some of these conditions were shown to have an effect of a decrease in aggression. For example, the quoted result that subjects who had been annoyed by the experimenter reacted with more hostility toward the provocateur after watching "justified aggression" than those who were given an "unjustified aggression" to view (cf. Berkowitz, 1984) thus also contains the result of a decrease in aggression when experiencing feelings of injustice or when feeling for the victim. Such results were not only obtained under laboratory-experimental conditions with students, but also in field experiments with incarcerated male youth (Parke, Berkowitz, Leyens, West, & Sebastian, 1977). When aggression is experienced as aesthetically aversive or counterproductive in terms of instrumental and value rationality, this can lead to the aggressor being blamed for the victim's suffering. So here we have a different quality of aggression inhibition than the fear of potential punishment (or also revenge, counter-aggression, etc: cf. a.o. Bandura, 1965; Berkowitz, 1984; Selg, 1992) which has been investigated particularly in the tradition of Bandura. Here it is much more a case of an inhibition of aggression based on moral feelings such as compassion, guilt, shame, outrage and the like (cf. a.o. Bandura, 1994; Berkowitz, 1984; Kornadt, 1982). On an operational level, such a moral inhibition of aggression should constitute the starting point for the elaboration of catharsis' purification pole.

A direct empirical testing of this moral aggression inhibition in the context of the classical concept of catharsis has been carried out by Bönke (1989). The study is based on three plays, which present the central versions of dealing with human pain, entanglements, aggression, and so on, according to the Aristotelian conception of catharsis. The first theme is constructive conflict resolution in the sense of reconciliation of antagonistic needs, structures, and so forth ("The Orestia" by Aeschylus). The second is the inextricable "becoming guiltlessly guilty" ("Der Nusser" by Franz-Xaver Kroetz—a German fate of Job set in post WWII) and, last but not least, the unsolvable misery of the individual and society being helpless in the face of violence suddenly appearing from without (Klaus Pohl's "Hunsrueck"—the amok slaughter of four foreign guests, presented in an absurd theater format). Contrasting predictions regarding aggression readiness and (moral) inhibition of aggression were deducted for the reception of these different dramatic presentations of fear and terror. The reception of "The Orestia" (as the prototype for the purification pole) should directly lead to a reduction in the willingness to engage in aggression, and an increase in moral aggression inhibition. From the presentation of fear and terror on an individual level without the possibility of resolution, which reminds one of Job's fate ("Der Nusser"), the result should be an increase in aggression readiness and no change in aggression inhibition. With the superindividual storm of violence lacking the possibility of resolution, the prediction is for a reduction in moral aggression inhibition, potentially linked to an increase in aggression readiness, as well. The study was a field experiment in which groups of 24 students at the Ruhr University Bochum (equally divided between female and male) attended the respective plays (added to this were control groups which either visited the zoo or a lesson at the adult education center). The willingness to engage in aggression and the presence of (moral) aggression inhibition was measured with the aggression-TAT, that is the content analysis of stories that the subjects were asked to tell in response to standardized pictures with thematic content relating to aggression. The pre- and postmeasurements (one week prior to the performance, directly before and after the performance, and one week after the performance) correspond to the predictions deducted from the catharsis conception and the results are statistically significant. Hereby, the possibility of a cathartic purification qua reduction in aggression readiness through moral aggression inhibition, caused by medial presentation(s), has received empirical confirmation. What is essential for elaborating the purification catharsis is to extend this empirical basis (already for the current focus of research on aggression readiness) to different media, medial content, and formal structures of media, and this also with respect to different reception groups, research approaches, and so forth. In this way, the optimal medial and situational conditions for a moral purification through media communication can be clarified. Once this is accomplished for the area of aggression, the classical conception of catharsis contains a multitude of possibilities for researching other problem areas regarding moral purification.

So we see that at least the outlines of a psychological explication of the purification-catharsis are within reach. For the purification pole of catharsis, it is the humane possibility of empathic feeling with and for the suffering of the other, as well as moral protest that are central starting points. These can be seen as a contrast to the hedonistic coexperiencing of aggression, which is modeled by the stimulation thesis. As the inhibition of aggression is not a matter of fear of punishment or retribution (as it is in the purgation-catharsis) but rather an ennoblement in the sense of a—morally more mature—reevaluation, the previous limiting of research to the phenomenon of aggression can in fact present a heuristic for the complete catharsis concept. Accordingly, purification must represent an emotional-motivational (re)learning done on the basis of value-oriented confrontation and subsequent reorganization of one's own wishes and desires towards a morally more advanced identity. In a pre/postcomparison, this developmental quality shows itself in a stronger integration of cognition and emotion, as well as in the motivating hope to succeed in changing morally problematic aspects of experiencing one's self and the world.

This all implies emotional learning that psycho-physiological or biological theories (cf. e.g., Meyer & Reisenzein, 1996) cannot do justice to. Rather it is "cognitive theories of emotion," or, as they are recently called "cognitive appraisal theories" (cf. Scherer, Schorr & Johnstone, 2001), which can be drawn upon and differentiated in order to reconstruct the relationship between beliefs, value orientations (lasting emotional attitudes), current emotions (self-referential appraisals), and motivations (desires and volitions) (cf. Scheele, 1990, 2003; Scherer et al., 2001). Only a such complex modeling of emotions will be able to adequately portray the goal of the integration of cognition and emotion contained in the purification concept. Relevant experiences and results in clinical psychology show that this integration cannot be achieved through cognitive work without emotional engagement, or through emotional reliving without cognitive work. Both are necessary for successfully coping with distressing contents, such as those that are experienced as (emotionally) meaningful to the self and (cognitively) understood (cf. Boesch, 1976; Bohart, 1980; Epstein, 1984; Nichols & Efran, 1985). It is only through such cognitive-emotional confrontation with moral problems that a learning effect is achieved that represents a change in the self-relevant area of values. This change consists of a correspondence between the solution to the medial experienced (moral) problem and the subjectively more highly appraised value concepts of the individual. The achieved, morally higher concept of the self and of the situation makes the individual feel better than before—a feeling as if "the whole spiritual existence of the person has experienced a lasting improvement" (Schadewaldt, 1955, p. 148; Gründer, 1968/1991). This change is fully in accordance with Lessing's reconstruction of the aristotelian catharsis as purification (1768/1988, 74th–78th part).

PURIFICATION-CATHARSIS AS A TASK OF AN ENCOMPASSING ENTERTAINMENT RESEARCH?

Now how can one combine this (moral) developmental goal with the (aesthetic) entertainment process? The empirical research on therapeutic process, in which cognitive-motivational goals of development likewise play an essential role, offers one possible starting point. Basically what is at stake here is the connection between the ideal-self and the acceptance of reality, which are necessary for self-critical reconstruction efforts (cf. Symonds, 1954). According to client centered psychotherapeutic modeling, such a self-critical development requires on the one hand an existential self-acceptance that is capable of giving a sense of security, but on the other hand also a self-distancing that makes possible "processes of reappraisal and newly ordering one's own behavior" (Boesch, 1976, p. 408). This conception of self-accepting self-distancing corresponds to the concept of "aesthetic distance" (Bullough, 1912) in the reception of medial presentations, which consists of a connection between receptive involvement and receptive distance (cf. Vorderer, 1992, p. 73ff.). Scheff (developing further Bullough's concept of distancing) has explicated the "optimal distance" as the synthesis of "deep emotional resonance" with creative reflection on values (Scheff, 1983, p. 66ff.). Aesthetic distance is optimal when two conditions are fulfilled: Too little distancing must be avoided, as it makes the metalevel of reflection no longer possible due to excessive involvement. At the same time, too much distancing must be avoided, where (morally) problematic situations are only reacted to with "numbness and/or confusion" (p. 69).

This conception of aesthetic distance obviously has consequences for the allied conception of entertainment. The matching conceptualization of entertainment in this regard should be neither too narrowly nor too exclusively focused on positive emotions. Both of these problematic aspects can be seen and clarified quite well in two widely dominant conceptions from Zillmann: the mood-management theory and the affective disposition theory (cf. e.g., Zillmann, 1988, 1994, 1996). The mood-management theory postulates that media reception is realized above all with the goal perspective of experiencing states that are as pleasurable as possible. This approach is not particularly suited to purifying reception processes, as it is aimed 'merely' at the relieving and amusing variants of media as prototypes for entertainment. In contrast, at first glance the affective disposition theory seems more applicable. In a differentiated manner, this theory models affective attitudes of the audience regarding the acting characters in the media presentation that are based on moral judgments. According to the theory, moral acceptance forms the basis for an affective correspondence with the protagonist(s) (empathetically hoping, rejoicing, worrying, grieving, etc., with them), whereas moral disapprobation forms counterempathetic reception (with ill will, gloating about the misfortune of the relevant characters, etc.: Zillmann, 1994). Here, however, the moral judgments have merely an instrumental function of generating sympathy and antipathy towards the characters. They do not have the function of defining the problem. In addition, these potentially contrary (partial) emotional processes are modeled above all in regard to the final resolution into a pleasant emotional state. However, a comprehensive conception of entertainment must also include the complex and mixed feelings, in which positive and negative emotional qualities are not resolved too soon but, in accordance to the concept of aesthetic distance, rather are endured by the recipients, and, possibly, even savored. Thus, long lasting effects that are central to the personality become possible.

Therefore, such complex and mixed emotional states and processes are fruitful for the explication of a comprehensive concept of entertainment (cf. Vorderer & Weber, 2003). An example for this is the "seemingly paradoxical enjoyment of sad entertainment" (Klimmt & Vorderer, 2004, p. 13) that is discussed as a paradigmatic case of the valid reconstruction of the observed breath of entertainment effects (cf. Miall & Kuiken, 2002). From our point of view, the purification conception of catharsis can present a further prototypical starting point for comprehensive entertainment research. This concept includes aspects of emotional and cognitive processing that could overcome the shortcomings of a too narrow conceptualization of entertainment. Namely, these are complex and mixed emotions that (in the optimal case) contain moral judgments that do not function instrumentally but rather represent problems that are relevant to the self. Accordingly, they are not to be resolved too quickly within the reception process into a pleasurable emotional final state. Instead, they should lead to a long-lasting solution that changes the individual self, its values, and moral actions. Such comprehensive modeling not only overcomes the hedonism of the narrow entertainment concept, but also the model of homeostasis to which it is linked. The outlined hedonism-conception implies homeostatic feedback loops in which too much and too little stimulation are seen merely as temporary phases toward achieving the optimal, medium level of arousal or emotional well-being respectively. However, in this way it is not possible to model psychological growth. It can, however, be included within an unreduced conception of entertainment. Users of media do not have to seek out "heavy material" merely in the hope of achieving a homeostatic-hedonistic resolution. Instead, a constructive solution is also possible with the (multilevel) model of reflection, as given in the moral comparison of media presentation and real self-world relations. That is to say, under certain circumstances, we consciously choose (morally) painful media experiences in order to achieve the satisfaction of gaining in understanding. When simulating actions in the context of media, we choose those (potentially problematic) aspects of the world and our reactions to them that allow for growth (Nussbaum, 2003, p. 244ff.; Rorty, 1989). This manifests itself in *self-modifying feelings* that restructure the reader's understanding of the textual narrative and, simultaneously, the reader's sense of self"

(Miall & Kuiken, 2002, p. 223, also regarding the empirical proof of such self-modifying feelings during literary reading; italics in the original).

Thus, a conception of entertainment comes into view in which both the aesthetic and the moral elements are unified. The aesthetic here is not only related to form, but also to the content, that is the transforming distance of the medial presentation to (everyday) reality. The narrow entertainment concept, with its focus on relief and amusement, primarily ascribes a hedonistic (pleasure) function to the aesthetic presentation, and if need be, as a resolution of negative emotions that were experienced in the meantime and only striven for because of the inherent promise of their resolution. However, if these "negative emotions" consist of moral burdens and lead to self-modifying feelings, the aesthetic presentation gains an additional function that goes further and more deeply. This function becomes evident when one sees the aesthetic-medial presentation in contradistinction to pure theoretical-moral argumentation (tractates). Although such tractates are indispensable for reasoning on moral norms, life experience is needed for the relevance to actions in the everyday world. This is the position that Nussbaum represents in her neo-Aristotelian aesthetic rehabilitation of ethics, exemplified by Greek tragedy but applying to all medial presentations (Nussbaum, 2001, 2003; Rorty, 1989). Here, the individual concepts and depictions "of a good and praiseworthy living" (*eudemonia*) are not only acquired through cognitive and abstract learning, but, above all, through emotional learning on the basis of experience (including symbolic experience) (Nussbaum, 2001, Chapter 10, interlude 2). The vividness of the medial presentation offers an approximation of such life experiences which may cause sensitivity to human suffering and solidarity with those who suffer (Nussbaum, 2001, 2003; Rorty, 1989). Thus, all those (burdensome) moral feelings (such as grief, desperation, guilt, regret, etc.: Nussbaum, 2001, Chapter 3) which make possible the further development of the self (according to purification catharsis) belong to the aesthetic experience—at least as an existential possibility. Comprehensive entertainment research should include this far-reaching constructive function of the aesthetic in its theoretical modeling for several reasons: First of all, most likely it plays a constitutive role in media reception for certain users. Furthermore, it also facilitates the reconstruction of a morally (more) worthy form of entertainment. This theoretic modeling includes identifying and validating the relevant conditions on the part of the recipients, as well as on the part of the contents and forms of medial presentation. This is what we sought to begin to do with the explication of purification catharsis aforementioned.

When one lays aside the postulate of value neutrality for scientific statements and thus, also, for theoretical modeling of central concepts such as catharsis and entertainment—as we argued at the outset—this even implies the question of whether it is not morally incumbent upon science to elaborate a theoretical model of entertainment in which precisely the possibilities for moral purification through media entertainment may be reconstructed.

ACKNOWLEDGMENTS

[1] We would like to thank Ms. Dipl. Psych. Johanna Vollhardt for her support in translating this chapter into English.

REFERENCES

Bandura, A. (1965). Influence of models' reinforcement contingencies on the acquisition of imitative responses. *Journal of Personality and Social Psychology, 1*, 589–595.

Bandura, A. (1969). *Principles of behavior modification.* New York: Holt, Rinehart & Winston.

Bandura, A. (1973). *Aggression: A social learning analysis.* Englewood Cliffs, NJ: Prentice Hall.

Bandura, A. (1974). Behavior theory and the models of man. *American Psychologist, 29*, 859–869.

Bandura, A. (1976). Die Analyse von Modellierungsprozessen [Analysis of modeling processes]. In A. Bandura (Ed.), *Lernen am Modell. Ansätze zu einer sozial-kognitiven Lerntheorie* (pp. 9–67). Stuttgart, Germany: Klett.

Bandura, A. (1986). *Social foundations of thought and action: A social cognitive theory.* Englewood Cliffs, NJ: Prentice Hall.

Bandura, A. (1989). Die sozial-kognitive Theorie der Massenkommunikation [Social cognitive theory of mass communication]. In J. Groebel & P. Winterhoff-Spurk (Eds.), *Empirische Medienpsychologie* (pp. 7–32). München, Germany: Psychologie Verlags Union.

Bandura, A. (1994). Social cognitive theory of mass communication. In J. Bryant & D. Zillmann (Eds.), *Media effects: Advances in theory and research* (pp. 61–90). Hillsdale, NJ: Erlbaum.

Bandura, A. (2000). Die Sozial-Kognitive Theorie der Massenkommunikation [Social cognitive theory of mass communication]. In A. Schorr (Ed.), *Publikums- und Wirkungsforschung. Ein Reader* (pp. 153–180). Opladen, Germany: Westdeutscher Verlag.

Baron, R. A., & Richardson, D. R. (1994). *Human aggression* (2nd ed.). New York: Plenum.

Bergler, R., & Six, U. (1979). *Psychologie des Fernsehens. Wirkungsmodelle und Wirkungseffekte unter besonderer Berücksichtigung der Wirkung auf Kinder und Jugendliche* [Psychology of television: Models and effects of impact, most notably on children and adolescents]. Bern, Switzerland: Huber.

Berkowitz, L. (1984). Some effects of thoughts on anti- and prosocial influences of media events: A cognitive-neoassociation analysis. *Psychological Bulletin, 95,* 410–427.

Berkowitz, L. (1993). *Aggression. Its causes, consequences and control.* New York: McGraw-Hill.

Bernays, J. (1858/1970). *Grundzüge der verlorenen Abhandlung des Aristoteles über Wirkung der Tragödie* [Main features of Aristotle's lost essay about the effect of tragedy]. Hildesheim, Germany: Olms.

Bierhoff, H. W., & Wagner, U. (1998). Aggression: Definition, Theorie und Themen [Aggression: Definition, theory, and subjects]. In H. W. Bierhoff & U. Wagner (Eds.), *Aggression und Gewalt. Phänomene, Ursachen und Interventionen* (pp. 2–25). Stuttgart, Germany: Kohlhammer.

Blasi, A. (1983). Moral cognition and moral action: A theoretical perspective. *Developmental Review, 3,* 178–210.

Boesch, E. E. (1976). *Psychopathologie des Alltags. Zur Ökopsychologie des Handelns und seiner Störungen* [Abnormal psychology of everyday life. Toward ecopsychology of action and its disruptions]. Bern, Switzerland: Huber.

Bönke, H. (1989). *Der kathartische Effekt des antiken Dramas auf den modernen Menschen* [The cathartic effect of ancient drama on the modern person]. Unpublished masters' thesis, Ruhr-Universität Bochum, Bochum, Germany.

Bohart, A. C. (1980). Toward a cognitive theory of catharsis. *Psychotherapy: Theory, Research and Practice, 17,* 192–201.

Breuer, J., & Freud, S. (1893/1955). On the psychical mechanism of hysterical phenomena: Preliminary communication. In J. Strachey (Ed. & Trans.), *The standard edition of the complete psychological works of Sigmund Freud* (Vol. 2, pp. 3–17). London: Hogarth Press.

Bullough, E. (1912). 'Psychical distance' as a factor in art and an aesthetic principle. *British Journal of Psychology, 5,* 87–118.

Bushman, B. J. (2002). Does venting anger feed or extinguish the flame? Catharsis, rumination, distraction, anger and aggressive responding. *Personality and Social Psychology Bulletin, 28,* 724–731.

Colby, A., & Kohlberg, L. (1978). Das moralische Urteil: Der kognitionszentrierte entwicklungspsychologische Ansatz [Moral judgement: The cognitive developmental approach]. In G. Steiner (Ed.), *Die Psychologie des 20. Jahrhunderts* (Vol. 7, pp. 348–365). Zürich, Switzerland: Kindler.

Dollard, J., Doob, L., Miller, N. E., Mowrer, H. O., & Sears, R. R. (1939). *Frustration and aggression.* New Haven, CT: Yale University Press.

Eckensberger, L. H., & Emminghaus, W. B. (1982). Moralisches Urteil und Aggression: Zur Systematisierung und Präzisierung des Aggressionskonzeptes sowie einiger empirischer Befunde [Moral judgement and aggression: Toward the systematization and specification of the concept of aggression as well as some empirical results]. In R. Hilke & W. Kempf (Eds.), *Aggression. Naturwissenschaftliche und kulturwissenschaftliche Perspektiven der Aggressionsforschung* (pp. 208–280). Bern, Switzerland: Huber.

Epstein, S. (1984). Controversial issues in emotion theory. *Review of Personality and Social Psychology, 5,* 64–88.

Eron, L. D. (1986). Interventions to mitigate the psychological effects of media violence on aggressive behavior. *Journal of Social Issues, 42,* 155–169.

Eron, L. D., & Huesmann, L. R. (1980). Adolescent aggression and television. *Annals of the New York Academy of Science, 347,* 319–331.

Feshbach, S. (1961). The stimulating versus cathartic effects of a vicarious aggressive activity. *Journal of Abnormal and Social Psychology, 63,* 381–385.

Feshbach, S. (1964). The function of aggression and the regulation of aggressive drive. *Psychological Review, 71,* 257–272.

Feshbach, S. (1989). Fernsehen und antisoziales Verhalten. Perspektiven für Forschung und Gesellschaft [Watching television and antisocial behavior. Perspectives of research and society]. In J. Groebel & P. Winterhoff-Spurk (Eds.), *Empirische Medienpsychologie* (pp. 65–75). München, Germany: Psychologie Verlags Union.

Flashar, H. (1984). Die Poetik des Aristoteles und die griechische Tragödie [Aristotle's poetics and greek tragedy]. *Poetica, 16*, 1–23.

Geen, R. G., & Quanty, M. B. (1977). The catharsis of aggression: An evaluation of a hypothesis. In L. Berkowitz (Ed.), *Advances in experimental social psychology* (Vol. 10, pp. 1–37). New York: Academic Press.

Geen, R. G., & Thomas, S. L. (1986). The immediate effects of media violence on behavior. *Journal of Social Issues, 42*, 7–27.

Gleich, U. (2004). Medien und Gewalt [Media and violence]. In R. Mangold, P. Vorderer, & G. Bente (Eds.), *Lehrbuch der Medienpsychologie* (pp. 587–618). Göttingen, Germany: Hogrefe.

Goranson, R. E. (1970). Media violence and aggressive behavior: A review of experimental research. In L. Berkowitz (Ed.), *Advances in experimental social psychology* (Vol. 5, pp. 2–31). New York: Academic Press.

Groeben, N. (1972). Literaturpsychologie [Psychology of literature]. Stuttgart, Germany: Kohlhammer.

Groeben, N. (1999). Fazit: Die metatheoretischen Merkmale einer sozialwissenschaftlichen Psychologie [Conclusion: The metatheoretical attributes of social science psychology]. In N. Groeben (Ed.), *Zur Programmatik einer sozial-wissenschaftlichen Psychologie: Bd. I. Metatheoretische Perspektiven: 2. Halbbd. Theoriehistorie, Praxisrelevanz, Interdisziplinarität, Methodenintegration* (pp. 311–404). Münster, Germany: Aschendorff.

Groeben, N., & Scheele, B. (1977). *Argumente für eine Psychologie des reflexiven Subjekts* [Arguments for a psychology of the reflexive subject]. Darmstadt, Germany: Steinkopff.

Groeben, N., & Vorderer, P. (1988). *Leserpsychologie. Lesemotivation—Lektürewirkung* [Psychology of reading. Reading motivation—reading effects]. Münster, Germany: Aschendorff.

Gründer, K. (1968/1991). Jacob Bernays und der Streit um die Katharsis [Jacob Bernays and the argument about catharsis]. In M. Luserke (Ed.), *Die Aristotelische Katharsis. Dokumente ihrer Deutung im 19. und 20. Jahrhundert* (pp. 352–385). Hildesheim, Germany: Olms. (Reprinted from *Epirrhosis. Festgabe für Carl Schmitt*, Vol. 2, pp. 495–528, by H. Barion, E.-W. Böckenförde, E. Forsthoff, & W. Weber (Eds.), 1968, Berlin, Germany: Duncker & Humblot).

Hanly, C. M. (1986). Psychoanalytic aesthetics: A defense and an elaboration. *Psychoanalytic Quarterly, 55*, 1–22.

Hartmann, D. P. (1969). Influence of symbolically modeled instrumental aggression and pain cues on aggressive behavior. *Journal of Personality and Social Psychology, 11*, 280–288.

Heckhausen, H. (1989). *Motivation und Handeln* [Motivation and action] (2nd ed.). Berlin, Germany: Springer.

Herkner, W. (1991). *Lehrbuch Sozialpsychologie* [Textbook in social psychology] (5th ed.). Bern, Switzerland: Huber.

Herzog, W. (1991). *Das moralische Subjekt. Pädagogische Intuition und psychologische Theorie* [The moral subject. Educational intuition and psychological theory]. Bern, Switzerland: Huber.

Huesmann, L. R. (1986). Psychological processes promoting the relation between exposure to media violence and aggressive behavior by the viewer. *Journal of Social Issues, 42*, 125–139.

Huesmann, L. R., & Malamuth, N. M. (1986). Media violence and antisocial behavior: An overview. *Journal of Social Issues, 42*, 1–6.

Janov, A. (1970). *The primal scream*. New York: Putnam.

Jo, E., & Berkowitz, L. (1994). A priming effect analysis of media influences: An update. In J. Bryant & D. Zillmann (Eds.), *Media effects: Advances in theory and research* (pp. 43–60). Hillsdale, NJ: Lawrence Erlbaum Associates.

Klimmt, C., & Vorderer, P. (2004). Unterhaltung als unmittelbare Funktion des Lesens [Amusement as a direct function of reading]. In N. Groeben & B. Hurrelmann (Eds.), *Lesesozialisation in der Mediengesellschaft. Ein Überblick* (pp. 36–60). Weinheim, Germany: Juventa.

Kornadt, H.-J. (1974). Toward a motivational theory of aggression and aggression inhibition: Some considerations about an aggression motive and their application to TAT and catharsis. In J. deWit & W. W. Hartup (Eds.), *Determinants and origins of aggressive behavior* (pp. 567–577). Den Haag, The Netherlands: Mouton.

Kornadt, H.-J. (1982). Grundzüge einer Motivationstheorie der Aggression [Main features of a motivation theory of aggression]. In R. Hilke & W. Kempf (Eds.), *Aggression. Naturwissenschaftliche und kulturwissenschaftliche Perspektiven der Aggressionsforschung* (pp. 86–111). Bern, Switzerland: Huber.

Kornadt, H.-J., & Zumkley, H. (1992). Ist die Katharsis-Hypothese endgültig widerlegt? [Has the catharsis theory been conclusively disproved?]. In H.-J. Kornadt (Ed.), *Aggression und Frustration als psychologisches Problem* (Vol. 2, pp. 156–223). Darmstadt, Germany: Wissenschaftliche Buchgesellschaft.

Kroetz, F.-X. (1986). Der Nusser. Uraufführung [world premiere].

Kunczik, M. (1982). Aggression und Gewalt [Aggression and violence]. In H. J. Kagelmann & G. Wenninger (Eds.), *Medienpsychologie. Ein Handbuch in Schlüsselbegriffen* (pp. 1–8). München, Germany: Urban & Schwarzenberg.

Langholf, V. (1990). Die "kathartische Methode." Klassische Philologie, literarische Tradition und Wissenschaftstheorien in der Frühgeschichte der Psychoanalyse [The cathartic method. Classical philology, literary tradition, and philosophy of science in the early developments of psychoanalysis]. *Medizinhistorisches Journal, 25*, 5–39.

Lessing, G. E. (1768/1988). *Gesammelte Werke in 5 Bänden* [Collected works in 5 volumes], Bd. 4. Hamburgische Dramaturgie. Berlin, Germany: Aufbau Verlag.

Leuzinger, P. (1997). *Katharsis. Zur Vorgeschichte eines therapeutischen Mechanismus und seiner Weiterentwicklung bei J. Breuer und in S. Freuds Psychoanalyse* [Catharsis. On the prehistory of a therapeutic mechanism and its advancement. In J. Breuer and S. Freud's psychoanalysis]. Opladen, Germany: Westdeutscher Verlag.

Lorenz, K. (1963). *Das sogenannte Böse. Zur Naturgeschichte der Aggression* [On aggression]. Vienna: Borotha-Schoeler.

Lukesch, H., & Schauf, M. (1990). Können Filme stellvertretende Aggressionskatharsis bewirken? [Can movies cause vicarious aggression catharsis?]. *Psychologie in Erziehung und Unterricht, 37*, 38–46.

Luserke, M. (Ed.). (1991). *Die Aristotelische Katharsis. Dokumente ihrer Deutung im 19. und 20. Jahrhundert* [The peripatetic catharsis. Documents of interpretation in the 19th and 20th century]. Hildesheim, Germany: Olms.

Maill, D. S., & Kuiken, D. (2002). A feeling for fiction: Becoming what we behold. *Poetics, 30*, 221–241.

Meyer, U.-W., & Reisenzein, R. (1996). [Emotion]. In G. Strube together with B. Becker, C. Freksa, U. Hahn, K. Opwis, & G. Palm (Eds.), *Wörterbuch der Kognitionswissenschaft* (pp. 139–141). Stuttgart: Klett-Cotta.

Möller, H.-J. (1981). Katharsis [Catharsis]. In. H.-J. Möller (Ed.), *Kritische Stichwörter zur Psychotherapie* (pp. 184–192). München, Germany: Fink.

Montada, L. (2002). Moralische Entwicklung und moralische Sozialisation [Moral development and moral socialization]. In R. Oerter & L. Montada (Eds.), *Entwicklungspsychologie* (5th ed., pp. 619–647). Weinheim, Germany: Beltz.

Mummendey, A. (1983). Aggressives Verhalten [Aggressive behavior]. In H. Thomas (Ed.), *Enzyklopädie der Psychologie: Themenbereich C Theorie und Forschung, Serie IV Motivation und Emotion, Band 2 Psychologie der Motive* (pp. 321–439). Göttingen, Germany: Hogrefe.

Mummendey, A., & Otten, S. (2002). Aggressives Verhalten [Aggressive behavior]. In W. Stroebe, K. Jonas, & M. Hewstone (Eds.), *Sozialpsychologie. Eine Einführung* (4th ed., pp. 350–380). Berlin, Germany: Springer.

Nichols, M. P., & Efran, J. S. (1985). Catharsis in psychotherapy: A new perspective. *Psychotherapy, 22*, 46–58.

Nichols, M. P., & Zax, M. (1977). *Catharsis in psychotherapy.* New York: Gardner.

Nussbaum, M. C. (2001). *The fragility of goodness. Luck and ethics in greek tragedy and philosophy* (Rev. ed.). Cambridge, UK: Cambridge University Press.

Nussbaum, M. C. (2003). *Upheavals of thought. The intelligence of emotions.* Cambridge, UK: Cambridge University Press.

Parke, R.D., Berkowitz, L., Leyens, J. P., West, S., & Sebastian, R. J. (1977). Some effects of violent and nonviolent movies on the behavior of juvenile deliquents. In L. Berkowitz (Ed.), *Advances in experimental social psychology* (Vol. 10, pp. 135–172). New York: Academic Press.

Peper, D. (1981). *Aggressive Motivation im Sport: Literaturanalyse, Theoriebildung und empirische Felduntersuchung zum Katharsis-Problem* [Aggressive motivation in sport: Analysis of the literature, theory formation and empirical field inquiry about the catharsis problem]. Ahrensburg, Germany: Czwalina.

Pohl, K. (1987). Hunsrück. Uraufführung [world premiere].

Pohlenz, M. (1956). Furcht und Mitleid? [Pity and fear]. *Hermes, 84*, 49–74.

Rorty, R. (1989). *Contingency, irony, and solidarity.* Cambridge, UK: Cambridge University Press.

Rule, G. B., & Ferguson, T. J. (1986). The effects of media violence on attitudes, emotions, and cognitions. *Journal of Social Issues, 42*, 29–50.

Schadewaldt, W. (1955). Furcht und Mitleid? Zur Deutung des Aristotelischen Tragödienansatzes [Pity and fear? On the interpretation of the peripatetic approach to tragedy]. *Hermes, 83*, 129–171.

Scheele, B. (1990). *Emotionen als bedürfnisrelevante Bewertungszustände. Grundriß einer epistemologischen Emotionstheorie* [Emotions as need-relevant appraisal states. Outline of an epistemological theory of emotions]. Tübingen, Germany: Francke.

Scheele, B. (1999). Theoriehistorische Kontinuität: Lernen von Aggression oder Möglichkeiten zur Katharsis?! [Theoretical continuity: Learning from aggression or possibilities for catharsis?]. In N. Groeben (Ed.), *Zur Programmatik einer sozialwissenschaftlichen Psychologie. Bd. 1: Metatheoretische Perspektiven. 2. Halbbd.: Theoriehistorie, Praxisrelevanz, Interdisziplinarität, Methodenintegration* (pp. 1–83). Münster, Germany: Aschendorff.

Scheele, B. (2001). Back from the grave: Reinstating the catharsis concept in the psychology of reception. In D. Schram & G. Steen (Eds.), *The psychology and sociology of literature: In honor of Elrud Ibsch* (pp. 201–224). Amsterdam: Benjamins.

Scheele, B. (2003). Rationale Gefühle [Rational feelings]. In N. Groeben (Ed.), *Zur Programmatik einer sozialwissenschaftlichen Psychologie. Bd. II: Objekttheoretische Perspektiven. 2. Halbbd.: Situationsbezug, Reflexivität, Rationalität, Theorieintegration* (pp. 233–272). Münster, Germany: Aschendorff.

Scheff, T. J. (1983). *Explosion der Gefühle. Über die kulturelle und therapeutische Bedeutung kathartischen Erlebens* [Catharsis in healing, ritual, and drama]. Weinheim, Germany: Beltz. (Original work published 1979)

Scheff, T. J., & Bushnell, D. D. (1984). A theory of catharsis. *Journal of Research in Personality, 18*, 238–264.

Scherer, K. R., Schorr, A., & Johnstone, T. (2001). Appraisal processes in emotion. Theory, methods, research. New York: Oxford University Press.

Schiller, F. (1801). Die Schaubühne als eine moralische Anstalt betrachtet [The stage regarded as a moral institution]. *Kleinere prosaische Schriften, 4. Teil,* 719–721.

Schneider, K., & Schmalt, H. D. (1994). *Motivation* [Motivation] (2nd ed.). Stuttgart, Germany: Kohlhammer.

Selg, H. (1992). Ärger und Aggression [Anger and aggression]. In U. Mees (Ed.), *Psychologie des Ärgers* (pp. 190–205). Göttingen, Germany: Hogrefe.

Selg, H., Mees, U., & Berg, D. (1997). *Psychologie der Aggressivität* [Psychology of aggression] (2nd ed.). Göttingen, Germany: Hogrefe.

Symonds, P. (1954). A comprehensive theory of psychotherapy. *American Journal of Orthopsychiatry 24*, 697–712.

Vorderer, P. (1992). *Fernsehen als Handlung. Fernsehfilmrezeption aus motivationspsychologischer Perspektive* [Television as action. Watching TV movies from a motivational perspective]. Berlin, Germany: Edition Sigma.

Vorderer, P. (2004). Unterhaltung [Entertainment]. In R. Mangold, P. Vorderer, & G. Bente (Eds.), *Lehrbuch der Medienpsychologie* (pp. 543–564). Göttingen, Germany: Hogrefe.

Vorderer, P., & Weber, R. (2003). Unterhaltung als kommunikationswissenschaftliches Problem. Ansätze einer konnektionistischen Modellierung [Entertainment as a problem for communications. Towards a connectionist model]. In W. Frueh & H.-J. Stiehler (Eds.), *Theorie der Unterhaltung. Ein interdisziplinärer Diskurs* (pp. 136–159). Köln, Germany: Halem.

Zillmann, D. (1988). Mood Management: Using entertainment to full advantage. In L. Donohew, H. E. Sypher, & E. T. Higgins (Eds.), *Communication, social cognition, and affect* (pp. 147–171). Hillsdale, NJ: Lawrence Erlbaum Associates.

Zillmann, D. (1994). Mechanisms of emotional involvement with drama. *Poetics, 23*, 33–51.

Zillmann, D. (1996). The psychology of suspense in dramatic exposition. In P. Vorderer, H. J. Wulff, & M. Friedrichsen (Eds.), *Suspense: Conceptualizations, theoretical analyses, and empirical explorations* (pp. 199–231). Mahwah, NJ: Lawrence Erlbaum Associates.

Zillmann, D., Bryant, J., & Sapolsky, B. S. (1979). The enjoyment of watching sport contests. In J. H. Goldstein (Ed.), *Sports, games, and play* (pp. 297–335). Hillsdale, NJ: Lawrence Erlbaum Associates.

Zumkley, H. (1978). *Aggression & Katharsis* [Aggression & catharsis]. Göttingen, Germany: Hogrefe.

24

An Evolutionary Perspective on Entertainment

Peter Ohler
University of Technology Chemnitz

Gerhild Nieding
University of Wüerzburg

In media theory and communications research it is becoming popular to refer to evolutionary arguments when explaining the entertainment function of the media (e.g., Steen & Owens, 2001; Schwender, 2001; see the chapter by Vorderer et al., this volume). But it still remains to be explained whether our preference for being entertained (and to entertain others) resides in an adaptive function of the species *Homo sapiens sapiens* or whether it is only an evolutionary by-product of other adaptations. However, even if this preference for entertainment is an adaptive function of our species, it remains unsolved as to which selection pressures produced this adaptation. Could this culturally invariant preference for entertainment (Barkow, 2001) be caused by the forces of natural selection (Darwin, 1859/1995) or is it tied to the more sophisticated forces of sexual selection (Darwin, 1871), which seems more plausible? In this chapter we will take a closer look at the lines of argument that are involved in this discussion. We will briefly introduce the principles of evolutionary biology including some remarks on sexual selection. This is followed by a short outline of evolutionary psychology. Finally, three evolutionary approaches concerning the evolutionary origins of entertainment are presented.

PRINCIPLES OF EVOLUTIONARY BIOLOGY

The classical assumption of the verity of natural selection by Charles Darwin (1859/1995) is still a main basis of evolutionary psychology. Darwin's argument includes the following steps: 1) the number of individuals of a species grows faster than the necessary resources that are available (principle of Malthus, 1826); 2) individuals of a species vary in structural and behavioral features; 3) the variations can be inherited; and 4) if individuals of a species acquire a competitive advantage in the acquisition of resources caused by variation, this will increase their chance to reproduce. It follows that the variation will manifest itself gradually in the population or a subpopulation of that species. In modern evolutionary biology the reproductive success of an individual, including the ability to survive until reproduction, is called fitness. Darwin's (1859/1995) principle of natural selection is equivalent to reproductive fitness, sometimes also called Darwinian fitness.

An important modification of the classical approach to fitness (which has certain shortcomings) was developed by Hamilton (1964a, 1964b) and his theory of inclusive fitness. Inclusive fitness is the sum of direct fitness, which is the reproductive success of an organism, and so-called indirect fitness, which is the sum of the reproductive success of an organism's genetic relatives. It follows that not only is the reproductive success of the individual itself relevant for the survival of the individual's genes, but also the support that influences the reproductive success of its kin.

THE THEORY OF SEXUAL SELECTION

Modern theorists distinguish sexual selection from natural selection (Miller, 2000). Natural selection favors individuals with a high fitness to survive and reproduce until the end of their reproductive period, whereas sexual selection favors individuals with a high fitness to pick sexual partners with the best genes for reproduction. The "parental investment theory" of Trivers (1971, 1972) explains the conditions of sexual selection that result in different short-term and long-term mating strategies of males and females (Buss, 1999; Buss & Schmitt, 1993). This theory, which is an elaboration of Darwin's (1871) theory of sexual selection, predicts that the sex that invests more resources in the rearing of offspring should be more selective in choosing a mate (e.g., human females), whereas the other sex should be more competitive with members of their own sex (e.g., human males) in accessing members of the opposite sex.

Sexual selection differs from natural selection in that it causes changes in the genetic makeup of a species much faster. The mechanism that produces such fast changes was first described by Fisher (1930) and is called *runaway sexual selection*, and the changes that are produced are called *sexual ornaments*. This concept is best explained by referring to a species like the ancestor of modern peacocks.

Within that proto-peacock population, different males possessed different tail lengths and they also differed in the colorfulness of their tails, but there was not a significant difference between males and females. Because the males invested nothing in raising the offspring, the sexual selection was strictly dominated by female choice. If in such a situation only one peahen was born with a curious mutation, a preference for males with longer and more colorful tails, the process of a runaway sexual selection may start. This peahen would only mate with males with slightly longer and more colorful tails. If the paternal feature of long, colorful tails is inherited in her sons and the preference for such males is also inherited in her daughters, all their sons will possess above-average tails and all their daughters will possess the preference of their mother. As a result, in the new generation, the genes for long-colorful tails and the genes for the sexual preference in hens will have spread in the population at least in a small amount.

The most important insight of Fisher (1930) was that this process of spreading specific genes will continue and accelerate dramatically via a positive feedback loop (Iwasa & Pomiankowski, 1995). As not only long and colorful tails in males, but also the female preference, is inherited a genetic correlation between the two traits is established. Sons will not only produce sons with longer and more colorful tails, but also daughters with that preference, and daughters will also produce sons with the sexual traits and daughters with the sexual preference. In the next generations more females will exist with the sexual preference and longer and more colorful tails will increase the sexual attractiveness of males. Therefore, they will mate at an above-average rate and the sexual trait and the preference will spread in the population until it is fixed in the genetic pool of the species. Through this process, the tail of males will become much longer and much more colorful until the species shows clear sexual dimorphisms[1] in the tails of males and females (as is the case in today's peacocks). The tails of the male have become a sexual ornament.

The process only stops at a certain level when the disadvantages caused by natural selection (e.g., energy investment for the trait; the ability to flee from predators) counteract the advantages in mating opportunities.

These principles of evolutionary biology suffice to construct an evolutionary psychology of the human cognitive architecture as well as the emotional and motivational make-up of our species.

A SHORT OUTLINE
OF EVOLUTIONARY PSYCHOLOGY

Evolutionary psychology (EP), sometimes also called Darwinian psychology (Plotkin, 1994), is a synthesis of modern evolutionary biology and modern (cognitive) psychology. In the version of Cosmides and Tooby (1997), EP can be seen as the description of information-processing-mechanisms designed by natural selection. The interplay of these mechanisms constitutes the cognitive architecture of the human mind.

Cosmides and Tooby (1997) argue that the human brain works like a bio-computer and its neural circuits enable the production of behavior, which is adapted to a specific environment. The neural circuits are designed by the processes of natural selection. They are solutions of adaptive problems. Adaptive problems are those problems that occur very often in the evolutionary history of a species. They are evolutionary recurrent situations (Cosmides & Tooby, 1994). Only the products of some (integrated) high-level neural circuits can be processed consciously, the majority of processes operate automatically. Different neural circuits are specialized to solve different adaptive problems. Solutions can only be achieved by very specific and functionally distinct mechanisms (Cosmides & Tooby, 1994; Tooby & Cosmides, 1992). These specialized modules increase the inclusive fitness, but something that fits one domain does not necessarily fit another domain. To solve this problem, many distinctive domain-specific modules evolved.

On such a basis we can distinguish two principal types of evolutionary psychology (EP). The classical form of EP assumes that the cognitive architecture and socio-emotional make-up of *Homo sapiens sapiens* reached its contemporary form in an *environment of evolutionary adaptedness* (EEA)[2] that existed between 200,000 and 100,000 years ago in the savannahs of Africa (Cann, Stoneking & Wilson, 1987), when the species lived as hunters and gatherers (e.g., Barkow, Cosmides, & Tooby, 1992). This approach takes for granted that the later cultural improvements of the species (e.g., parietal art and portable art, c.f. Conkey, 1999) since circa 40,000 years ago are entirely the result of the cognitive architecture of modern humans. The cognitive architecture, which secured the survival of hunters and gatherers in the Pleistocene is

---

[1]Sexual dimorphism includes all differences in size, shape, color etc. in the males and females of a species. The different size and color of the tails in male and female peacocks is an example of sexual dimorphism.

[2]The EEA "is not a specific time or place. It is the statistical composite of selection pressures that caused the design of an adaptation" (Cosmides & Tooby, 1997). The constant terrestrial illumination conditions that formed the design of the verebrate eye are part of the EEA of *Homo sapiens sapiens,* as are the selection pressures that caused the mental adaptations in the hominid line in the last two million years.

Being confronted with the adaptive problems to find food, to find mates, to detect and avoid predators, to cooperate with the other members of the group, to detect cheaters, and so on in the savannahs of Africa ca. 100,000 years ago caused the fine-tuning of the modern human mind.

still at work today. There was not enough time in evolutionary dimensions (variation qua muta-tion, inheritance, and selection) to change the cognitive architecture of *Homo sapiens sapiens* fundamentally. Therefore, in this approach, culture does not influence the way we think.

The other type of EP favors a co-evolutionary approach, assuming a close linkage between brain and culture that accelerated human evolution in the last 40,000/50,000 years (Donald, 1991, 2002). The highly social mind of modern humans is able to use external representations (e.g., parietal art, development of writing) as a cultural strategy that improves remembering and problem solving. Human anatomy and behavior seem to have evolved slowly before 50,000 years ago; afterwards, anatomy remained unchanged, but behavior (including cultural traditions and use of some first media) accelerated dramatically. Perhaps this change was caused by a mutation (Klein & Blake, 2002).

EVOLUTION AND ENTERTAINMENT

Three main theoretical positions concerning the evolutionary origins of entertainment can be distinguished. The leisure-time approach holds that the human preference for entertainment is not an adaptive function but rather an evolutionary by-product of other adaptations of the species. Two other positions argue that the preference for entertainment is indeed an adaptive function. One approach maintains that entertainment is a function in the process of sexual selection in humans (Miller, 2000), whereas the other position consists of varying approaches that share the idea that entertainment is a prolongation of play (e.g., Ohler & Nieding, in press a; Steen & Owens, 2001; Vorderer, 2001).

The Leisure-Time Approach

Zillmann (2000) argues that entertainment is the consequence of the leisure time left after survival needs are satisfied. He proposes that only in the hominid line was leisure time ac-cumulated to an extent that allowed the development of entertainment. Many species are permanently forced to expend never-ending efforts to survive, although some predator species possess leisure time, but waste it by napping, and still other species use leisure time for play activities. In Zillmann's concept, animal play seems to have the function of providing training skills that are necessary and useful for adult individuals (Groos, 1899; Smith, 1982). Play, in this concept, is equivalent to the practice of behaviors that are relevant for survival.

Tool use, social organization, and the division of labor of australopithecines did not result in enough leisure time for pleasure activities. Perhaps, before the appearance of *Homo erectus*, no species in the hominid line existed that was not highly absorbed in working for subsistence. Somewhere in the hominid line, organized hunting, tool use, domestication of fire, and im-proved communication skills reached a level of elaborateness that allowed a species with this combination of features to quickly acquire the resources necessary for subsistence. As a result, they had a considerable amount of leisure time.

All activities of entertainment, starting with pleistocene campfires, ancient partying, ancient Greek theatre, and so on, serve the same function. They help individuals spend their leisure time. The same is the case with modern forms of entertainment, such as watching TV, or playing interactive computer games. All these entertainment activities have some structures and functions in common (e.g., they are self-rewarding for the actors/recipients/players). This should be expected if entertainment activities fulfill an adaptive function. Adaptations, such as mating behavior or play, are highly constant across different historical periods and cultures. This should not necessarily be the case with evolutionary by-products. If entertainment is only an evolutionary by-product of other adaptations, as the leisure-time approach stipulates, then other forms of culturally and historically divergent activities to bridge leisure time should have emerged. But this is not the case, which is a fundamental weakness of the leisure-time approach.

Entertainment as Sexual Selection

Runaway Sexual Selection in Humans

Miller (2000) uses the argument of *runaway sexual selection* developed by Fisher (1930) to reconstruct the evolution of brain size in the hominid line with this theory. The brain size in the hominid line doubled, at least, in the last 1,7 Million years (Holloway, 1999) and the entire set of specific human features such as language, creativity, humor, music, and art developed during this period of brain growth. One necessary pattern for runaway sexual selection is polygyny: At least some sexually mature males mate regularly with more than one sexually mature female. Thus, males of an undetermined hominid species, for instance, of the family *homo erectus*, varied in the feature "creative intelligence." Some females, perhaps via a mutation, developed a preference for intelligent creativity. Males with a genetically heritable higher value of intelligent creativity attract more sexual partners, mate more than average, and therefore produce more offspring. If the paternal feature of an above-average creative intelligence is inherited in every male offspring, and the preference for such males is also inherited in every daughter, all sons will posses above-average creative intelligence and all daughters will possess the taste of their mother. Analogous to the peacock tail, the *sexual trait*—exhibition of creative intelligence—and the *sexual preference*—with a highly sensitive perceptual-cognitive system to distinguish differences in exhibitions of creative intelligence—will spread in the population until they are fixed in the genetic pool of the species. In this view, the brain, especially of males, is solely a sexual ornament, like the peacock tail, and is in no way an instrument for adaptation. The female brain serves at least one adaptive function: to choose the most entertaining male out of the group of males that are attracted to her.

The growth of creative intelligence (males) and the growth of cognitive functions to perceive and evaluate this intelligence (females) are related with the brain growth in the sexes. This means that the brain growth in the hominid line is the result of a runaway sexual selection driven by a positive feedback process.

According to Miller (2000) this can only be one part of the big picture. The astonishing brain growth in the hominid line is too slow to be the product of a runaway process. The sexual dimorphism in human males and females—in brain size and its function, and in anatomical features—is too small to be the result of a runaway process. And, as far as we know, our ancestors were moderately polygynous, but much less so than peacocks, elephant seals, and gorillas.

Creative Intelligence as Fitness Indicator

Miller (2000) argues that human creative intelligence is a *fitness indicator*. Fitness indicators are "adaptations that evolved to advertise an individual's fitness during courtship and mating ..." (Miller, 2000, p. 439). Sexual ornaments are fitness indicators. A peacock with a splendid tail shows to the peahens how strong and healthy he is and that he can afford the resources for this wasteful splendor. The performance of behaviors that are too costly for individuals with lower fitness can also function as fitness indicators. Zahevi (1975) introduced the *handicap principle* to explain the reliability of fitness indicators. They must create a high cost, such that an individual that is in a worse condition cannot afford it. Furthermore, it must be impossible for individuals to fake them.

The application of this idea to the human brain is termed the "healthy brain theory" by Miller (2000, p. 104). Our extravagantly large brains can only produce the diverse signals of creative intelligence when they are in a good shape. According to Miller (2000) "the healthy brain theory proposes that our minds are clusters of fitness indicators: persuasive salesmen like art, music, and humor, that do their best work in courtship, where the most important deals are made" (p. 105).

The Ornamental Mind Theory

Approaches in evolutionary psychology (e.g., Buss, 1999; Pinker, 1997) mostly focus on physical features of the face or body (Gangestad & Thornhill, 1997; Singh & Young, 1995) when looking for fitness indicators that signal health and fertility. Whereas it is widely accepted that human mating is related to an entire set of mental, social, and physical features of individuals (Buss,1999), it is the merit of Miller (2000) to emphasize that "many of our psychological traits may have evolved as fitness indicators too" (p. 106).

The central idea of his *ornamental mind theory* is that the human mind evolved through sexual choice—probably starting in the course of a runaway process—and ended up as a fitness indicator. The human mind consists of a *set of entertainment systems* designed to appeal to other minds during courtship. Contrary to Cosmides and Tooby (1994), who compare the adapted mind (with its distinct modules) with a Swiss army knife (with its distinct tools), Miller (2000) compares it with an amusement park. To formulate the thesis a bit metaphorically: in the cradle of human culture lies an entertainer.

This theory does not expect as much difference between the two sexes as the runaway selection approach does. The peacock only needs to possess a splendid tail, he does not need the peahens discriminative perceptual system for the appreciation of favored tails. When sexual ornaments are not grown but are produced (as output of the ornamental brain) the situation is different. In order to tell an entertaining story or to produce an admirable piece of art, it is necessary to embody the aesthetic discrimination of the recipients. Therefore this theory predicts much higher similarities between the sexes in their aesthetic taste than the runaway approach. However, it is similar to other approaches in its prediction of a higher output of entertainment performances by males.

The human preference for entertainment and the creative intelligence that is necessary to be entertaining evolved according to Miller (2000) "through sexual selection as an anti-boredom device" (p. 412). In modern human societies, industries that support the entertainment of our mates exist, but this cannot replace the individual's entertainment efforts in the courtship game (Miller, 2000):

> We can (indirectly) pay Hollywood scriptwriters to make our intended romantic partners laugh. But our ancestors could not do this, and even now it does not suffice. If we prove boring during the conversation after the film, our dates may say they had a lovely time, but let's be just friends. You can't buy love. You have to inspire it, partly through humor, the premier arena for advertising your creativity (p. 418).

Entertainment as Play

Stephenson (1988) argues that the reception of mass media should be modeled as a subjective play of the recipients. Vorderer (2001) suggests that entertainment experiences share many aspects with play-experiences and that therefore entertainment could be reconstructed as a form of play. Steen and Owens (2001) propose that culturally, media entertainment is a more elaborate form of pretense that phylogenetically evolved as a cognitive adaptation for pretense play. Ohler and Nieding (in press a) believe that the creative explosion of modern humans that took place within dispersed populations some 50,000 years ago could be modeled as an interplay of an encapsulated play-module and the possibilities of the media, which were discovered in those times (e.g., parietal art, portable art). If such ideas are valid, play really is a phylogenetic cornerstone for modern media entertainment. Therefore, a closer look on the evolution of play behavior is necessary.

A Theory of Play in Animals and Humans

The starting point of evolutionary considerations about play is the phenomenon that not only (young) individuals in the species *Homo sapiens sapiens* show play behavior, but also other species (Bekoff & Byers, 1998, Burghardt, 1998). This means that the behavior system "play" is much older than modern humans, it had already existed for millions of years when the first hominids appeared.

The basic assumption of our behavior-diversification proto-cognition theory of play (BD-PC theory; Ohler, 2000; Ohler & Nieding, in press a) is that play behavior was selected in evolution because of its potential to generate behavior variants. In a vertebrate species a mutation appeared, which caused the very curious behavior to produce random sequences of acts across the borders of behavior systems (e.g., fight, flight, hunting, nutrition, reproduction; Bekoff, 1995). Individuals of this species who possess this behavioral feature are able to retrieve a repertoire of behaviors more effectively than other animals of that species. If this species lives in an EEA with variable changing niches—but with changes that can be reacted to with responses on a behavioral level—these individuals have an advantage in fitness; they show higher reproductive success. Over many generations, play behavior will manifest itself in the genetic pool of that species.

The neural circuits and control-mechanisms, which generate play behavior, are now part of the psychological architecture of that species. All individuals now possess a distinctive play module (Cosmides & Tooby, 1994). It is always triggered when specific environmental cues, resembling the adaptive problems in the EEA, occur. As a result, those individuals will execute the typical combinatory behavior sequences transcending the borders of behavior systems.

We will make a jump in the evolutionary period of time. The evolved play mechanism is preserved even in species with more highly developed cognitive systems, which can be observed in mammals like canines (Bekoff, 1998, 1995). The play module is no longer—in contrast to the situation of the original first players—restricted to circuits controlling behavior, but can also operate on cognitive modules.

Up to that point in the evolution of representational systems, no individual of any species was capable of representing imagined entities. All species were only capable of operating cognitively on the basis of primary representations (Leslie, 1987; Perner, 1991). Primary representations are always tied directly or indirectly (via cues still present) to the perception of the representing organism. The contents of primary mental representations are always related to entities that are available in the present situation of the representing organism.

Then a species occurred—probably either a common ancestor of the families *hominidae* and *pongidae* (of humans and great apes) or an ancestor in the hominid line—that possessed a cognitive architecture with a capacity allowing a qualitatively new type of mental representation. In this situation, the play module would enable a quantum leap in the functioning of the representational system. When these individuals activated primary representations that triggered the play module, this would work in its well-established fashion. It would force every unit in the activated part of the cognitive system to combine with every other unit. This systematic combination of units in primary representations lead to the emergence of a semiotic function that was not realized before. A new relation between elements is thus established: A mental element is able to represent another mental element. Now, entities can be represented that are only imagined. Secondary representations (Povinelli, 1998; Perner, 1991) are the prerequisite for all cognitive operations that allow hypothetical and/or counterfactual thinking (Mitchell & Riggs, 2000).

This means that the play module is responsible for the quantum leap to the first secondary mental representations in phylogenesis, which some authors think is the fundamental difference between humans and all other primate species (Gärdenfors, 1995; Povinelli, 1998).

The capacities of the new representation system in its original state were surely very small. Only rudimentary anticipatory planning was possible. However, for the individuals possessing such a system, suddenly the possibility emerged for them to play cognitively with behavior alternatives and not just to pick alternatives at random and to act blindly based on trial and error. If this allowed an individual to avoid at least one tragedy from birth to reproduction, the direct fitness of that individual was increased enormously. Again, over many generations, this feature would manifest itself in the genetic pool of that species, and all members of the population would have secondary mental representations at their disposal.

This would mean that early pretense play, which appears at around 12–13 months in humans, should be the ontogenetically first psychological domain, in which secondary mental representations are realized. Even in ontogenetically later play forms, for instance, constructive play, social play, and games with rules (e.g., computer and video games), the combinatory diversified depth structure of activities triggered by the play module remain in function. But the different play forms express different surface features when they are executed (Ohler, 2000).

Play and Entertainment

This theory of the evolution of play can also be applied to model aspects of entertainment experiences in the context of the use of media. If a general definition of media can be given at all, then the shortest and most concise definition would be: Media are *external representational systems* that are organized via symbol systems (Ohler & Nieding, in press b). The symbol systems dominantly organizing different media can be ordered on an axis with two poles. One pole is characterized by very abstract signs that are used in written language (e.g., novels) and the other pole by perceptual symbol systems (Barsalou, 1999). The current new media that are most extremly organized through perceptual symbol systems are immersive virtual reality environments (Biocca, 1997). TV, film, and computer games are dominantly organized by symbol systems that lie somewhere between the extremes of the symbol systems organizing novels and virtual reality.

The emergence of media circa 40,000/50,000 years ago (e.g., parietal art, portable art) allowed the use of these external representations to produce selective presentations of aspects of the physical and social environment. Primary mental representations profit from this possiblity in that they are relieved in their permanent work of information reduction. Furthermore, an entirely new playground for secondary mental representations emerged. The play module that allowed the first pretend activities becomes at this point in evolution a decisive part of a cognitive system that allowed the establishment of mimetic and mythic cultures in the hominid line (Donald, 1991, 2002). The individuals in these cultures were able to mentally construct narrations with very complex plot lines and to share them in a process of oral communication. All pretend and make-believe activities up to this time still needed a massive amount of mental work, though.

This situation changed dramatically with the emergence of the first media. With media, a new sophisticated interplay between external and secondary mental representations is possible. The media offer frameworks that reduce the workload to represent imagined worlds and, in doing this, they take enormous pressure off the modules that are engaged in imaginative processes. A playful management of expectation horizons (Ohler & Nieding, 1996) is possible; if someone does not need to keep every element of the world of discourse in his mind, he/she may gain degrees of freedom to think about: "What would happen if?"[3]

---

[3] Variability is the invariant key component of the play module that remains always stable despite new ontogenetically emerging play forms. In a playful mode, users of computer games employ a broader range of strategies (Ohler & Nieding, in press a).

Different media allow the externalization of various elements of their worlds of discourse. This is precisely the deeper reason why some media are experienced as cognitively more demanding than others (Salomon, 1984). Despite these differences, all media are similar in that they are not only able to present a possible world (Bruner, 1987; see also the chapter by Vorderer et al., this volume) but also in that they organize this possible world concerning the media-specific symbol systems. This allows the imaginative parts of the cognitive system to undergo playful endeavours of different kinds, depending on the medium and on other aspects of the possible world like genre and content. This is at least one important cognitve basis for the experience of media entertainment.

If the frameworks of the media allow the recipients to shift their cognitive systems into a more playful mode, which means the activation of the play module that is very old in evolutionary terms, this should be self-rewarding since most evolved functions are self-rewarding. This is one motivational basis for the pleasure recipients experience in media entertainment.

CONCLUSION

Three different approaches concerning the evolution of entertainment are presented in this chapter. One approach suggests that entertainment is an evolutionary by-product (leisure-time approach) and the other two approaches state that entertainment serves as an adaptive function: One uses the explanation pattern of sexual selection (the ornamental mind theory), whereas the other starts with natural selection of a play module and tries to reconstruct the destiny of this module in later stages of anthropogenesis.

When different approaches try to explain a single phenomenon, it is often the case that none of the approaches alone are able to grasp the big picture. Leisure time is a prerequisite for the enactment of such elaborate courtship patterns that are necessary for the sexual selection of an ornamental mind. Leisure time, though, is also necessary for play that is not restricted to the young of a species (which is one of the pecularities of the human race). The mating mind, as Miller (2000) describes it, is a playful mind. Therefore, it is highly possible that something like a play module is also triggered in the execution of typical human courtship behaviors. As soon as classical play approaches [that assume that play is physical exercise (Smith, 1982) or pre-exercise (Groos, 1899, 1901) for adult skills] are replaced by models like the BDPC-theory, former assumptions will also vanish. Is it really natural selection that formed the design features of play, or may it even be the case that selection pressures to find a mate are responsible for the curious play behavior? BD-PC-theory would agree with such a possibility (Ohler, 2000).

While the big picture is still unclear, it makes sense to examine a seemingly very culturally determined phenomenon like media entertainment with the tools of evolutionary approaches.

REFERENCES

Barkow, J. H., Cosmides, L., & Tooby, J. (Eds.). (1992). *The adapted mind. Evolutionary psychology and the generation of culture*. New York: Oxford University Press.

Barkow, J. H. (2001). Universalien und evolutionäre Psychologie [Universals and evolutionary psychology]. In: P. M. Hejl (Ed.), *Universalien und Konstruktivismus. Delfin 2000* (pp. 126–138). Frankfurt, Germany: Suhrkamp.

Barsalou, L. W. (1999). Perceptual symbol systems. *Behavioral and Brain Sciences, 22*, 577–660.

Bekoff, M. (1995). Play signals as punctuation: The nature of social play in canids. *Behaviour, 132*, 419–429.

Bekoff, M. (1998). Playing with play: What can we learn about cognition, negotiation, and evolution? In D. D. Cummins & C. Allen (Eds.), *The evolution of mind* (pp. 162–182). Oxford, England: Oxford University Press.

Bekoff, M., & Byers, J. A. (Eds.). (1998). *Animal play: Evolutionary, comparative, and ecological perspectives*. Cambridge, England: Cambridge University Press.

Biocca, F. (1997). The cyborg's dilemma: Progressive embodiment in virtual environments. *Journal of Computer Mediated Communication [Online serial], 3*(2).

Bruner, J. (1987). *Actual minds, possible worlds.* Cambridge, MA: Harvard University Press.

Burghardt, G. M. (1998). Play. In G. Greenberg & M. M. Haraway (Eds.), *Comparative psychology. A handbook* (pp. 725–735). New York: Garland Publishing.

Buss, D. M. (1999). *Evolutionary psychology. The new science of the mind.* Boston: Allyn and Bacon.

Buss, D. M., & Schmitt, D. P. (1993). Sexual strategies theory: An evolutionary perspective on human mating. *Psychological Review, 100,* 204–232.

Cann, R. L., Stoneking, M., & Wilson, A. C. (1987). Mitochondrial DNA and human evolution. *Nature, 325,* 31–36.

Conkey, M. W. (1999). A history of the interpretation of European 'palaeolythic art': magic, mythogram, and metaphors for modernity. In A. Lock & C. R. Peters (Eds.), *Handbook of human symbolic evolution* (pp. 288–349). Oxford, England: Blackwell Publishers.

Cosmides, L., & Tooby, J. (1994). Origins of domain specifity: The evolution of functional organization. In L. A. Hirschfeld & S. A. Gelman (Eds.), *Mapping the mind. Domain specifity in cognition and culture* (pp. 85–116). New York: Cambridge University Press.

Cosmides, L., & Tooby, J. (1997). *Evolutionary Psychology: A Primer.* Retrieved July 21, 2001, from http://www.psych.ucsb.edu/research/cep/primer.html

Darwin, C. (1871). *The descent of man, and selection in relation to sex. 2 vols.* London: John Murray.

Darwin, C. (1859/1995). *On the origins of species. A facsimile of the first edition (14th ed.).* Cambridge, MA.: Harvard University Press.

Donald, M. (1991). *Origins of the modern mind. Three stages in the evolution of culture and cognition.* Cambridge, MA: Harvard University Press.

Donald, M. (2002). *A mind so rare: The evolution of human consciousness.* New York: Norton & Company.

Fisher, R. A. (1930). *The genetical theory of natural selection.* Oxford, England: Clarendon Press.

Gangestad, S. W., & Thornhill, R. (1997). Human sexual selection and developmental stability. In J. A. Simpson & D. T. Kenrick (Eds.), *Evolutionary social psychology* (pp. 169–195). Mahwah, NJ: Lawrence Erlbaum Associates.

Gärdenfors, P. (1995). Cued and detached representations in animal cognition. *Lund University Cognitive Studies (LUCS), 38.*

Groos, K. (1899). *The play of animals.* New York: Appleton.

Groos, K. (1901). *The play of man.* New York: Appleton.

Hamilton, W. D. (1964a). The genetical evolution of social behaviour. I. *Journal of Theoretical Biology, 7,* 1–16.

Hamilton, W. D. (1964b). The genetical evolution of social behaviour. II. *Journal of Theoretical Biology, 7,* 17–52.

Holloway, R. (1999). Evolution of the human brain. In A. Lock & C. R. Peters (Eds.), *Handbook of human symbolic evolution* (pp. 74–125). Oxford, England: Blackwell Publishers.

Iwasa, Y. & Pomiankowski, A. (1995). Continual change in mate preferences. *Nature, 377,* 420–422.

Klein, R. G., & Blake E. (2002). *The dawn of human culture. A bold new theory on what sparked the "big bang" of human consciousness.* New York: Wiley & Sons.

Leslie, A. M. (1987). Pretense and representation: The origins of "Theory of mind." *Psychological Review, 94,* 412–426.

Malthus, T. R. (1798/1826). *An essay on the principle of population; or, a view of its past and present effects on human happiness, with an inquiry into our prospects respecting the future removal or mitigation of the evils which it occasions. 2 vols.(6th ed.).* London: John Murray.

Miller, G. F. (2000). *The mating mind. How sexual choice shaped the evolution of human nature.* New York: Doubleday.

Mitchell, P., & Riggs, K. J. (Eds.). (2000). *Children's reasoning and the mind.* Hove, England: Psychology Press.

Ohler, P. (2000). *Spiel, Evolution, Kognition. Von den Ursprüngen des Spiels bis zu den Computerspielen* [Play, evolution, cognition. From the origins of play to the computer games]. Habilitationsschrift an der Technischen Universität Berlin, Germany.

Ohler, P., & Nieding, G. (1996). Cognitive modeling of suspense-inducing structures in narrative films. In P. Vorderer, H. J. Wulff, & M. Friedrichsen (Eds.), *Suspense. Conceptualizations, theoretical analyses and empirical explorations* (pp. 129-147). Hillsdale NJ: Lawrence Erlbaum Associates.

Ohler, P. & Nieding, G. (in press a). Why play? An evolutionary perspective. In J. Bryant & P. Vorderer (Eds), *Playing Computer Games: Motives, Responses, and Consequences* (Chap. 8). Hillsdale, NJ: Lawrence Erlbaum Associates.

Ohler, P. & Nieding, G. (in press b). Medienpsychologie [Media psychology]. In A. Schütz, H. Selg, & S. Lauterbacher (Eds.), *Einführung in die Psychologie* (Chap. 20). Stuttgart, Germany: Kohlhammer.

Perner, J. (1991). *Understanding the representational mind.* Cambridge, MA: The MIT Press.

Pinker, S. (1997). *How the mind works.* New York: Norton.

Plotkin, H. (1994). *Darwin machines and the nature of knowledge.* Cambridge, MA: Harvard University Press.

Povinelli, D. J. (1998). Can animals empathize? Maybe not [Electronic version]. *Scientific American, 9,* 67–75.

Salomon, G. (1984). Television is 'easy' and print is 'tough': The differential investment of mental effort in learning as a function of perceptions and attributions. *Journal of Educational Psychology, 4,* 647–658.

Singh, D., & Young, R. K. (1995). Body weight, waist-to-hip-ratio, breasts, and hips: Role in judgements of female attractiveness and desirability for relationships. *Ethology and Sociobiology, 16,* 483–507.

Schwender, C. (2001). *Medien und Emotionen. Evolutionspsychologische Bausteine einer Medientheorie* [Media and emotions. Evolutionary-psychological constituents of a media theory]. Wiesbaden, Germany: Deutscher Universitäts-Verlag.

Smith, P. K. (1982). Does play matter? Functional and evolutionary aspects of animal and human play. *The Behavioral and Brain Sciences, 5,* 139–184.

Steen, F. F., & Owens, S. (2001). Evolution's pedagogy: An adaptationist model of pretense and entertainment. *Journal of Cognition and Culture, 1,* 289–321.

Stephenson, W. (1988). *The play theory of mass communication.* New Brunswick, NJ: Transaction Publishers.

Tomasello, M. (1999). *The cultural origins of human cognition.* Cambridge, MA: Harvard University Press.

Tooby, J., & Cosmides, L. (1992). The psychological foundations of culture. In J. H. Barkow, L. Cosmides, & J. Tooby (Eds.), *The adapted mind. Evolutionary psychology and the generation of culture* (pp. 19–136). New York: Oxford University Press.

Trivers, R. L. (1971). The evolution of reciprocal altruism. *Quaterly Review of Biology, 46,* 35–57.

Trivers, R. L. (1972). Parental investment and sexual selection. In B. Campbell (Ed.), *Sexual selection and the descent of man: 1871-1971* (pp. 136–179). Chicago: Aldine.

Vorderer, P. (2001). It's all entertainment–sure. But what exactly is entertainment? Communication research, media psychology, and the explanation of entertainment experiences. *Poetics, 29,* 247–261.

Zahevi, A. (1975). Mate selection—A selection for a handicap. *Journal of Theoretical Biology, 53,* 205–214.

Zillmann, D. (2000). The coming of media entertainment. In D. Zillmann & P. Vorderer (Eds.), *Media entertainment. The psychology of ist appeal* (pp. 1–20). Mahwah, NJ: Lawrence Erlbaum Associates.

Author Index

Subject Index

CPSIA information can be obtained
at www.ICGtesting.com
Printed in the USA
BVHW061838060219
539568BV00008B/88/P